Restoring the Pacific Northwest

The Art and Science of Ecological Restoration in Cascadia

Edited by
Dean Apostol and Marcia Sinclair

SOCIETY FOR ECOLOGICAL RESTORATION INTERNATIONAL

Washington · Covelo · London

Library of Congress Cataloging-in-Publication Data.

Restoring the Pacific Northwest : the art and science of ecological restoration in Cascadia / edited by Dean Apostol and Marcia Sinclair ; foreword by Eric Higgs ; Society for Ecological Restoration International.
p. cm.
Includes index.
ISBN 1-55963-077-9 (cloth : alk. paper) — ISBN 1-55963-078-7 (pbk. : alk. paper)
1. Restoration ecology—Northwest, Pacific. I. Apostol, Dean. II. Sinclair, Marcia. III. Society for Ecological Restoration International.
QH104.5.N6R47 2006
639.909795—dc22
2006009303

British Cataloguing-in-Publication data available.

Printed on recycled, acid-free paper 

Composition by Lisa Beck

Manufactured in the United States of America
10 9 8 7 6 5 4 3 2 1

This book is dedicated to seven generations of conservationists and restorationists. The present generation, partially reflected in the contributors to and supporters of this book, stands on the shoulders of our elders and ancestors of the past three generations, who recognized the unique qualities of the Northwest and its ecosystems and fought the good fight for its conservation. We learned from you, remember you, and honor you.

The next three generations will build on our work and create a new, restorative culture that is reflected in agriculture, forestry, art, music, architecture, politics, and an ecosystem that retains the full complement of native biodiversity that existed here historically. If we do our job well, they will learn from us, remember us, and honor us.

For Ann Lennartz

When this book was still a somewhat amorphous idea, a mutual acquaintance recommended I call Ann Lennartz to ask if she would consider providing financial support. Although we had never met, within a few minutes we were talking as if old friends, sharing the coincidence of both being Iowa State grads. This single warm conversation was enough for Ann to write a check large enough to get the ball rolling. She also pitched in later with a challenge grant when we were half done, the cupboard was bare, and we were teetering on the brink of collapse. Both times, and in subsequent conversations, when I thanked her for her generosity and trust she simply said, "This book is needed." Ann passed away recently, shortly before we were able to place a copy in her hands. But she knew it was near completion and said she was looking forward to buying a large number of copies to give away as gifts. Ann, you have given the gift of this book to the entire community of restorationists in the Northwest and far beyond. Over the years I know you have given many gifts of hope and we are all deeply and forever indebted to you.

CONTENTS

FOREWORD

Eric Higgs

The Pacific Northwest, as it is called in the United States, is a hotspot. Not just a hotspot in the conventional ecological sense as a haven for species at risk but also a place of boggling diversity in restoration projects. Something about this landscape inspires the promise of an ecological society and the fierce dedication to restorative ideals. We see it in the genius of the community-rooted Mattole watershed project in northern California, the restoration of threatened Garry oak habitat in urban Victoria, British Columbia, the rekindled historical fire regimes in the dry interior forests, and so many more examples. The early successes of the Northwest and British Columbia chapters of the Society for Ecological Restoration International, including hosting two international restoration conferences (Seattle 1995, Victoria 2004) and many well-attended regional meetings, are testaments to the strength of the movement and the capacity of the restoration community here.

Where is here? Dean Apostol and Marcia Sinclair have taken a wide embrace in their definition of the region. For many in the United States, *Restoring the Pacific Northwest* may connote the dripping conifer forests of Oregon and Washington, a *Snow Falling on Cedars* image. But there are two reasons to mistrust this impression. A regional ecological embrace follows the sinewy form of watersheds. In this case those rivulets, creeks, streams, and rivers, and all the lands they nourish, flow through a vast region starting in Alaska and Yukon in the north, almost all of British Columbia, Washington, Oregon, and Idaho, as far east as central Montana, and all the way down to northern California, entering the Pacific Ocean and returning. We think of flow in only one direction, but recent studies illustrate the movement of nutrients via salmon migration hundreds of miles inland. Northwest rivers are legends and include (from north to south) the Copper, Alsek, Stikine, Skeena, Nechako, Fraser, Columbia, Snake, Willamette, Umpqua, Rogue, Klamath, and Mattole. Along the coast we find the archetypal temperate rainforest and estuarine ecosystems, but moving inland to the plateau we find dry inland forests of ponderosa pine and Douglas fir, arid sagebrush zones and deserts, rich wetlands, prairies, and sublime subalpine and alpine ecosystems. To think of this region as a coherent whole takes a creative leap, but such ambition teaches us an important lesson: We are connected socially, culturally, and ecologically by the waters that flow to the north Pacific and, to a great extent, by the Pacific salmon.

The second reason the title should give us pause is the difference political borders make in how we understand a landscape. Americans think of this region as the Pacific *Northwest*, but Canadians think of it as the *West Coast*. I write from Victoria, a city at the extreme *southwest* of Canada, which makes the idea of Oregon and Washington as the Northwest difficult to comprehend. For someone living on the other side of the Pacific basin (in Japan, for example), the landscape described in this book would be the Pacific *Northeast*. It gets confusing! Others have grappled with this issue and invented *Cascadia*, a name that evokes the volcanic mountains that have largely shaped the character of the land. Yet so far this bioregional moniker has not overcome the inertia of centuries-old political boundaries and vernacular associations. I don't intend to rewrite or relabel the region, but I want at the very least to sow seeds of doubt, complicate the matter, and put on record that the geographic naming issue won't be easily solved. For every American who reads this book, please imagine what it is like to identify with a region whose name for Canadians in particular makes little sense. Still, whatever we choose to call this place, we all should know what we are referring to ecologically and spiritually.

This region has inspired restoration projects because of its diversity, beauty, and abundance. Having lived for more than a decade in the harsh high plains of Alberta, I pause for a moment every time the plane touches down in my recently adopted Victoria to breathe the floral, humid air. Fruit hangs everywhere in the summer, and until recently the ocean has been rich in healthful food (marine ecosystem restoration is an emerging concern). The ecological diversity is reflected by cultural diversity, with dozens of traditional indigenous languages and local adaptations. And beautiful? Every landscape has its admirers, but it is no coincidence that great Canadian man of letters Stephen Leacock described British Columbia as "surrounded on one side by ocean and the other three by envy." Such beauty has inspired everything from sober second thoughts to direct action against activities that despoil abundance, beauty, and diversity. The pitched battles to protect old-growth coastal forests, and countless campaigns to preserve areas of local, regional, national, and international significance, are part of our culture. So, too, is restoration emerging as part of our outlook, which is why this book arose in the first place.

Restoring the Pacific Northwest is a perfect fit for the Society for Ecological Restoration International's partnership with Island Press: a series on *The Science and Practice of Ecological Restoration*. The intent of the series is to provide practical books informed by good science and scholarship on all types of restoration. The publication of regionally focused books falls squarely in the purview, and this particular one will brightly illuminate the path for others to follow. As the restoration movement grows, we will someday have dozens of books that will gather the best intelligence from regions across the world and inspire distinctive cultures of restoration appropriate to those places.

The initial title of this book, *An Encyclopedia of Pacific Northwest Restoration*, made an audacious claim to presenting comprehensive knowledge of restoration in the Northwest. As will become clear to the reader, this ambition had to be scaled back because restoration practice and knowledge have become too comprehensive to cover in a single volume. It should be viewed instead as a territorial marker of ecological restoration: *In this particular geography, this is what we in the restoration field are attempting to do*. As such, it is a critical step, intended to lay the foundation for many more detailed books. The many contributors to this book are a *Who's Who* of restoration in the region. They deserve gratitude for taking the time to share their work, in most cases without compensation. If we have been successful, updates will follow, more detailed books on subregional ecosystems will be written, and the project to generate and share restoration information will grow throughout this enormous region. That it is understood as a dynamic project respects not only the nearly impossible diversity of restoration initiatives but also the changing character of restoration techniques, insights, legislation and policies, and resolve.

What actions and thoughts will this book inspire? I hope it will galvanize the power of restorationists throughout the region, providing them with a stronger identification and context for the importance of their work and the myriad connections that support the work they are doing. It will add weight to arguments that restoration has arrived, that it is a crucial conservation tool, and that it will provide an important boost to students who are hungry to see more offerings of restoration education. My fondest wish is that in the decade that follows the publication of this book we will witness the development and spread of integrated restoration education. What better prospect is there for a life well worked than to engage in ecological restoration, the most hopeful of conservation endeavors?

PREFACE

In early 2003, when Dean Apostol raised the idea to the Rivers Foundation of putting together a multiauthor volume that would present the state of ecological restoration in the Pacific Northwest, we were happy to help him in any way that we could. Three years later, he and Marcia Sinclair have skillfully shepherded the book to completion, and you are holding the wonderful result in your hands. This inspiring and practical volume brings together many of the leading scientists and practitioners who together have logged thousands of hours in the diverse landscapes and watersheds from northern California to Alaska and east to the high deserts, woodlands, and grasslands. I am confident you will find the information in these pages to be useful, illuminating the breadth and depth of the range of ecosystems in the region and the many challenges of restoration.

Restoring the Pacific Northwest: The Art and Science of Ecological Restoration in Cascadia gathers and presents the best examples of state-of-the-art restoration techniques and projects. It is an encyclopedic overview and an invaluable reference not just for restorationists and students working in the Pacific Northwest but for practitioners across North America and around the world.

Please read it, use it, and help us to build on its foundation to restore this hauntingly beautiful region called Cascadia.

Peter Lavigne, President
Rivers Foundation of the Americas

ACKNOWLEDGMENTS

A book of this scope could never have been completed by two people. It is hard to know where to start or end with thanking those who had a hand in its development. First, credit should go to Stephen Packard, Cornelia Mutel, and all those who created the *Tallgrass Restoration Handbook* in 2005 (Island Press). That book, which brought leading practitioners together to assemble information about Midwestern ecosystem restoration, was the original inspiration for this effort. Midwestern restoration colleagues may (or should) envy our high mountains, mild climate, and rugged coastline, but we in the upper left corner of North America will forever envy their role in initiating the modern ecological restoration movement.

Sarah Vickerman and her staff at the Defenders of Wildlife West Coast office provided helpful tips on multicontributor book management and budgeting, having just completed *Oregon's Living Landscape*. This advice was crucial in helping us understand more fully what we were getting into and proved to be remarkably accurate in the end.

Without funding, there would have been no book. Ann Lennartz provided a startup grant to a fellow Iowa State alumnus whom she had never met, based on a brief phone conversation, an e-mail outline of an idea, and a single reference (Peter Goldman of the Washington Forest Law Center). She pitched in again with a challenge grant when subsequent fundraising fell short and got us across the finish line. Local foundations, firms, agencies, and other individuals added funding to the collective pot in both small and larger amounts. Included among these are the Bullitt Foundation of Seattle, the Spirit Mountain Community Fund, Trout Unlimited, Oregon Department of State Lands, Portland Metro Regional Government Parks and Greenspaces, Walker Macy Landscape Architects, the Society of Wetland Scientists Northwest Chapter, and Vigil-Agrimis. Greenworks, PC, a Portland landscape architecture firm, provided very timely support in helping to organize and format book graphics. A special thanks to Andrea Cameron, Jamie English, David Elkin, and Brian Wethington for their technical skills and to Mike Faha for hiring talented people and allowing us to borrow them for a few days.

Two organizations, the Society for Ecological Restoration Northwest Chapter (SERNW) and the Rivers Foundation of the Americas (RFA), provided fiscal oversight and assistance with fundraising. SERNW, in particular board chair Steve Link, treasurer Leslie Ryan-Connelly, and staff members Melissa Keigley and Nancy Hahn, supported the first phase. RFA helped the book reach a successful conclusion. The work of both organizations included consultation, coordination, assistance in grant preparation, fund management, and reporting back to those who provided funding. A large number of individuals and some businesses contributed funding to the RFA that was used in part to fulfill a challenge grant match. They are thanked here and listed in full in the "Supporters and Partners" section at the end of the book.

Island Press editors Barbara Dean and Barbara Youngblood are two understanding and nurturing professionals who have honed the art of providing the right blend of timely information, gentle nudging, shoulders to occasionally cry on, sage advice, strategic deadline extensions, and an occasional kick in the rear to see us through the maze of book publishing. We commend them to any potential writer contemplating a natural resource–related book quest.

Don Falk and James Aronson, the Science and Practice of Ecological Restoration series editors representing the Society for Ecological Restoration International, provided detailed (tough to swallow at times, but crucial) peer critiques of early ideas and draft chapters, particularly Chapters 1 and 2. Any slippage from the empirical high road they laid out is our fault, not theirs.

The fifty-seven contributors who researched and authored most of this book are true heroes. Brief biographies of all of them are provided at the end this book, and we encourage the reader to review them. For the most part, they worked on their own time and received not a nickel for their efforts. They are all dedicated restoration practitioners who were eager to share what they have learned through many years of work. It has been a privilege and honor working with them. We have met many personally and hope to meet the rest at some point before we all become compost.

A number of additional restoration practitioners read chapters and provided feedback on individual chapters. Most of them are listed at the end of the chapters they worked on, but in the rush to complete this book it is likely that we missed more than one, and we apologize for that.

Thousands of individuals toil daily in the world of ecological restoration in the Northwest, from middle school students who pull ivy out of urban woodlands to watershed councils and friends of a thousand small places. Many would not think to call themselves restoration practitioners because their involvement is only part-time and usually is unpaid, but with them lies the fate of the land. Dave Corkran of Catlin Gable School, Dan Evans of Davinci Middle School, and Tom Hinton of Madison High School are among a growing number of teachers who have inspired hundreds of Northwest students to learn about their local ecology and get involved. In the process they have begun to build an ethic and culture of restoration and stewardship that will be much longer lasting and effective than this book could hope to be.

My co-editor, Marcia Sinclair, encouraged me to pursue the initial idea of this book. She has seen many ideas pop up in my head, but fortunately most of them dissipate before consuming 5 years of our lives. She put up with a very cranky Greek for several weeks of struggle toward the deadline, interjected timely editing suggestions, and with her colleagues produced a great chapter that raised the graphics bar for the rest of the book. Marcia also pitched in and used her extensive skills to help format, edit, and organize the entire book in the last few weeks. My teenage son, Simon, gracefully put up with a neglectful dad in the summer of 2005. As I write this he is off building trails, swatting bugs, avoiding large mammal encounters, and learning the Latin names of plants in Alaska, a next-generation nature nerd preparing to take his turn with the conservation baton.

Lastly, I want to offer a special acknowledgment to the man who first got this former Chicagoland street punk interested in the idea of ecological restoration: Distinguished Professor Emeritus Robert Dyas of Iowa State University. His enthusiasm for the remnant prairies and native woodlands of his adopted state infected his students with a bug that many of us have never been able to shake. He opened us to the world of Jens Jensen and Aldo Leopold and made the bland Midwest and the wider world seem filled with limitless opportunities to go forth and do good work. Marcia and I hope this book can provide similar inspiration for others.

INTRODUCTION

Dean Apostol

Restoring the Pacific Northwest represents a remarkable collaboration between writers, practitioners, foundations, agencies, businesses, and individual supporters, all with a common interest in using the art and science of ecological restoration in the service of conservation. This book came to life as an idea linked to a regional conference being planned by the Society for Ecological Restoration Northwest Chapter (SERNW). The initial idea was to gather leading practitioners to document the state of the art of ecological restoration in the region and thus help advance its practice. This reflects the stage of development of restoration, which is still quite a young field and is advancing primarily through the efforts of field practitioners rather than academic researchers. Practitioners are very busy saving the planet and do not have much time for research and writing, so much of the best technical knowledge is locked up in the heads and field notes of biologists, ecologists, botanists, landscape architects, and laborers who toil away on restoration projects. When practitioners get together at conferences and symposia, they share this information and learn from each other, but their knowledge is rarely disseminated more widely. Much good restoration information also exists in "gray literature" (non–peer reviewed) technical reports and memos, also not very accessible to a wider audience.

Conference presenters were asked to participate in development of a book that would provide an encyclopedia of regional restoration practice. A similar effort had been made at the second Society for Ecological Restoration conference in Chicago in 1990. There, a multiday session on tallgrass prairie restoration picked the brains of leading practitioners, who shared the tricks they had learned over decades of practice. The event was recorded, with a writer in residence. Several years later Island Press produced the *The Tallgrass Restoration Handbook* (edited by Stephen Packard and Cornelia Mutel, 1997, with a new edition in 2005), now well on its way to becoming a classic of restoration literature.

It soon became apparent that the encyclopedia project could not be organized and funded in time to take advantage of the conference as initially envisioned. Nevertheless, funding gradually was secured, contributors were recruited, a prospectus was drafted, and a book began to take shape. As the project progressed, SERNW decided to hand off fiscal oversight to the Rivers Foundation of the

Americas, which assisted seeing the effort through. The result represents what we hope will be a milestone: a first gathering of many of the leading lights of ecological restoration in the Pacific Northwest, with their best ideas and experiences captured in the book you now hold in your hands.

Book Organization

The focus of this book is ecological restoration in the "Pacific Northwest," an imprecise geographic term for a region that has been described and defined in many different ways. Some use it to refer only to the rainy parts of our region lying west of the Cascades, British Columbia Coast Range mountains, and southeast Alaska—the land of mysterious giant, dripping conifers where the bulk of regional population resides. The U.S. Forest Service defines the Pacific Northwest administratively as limited to Oregon and Washington. Other agencies and geographers use the term to include Idaho, western Montana, and sometimes northern California, British Columbia, and Alaska. As Eric Higgs points out in the foreword, "Pacific Northwest" is a particularly problematic term for Canadians, whose idea of north is a bit different from that of Americans. An obvious choice for this book, given its origins, was to simply encompass the boundaries of the SERNW. This range included both sides of the mountains, the entire Columbia Basin, and some of northern California and southeast Alaska. But the complication was the hole in the region formed by the creation of a separate British Columbia chapter, which divides Alaska from the Northwestern states. So following in the American imperial tradition, we decided to annex much of British Columbia, although we did ask their permission. Color Plate 1 expresses the final geography chosen for this book quite beautifully and accurately. It defines the Pacific Northwest as all the watersheds that flow to the Pacific Ocean through North America's temperate rainforest zone. This definition extends the range east to the continental divide in Montana. *Cascadia* is included in the subtitle because it provides a more nation-neutral bioregional description of roughly the same territory. Granted, the Cascade Mountains are only a part of the whole, yet the volcanoes that crown them represent the dynamism of nature here, the beauty of this land, and provide a clear compass point and enduring image of the region for residents, visitors, and those far away who know of us only through photographs and travel writing.

The Pacific Northwest encompasses overlapping and competing ecosystems and cultures. There has always been, at least since European settlement, a tension and perhaps jealousy between those living east of the mountains, in the dry rain shadow with its more continental climate, and those living to the west, in the moist maritime zone. Those of us in the wet west look with envy to the dry east in the middle of our dreary winters, but in times of searing drought, or perhaps during the search for a decent double mocha latte, the reverse happens. Recently, the mountain divide has also become something of a political divide, as red and blue America fight it out every 4 years, even while the more peaceful

Canadians legalize same-sex marriage and pot smoking. Tension and envy, we are happy to report, are not apparent among ecological restorationists living in different parts of this region. We are too poor, powerless, and few in number to fight much among ourselves. Also, we are too busy in our mission to prevent further loss of biodiversity and ecosystem decline. Our overriding goal in setting the geography and exploring restoration within it has been to provide the region's first general text on the art, science, and practice of ecological restoration. Thus, this book attempts to cover the large sweep of Northwest ecosystems and leaves the finer details to others.

The book is divided into three sections. The shortest, Part I, "The Big Picture," contains Chapter 1, "Northwest Environmental Geography and History," and Chapter 2, "Ecological Restoration." Chapter 1 provides a broad overview in order to set the stage for a more detailed understanding of regional restoration. Chapter 2 is primarily a summary of general principles of ecological restoration that are applied regionally and internationally, with a focus on terms, definitions, and approaches developed by the Society for Ecological Restoration International. Together these two chapters address issues that otherwise would have been repeated in most or all other chapters.

Part II, "Pacific Northwest Ecosystems," lumps a very large and diverse region into nine major ecosystem types. These only partly correspond to more systematized classifications. Ecosystem classifications are always somewhat subjective and depend on the expected end use. For example, Robert Bailey's ecoregional classification system for the United States includes domains, divisions, provinces, and sections at increasing levels of detail. The Bailey system lumps conifer forests and alpine meadows together, but we have separate chapters for each (see www.fs.fed.us/land/ecosysmgmt/ecoreg1_home). It also includes a "Pacific lowland mixed conifer forest" zone to describe the Georgia Trough (British Columbia) to Willamette Valley (Oregon) area that includes prairie and oak habitats and a large number of wetlands and riparian woodlands. Bailey's intermountain semidesert province corresponds quite well with our Sagebrush Steppe chapter.

British Columbia's bio-geoclimatic classification system combines vegetation, soils, and climate, whereas their ecoregion classification integrates terrestrial and marine environments. This system breaks things down much more finely than does Bailey and may be one of the most sophisticated ecosystem classifications in North America, if not the world (www.gov.bc.ca/ecology/ecoregions/ecoclass). Bio-geoclimatic zones are named for climax plant communities, with gradients for subzones that go from dry to wet, hot to cold, and degree of maritime influence.

The U.S. Forest Service and The Nature Conservancy both rely on plant associations to describe ecosystems, starting with dominant overstories and working down through understory variations. The Forest Service has created hundreds of classifications across its domain. For example, in Mt. Hood National Forest there are twenty classifications for understory associations in the western hemlock (*Tsuga heterophylla*) zone (Halverson et al. 1986). The Forest Service uses classifications as environmental indicators, initially to guide logging and reforestation but now for many purposes, including restoration.

Thus, Chapter 5, "Old-Growth Forests," can be applied to dozens of variations found west of the Cascades and British Columbia Coast ranges and potentially inclusive of inland maritime forests in Idaho, Montana, and British Columbia. The nine ecosystems encompassed in this section are intended to apply to most ecological restoration projects across the region. Unfortunately, some ecosystems that are the subject of restoration attention, such as coastal sand dunes, have been left out. Each of the nine merits a full book in order to be treated in sufficient detail. One of these ecosystems, interior forests, already has been published as a full book, *Mimicking Nature's Fire* (by Stephen Arno and Carl Fiedler, 2005), which covers a much larger region than the Pacific Northwest.

The nine ecosystem types profiled are bunchgrass prairies, oak woodlands and savannas, old-growth conifer forests, riparian woodlands, freshwater wetlands, tidal wetlands, interior forests, sagebrush steppe, and mountain ecosystems. Broad as it is, this list misses much of the restoration story, so we also chose to address restoration topics or challenges that often involve more than one ecosystem or that may influence restoration in any. These are profiled in Part III, "Crossing Boundaries," and include urban natural areas, streams, watersheds, wildlife, invasive species, and traditional ecological knowledge. Through these chapters, a larger ecological restoration narrative is told, to a great extent more cultural and less technical. These topics tend to cut across or encompass the ecosystem types profiled in Part II. For example, many Northwest streams originate in mountain ecosystems, flow down through interior or old-growth forests, then are bordered by riparian woodlands and empty into an estuary. Their watershed context may include prairies, oak woodlands, and urban areas. Multiple invasive species may be encountered along the way. Wildlife reintroduction and barrier removal contribute to conservation efforts. And part of the restoration planning may rely on traditional ecological knowledge.

As Eric Higgs expresses so well in his Foreword, all of the contributors to this book have high hopes that it will advance the field of restoration in the region and perhaps elsewhere. By tackling Northwest restoration in a comprehensive way, across a range of ecosystem types and topics, this book will help practitioners, policymakers, agencies, communities, landowners, and students connect more dots as they contemplate a restoration project, program, business, or career. We see it not as the first word, which has already been spoken through many of the efforts documented in these pages, and certainly not as the last word because there is so much to be learned and written. Instead, we see is as a foundation on which to build. Please let us know if we have succeeded.

References

Arno, S. F. and C. E. Fiedler. 2005. *Restoring Nature's Fire*. Island Press, Washington, DC.

Halverson, N. M., C. Topik, and R. Van Vickle. 1986. *Plant Association and Management Guide for the Western Hemlock Zone, Mt. Hood National Forest*. U.S. Department of Agriculture Forest Service, Pacific Northwest Region, Portland, OR.

Packard, S. and C. F. Mutel. 2005. *The Tallgrass Restoration Handbook for Prairies, Savannas, and Woodlands*. Island Press, Washington, DC.

PART I

The Big Picture

Part I consists of just two chapters, but they lay a foundation for all that will follow. The first, titled "Northwest Environmental Geography and History," provides a regional overview, a discussion of biogeography and environmental history, a summary of restoration practice, and a brief discussion of the state of the art of ecological restoration.

These subjects are all worthy of book-length treatments. The intent is not to be comprehensive but to provide a context for readers who lack detailed knowledge about the regional environment of the Pacific Northwest. Readers who have more detailed knowledge than this author might want to skip ahead.

Chapter 2, "Ecological Restoration," is a general introduction to the development of restoration practice and includes summaries of some of the key concepts that international leaders of the field, particularly those in the Society for Ecological Restoration International, have generated in the past few years. These include the current definition of *ecological restoration*, discussion of the expected attributes of restored ecosystems, and the reasons why restoration is needed. This discussion will be of particular value to policymakers and restoration advocates. Too often the word *restoration* is thrown around casually. I recall one visit to a wildlife refuge where a former farm field, planted with a few oak saplings and seeded to native grasses, was presented as a "restored" white oak savanna. I tried to hold my tongue but couldn't. Readers who have not been steeped in the past 15–20 years of restoration conferences, journal articles, debates about definitions, and intellectual development in the field also will benefit from this chapter. The process of ecological restoration, from goal setting and project planning through monitoring and adaptive management, is discussed in many subsequent chapters but explored here in greater depth.

Chapter 1

Northwest Environmental Geography and History

Dean Apostol

It is not bragging to claim that the Pacific Northwest is one of the world's most spectacular regions. Our mountains and glaciers would make the Swiss envious, and our jagged, rocky coast washed by crashing ocean surf is the equal of New Zealand, Norway, or western Ireland. Our old-growth conifer forests have some of the world's tallest trees and highest levels of biomass. The vast sagebrush steppe is a land of national park–scale superlatives. One of the least populated places in North America, it boasts the deepest canyon (Hell's Canyon of the Snake River) and largest natural fault (Steens Mountain) on the continent.

Our human history and cultural development are equally impressive and fascinating. Northwest Indians attained a unique and very sophisticated level of art and culture that reflected the material abundance of the land and sea. The journals of Lewis and Clark reveal the land as it was before Euro-Americans set about changing it. Pioneers on the Oregon Trail bypassed nearly 3,000 miles of central continent to reach Oregon Country, rich in fish, farmland, and forest. Today's farmers continue to cultivate deep, rich alluvial Willamette Valley soils, reaping harvests of grain, fruit, and vegetables. The sparse soils of the Oregon Coast Range and eastern Washington produce some of the highest-quality wine grapes anywhere.

Much has been written about the geography and history of the Pacific Northwest, and this chapter can offer only a modest summary. As illustrated in Plate 1 in the color insert, Cascadia spans the middle to northerly temperate latitudes, from around 40 degrees south (northern California) to nearly 60 degrees north (the southern mainland of Alaska). Marine air over the northern Pacific Ocean fights a timeless war with continental air masses, each taking charge at different times of the year. West of the Cascade and Coastal mountain ranges, the Pacific usually has the upper hand, while to the east the drier continental system rules. Lands in the south, particularly the Siskiyou–Klamath Mountain subregion, are much drier than the north, especially in summer, when a blessed high-pressure system parks itself over the ocean off the Oregon coast. The climate is *maritime* in the north, increasingly *Mediterranean* in the south, and *continental* in the east, with a great number of intermediate zones and microclimates between them (Goble and Hirt 1999, Franklin and Dyrness 1973).

The Pacific Northwest is also shaped by three large geologic forces: tectonics, volcanoes, and glaciers. Ocean plates grind under the continent, shoving, melting, and lifting rock. The result is a geologically young land, formed of materials drawn from the ocean depths and reborn through volcanic action. Significant amounts of exotic terrain collided with the continent from great distances over a period of 200 million years and formed parts of the lands west of Idaho (Alt and Hyndman 1995). Generally the terrain in the east is older than that farther west. The rocks of the Klamath–Siskiyou Mountains in the southwestern part of the region are the exception, having arrived more than 200 million years ago, now jumbled into a chaotic heap (Trail 1998).

Elevation ranges are substantial, from sea level to more than 4,000 meters. Southeastern Alaska, the west coast of British Columbia, and the Olympic Peninsula include the wettest places in North America, whereas the sagebrush steppe, in the rain shadow of the Cascade Mountains, is characterized by arid plains. Diverse ecosystems are distributed across this terrain, responding to elevation, rainfall, underlying geology, soils, aspect, and cultural influences. An island mountain archipelago reaches south from Alaska, down the British Columbia coast, and through Puget Sound. Vancouver Island, at more than 32,000 square kilometers, is the largest island along North America's Pacific Coast. The Queen Charlotte Islands, or Haida Gwai, have been called the Canadian Galápagos, reflecting their remoteness from the main continent. Over the past 2 million years, successive advances and retreats of glaciers have also carved and shaped the landscape. The last retreat of the continental ice sheets was only some 12,000 years ago (the ice was 5,000 feet thick at the Washington–British Columbia border), and the Northwest remains a land with many glaciers.

Ecosystem Biogeography

Topographic complexity and proximity to the northern Pacific Ocean combine to create very diverse assemblages of plants and animals. The Northwest has tremendous landscape diversity over a fairly small area, a function of numerous mountain chains and quite variable precipitation, including some of the wettest and driest areas on the continent. Drive 200 miles in any direction from nearly any point in the region and you will experience significant ecosystem change, possibly more than anywhere else on the North American continent. Plant communities tend to run in north–south rather than east–west gradients, reflecting the orientation of major mountain ranges. West of the major mountains is the largest temperate rainforest in the world (Franklin and Dyrness 1973). In southeast Alaska, where the climate is cool and very wet even in summer, Sitka spruce (*Picea sitchensis*), western hemlock (*Tsuga heterophylla*), and yellow cedar (*Chamaecyparis nootkatensis*) are the dominant overstory trees. Perched peat bogs, or muskegs, with lodgepole pine (*Pinus contorta*) are also increasingly common in the north (Pojar and MacKinnon 1994). Rivers and streams teem with salmon. Farther south, in British Columbia, western redcedar (*Thuja plicata*) becomes increasingly dominant, and Douglas fir (*Pseudotsuga menziesii*) occupies more inland sites. Farther south along the coast, summer air temperatures warm, and as a consequence dense fogs form. Redwoods (*Sequoia sempervirons*) become a key component of the forest. Prairies, or grass balds, increase in frequency on coastal headlands farther south as well, adding diversity to the extensive forest matrix (Franklin and Dyrness 1973; Chapter 5).

Inland valleys, particularly the Georgia Straight–Puget Trough of Washington and the Willamette Valley of Oregon, are in the rain shadow of coastal mountains and therefore are much drier, which has allowed them to support bunchgrass prairies and Garry oak (*Quercus garryana*) woodland ecosystems (Chapters 3 and 4). Farther east the mountains rise to heights well above the tree line, with rich meadows, huckleberry fields, and subalpine parklands forming below and around the many glaciers (Chapter 11). North of Mt. Rainier, the tree line drops lower and parklands become more extensive than in the south (Chapter 11). East of the mountains a second rain shadow, a characteristic of the north–south mountain orientation, causes the forests to quickly change from hemlock to fir to pine and eventually to open up onto the vast sagebrush steppe of the interior Northwest (Chapters 9 and 10).

This geography includes the most extensive network of salmon-bearing streams on the planet (Chapter 13). All the major mammals of the North American continent still find homes in the region: grizzlies, wolves, lynx, cougars, bison, moose, elk, and many more (Chapter 15). It is a rich and beautiful environment that continues to attract tourists and immigrants from around the world, even while holding its native born close by.

A Very Brief Environmental and Ecological History

Although there is some dispute over exact dates, the evidence clearly shows that humans have been part of the Pacific Northwest landscape for at least 10,000 years and probably longer. "Kennewick Man," whose remains were discovered along the Columbia River shore, has been dated at more than 8,000 years old (Burke Museum 2005). The last retreat of glaciers 12,000–20,000 years ago allowed development of forests over much of the western part of the region, although the composition and structure we are familiar with today settled into place much later (Schoonmaker et al. 1997, Goble and Hirt 1999). Retreat of the glaciers was followed by a warm, dry climate that favored the northward spread of oaks and the westward movement of Douglas fir. Sagebrush steppe vegetation reached much farther west than at present, all the way to the western end of the Fraser River Valley. Then the climate cooled and became wetter, favoring hemlock and cedar and resulting in shrinkage of the range of oaks. There is speculation, but not physical evidence, that this may be the time when Northwest Indian people developed fire management of prairies and oak woodlands in interior valleys, which allowed them to persist even after a cooling of the climate should have resulted in their overtake by forest.

The pattern and distribution of forests, oak woodlands, prairies, streams, steppe, and wetlands that Lewis and Clark traveled through in the early nineteenth century had been in place only for 4,000–5,000 years. Forests were part of an ever-shifting mosaic that responded to periods of drought and large wildfires (Agee 1993). Rivers flooded, changed course, and created dynamic riparian zones, sometimes with enormous log jams (Ecotrust 2002). But the basic distribution of major vegetation communities was fixed, with only small further shifts at the margins.

Indian people interacted with this pattern and affected it in many ways. Level terraces along rivers and estuaries were cleared and occupied as village sites. Northwest Indians were skilled woodworkers and harvested trees and planks for canoes, building materials, tools, and weapons. Brush was gathered for firewood, basketry, and other uses. Western redcedar (*Thuja plicata*) was the "tree of life," used for everything from roofing to baby diapers and menstrual pads. Most importantly, almost every major ecosystem type in the region was shaped in part by Indian fire (Boyd 2000). Prairies and oak and pine woodlands were burned on a frequent basis, with forests underburned less frequently. Small clearings were made to attract game animals, even in the far northern coastal areas. Huckleberry patches and travel corridors in the mountains were also burned. There is little question that all of this burning had a profound effect on regional ecosystems (Chapter 17).

Plants were gathered in great numbers. Some, such as camas, may have been transplanted deliberately and managed to increase abundance (Boyd 2000). Salmon was the main food source for people far into the interior of the Columbia Basin. Harvest techniques included construction of weirs to funnel fish into shallow areas where they could be trapped and harvested more easily.

Development of sea trading by captains Vancouver, Gray, and others, Lewis and Clark's journey, the development of the beaver trade, missionaries, and thousands of pioneers seeking new land initiated profound social and environmental changes in the region. Diseases such as measles and smallpox reduced Indian populations by as much as

90% in some areas, beginning in the late eighteenth century (Schoonmaker et al. 1997). This catastrophe probably disrupted burning cycles and caused abandonment of villages, contributing to the pioneer impression that the Northwest was only lightly populated or unsettled (Robbins 1997).

Widespread trapping of beavers by agents of Hudson's Bay Company, designed in part to create a "beaver desert" that would discourage competitors, had profound effects on streams and wetlands all across the region (Lichatowich 1999). Only recently have ecologists begun to appreciate the critical role beavers play in sustaining complex aquatic and riparian habitats. Farmers settled and plowed the most fertile prairie soils first, then quickly spread to oak woodlands and prairie margins. Forests were cleared, cities were platted and built at the heads of deepwater navigation, and dredging of the Willamette and other rivers converted highly complex, braided systems to simple channels that could accommodate large vessels. By 1895, an estimated 50% of the bottomland hardwood forests of the Willamette Valley had been converted to agriculture, and the riparian conifers were almost completely gone (Hulse et al. 2000). Dikes were built along estuaries and lowlands, with wetlands ditched and drained. Intensive logging began along streams and rivers in the lower mountain reaches. On smaller streams, such as those in the Oregon Coast Range, splash dams temporarily backed water up to corral logs, which were then dynamited, releasing a torrent downstream that tore out natural log jams and sluiced riverbeds down to bedrock (Ecotrust 2002).

Early Northwest Euro-American settlers, loggers, and town builders for the most part had little understanding and less regard for native ecosystems or native people. They saw the region as a vast wilderness and saw their job as taming it and bringing it to heel. Government surveyors laid out a straight-lined grid of townships, sections, and range, initially stopping only where the land was considered unsuitable for farming or town building. The few remaining Indians were herded off to remote reservations on land white people did not want, at least at first. Over time most of these lands were confiscated as well. Fish, particularly salmon, were harvested with no thought of sustaining them, and population declines were well documented by the late nineteenth century (Lichatowich 1999).

Indian and pioneer trails became muddy farm-to-market roads, some surfaced with planks. Plank roads were later straightened and finally paved. Logging moved deeper into the forests along temporary railroads, then migrated upslope as trucks became the preferred transportation mode. The U.S. Forest Service, first established as a guardian of watersheds, got into the logging business in a big way in the 1950s to help feed the postwar housing boom and make up for depleted private forests. Logging revenues were used to build a powerful agency. British Columbia forests, 90% of which are owned by the Crown, have followed a similar path, although they trailed a few decades behind the U.S. logging curve. Clearcutting, at first shunned by foresters trained in European methods and a timber industry interested only in the highest-value trees, became the preferred harvest technique. It proved to be more efficient than selective cutting and allowed quick reforestation with sun-loving Douglas fir, the fastest-growing and most valuable tree in the much of the region. East of the mountains in the pine woodlands Indian fire was stopped, natural fires suppressed, and selective logging of old-growth pine initiated. Shade-tolerant fir trees quickly occupied the ground and filled in, and at first foresters were delighted (Arno and Fiedler 2005; Chapter 9).

High mountain meadows were grazed by armies of sheep and herded great distances to market. Millions of cattle were let loose on the unfenced, previously lightly grazed sagebrush steppe, resulting in a severe loss of native bunchgrasses (Chapter 10). Invasive species, some brought in accidentally, others on purpose, hitched rides into the Northwest with the new settlers. Hundreds of large and small dams were built for hydropower and flood control. At first they were mainly in upper tributaries, but some were near tidal zones, such as the Elwha on the Olympic Peninsula (Lichatowich 1999).

Ecological losses have only recently begun to be tallied. Old-growth conifer forests west of the Cascades, the most studied ecosystem in the Northwest, now cover only 10–15% of the presettlement extent in Oregon and Washington, with higher amounts remaining in British Columbia and southeast Alaska (Chapter 6). Interior pine forests have all been degraded by fire suppression, grazing, and logging. Tens of thousands of miles of roads, most very poorly built, prone to failure, with fish-blocking culverts, are laced through forested mountains. Bunchgrass prairies west of the Cascades are down to a tiny fragile fraction of their original extent (Chapter 3). Oak woodlands, though still occupying roughly the same area, have been severely disrupted by fire suppression and grazing, their understory communities most highly affected (Chapter 4). Freshwater wetlands have been reduced by 50%, or even more in densely settled places such as the Willamette Valley, the Puget Sound, and along the Fraser River (Chapter 7). Tidal wetlands have been substantially reduced along the Oregon and Washington coasts (Chapter 8). Riparian woodlands have been affected all across the region, particularly in lowland agricultural valleys and urban areas. The Columbia and Snake rivers, once the world's greatest producers of salmon, have become a series of warm water lakes except for a short stretch that runs past one of the most environmentally contaminated lands in North America, the nuclear reservation at Hanford, Washington. Fifty Columbia River salmon stocks have become extinct, and the remaining twenty-five are on a very expensive life support system (Lichatowich 1999). Urban streams have been buried in culverts, channelized, subjected to huge pulses of water from impervious surfaces, and polluted with oil, gasoline, and assorted street gunk (Booth 1991).

Despite the highly touted reputation of the Pacific Northwest across much of the continent and world as an "Ecotopia," the sad reality is that only fragments of natural ecosystems are left. This region has been logged, farmed, urbanized, roaded, grazed, drained, and dammed a lot more than many realize. Most conservationists' attention has been on preserving the remaining natural fragments of old-growth forests and roadless areas. Watershed, stream, and riparian protection have received increasing attention as a consequence of salmon decline. But the case has just begun to be made for spending significant resources on ecological restoration. Local economies and governments still struggle to recover from the dot-com bust and relentless antitax campaigns. Federal and provincial land management agencies, responsible for more than half the land in the region, have been decimated by staff reductions resulting directly from reduced logging revenues. Additionally, they are not yet staffed for a restoration mission, and the existing policy framework provides inconsistent direction that bounces from serving the natural resource extraction beast to protecting and restoring ecosystems. State and local agencies also lack staff expertise and reliable funding sources for restoration, although programs are maturing across the region.

Northwest Restoration Summary

Salmon, Rivers, and Watersheds

Most regional policy and funding attention has been on restoration of streams and watersheds, aimed primarily at improving conditions for wild salmon and trout. The results of more than two decades of restorative work are mixed. Although fifty populations of Columbia Basin salmon are extinct, salmon have returned to the Umatilla River in eastern Oregon after 70 years of absence, a result of a cooperative restoration project led by the Umatilla Indians that allowed water formerly diverted for irrigation to remain in the river. To do this, a deal was made to substitute Columbia River water to irrigators (Umatilla Indian Reservation Web Site). Although twenty-six separate Columbia Basin wild salmon and steelhead populations have been listed as threatened or endangered since 1991, no further extinctions have been recorded since the 1980s. Nevertheless, twenty-five of the twenty-six are continuing to decline in population (NOAA 2005, Sheets 2004). Increases widely reported in the media over the past few years have been primarily hatchery-bred, not wild fish.

It is estimated that historically, 10–14 million native adult salmon returned to the Columbia Basin each year. Natural salmon runs are now less than 5% of these historic levels (Sheets 2004). Dams have been identified as the main cause of this decline, although harvest levels, set too high because of an abundance of hatchery fish, also have played an important role. In the Siuslaw Basin on the central Oregon coast, reliable estimates indicate that historically more than 200,000 coho salmon returned in an average year. Now the number is around 4,000, despite the general absence of dams on the Siuslaw. High ocean harvest levels, habitat loss from logging, valley bottom farming, and the historic legacy of splash dams reduced coho habitat and abundance in the Siuslaw. Chinook salmon, which spend less time in degraded rivers and more in the ocean, remain at or near historic levels (Ecotrust 2002).

Salmon populations in the Columbia Basin are approximately where they were in 1986: two and half million adult fish, 80% of which are of hatchery origin (Sheets 2004). Bonneville Power has been spending an average of US$250 million per year on salmon recovery, of which only 15% goes for habitat restoration. This level of funding is $147 million below what the Northwest Power and Conservation Council recommended over the 2001–2003 period (Sheets 2004). In a 2001 review of the four major salmon recovery programs the Independent Scientific Advisory Board concluded that the probability of successful recovery, even assuming faithful implementation, was low (ISAB 2001).

Nevertheless, a lot has been learned about stream habitat restoration in the past two decades. In particular, the role of complex structure provided by large wood and the importance of periodic floods to reshape channels have helped biologists move from trying to engineer habitat (rocks, cables, gabions fixed in place) to working more with natural processes. A better understanding of the relationship between sediment delivery, storage, and transport, which some fishery ecologists call the stream digestion process, also helps (Ecotrust 2002). This has led to more attention on controlling upland sediment sources than on fixing downstream habitat. In some areas, restoration results have been very positive, with some stream systems such as Kennedy Flats on Vancouver Island showing impressive recovery after a decade of active restoration. In others, such as the Mattole River on northern California's "lost coast," nearly two decades of intensive and extensive efforts appear only to have stabilized the aquatic ecosystem, but a culture of restoration has become firmly rooted (Chapter 14). Region wide, we may be nearing the bottom of the trough, and the suite of new protection policies, combined with restorative action, may stem further decline or extinctions but may still not be near enough to result in significant recovery.

Forests and Woodlands

The first significant shock to the perception of the Northwest as an ecological Shangri-La was delivered by the listing of the northern spotted owl as an endangered species in 1990. Up to that point, old-growth conifer forests were seen as timber bank accounts to pay for schools and roads and to provide jobs in rural areas. It took a prolonged and concerted effort by a few stubborn local environmental organizations, eventually supported by national ones, to call attention to what was being lost. It also took the pathbreaking research of Dr. Jerry Franklin and his colleagues, affiliated with the Pacific Northwest Research Station and the H. J. Andrews Experimental Forest, to discover the unique and complex ecology of old growth. By the time President Clinton initiated the Northwest Forest Plan, old-growth forests in Washington,

Oregon, and northern California were down to a fraction of their presettlement range. Vancouver Island in British Columbia also has very little old growth left. "Restoration" of old growth at first appears to be an oxymoron because we think of these as completely natural or primeval ecosystems that have no human influence. Yet foresters at the Forest Service and Bureau of Land Management are now tackling the restoration challenge using variable thinning in young forest plantations to begin to build back structural and ecological complexity that the old forests exhibit.

Restoring *interior Northwest forests*, primarily Ponderosa pine, is a very different challenge. Foresters have slowly come to appreciate the critical role that frequent fire played in keeping these ecosystems healthy and in many cases are now combining selective logging with fire to recreate the open, parklike structure that pine forests once had.

Much recent effort has gone into restoring *riparian woodlands* in forested, agricultural, and urban landscapes and increasing regulatory protection of remaining riparian forests. It appears that progress is being made. Riparian areas are hospitable environments for plants, and one can start some species such as cottonwood and willows by simply jamming live stakes into the soil. A key challenge in riparian areas is competition from invasive species. Japanese knotweed in particular appears to have exploded in its distribution along regional streams over the past few years.

Oak woodlands are generally thought to be in poor condition, although there are no reliable inventories that document the full extent. Oak advocates generally believe that 100% of Oregon white and Garry oak savannas from British Columbia to the southern end of the Willamette Valley have been lost or significantly degraded by farming, urban development, and fire suppression. In many cases the grand old oaks are still there, but the understory communities have been lost. Fortunately, a budding oak restoration community has sprung up, led by an informally organized, somewhat anarchic Oregon oak woodland group in the Willamette Valley and a more disciplined Garry Oak Ecosystem Recovery Team in British Columbia. These groups are populated by nature nerds from local and federal agencies, progressive landowners, and the odd independent consultant. Oak restoration efforts are taking place across the entire ecological range, from northern California up through Oregon and Washington, reaching into Vancouver Island and spreading east through the Fraser River Valley and Columbia Gorge.

Freshwater wetlands have been the target of restoration, or at least mitigation work, for decades. What started as creation of simple "duck donuts" has evolved into complex approaches working with site hydrology and multiple plant communities. Restoring wetlands is still very challenging and will remain so, but there is no question that both the art and the science have improved. There remains the question of whether we are gaining or losing acreage because most wetland restoration is still compensating for ongoing wetland destruction.

Restoring *tidal wetlands* is a more recent undertaking, driven largely by efforts to protect salmon. Dike breaching or removal has proved quite successful at initiating restoration, but in most cases it takes years of tidal action for marsh systems to rebound.

The *sagebrush steppe* ecosystem of the Northwest interior has been the focus of restoration work by the Bureau of Land Management, Fish and Wildlife Service, and other agencies. There have been no comprehensive reviews on the status of restoration across the region, so it is difficult to know whether overall degradation has stabilized or decline continues. Restorative efforts are showing success on a site-by-site basis, but we are probably still losing ground to the cheatgrass–wildfire cycle (S. Link, personal communication, 2005).

State of the Art

One initial goal of this book, to describe the state of the art of ecological restoration, turned out to be impossible to meet. The practice of regional restoration is gaining such strong momentum, with new projects being planned and implemented every day, that the learning curve is too steep to capture. Thus a book like this one, more than 5 years in the making, finds itself chasing an ever-growing catalogue of projects and a rapidly expanding knowledge base. A secondary goal, which we believe has been better achieved, was to profile the wide range of restoration projects in progress, embed them in a geographic context, and raise the bar on restoration practice. In case studies presented throughout Parts II and III, we have gathered what we believe to be the best of the best. This book should encourage students, practitioners, and policymakers engaged in ecological restoration to reach high and not settle for mediocrity. We believe that every ecosystem profiled in this book can be restored or, rather, put on the path to restoring itself. We are just beginning to reach a point of knowledge, civic attention, and funding sufficient to support real gains. Governments, communities, and private landowners all across the region are working at ecological restoration, and if this effort is able to prevent further loss, the next generation will begin to make substantive gains.

References

Agee, J. A. 1993. *Fire Ecology of the Pacific Northwest*. Island Press, Washington, DC.

Alt, D. and D. W. Hyndman. 1995. *Northwest Exposures: A Geologic Story of the North West*. Mountain Press, Missoula, MT.

Arno, S. and C. Fiedler. 2005. *Mimicking Nature's Fire*. Island Press, Washington, DC.

Booth, D. 1991. Urbanization and the natural drainage system: impacts, solutions and prognosis. *Northwest Environmental Journal* 7: 93–118.

Boyd, R. 2000. *Indians, Fire, and the Pacific Northwest*. Oregon State University, Corvallis.

Burke Museum Web Site: www.washington.edu/burkemuseum/kman/kman_home.

Ecotrust. 2002. *Siuslaw River Watershed Assessment*. Unpublished report available at www.inforain.org/dataresources/.

Franklin, J. F. and C. T. Dyrness. 1973. *Natural Vegetation of Oregon and Washington*. Pacific Northwest Forest and Range Experiment Station. GTR PNW-8. USDA Forest Service, Portland, OR.

Goble, D. D. and P. W. Hirt. 1999. *Northwest Lands, Northwest Peoples*. University of Washington Press, Seattle.

Hulse, D., S. Gregory, and J. Baker. 2000. *Willamette River Basin Atlas, Trajectories of Environmental and Ecological Change*. Pacific Northwest Ecosystem Research Consortium. Corvallis: Oregon State University. Available at www.fsl.orst.edu/pnwerc/wrb/Atlas_web_compressed/PDFtoc.html.

ISAB (Independent Scientific Advisory Board). 2001. *A Review of Salmon Recovery Strategies for the Columbia Basin*. Available at www.nwcouncil.org/library/isab2001.

Lichatowich, J. 1999. *Salmon without Rivers: A History of the Pacific Salmon Crisis*. Island Press, Washington, DC.

NOAA. 2005. National Marine Fisheries salmon status reviews and status review updates. Available at www.nwr.noaa.gov/1salmon/salmesa/pubs.

Pojar, J. and A. MacKinnon. 1994. *Plants of the Pacific Northwest Coast*. British Columbia Ministry of Forests and Lone Pine Publishing, Vancouver.

Robbins, W. G. 1997. *Landscapes of Promise: The Oregon Story 1800–1940*. University of Washington Press, Seattle.

Schoonmaker, P. K., B. von Hagen, and E. C. Wolf. 1997. *The Rain Forests of Home: Profile of a North American Bioregion*. Island Press, Washington, DC.

Sheets, E. W. 2004. Restoring salmon: how are we doing? *Open Spaces. Views from the Northwest* 7(1). Portland, OR. Available at www.open-spaces.com/issue-v7n1.php.

Trail, P. 1998. Recognizing paradise: the world discovers the Klamath–Siskiyou ecoregion. *Jefferson Monthly Magazine* 22(2).

Umatilla Indian Reservation Web Site. n.d. *Salmon Success in the Umatilla River*. Available at www.umatilla.nsn.us/umariver.html.

Chapter 2

Ecological Restoration

Dean Apostol

The art and science of ecological restoration have come a long way from their origins. Historic threads of restoration include land management by traditional people, the writings of George Perkins Marsh (*Man and Nature*), reforestation and soil conservation in Europe and North America in the late nineteenth and early twentieth centuries (including the work of Arthur Sampson in Oregon's Wallowa Mountains), the naturalist school of landscape architecture inspired by Jens Jensen in the upper Midwest, Aldo Leopold and Theodore Sperry's efforts to transplant prairie sod at the University of Wisconsin in the 1930s, wetland mitigation projects spawned by the U.S. Federal Clean Water Act in the 1970s, and multiple stream and salmon enhancement projects over many decades in the Northwest (Berger 1990, Hall 1997). The Society for Ecological Restoration (SER) knit many of these threads together in 1987, has since grown to include members from thirty-seven nations, and has added the modifier "International" to its name (SERI).

Many of the contributors to this volume expressed a desire to include background information on ecological restoration as a way of qualifying their discussions. Rather than have each team do this, we decided to create this chapter to address broad restoration definitions and issues that are consistent across ecosystem types and projects. Most of the information, unless noted otherwise, is drawn directly from SER online publications, particularly the *SER International Primer* (2004) and the *Guidelines for Developing and Managing Ecological Restoration Projects* (Clewell et al. 2000). The reader of this book is encouraged to read these for greater detail and insights. As online documents, they may be updated from time to time because restoration research and practice are evolving rapidly.

Defining Ecological Restoration

After more than a decade of debate, SERI settled on the following definition: *Ecological restoration is the process of assisting the recovery of an ecosystem that has been degraded, damaged, or destroyed.* This is an elegant definition that was wrestled to the mat by a number of dedicated restoration practitioners and scholars (Higgs 2003). It is important to note that *restoration ecology* is defined as the science that provides concepts, models, and methods, whereas *ecological restoration* is the practice that puts knowledge in place. Because restoration is still young, overlaps a number of fields, and has not developed an extensive academic infrastructure, there is a lot of synergy between practice and science. Unlike in other fields, it is often the practitioners who are leading the science and academics by documenting, experimenting, and adapting rather than the other way around.

In the late 1980s the debate over defining restoration centered on the balance between *mitigation* and *restoration*. Midwest prairie and oak ecologists and legions of volunteers had been at work since the 1970s trying to replicate as closely as possible the historic composition and structure of ecosystems that were down to their last small remnants (Stevens 1995). In the meantime, the Clean Water Act spawned a wetland mitigation industry that had bureaucratic, commercial, and professional interests and a lot of money at stake. The motivations, resources, and expectations of these two camps were and still are quite different, so the argument over what constitutes restoration became political and almost theological as well as scientific. The more people grappled with the question of what restoration is, the more complicated the debate became (Figures 2.1 and 2.2).

FIGURE 2.2. A wetland mitigation project in Portland, Oregon has value but falls short of the *restoration* definition. (*Photo by Dean Apostol*)

To qualify as restoration, did a project have to return a site to an exact preexisting composition and structure? Some argued for this, but others argued that it was an impossible and inappropriate goal (Clewell et al. 2000). Which preexisting ecosystem should we choose? Many have argued for ecosystems predating European contact, but as the SER became more international this idea quickly became quaint. At one SER conference a British member raised the question that because Europeans are by definition the indigenous people of Europe, how could they restore something to a precontact state? Americans and Canadians were at first taken aback but soon accepted the point.

FIGURE 2.1. Volunteers remove invasive Siberian elms from a Chicago prairie. (*Photo by Dean Apostol*)

The concept of dynamic ecological trajectories also complicated the issue. Older ideas of ecology centered on stable climax ecosystems had been largely displaced by dynamic equilibrium theories, in which disturbance and change are the norm, stability the exception. This implies that the state of an ecosystem in the past cannot be replicated precisely, simply because it has changed over the past 100 or 1,000 years. Some argued for going back as close as possible to the point where the ecosystem suffered a radical change, then working with natural processes to allow it to continue its prior evolutionary path. This clarification helped, but still left out a lot. Some important ecosystems were closely tied to human culture. Others were and are threatened by climate change. Mined areas often are so altered that they cannot be returned to any preexisting state because the topography has been radically modified and topsoil lost.

Motivation, process, and product also became important parts of the discussion. The "why" of restoration was a big subject of debate and still is. For example, suppose one is restoring a wetland in one location in order to make up for the planned destruction of one in another location. Even if a perfectly proportional project could be achieved, one could question its value, given the large amount of resources spent to basically break even. Some activities, such as Midwest prairie restoration, were inherently additive. That is, the motivation and results were not driven by a desire to destroy more prairie and replace it elsewhere. All inventoried prairie habitats, at least in densely settle places such as Illinois, have already been protected, so new restorations add acres. But by contrast, wetlands are still widespread and stand in the way of further development, particularly in ports and along the routes of highways that are to be widened.

In the Pacific Northwest, the link between commercial forestry and restoration has raised the motivation issue. There is widespread support for thinning fire-prone dry forests, particularly east of the Cascades and in the Siskiyou Mountains, in order to reduce fire intensity and restore the open woodland character that existed before fire suppression. But many environmentalists have argued that there should be no commercial gain from restoration. This issue arose because restoring fire-prone forests, riparian woodlands, and even old-growth forests in some cases generates commercially valuable products. Such gains may become even more common if biogenerating power plants come on line that can use small-diameter wood and brush.

Early on, the definition of ecological restoration kept growing larger and more cumbersome. As each new issue came up, words or phrases were tacked on to account for it. But over time, patient wordsmiths were able to shrink the number of words while broadening the definition to encompass a wide range of projects and to get to the heart of motivation. The point of restoration has now become the *assisted recovery* of an ecosystem—brilliant because nature has impressive self-healing powers and often needs only a nudge in the right direction.

Mitigation sometimes can be but most often is not the same as ecological restoration. It describes projects that are designed to compensate for planned or preexisting environmental damage and is generally associated with land, port, or highway development projects that damage or destroy existing wetlands. A few mitigation projects rise to the higher standards of restoration, but most do not. *Ecosystem fabrication* and *environmental engineering* describe projects in which a designed system is put in place that may or may not be associated with a former condition or reference ecosystem. For example, a designed wetland and bioswale complex may be built to help manage stormwater from a new building or campus. It may provide important ecological functions that protect downstream aquatic habitats, have mostly native plants, and be environmentally beneficial as compared with traditional landscaping (Figure 2.3). But it rarely qualifies as restoration.

FIGURE 2.3. A bioswale designed to reduce impacts from impervious surfaces to local streams. (*Photo by Marcia Sinclair*)

Process-Based Restoration

A related, very important concept in ecological restoration has been the shift from structure-based to process-based restoration. In the 1980s, most restoration was strongly related to horticulture. It involved mainly growing and outplanting native vegetation. Streams were restored with gabions, boulders, and logs carefully placed to provide fixed habitats. But over time, it became clear that ecological processes often hold the key to successful restoration. It makes little sense to plant a prairie unless fire or a suitable substitute can be harnessed to keep woody vegetation at bay. If continuing erosion overwhelms the ability of a stream to process sediment, then what is the point of building structures? Stream ecologists now work to identify barriers that prevent the system from restoring itself, including revetments, diversions, levees, culverts, and roads that trigger sedimentation (Figure 2.4). Direct structure improvements such as additions of large wood are still done, but increasingly these are viewed as temporary measures used until the stream can fend for itself.

Agreeing on What a Restored Ecosystem Is

Even though SERI finally was able to create a simple, short, and elegant definition, they could not stop at that point. The *SER International Primer* runs to fifteen pages, and a second companion document, *Guidelines for Developing and Managing Ecological Restoration Projects*, identifies fifty-one steps, all the way from initial concept to publishing. One key issue is what should be expected from a restored ecosystem. The *Primer* lists nine expected attributes. First, native plant assemblages should resemble reference ecosystems in structure and composition. Second, the ecosystem should have only native species, except when the reference sys-

FIGURE 2.4. A breached dike in the Siuslaw River estuary near the central Oregon coast allows a tidal wetland to begin restoring itself. (*Photo courtesy of the Karnowsky Creek Restoration Team, Siuslaw Watershed Council*)

TABLE 2.1.

Attributes of restored ecosystems, with an oak woodland example.

Attribute	Description	Northwest Example
Plant assemblage	Similar to a defined reference, either a site or a written description.	Garry oak woodland.
Composition	All or nearly all native plants.	Oak overstory, groups of hazel, snowberry, oceanspray, Roemer's fescue grass matrix, camas, shooting star, wyethia, lupine, paintbrush forbs.
Functional groups	Trees, shrubs, grasses, forbs, lichens present in correct proportions.	See above.
Reproduction	System shows ability to self-reproduce.	Oak seedlings, bunchgrass spreading, forbs seeding in.
Development stage	The composition and structure are in a successional stage normal for this system.	Age and density of oaks strongly related. Older age should show larger overstory tree distance, with regeneration in clumps.
Ecosystem context	The restored system fits within a larger conservation framework as a patch, part of a matrix, or corridor.	Oak woodland is a stepping-stone habitat in the south Puget Sound oak–prairie complex.
Off-site security	System is protected from threats originating on other properties.	Scotch broom patches on neighbor's pasture are being removed through a cooperative agreement that uses frequent mowing and spot herbicides.
Resiliency	The restored system is ready to respond to disturbances within the normal range of its ecology.	Prescribed or wildfire is held in the understory.
Self-sustaining	System is able to replicate itself.	Oaks are regenerating, grasses and forbs spreading, new niches filled.

tem is a cultural landscape that includes nonnative and domesticated ones. Third, a restored ecosystem should include all the main functional groups or should be able to attain them through natural colonization. Fourth, it should be able to have self-sustaining populations.

Fifth, a restored ecosystem should exhibit functions that are in the normal range for its stage of development. Presumably this means it is on a historically consistent trajectory of growth and change. Sixth, the restored ecosystem should be integrated into a larger functioning landscape matrix or mosaic. This is a very important consideration in many Northwest restoration efforts, including oak and riparian woodlands, old-growth forests, and watersheds. The reader will find numerous examples of contextual restoration in this book. Seventh, restored ecosystems should be safe or secured from outside threats. For example, a restored prairie patch downwind of a major Scotch broom infestation might not be secure unless steps are taken to address the broom. Stream restorations in particular must pay attention to upslope sediment risks.

Eighth, a restored system should be resilient. Ecosystem resiliency is its ability to return to a predisturbance state in terms of key elements (i.e., composition) and processes, such as nutrient or carbon cycling. Finally, a restored ecosystem should be self-sustaining, at least at a similar level to reference ecosystems. Some fluctuation is expected within a normal range (Table 2.1).

It is rare for any restored ecosystem to fully exhibit all nine of these attributes. The best examples may be found in restorations that are in already fairly intact and healthy landscapes that need only a slight nudge or assistance. (This theme recurs throughout this book: The more degraded an ecosystem is, the harder it is to restore it.) But these nine establish a high ideal that equates restored systems to "natural," or reference, ones and thus can serve as a guiding star to restorationists. The *Primer* points out that restored ecosystems add to the natural capital of a landscape. This term makes some uneasy because it introduces economic valuation to what many think should be a purely altruistic pursuit. But a restored upland forest may intercept and filter rainwater as well as a reference forest and thus be an important asset in protecting a community water supply. A restored prairie that expands the range of an endangered butterfly has enhanced local biodiversity and improved the landscape aesthetic, which may also improve local economics. On the other hand, most restorations to date have not achieved the same levels of ecological function as reference ecosystems. Whether this is a matter of flawed technique or simply not enough time is unknown.

The Need for Restoration

The SER *Primer* lists four terms that describe degrees of change in natural and cultural ecosystems that may trigger restoration. The first, *degradation*, is defined as gradual, subtle, slow-acting changes that compromise ecosystem integrity or health. Northwest examples include invasion of oak or pine woodlands by native fir trees, invasion of remnant prairies by Scotch broom or forests by English ivy, siltation of aquatic ecosystems from upslope logging or farming, and cheatgrass taking hold in an area of sagebrush steppe that has been overgrazed. In all of these cases, the basic ecosystem structure and composition are still there but have started to become compromised.

Given a long enough time, ecosystem degradation progresses to *damage*, a term used to describe more acute and obvious ecosystem changes such as fir trees overtopping the oaks, broom crowding out the native prairie grasses, and ivy causing trees to topple and smothering the native understory (Figure 2.5). Damaged ecosystems still have natural elements in place, but they are teetering on the brink. Unless corrective measures are taken soon, the ecosystem could be lost.

Destruction is the next downhill stage, in which there is nearly complete removal or loss of major ecosystem elements. It will take heroic and usually expensive efforts to reverse the negative impacts. Examples of destroyed ecosystems in this region abound. There are prairies at Fort Lewis, Washington, that have completely been lost to Scotch broom. Many urban streams are so channelized and simplified that restoration is nearly impossible. The mighty Columbia River is now a series of slackwater lakes from Bonneville to Canada in the north and Idaho in the east. Areas of extensive cheatgrass infestation in the sagebrush steppe have suffered repeated fires that have wiped out most native plants. A few fragments of natural ecology may persist, but only barely.

A *transformed* ecosystem is similar to a destroyed one in some respects, but it has been completely and deliberately converted to a different land use.

FIGURE 2.5. Invasive Scotch broom threatening remnant native prairie south of Puget Sound in Washington State. (*Photo by Peter Dunwiddie*)

In the Northwest, this could be an urban neighborhood occupying the site of a former old-growth conifer forest, a stream buried in a culvert with a highway above it, or a wheat field where there once was a prairie. Although the Northwest arguably has more ecologically healthy land than anywhere else in the lower forty-eight states and southern Canada, there are few areas left that have not been at least slightly degraded. Alpine, subalpine, old-growth conifer forests and perhaps sagebrush steppe may be the least degraded regional systems, along with streams and watersheds in the northern part of the region.

There is much debate, particularly in watershed restoration circles, about whether to focus limited time and money on protecting still intact systems, restoring slightly degraded ones, or addressing the worst first. A general principle is to secure the best systems first by protecting them from outside threats. For example, if there are a few areas of mature or old-growth forest in a watershed that may be logged, it is better to find a way to buy them or get conservation easements in place and then start replanting or addressing degraded or destroyed areas. An exception is a highly degraded area, perhaps a road built across a steep, unstable midslope, that could fail in the next rainstorm and wipe out miles of high-functioning aquatic habitat. It may be best to deal with the road first and then initiate the long process of securing the remnant forest. These are always difficult choices that entail risk analysis and creative thinking (Figure 2.6).

Restoration ecologists have coined the terms *reference ecosystem* and *reference landscape* to describe models or images of what a project is attempting to achieve. The nearest mature riparian woodland might serve as the model for a restoration project, assuming that physiographic conditions are similar enough. Reference ecosystems can be actual sites, written descriptions, or ideal archetypes. A restored system can never be an exact copy of the reference because it has to be its own entity. A reference acts as a carrot that compels the restoration horse forward. There is an art to divining reference ecosystems that includes detective work. The term *reference landscape* is used when one is restoring a larger area that includes two or more discrete ecosystems, say a prairie, an oak woodland, and a riparian forest. The reference ecosystem or landscape does not have to be a real place. It can be derived from written descriptions of a past ecosystem, much as Stephen Packard has done with oak savannas in Illinois (Stevens 1995). In the Pacific Northwest, early government land surveys (section line field notes and maps), military surveys, oral histories, pioneer journals, photos archived at historical museums, and the diaries of Lewis and Clark have all been conscripted into the service of restoration. Paleoecology, a field that studies fossil pollen trapped in peat bogs, fire scars on old stumps, and rodent middens has become an important restoration ally. This work has allowed us to take the clock back at least to times just after the glaciers' most recent retreat. Paleoecology may also be the key to understanding how global climate change will create new ecological trajectories and plant assemblages that restorationists increasingly will have to anticipate.

As mentioned earlier, ecological restoration has become much more process and function oriented over the past few years. This means that it is the dynamic rather than static elements of ecosystems that are the target. Fires, floods, and succession are three critical processes that regional restorationists often work with and around. The holy grail of restoration is to get an ecosystem to the point where it is self-sustaining.

For many years, ecologists described systems in terms of stable climaxes, meaning composition and structure that could hold in place indefinitely. Now, stable ecosystems are described as those that can maintain a *dynamic equilibrium*. This means that they can absorb stresses or disturbances and shrug them off. They are not transformed; they can adapt as long as they have all their parts and the disturbance is within the range of their ecological history. Regionally, many ecosystems have been pushed to a dangerous point with regard to stability.

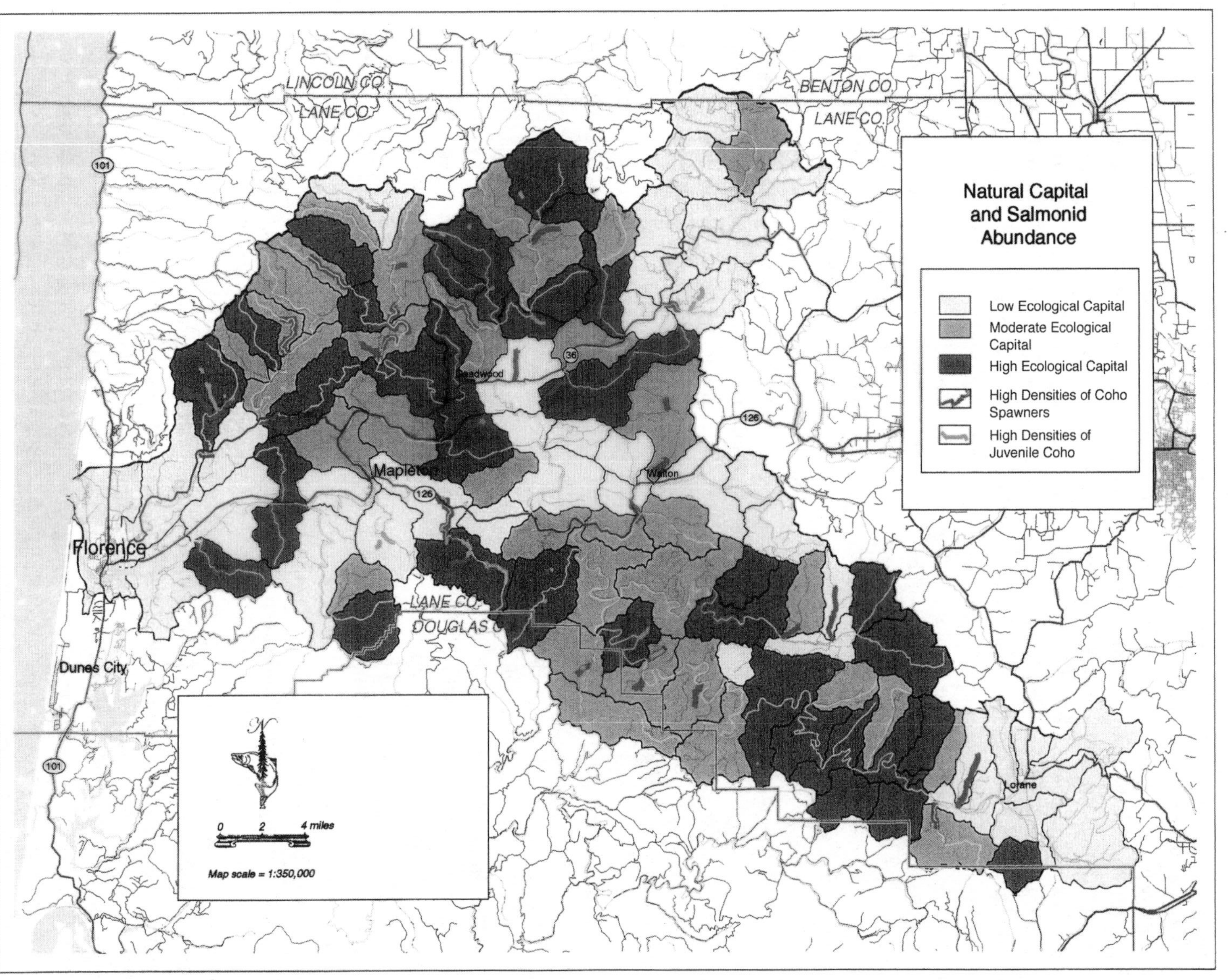

FIGURE 2.6. Natural capital map shows a positive relationship between landscape condition and salmon abundance in the Siuslaw watershed and suggests priority catchments for conservation and restoration. (*Map courtesy of Ecotrust*)

Old-growth forests are far below their lowest estimated point of extent, and remnants are very fragmented and concentrated at higher elevations. Interior forests have very high fuel loads that they never experienced in the past. Fires fueled by cheatgrass in the sagebrush steppe create conditions that local plants cannot survive. Streams have been stripped of wood, confined into channelized ditches, or transformed into a series of lakes. The multiple listings of endangered and threatened species in all of our ecosystems are largely a consequence of induced instability.

Setting Goals and Objectives

Ecological restoration is by definition a goal-oriented process. As such, it involves planning, research, project management, public involvement, budgeting, and all the skills typically associated with applied ecological disciplines. In the absence of clearly stated, achievable goals, there is no way to know whether a project will ever reach a successful conclusion. The early stages of restoration projects can be exasperating to hands-on doers who want to get out there and hack down the blackberries or plant riparian zones.

In general, goals should be stated broadly and focused on restoration of composition and structure, species, functions, or ecosystem processes. Good project planning should consider all of these at some level. For example, suppose the project is to restore old-growth conifer forest in an area of young monoculture plantations. The reference system might be a nearby lowland Douglas fir–hemlock patch of old growth. Desired functions might include rainfall capture, habitat for multiple species, and carbon accumulation. An initial composition and structure would seek to create a "natural" composition and pattern that would exist in a young forest on a trajectory toward old growth. This may mean introducing an intermediate disturbance that removes 30% or more of the existing trees, in a pattern that reflects a windstorm or fire (process).

Objectives are more detailed statements that can be measured at some level. For example, a broad goal may be to set a 40-year-old conifer plantation on a path toward development of old-growth forest structure. A companion objective could be to achieve 90% of the plant composition diversity one would expect in a natural 50-year-old stand within 10 years of a treatment. This diversity could be expressed as a species list, with rough proportions for overstory and understory plants, using a reference site as the model.

Restoration Planning

Restoration plans can include written descriptions, plant lists, removal targets for invasive species, drawings, maps, specifications, timetables, and budgets. There is no one right way to do a plan, but there are probably wrong ways. The selected approach, level of detail, and form of a plan depend on the experience and resources of the landowner or managers and on the project leader. Restoration project budgets can range from a few thousand dollars up into the millions. Clearly, the higher the budget and more complex the project, the greater the rigor of planning needed.

Ecological restoration is a multidisciplinary or interdisciplinary field, and this diversity also influences planning processes and products (Figure 2.7). For example, landscape architects rely mostly on drawings, plans, and specifications, whereas ecologists favor written descriptions of desired conditions. Clewell et al. (2000) divide restoration planning into conceptual, preliminary, and implementation stages. At the conceptual stage, the basic rationale for restoration and a general strategy are put forward. Typical elements of a concept plan include identifying site location, boundaries, ownership, analysis of restoration need, and description of what is to be restored (e.g., a riparian woodland). Clewell also suggests that at this stage there should be a description of the type of restoration envisioned (repair, creation, or replacement of an

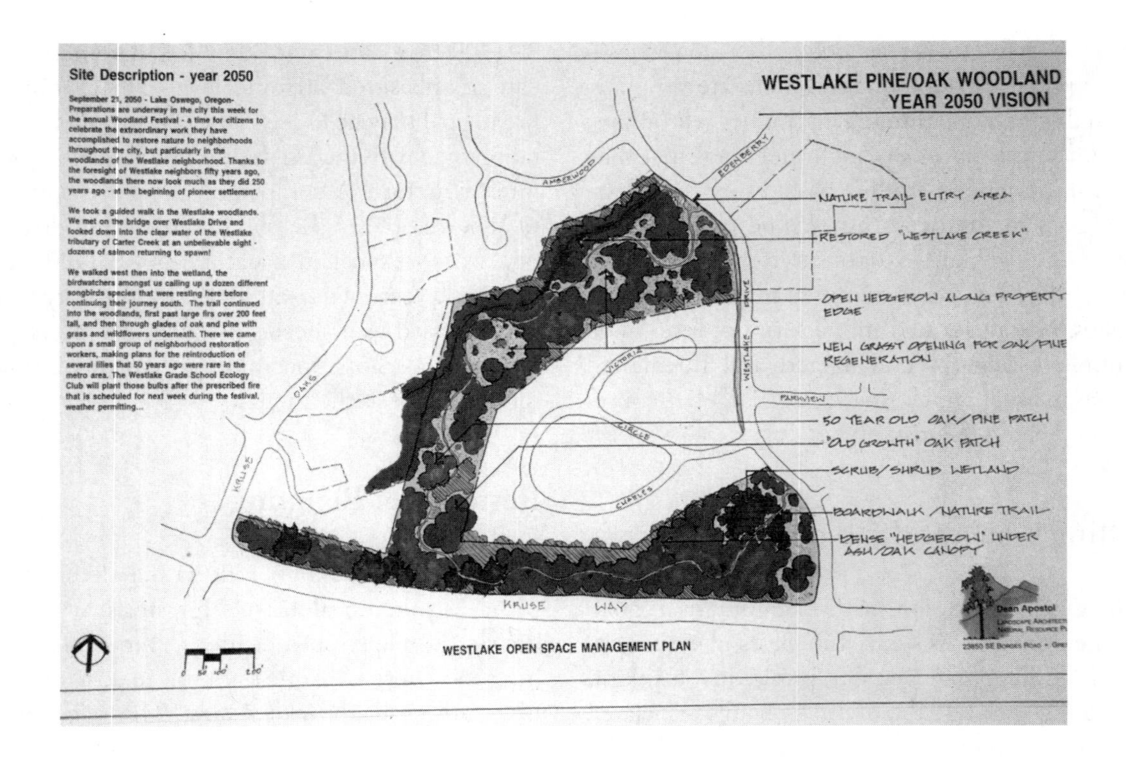

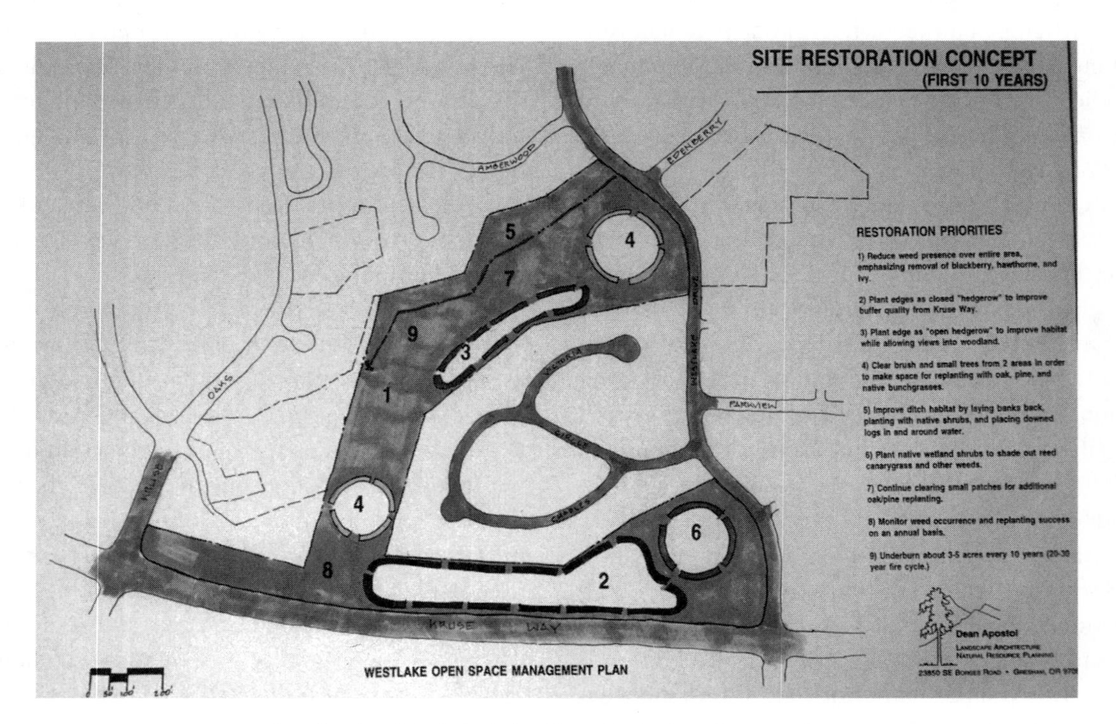

FIGURE 2.7. A restoration plan for the Westlake Woodland in Lake Oswego, Oregon communicates long-term vision (50 years) and shorter-range strategies (10 years). (*By Dean Apostol*)

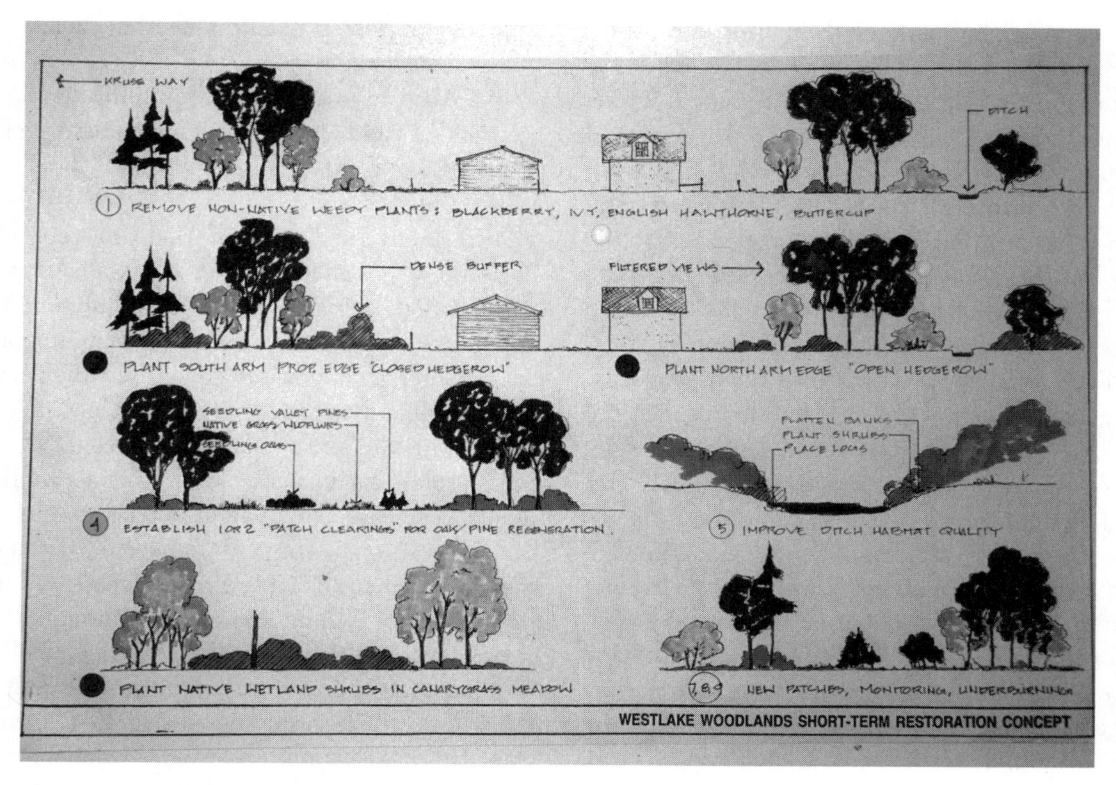

FIGURE 2.7. (*Continued*)

ecosystem). Additional concept plan elements include restoration goals, a description of physical conditions, identification of stressors, and possible funding sources.

Once the decision is made to proceed, one moves to more detailed planning and implementation strategies. If one has not already been chosen, a qualified project manager is appointed who has the skills and experience to shepherd the project to a successful conclusion. Clewell goes into step-by-step details, such as setting specific objectives, budgeting, and listing tasks. One concern is that he suggests waiting to engage in public involvement until step 31, way too late for any project that might be controversial (i.e., chemical use, logging, significant aesthetic impacts, prescribed burning, and removal of roads, to name just a few).

Public Involvement

The best way to think about public involvement is to use the golden rule: "Do unto others as you would have them do unto you." When would you want to first learn about a significant change to a landscape you live near, recreate in, or care about for any reason? Ecological restorationists have a hard time believing that their work can be controversial because they are repairing damaged ecosystems. Who can be against that? It turns out that plenty of people can be against it. Even though most people tend to support restoration in principle, many may not support specific techniques commonly used by restorationists (Barro and Bright 1998). Aggressive weed removal or thinning projects can radically change the visual character of a site. Cutting trees, particularly mature native ones, is nearly always resisted by someone. Herbicide use causes great concern among local residents. Placing large logs in streams can block kayakers and even endanger them. Closing a road may block access to a favored place. Make no mistake, restoration is often controversial. Education in public involvement is acquired primarily through experience. To avoid learning in the school of hard knocks, there are a few simple suggestions.

First, be strategic. This means identifying early on everyone who might be interested in the project. Make a list and check it twice. Group publics into categories, such as environmental interests, anglers and hunters, and local landowners. Recognize that an individual might belong to more than one category. Second, create a short, concise, and honest description of the project so that any layperson can understand what you are proposing and why very quickly. Third, understand why you are engaging the community. If you are working on public land, then those you are talking with own that land, meaning that they have every right to be engaged and to tell you what they like or do not like about what you are proposing. Listen to them and do not get defensive. Often people who have lived in the area for a long time may have valuable information to offer about a site.

Fourth, before you call a meeting or write an article for the local paper, it is helpful to develop key messages or "talking points." Though often misused and overused in public relations, these are very useful tools. Examples include the following:

- "We are considering a project to restore a healthy riparian ecosystem on Bailey Creek, in Cascadia County Park."
- "This creek once had healthy populations of salmon and trout but has become degraded over the years by land development, including at the park."
- "This project is part of a larger watershed restoration effort that includes many private landowners, state and federal agencies, and Cascadia County."
- "We want to know what you think about this project and whether you may want to get further involved."

It is useful to ask yourself what you want from those you are contacting. You may want support, participation, permission, or contributions. You may just want them to decide to not fight the project. Another good idea is to go to where the people are first rather than expecting them to come to you. This means attending organization meetings, placing articles or announcements in appropriate media, and using e-mail and the telephone. A big public meeting may or may not be needed. If a big meeting is scheduled, it is wise to prepare well and have a neutral facilitator to manage things.

Finally, restoration projects should include both an evaluation of the public process and an evaluation of the restoration itself. Did it go as expected? Did people understand what was proposed? Did they at least think the process was fair, even if they did not like the outcome? For more information on the art of public involvement, including training programs, contact the International Association for Public Participation (www.iap2.org/index.cfm); regional chapters in the Pacific Northwest include those of British Columbia, Cascade, northern California, and Puget Sound. Public involvement is increasingly a specialized professional field, and restorationists should not wade into troubled waters naively.

Monitoring and Adaptive Management

Before project implementation has begun, a monitoring program must be thought through and put in place. Without monitoring we cannot learn, and without learning we cannot adapt restoration practices. The essence of monitoring is to understand how well a restoration is working by collecting new data to bridge gaps with our existing knowledge base.

Noss and Cooperrider (1994) define monitoring as the periodic measurement or observation of a process or object. Monitoring has been common in natural resource management for many decades. Foresters have long monitored the growth and development of plantations, often keeping detailed records. Wildlife managers have always monitored game species, through bag numbers and the general health of animals. Ecologists have used transects and quadrats to measure changes in plant community composition and structure. There are three basic types of monitoring, all of which can be applied to ecological restoration projects.

The first of these is *implementation monitoring*, which is simple and straightforward. It answers the basic question of whether one did what one said one would. A restoration plan might entail removing exotic species and replanting with natives over a given area. Implementation monitoring is a basic checklist that documents what was actually done. Upper managers and politicians who provide funds or direction on annual cycles like implementation monitoring because it lets them know that a project was accomplished. Some implementation monitoring is done to meet rules and regulations or certification standards and to ensure legal compliance.

Effectiveness monitoring is more challenging. It asks, "Did the project actually work as intended?" For example, an interior forest restoration might have called for thinning and underburning a mixed fir, oak, and pine patch with the intent of restoring an open pine–oak woodland in order to improve habitat for western gray squirrels. Effectiveness measures could include whether the desired conditions, based on a reference ecosystem, have been created. These could include tree growth, composition, expected regeneration of native understory vegetation, and the presence of squirrels. Effectiveness monitoring often must wait until some time after implementation. For example, understory regeneration might not be expected to develop until several years after the initial work has been done. Even if a restoration was implemented precisely according to plan, where effectiveness is poor, the restoration has to be considered less than fully successful.

Validation monitoring adds a third level of complexity. It asks whether the results were a consequence of the actions taken and whether the assumptions and models used in developing the plan were correct (Noss and Cooperrider 1994). For example, suppose we had assumed that if we successfully restored the open pine–oak woodland just described, then western gray squirrels would take up residence within a certain period of time. The project is completed (implementation monitoring), and the desired forest structure and composition have been achieved (effectiveness monitoring), but after 10 years there are no squirrels. The problem could lie at a different ecological scale. For example, it may be that the nearest squirrel populations are too far away, or there are barriers to dispersal. There could be competition from nonnative squirrels that happen to like the same habitat conditions. It may also be the case that what we assumed about gray squirrel habitat was incorrect. Or perhaps some unnoticed element of the ecosystem is still missing from the restored patch.

To avoid becoming a pointless exercise in data gathering, monitoring must be well designed. If the wrong indicators are chosen for any objective, or the inappropriate elements measured, or the scale is too large or small, then the data will throw one off. To be effective, a monitoring program must have a clear vision of what it is intended to achieve, there must be a true commitment to implementation on the part of local managers, and the information gained must be incorporated into future planning. This is all much easier said than done when even the budget for implementation is too small by half.

Good record keeping is part of the picture, of course. A simple spreadsheet that includes actions taken, implementation dates, expected results, and monitoring dates and results is helpful. In this way the monitoring program is clearly linked to the original restoration aims of the plan. It is also helpful if external auditing is done, assuming it is clear and honest in its findings. There could be a comment section noting implications and suggesting remedial actions. A benefit of this type of record keeping is that as employees come and go there is a record to help everyone understand what point the project has reached and where it should go next.

If a restoration plan is implemented in phases that are spaced at multiyear intervals, the main monitoring periods could coincide but precede the next action by 1 year, so that the plan is reviewed and revised in time to adjust the next phase of implementation. Where implementation phases are at longer intervals, such as 10 or more years in the case of an old-growth forest restoration, monitoring may be needed more frequently between phases and a regular timing developed. Broadly speaking, implementation monitoring is carried out soon after the restoration action, effectiveness monitoring follows at a suitable time interval, and evaluation monitoring is conducted later still. The feedback then goes to area managers, who may use it to help design other restoration projects or to adjust their next planned intervention on the subject site.

Many useful tools are supplementing or replacing old, tattered field ecology notebooks, including hand-held computers linked to global positioning systems that tie data to a specific place in real time. This allows for easy transfer to geographic information system layers. Older methods, particularly aerial and site photographs, are still very important and should not be discarded.

Depending on the scale and complexity of a restoration project, it may be necessary to involve a full multidisciplinary team to carry out various monitoring elements or tasks because the range of factors involved in the original design may warrant specialized expertise.

Adaptive management involves a coordinated relationship between initial planning, monitoring, evaluation, and adjustment. It is a logical evolution of monitoring that introduces active feedback of findings. All ecological restoration is still essentially experimental (Noss and Cooperrider 1994). With imperfect knowledge, we rely on expert opinion and assumptions to predict results. Because it is not practical to establish huge experimental restoration programs for every Northwest ecosystem, land managers have to use other methods to increase the overall knowledge base and expertise. To an extent this can be accomplished by approaching monitoring as an experiment and a unique chance to learn and improve.

The key in adaptive management is to include research content as part of monitoring. For example, in the hypothetical pine–oak forest restoration project described earlier, validation monitoring is set up so that a research question about the use of the woodlands by gray squirrels and an experimental design are included from the outset. Then, if the expected results fail to appear and the conclusions suggest that the restoration prescription erred, the project can be revised and adjusted at the next phase or on the next project site. It should be noted that it is always easier to remove additional trees later than to grow new ones, so if the canopy density that squirrels need was unknown, then the initial thinning should err on the light rather than heavy side.

Restoration Certification

SER has worked hard to clear up definitions and establish ground rules for practice. An important and very controversial debate has centered on the idea of certification. Independent certification has become a very important tool in ensuring that a project, product, or land management strategy is progressive and sustainable. For example, the Forest Stewardship Council has strict certification standards for sustainable forestry that are applied worldwide. Organic agriculture is certified by the U.S. Department of Agriculture. The Wildlife

Federation certifies backyard habitats. There are also a number of "salmon-safe" certification programs in the Northwest.

Ecological restoration lacks specific legal standing, so almost anyone can claim that a site has been "restored." Restoration certification has been proposed as a way to establish universal standards. As with other certification programs, the practitioner, the site, or both could be certified. But because ecological restoration occurs over many ecosystems and incorporates a wide range of professional disciplines and volunteer organizations, professional certification has been problematic. Being proficient at restoring sagebrush steppe ecosystems, for example, would not make one the best choice for planning an old-growth conifer restoration.

The alternative, certifying the restored site itself, has gained greater support in SERI (E. Higgs, personal communication, 2005). The basic idea is that composition and structure standards could be created for various ecosystems, and once a restored site has reached this level it could be certified. Or, if a set of steps is followed and well documented, perhaps this would suffice, even if the product falls short. In this way, the bar for restoration practice can be raised, but low-budget volunteer and nonprofessional efforts could be honored alongside more expensive, professionally executed ones.

Summary

The practice of ecological restoration is still quite young, and the field is evolving rapidly. Restoration scholars, designers, and field implementers engage in intense debates over the meaning of words, the appropriate practices, the value of mitigation, and many other topics. Restored ecosystems, once dismissed as unnatural—mere cultural artifacts (Katz 1992)—are coming closer to fulfilling the ecosystem attributes of reference systems. Practitioners have a long way to go, but with the terminology debates behind us (it is hoped), this is the time to continue to improve and perhaps someday perfect the art, craft, and scholarship of ecological restoration.

References

Barro, S. C. and A. D. Bright. 1998. Public views on ecological restoration. *Restoration and Management Notes* 16(1, Summer): 59–65. University of Wisconsin Arboretum.

Berger, J. J. 1990. *Environmental Restoration: Science and Strategies for Restoring the Earth*. Island Press, Washington, DC.

Clewell, A. F. 2000. Restoring for natural authenticity. *Ecological Restoration* 18(4, Winter): 216–217.

Clewell, A., J. Reiger, and J. Munro. 2000. *Guidelines for Developing and Managing Ecological Restoration Projects*. Society for Ecological Restoration, Tucson, AZ. Available at www.ser.org.

Hall, M. 1997. Co-workers with nature: the deeper roots of restoration. *Restoration and Management Notes* 15(2, Winter). University of Wisconsin Arboretum.

Higgs, E. 2003. *Nature by Design: People, Natural Process, and Ecological Restoration*. MIT Press, Cambridge, MA.

Katz, E. 1992. The big lie: human restoration of nature. *Research in Philosophy and Technology* 12: 231–242.

Noss, R. F. and A. Y. Cooperrider. 1994. *Saving Nature's Legacy*. Island Press, Washington, DC.

Stevens, W. 1995. *Miracle under the Oaks: The Revival of Nature in America*. Pocket Books, New York.

Society for Ecological Restoration International Policy and Science Working Group. 2004. *The SER International Primer on Ecological Restoration*. Society for Ecological Restoration, Tucson, AZ. Available at www.ser.org.

PART II

Pacific Northwest Ecosystems

Part II is the heart of *Restoring the Pacific Northwest*. It consists of nine chapters that focus on ecological restoration in the major ecosystems of the region: bunchgrass prairies, oak woodlands and savannas, old-growth conifer forests, riparian woodlands, freshwater wetlands, tidal wetlands, pondorosa pine and interior forests, shrub steppe, and mountains. These are ecosystem types of convenience; they do not correspond to specific plant communities or published ecosystem classifications. In these chapters and in Part III, case studies provide detailed examples of restoration projects in progress.

In some cases, such as old-growth conifer, sagebrush steppe, and interior forests, they represent broad bio-geoclimatic plant zones that cover millions of acres. By contrast, bunchgrass prairies and oak woodlands occur as large or small patches within these larger zones and often are considered to be ecotonal or edaphic. Freshwater wetlands also are usually patch habitats but are highly variable in terms of composition and structure and are quite unevenly distributed, relating more to landform position and soils than to particular plant zones. Riparian woodlands are vital connecting corridors that course through the region, are closely associated with the aquatic condition of streams, wetlands, and lakes, and mediate between terrestrial and aquatic environments. Tidal wetlands are specialized ecosystems confined to coastal areas and the lower portion of major regional rivers. They include very diverse compositions and structures, from unvegetated but biologically rich mudflats to Sitka spruce swamps. Mountain ecosystems also include a diverse assemblage of plants, including wildflower meadows, huckleberry fields, and subalpine communities that occupy areas straddling Northwest timberlines.

In each of the nine ecosystem types, restoration issues, challenges, and techniques are similar. For example, old-growth forests, whether dominated by spruce, redwood, Douglas fir, or western hemlock, have similar structural properties and characteristics. All freshwater wetlands are very sensitive to changes in their hydrology, all tidal wetlands experience water fluctuations, usually on a daily basis, and all mountain ecosystems must cope with short growing seasons and deep snowpacks. Each ecosystem profiled deserves its own restoration book, and one (interior forests) already has one. All the contributor teams struggled to summarize and condense their knowledge and experience in the brief chapters allotted them. It is our hope and expectation that even readers who are familiar with or expert in the restoration of one of the ecosystem types profiled will find these chapters enlightening and of great use in broadening their knowledge.

Chapter 3

Bunchgrass Prairies

MARCIA SINCLAIR, ED ALVERSON, PATRICK DUNN, PETER DUNWIDDIE,
AND ELIZABETH GRAY

Prairie is a French word for "meadow" and was used by early explorers and settlers to describe open grassy areas in eastern and central North America. The term *prairie* describes an open habitat dominated by grasses and forbs, with little or no woody vegetation. In the Pacific Northwest, prairies are found at low elevations west of the Cascades, from the Willamette Valley of Oregon north to the Georgia Basin of southwest British Columbia. As defined here, the term does not include montane meadows, nor does it include grass steppe systems east of the Cascades that share some species but occur in a colder, continental climate, in contrast to the more moderate modified Mediterranean climate west of the mountains.

In the Pacific Northwest, approximately 350 species, subspecies, and varieties of plants are generally restricted to prairies and associated savanna and oak woodland habitats (E. Alverson and C. Chappell, unpublished data). The distribution of many of these species is centered in southwest Oregon and northern California, although some are predominantly found east of the Cascades. Thirty-two prairie-associated plant taxa are endemic or nearly endemic to Pacific Northwest prairies. Several mammals, birds, and insects are endemic to these habitats as well.

Historically, these Northwest prairies were not transient early successional stages of otherwise forested sites but persistent features on the landscape, maintained by frequent fires or by particularly stressful site conditions such as droughty soils. The continued existence of prairies for many centuries or millennia resulted in colonization and establishment of a diverse variety of long-lived perennial forbs and grasses, with an additional component of persistent annuals.

Three lines of evidence suggest that Northwest prairies evolved with high-frequency, low-intensity fires. First, monitoring of prairie vegetation after prescribed burns has shown that many native prairie species respond positively to these low-intensity, dry season fires (Pendergrass 1995, Streatfield and Frenkel 1997, Dunwiddie 2002). This suggests that these ecosystems evolved with fire and supports the idea that historically fire was an important presence in the prairie landscape. Evidence of Northwest prairie species' adaptations to fire include an abundance of taxa with bulbs, corms, or rhizomes, in which meristems are insulated by soil so buds are not damaged by the heat of the fire; increased vegetative growth and flower and seed production; and frequent establishment and survival of seedlings after burning. In addition, woody species such as Oregon white oak and snowberry, common in savannas and prairie ecotones, readily resprout after fire.

Second, in many areas, the suppression of fire by European settlers in the late nineteenth and twentieth centuries has resulted in the rapid expansion of trees and many native forest species into prairies. Clearly, these grassland habitats would not have been as extensive without frequent fires keeping the adjacent forests at bay (Figure 3.1). Third, anthropologists and ethnobotanists have gathered substantial evidence of the use of fire by Native Americans in Northwestern prairies. Although lightning ignitions are extremely rare in Northwest grasslands, Native American use of fire was a well-established cultural practice that was key to sustaining important food and medicinal plants. It is reasonable to surmise that these practices were widespread and of long duration (Boyd 1999, Turner 1999).

Prairies often are associated with savannas and open woodlands, particularly in the Willamette Valley but also in western Washington and southwest British Columbia. *Savanna* is a modern term that describes a community of scattered trees with grasses and forbs. Woodlands have a similar low density of trees but generally have a significant shrub understory. There is often a gradual transition from prairie to savannas and woodlands, with ecotones that share elements of adjacent communities. Oregon white oak, Douglas fir, and Ponderosa pine are the typical taxa that make up the arboreal component of these transition communities.

Although they are dominated by perennial forbs and grasses, quite a few annual plants persist in established prairies, filling in the gaps between the bunchgrasses. Such annuals can serve as early successional species in prairie restoration projects. Prairie perennials have varied life spans. Individual plants of species such as Roemer's fescue probably can live for decades, whereas others such as blue wild rye are short lived. (See Plates 2 and 3 in the color insert for oak and prairie images.)

Classification and Distribution of Northwest Prairie Ecosystems

West of the Cascades, prairies were most extensive in the Willamette Valley, with lesser acreages

FIGURE 3.1. Upland prairie and adjacent Oregon white oak (*Quercus garryana*) and Douglas fir (*Pseudotsuga menziesii*) woodland in the Willamette Valley, Oregon. (*Photo by Ed Alverson*)

occurring in the Puget Trough and southern Georgia Basin of British Columbia. However, a few prairies also could be found along the coast and in valley bottoms along the western slopes of the Cascades. In the Willamette Valley at the time of Euro-American settlement, there were about 1 million acres of prairie and another 500,000 acres of savanna (Figure 3.2). About one third of the prairie acreage was wet prairie, an estimate based on the extent of hydric soils (Christy and Alverson in prep.). In southwest Washington, mostly Clark and Lewis counties, there were approximately 50,000 acres of prairie and savanna (Caplow and Miller 2004). In the south Puget Sound region, historic prairies are estimated to have occupied 150,000 acres (Crawford and Hall 1997) (Figure 3.3). On the Saanich Peninsula of southwest Vancouver Island, estimates of the historic extent of prairie and savanna run around 25,000 acres (Garry Oak Ecosystem Recovery Team site). Smaller prairies were scattered throughout the area (Figure 3.4).

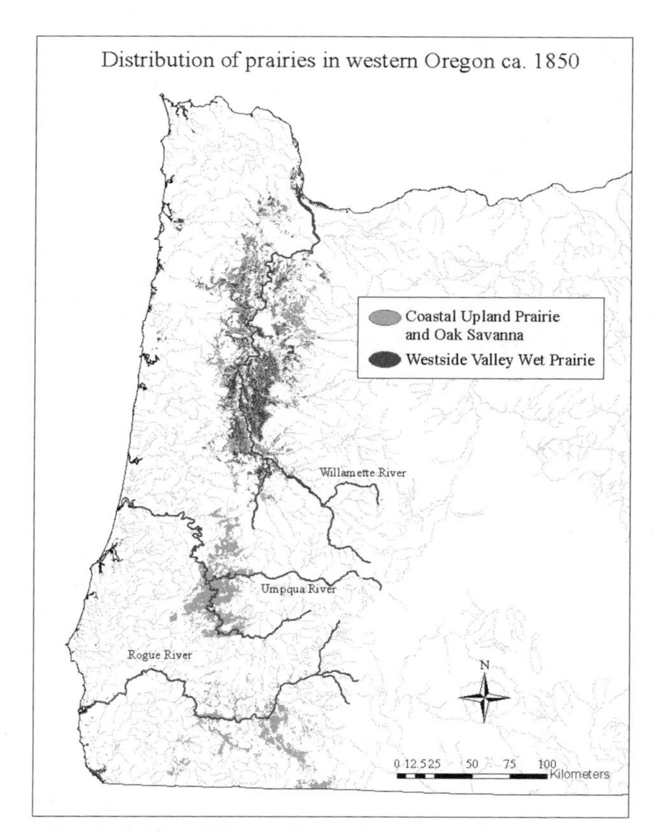

FIGURE 3.2. Historic extent of Willamette Valley prairies. (*Oregon Natural Heritage Information Center, Oregon State University*)

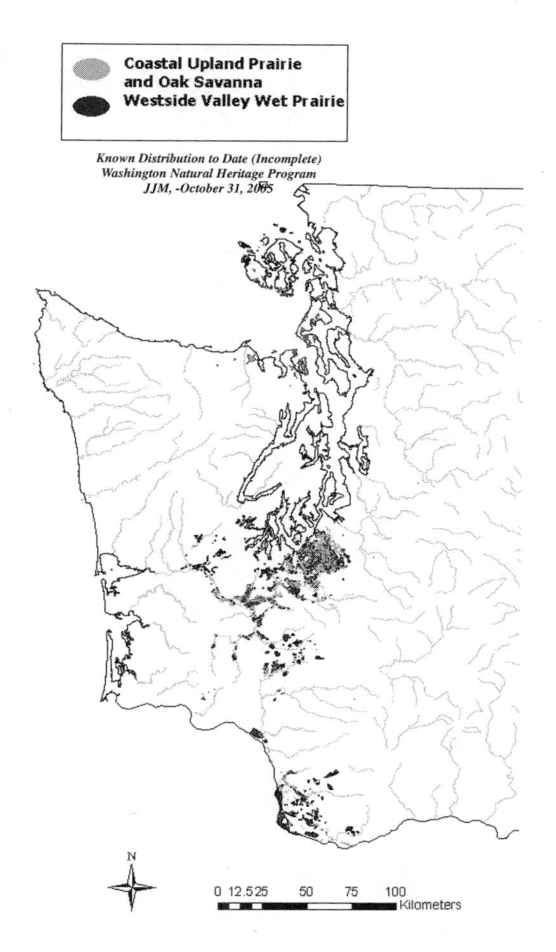

FIGURE 3.3. Historic extent of Puget Trough prairies. (*Washington Natural Heritage Program, Washington Department of Natural Resources*)

Today, most remaining prairie remnants are small fragments, less than 50 acres. There are only a few exceptions, such as the extensive gravelly prairies of Fort Lewis, Washington, and wet prairies in the southern Willamette Valley near Eugene and Corvallis.

Throughout the ecoregion there were areas that supported unique but related vegetation,

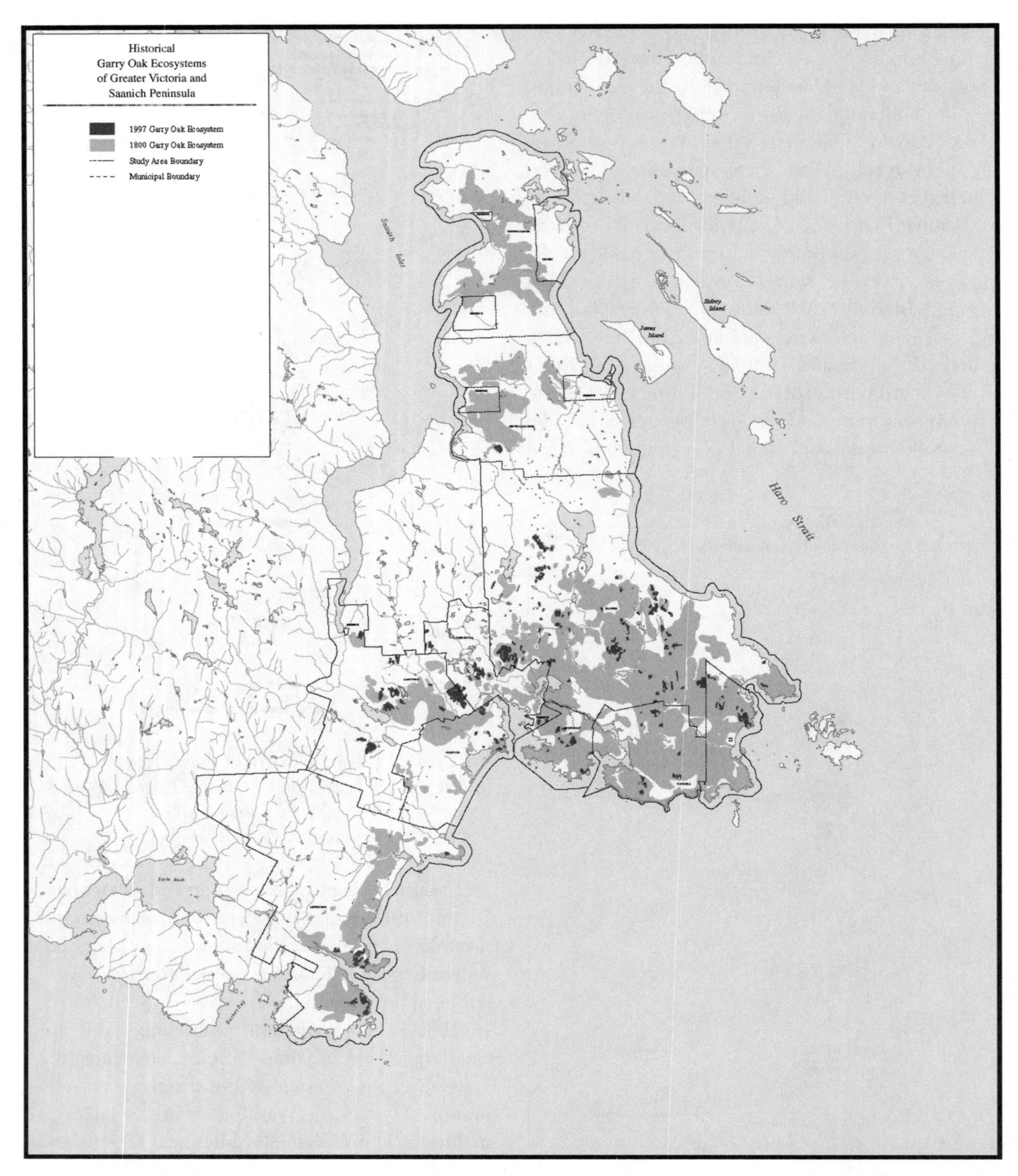

FIGURE 3.4. These maps of the Garry oak ecosystems in Victoria and the Saanich Peninsula (*above*) and Cowichan Valley and Saltspring Island (*facing page*) British Columbia compare the 1800 and 2003 distribution of this ecosystem, which serves as a surrogate for prairie and oak savanna. In this 200-year period, deep soil systems declined from 1824 hectares to 83. Shallow soil systems declined from 1301 hectares to 619. (*British Columbia Ministry of Water, Land and Air Protection, Victoria B.C. Courtesy Garry Oak Ecosystem Recovery Team*)

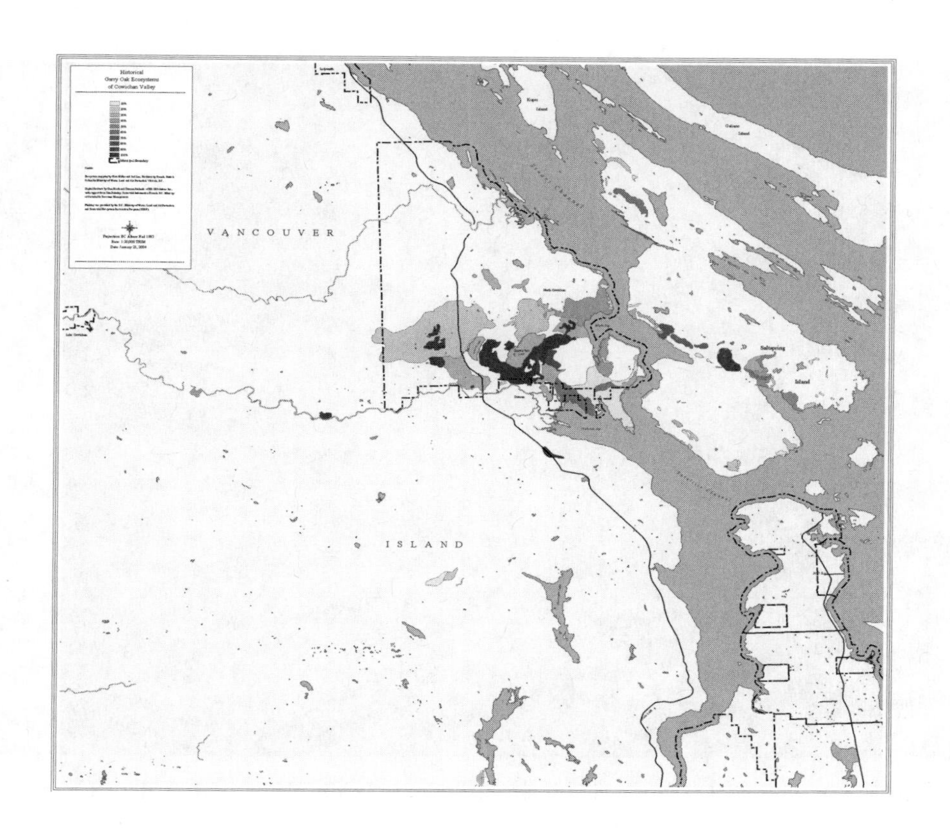
Historical
Garry Oak Ecosystems
of Cowichan Valley
VANCOUVER
ISLAND
Saltspring
Island

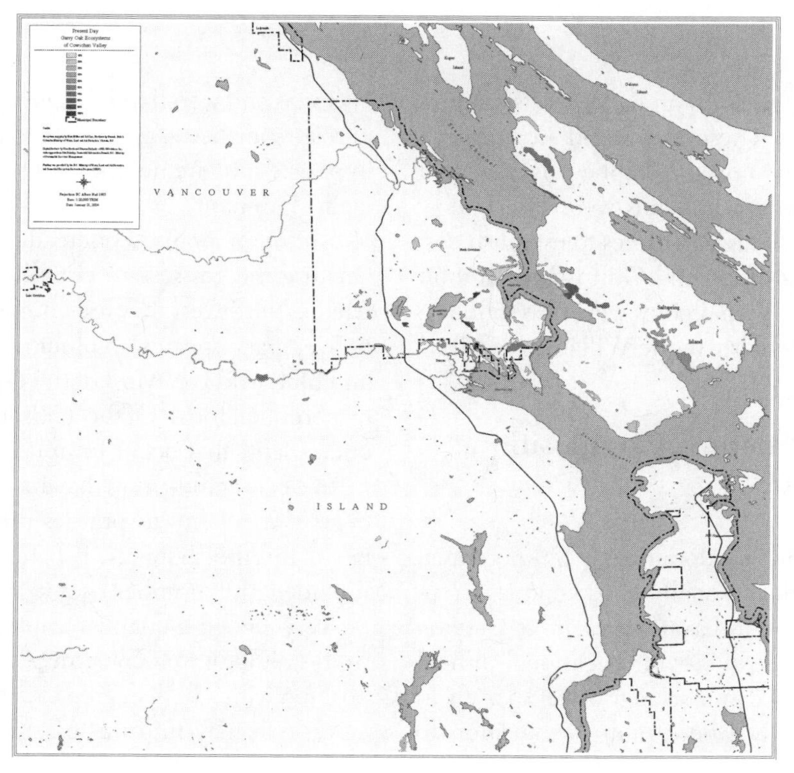
Present Day
Garry Oak Ecosystems
of Cowichan Valley
VANCOUVER
ISLAND
Saltspring
Island

FIGURE 3.5. Oregon white oak–California black oak savanna in the Willamette Valley, Oregon. (*Photo by Ed Alverson*)

occurring in shallow soils, on rocky bluffs, and on herbaceous balds. These are found in northern Puget Sound on north Whidbey Island and Fidalgo Island, where the soils were too shallow and droughty for Douglas fir trees to establish easily, and on the San Juan and Gulf islands and on Vancouver Island. Vernal pools were also found in wet prairies, particularly in the Willamette Valley.

Vegetation Structure, Composition, and Function

Prairie vegetation is dominated by perennial bunchgrasses with seasonally conspicuous perennial forbs. Annual forbs are present to varying degrees from site to site, with occasional annual grasses as well. Woody plants may occur sparsely in prairies, particularly where prairies transition to savannas and woodlands (Figure 3.5). Shrubs such as rose, snowberry, and poison oak may be present in prairies but are not thought to have been historically dominant.

Although prairies often are called grasslands because the grasses are visually dominant, in the Pacific Northwest they are composed of only a few native grass species, including Roemer's fescue and blue wild rye. Most of the diversity is provided by perennial forbs. Of the 350 species of native vascular plants that occur in upland and wet prairies in the ecoregion, most have a moderate to high degree of fidelity to prairies and are unlikely to occur in other habitats. If balds and outcrops are included, that number rises to 450 taxa.

Very few high-quality remnant prairies remain, so it is difficult to reconstruct their historical composition with a high degree of certainty. In unusual cases, one may find up to 30 native species in a

1-square-meter plot, but these are exceptional. Most sites have much lower species richness, and it is unclear to what extent this reflects losses caused by grazing, lack of fire, competition with nonnative taxa, or other alterations to historic ecosystem processes. Many perennial grasses and forbs are long lived once established. Most are reasonably tolerant of moisture stress but often are not highly competitive at colonizing bare ground.

Much less is known about the invertebrate fauna of the Northwest prairies. Coevolutionary relationships between native plants and invertebrates are an important feature and source of prairie biodiversity. For example, butterfly species such as the valley silverspot and Fender's blue rely on specific larval food plants. Similar host plant specificity may exist with native bees as well. But of the 350–400 species of native bees that once inhabited the prairies and savannas north of California, probably 80% are extirpated or exist only in minute relict populations. Thus, although a project's restoration goal may be to restore native biodiversity, the restored plant community may not be fully successful unless key native pollinators also are restored. It is likely that the destruction of prairies that has resulted in reduced native plant biodiversity has also resulted in a similar reduction of native insect biodiversity.

Variation and Commonalities

Upland prairies typically occur on well-drained silt loam or gravelly loam soils of low to moderate fertility. However, a subset of the native prairie flora is limited to more specialized conditions. Wet prairies, which are most extensive in the Willamette Valley, typically develop on clay soils. In winter, clay expands and forms an impervious layer that keeps water pooled on the surface. The resulting hydric soils generally are saturated with water but are not deeply inundated during most of the rainy season. These soils then become very dry in the summer, so plants that occur in wet prairies must be able to tolerate alternating wet and dry conditions. Vernal pools are depressions where water accumulates in the winter and early spring. They typically develop in soils that are a combination of clay and gravel. This most extreme prairie habitat has standing water in the wet season but turns dry and baked in the summer. The flora of vernal pools typically has a higher proportion of annuals (Figure 3.6).

In the Willamette Valley, prairies developed primarily on the deep fertile soils that were most valued for farming. Thus, most of the prairies were converted to agriculture soon after Euro-American settlement. By contrast, prairies in the Puget Sound region occurred primarily on well-drained, gravelly soils that are less productive for agriculture (Figure 3.7). Throughout the ecoregion, any of the sites not farmed were inevitably altered by grazing of domestic livestock. In the Georgia Basin, most prairie vegetation grows on shallow soils over bedrock, on south-facing bluffs, or where glacial activity left shallow soil and exposed rock. Such substrate characteristics are important constraints that determined the extent and distribution of historic prairies. Restorationists need to pay close attention to the variation in soils, hydrology, and topography of their sites in order to fully understand the context and feasibility of their restoration agendas in the historic landscape.

The prairies in the Pacific Northwest differ in several significant ways from their Midwestern counterparts. Climatically, the moist, mild winters and pronounced summer droughts that define the Northwestern prairies contrast markedly with the summer rains and frigid winters that characterize prairies in the Great Plains. As a result, the primary growing seasons in the Northwest are winter and spring, with grasses generally becoming dormant by late summer. Although Midwest tallgrass prairies receive roughly the same total precipitation, spring and summer are the primary growing seasons. Bunchgrasses are conspicuous in both areas, although sod formers also may be abundant in the Midwest (Figure 3.8). Overall species richness and

Figure 3.6. Popcorn flower blooms in a Willamette Valley vernal pool. (*Photo by Ed Alverson*)

Figure 3.7. Herbaceous balds are sites where bedrock is close to the surface. (*Photo by Ed Alverson*)

Figure 3.8. Pacific Northwest prairies are dominated by bunchgrasses such as Roemer's fescue (*Festuca roemeri*), in contrast to the sod-forming grasses of Midwestern prairies. (*Photo by Ed Alverson*)

productivity tend to be far lower in Pacific Northwest prairies than in those of the Midwest. Furthermore, tallgrass and shortgrass prairies evolved with extensive herds of bison and other ungulates, which appear to have been far less influential in Northwestern prairies. However, in the grasslands of both the Northwest and the Great Plains, fire appears to have played a major role in excluding trees and shrubs.

Most of the Midwestern tallgrass prairies are composed of warm season grasses that follow the C4 photosynthetic pathway, which allows these grasses to grow under higher temperatures and higher moisture stress. In contrast, Pacific Northwest grasses are cool season grasses that follow the C3 photosynthetic pathway and go into dormancy under conditions of high temperatures and drought. This distinction has a direct bearing on restoration and management of prairies. In the Midwest, the nonnative grasses are primarily cool season (C3) grasses. Mowing and fire can be timed to benefit natives or set back nonnatives. However, in the Pacific Northwest both native and nonnative grasses are cool season grasses, precluding use of such warm season–cool season restoration strategies.

Paleoecological History

About half of the geographic area occupied by Pacific Northwest prairies was covered with ice at the height of Pleistocene glaciation. Glaciation left a legacy of gravelly outwash plains in the southern Puget Trough and scoured areas of bare bedrock, particularly in the Georgia Basin. Both of those areas supported substantial prairie vegetation at the time of Euro-American settlement. The southern half of the area was not glaciated but was much colder than today.

A period of rapid climate warming brought about the retreat of the continental glaciers and continued into the early Holocene, 12,000 to 8,000 years ago. During this period, lodgepole pine colonized the glaciated areas (Leopold and Boyd 1999). At this time there was substantial vegetation change in Willamette Valley and other areas south of the boundary of continental glaciation as more temperate species replaced the colder, Pleistocene flora.

The hypsothermal interval between 4,500 and 9,500 years ago was a time of warm and dry climate with maximum solar input and highest temperatures. Pollen trapped in sediments in bogs and lake bottoms shows this as the period when oak pollen was most abundant. Oak is a proxy for prairie grasses and forbs, which generally don't show up in any quantity in the pollen records. It is likely that during this period prairies were much more extensive than at the time of Euro-American settlement. It is possible that there was a continuous band of prairie and savanna from the Willamette Valley north through the Puget Trough.

As the climate generally cooled after the hypsothermal interval, forests probably expanded into former prairie areas, leaving behind a series of prairie patches of various sizes. Many pollen records show that starting about 4,500 years ago the extent of oak habitats decreased as conifers increased (Leopold and Boyd 1999), which may have been accompanied by a decrease in the extent of prairies. Because Native Americans were almost certainly present across the region throughout much of the Pleistocene, burning is likely to have been an increasingly common practice to maintain these culturally significant open prairies, particularly as climates became cooler and wetter in the latter Holocene.

Native American Practices

Native Americans used fire for a variety of reasons, including harvesting of prairie vegetation and creation of migration routes, and thereby helped maintain prairies for thousands of years (Packard and Mutel 1997). At the time of initial Euro-American exploration and settlement, Native Americans were observed burning prairies and savannas in the Pacific Northwest. In early accounts, this burning was attributed to a variety of purposes. For example, David Douglas (1959) described a small tobacco plot that was prepared for planting by burning of cut brush over the garden plot. Although this example might not have landscape-scale implications, it shows a sophisticated attitude toward fire and the ability to use fire to meet a specific objective.

In other cases, fire may have been used to aid in seed gathering. Native Americans may have burned off tall grass under oaks before the acorns fell, making them easier to find. They also set fire to tarweed patches, burning the sticky tar but leaving the stalks and the upright involucres, or cup-like seedpods. Once these areas were burned off, they knocked the seeds into a basket with a stick. In the process, the seeds were toasted.

Brush burning improved visibility, providing advance warning of hostile attack. Heat and smoke prevented enemies from moving through an area. Fire was described as an aid to game hunting. Small patches burned in late summer or early fall grew lush new grass the next spring that attracted game animals. Hunters burned rings around unburned areas where deer or elk concentrated so they could easily be shot with bows and arrows (Boyd 1999).

A number of important food plants occurred primarily or exclusively in prairie or savanna habitats and undoubtedly were deliberately favored by Native American burning activities. Native people ate many lily family bulbs and corms, including chocolate lily (*Fritillaria affinis*), fawn lily (*Erythronium oregonum*) (Figure 3.9), and tiger lily (*Lilium columbianum*). Camas (*Camassia quamash* and *Camassia leichtlinii*) was particularly important for most tribes. In the spring after a fire, a higher proportion of camas plants flower than in unburned adjacent areas (Jancaitis 2001). It appears that plants from the lily family respond to fire physiologically by producing more flowers and possibly more seed.

A number of sunflower family species produce seeds that were ground for oil. Kalapuya people ate tarweed seeds (*Madia* spp.) and used balsamroot (*Balsamorhiza deltoidea*) or mule's ears (*Wyethia angustifolia*) oils for religious ceremonies. Roots of certain species of the carrot family, such as yampah (*Perideridia* spp.) and possibly desert parsley (*Lomatium* spp.) were important. Bracken fern (*Pteridium aquilinum*) rhizomes also were used as food. Early explorers described some prairies as completely filled with bracken fern. Farmers subsequently had difficulty ridding their fields of what had been a prized food of Native Americans.

It is possible that prairie species were dispersed by Native Americans moving from one site to another, either deliberately or accidentally as they

FIGURE 3.9. Although camas was a major component of the indigenous peoples' diet, they ate other bulbs and corms, including *(left)* chocolate lily (*Fritillaria affinis*) and *(right)* fawn lily (*Erythronium oregonum*). (*Photos by Ed Alverson*)

discarded food. Similarly, oaks from the Columbia River northward have a scattered distribution, and some are widely separated from other oak stands, perhaps reflecting historic dispersal by Native Americans. Oaks often are associated with Indian village sites in western Washington and southwest British Columbia (Storm 2002).

Although there is clear evidence that native people were very familiar with fire and capable of sophisticated application of fire, it is less clear how extensively and frequently it was applied at a landscape scale. Determining how often and in what configurations fire should be reintroduced in prairie restorations remains an important area for investigation.

Significance and Restoration Urgency

Pacific Northwest prairies have been greatly reduced from their presettlement extent, and most existing remnants have a reduced diversity of native biota because of livestock grazing, introduced species, and lack of fire. Estimates of remaining prairies vary from 10% of historic (presettlement) extent in the south Puget Sound region of Washington (Crawford and Hall 1997), to less than 5% of historic in southwest British Columbia (including oak savannas) (Garry Oak Ecosystem Recovery Team), to 1% of historic in the Willamette Valley (Wilson et al. 1995). Many of the smaller prairies that were located away from major prairie areas are now completely devoid of native prairie vegetation.

Because of the drastic degree of reduction of former prairies, many species that are largely restricted to prairie habitats are now at risk of extirpation or extinction. Many prairie plant species and a number of butterflies, birds, and mammals are formally listed as threatened or endangered species or otherwise recognized as at risk of extinction. For example, twenty-six globally at-risk plants are associated with prairies in the Willamette Valley–Puget Trough–Georgia Basin ecoregion (Floberg et al. 2004). Prairie restoration has the potential to aid endangered species recovery by increasing the viability of existing populations of rare species and assisting with the recovery of species that have been reduced to small fragments of historically larger populations.

Principles of Restoration

Emulating the Historic Landscape

Successful restoration of prairies entails knowledge of historical conditions, both natural and human induced. Often, restoration goals are based on knowledge of predisturbance ecosystem conditions. However, because prairies have been maintained by disturbance (e.g., fire) for thousands of years, goals for restoration should emphasize historic conditions and include disturbance regimes (Westman 1991). Restoration efforts should begin by gathering baseline data on soil, geology, topography, vegetation, and land use to understand site constraints and establish the landscape context that will help guide the setting of restoration goals. This is especially relevant for Puget Sound prairie restoration, where most of the original habitat has disappeared (Dunn and Ewing 1997).

Public land surveys and early explorers' botanical records are good sources of historic information on the prairie landscapes and vegetation and can provide useful guidance (Packard and Mutel 1997) (Figure 3.10). However, presettlement landscape information is so limited that prairie restoration must emphasize capturing the flavor of the historic conditions rather than the actual conditions themselves (Harker et al. 1999).

Undisturbed reference sites provide useful information on ecosystem characteristics and are helpful in identifying the range of composition and structure that characterized historic prairies. However, in the Pacific Northwest, such sites are rare or nonexistent. This presents unique challenges for prairie restoration in this region. By combining information on site characteristics, habitat heterogeneity, and historic conditions with data from reference sites, restorationists may begin to develop meaningful and appropriate goals that take into account the variability found in prairie systems (Hobbs and Norton 1996).

Restoration projects often use a combination of structural, functional, and social parameters that incorporate goals for effective restoration (Westman 1991). Dunn and Ewing (1997) list specific goals for prairie restoration and management, including native biodiversity preservation, protection of rare species and wildlife habitat, maintenance of vegetation structure and natural communities, control of invasive organisms, and promotion of self-sustaining habitats.

In general, restoration goals can be grouped into three major categories: species restoration, ecosystem function restoration, and ecosystem process restoration (Ehrenfeld 2000). There are advantages and disadvantages to each. A species emphasis will improve the chances for survival of certain organisms and may result in greater biodiversity. However, this method does not address landscape-level processes and the overall complexity of habitat needs. Populations fluctuate frequently, making it difficult to determine a representative level of species abundance (Harker et al. 1999).

Restoration goals that focus on ecosystem function emphasize small- and large-scale ecosystem processes with the underlying knowledge that species populations are variable. With this focus, biota and physical habitat are the parameters used

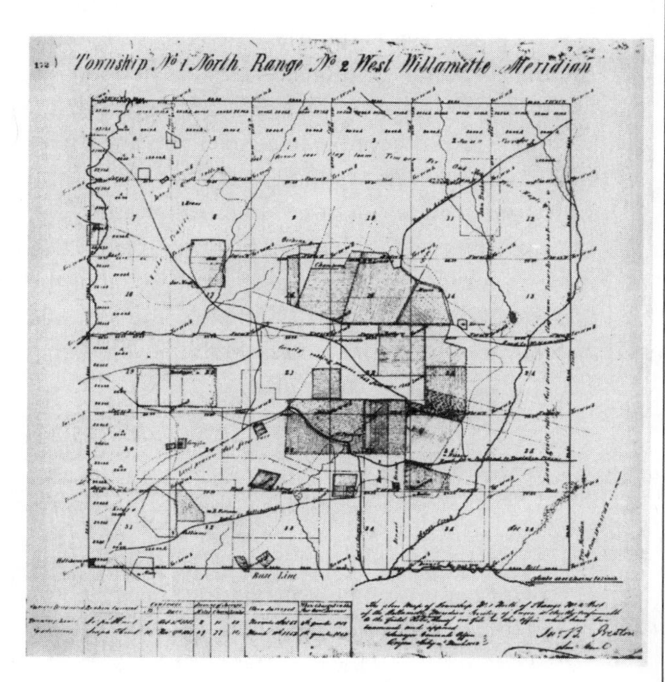

Township 8 North-Range 5 West, Willamette Meridian

North between sections 22 and 23

3.50	A road, course east and west
12.00	A swale or drain, 50 links wide, course east and west
30.00	Leave prairie and enter oak openings
40.00	Set quarter section post from which a W. oak 40 inches in diameter, bears 3½° W 214 links W. oak 36 inches in diameter, bears N 86½° E 332 links
80.00	Set post corner of sections 14, 15, 22, and 23, from which a W. oak 16 inches in diameter, bears N 53° E 175 links South 30 chains gently rolling prairie North 50 chains scattering oak timber

The bottoms along the Willamette are heavily timbered with fir, maple, ash, and a dense undergrowth of vine maple, hazel, and briars. There are numerous sloughs, which would make the township almost impossible to survey in winter (T13S-R4W).

FIGURE 3.10. *(a)* From 1851 to 1865 the Federal Land Office dispatched surveyors to map the Willamette Valley in preparation for settlement. *(b)* The surveyors' notes provide a valuable historic reference. (*Willamette River Basin Planning Atlas, Oregon State University Press*)

to set goals. Biota can be measured through species composition, gene frequencies, and biomass. Physical habitat can be measured by topographic features, soil structure, and nutrient flux (Westman 1991).

The third approach, emphasizing the maintenance of ecosystem processes, is advantageous for many reasons, although it is challenging to identify ecosystem definitions and functions necessary for restoring natural processes (Ehrenfeld 2000).

Filters Used in Setting Goals

Goal setting for restoration often uses both coarse and fine filter approaches. Coarse filters conserve representative ecological communities, which in turn protect associated species (Groves 2003). The coarse filter strategy is based on community-level conservation and has the potential to protect 85–90% of species found in any one particular ecosystem (Noss and Cooperrider 1994). In landscape-scale restoration, a fine filter strategy also is used to conserve the rare, endemic, or wide-ranging species that are missed by the coarse filter (Figure 3.11).

It is important to consider the entire landscape to determine the optimal system of reserves for conservation. Fragmentation of prairie landscapes, which interferes with landscape-scale processes such as fire, has led to significant species loss. Most ecosystems need multiple interconnected core

FIGURE 3.11. The federally threatened golden paintbrush (*Castilleja levisecta*) is a good example of a rare prairie species that would be addressed through a fine filter strategy. (*Photo by Ed Alverson*)

reserves to protect biodiversity. Ideally, the network of core reserves is surrounded by buffer zones that help to filter potential disturbance. Edge effects become less of a problem as the size of the preserve increases (Noss and Cooperrider 1994).

It is essential to provide a surrounding habitat mosaic for plant dispersal and an area for animal migration. A matrix design is more beneficial for species that disperse randomly (Noss and Cooperrider 1994). However, the characteristics of reserve size and shape are not equally advantageous for all species (Westman 1991). Hobbs and Norton (1996) indicate that details on reserve connectivity and size requirements depend on the landscape and specific goals. In prairies, the core design and the matrix design would each be beneficial in different ways. For example, a larger core habitat would be less susceptible to edge effects and therefore would be more resistant to exotic species invasion and coniferous tree encroachment. The matrix habitat would be beneficial for species that disperse randomly, such as butterflies and the spores of certain plants. Ultimately the restored habitat must be connected favorably to the larger overall landscape (Harker et al. 1999).

Faunal Restoration

Most prairie restoration projects emphasize restoring plant communities, considering species abundance and composition and the proportion of nonnatives to natives. However, when the restoration focuses solely on plants, the restored site may support appropriate prairie floral composition and structure but lack the faunal component (Allison 2004). This is partly because some animals need specific consideration of their life histories and plant associations. Therefore, it is important to consider restoration of the faunal component when determining objectives. It is likely that animal populations will need separate actions. For example, restored prairies often lack abundant insect populations. Insects that are linked to specific host plants, such as particular butterflies, are less resilient. Restoration of these insects entails knowledge of the specific insect–plant associations and may involve insect translocation (Packard and Mutel 1997).

Often considered indicator species, amphibians and reptiles are less mobile than many other animals. Therefore, amphibians and reptiles need larger reserves and connectivity and may need to be translocated to an established reserve site. Puget Sound herpetofauna historically included salamanders, newts, toads, frogs, turtles, lizards, and snakes, although many are now absent from the region because of both a reduction in prairie habitat and general human disturbance. For example, racers (*Coluber constrictor*) probably have been extirpated from western Washington by a reduction in prairie habitat and overall fragmentation (Dunn and Ewing 1997). Reintroduction of these species to restored prairie habitat will almost certainly entail human intervention.

Many bird species use prairies for food, shelter, and nesting. Edge effects increase as the prairie landscape becomes more fragmented, making breeding birds more vulnerable to predation. Many southern Puget Sound prairie birds such as MacGillivray's warbler have declining populations because of nest predation, cowbird parasitism, and prairie habitat loss (Dunn and Ewing 1997). In prairie systems, having the appropriate vegetation structure may be the most important restoration objective for birds. Restoring particular native plant species may be largely irrelevant from an avian perspective unless it affects their prey community composition (Vickery and Dunwiddie 1997).

How Much Restoration Is Enough?

In general, larger prairie landscapes are better. This means that on a landscape level there should be a network of sites spread over a range of gradients that include both natural habitat and areas restored to reduce long-term fragmentation. The reserve should be large enough to provide habitat for a variety of organisms that have different needs, ultimately enhancing their dispersal and movement. Larger reserves also contain higher genetic diversity (Bakker and Berendse 1999).

Size also depends on the goal of restoration; some species need larger areas than others. In general, larger prairie restoration areas provide greater habitat and food to animals and support more diverse populations and a wider array of plant species (Packard and Mutel 1997). In practice, however, size considerations may be largely moot in the Pacific Northwest. In most areas, development and habitat alteration have occurred to such an extent that restoration options may be extremely limited by patterns of ownership and historic and current land use practices.

Distinguishing Restoration from Management and Maintenance

Restoration is a process that involves returning habitat to a natural state, representative of historic conditions (Allison 2004). In prairies, restoration often involves large-scale native planting and may entail the translocation of associated animal species (Dunn and Ewing 1997; Packard and Mutel 1997). Once the prairie has been restored, the ecosystem must be managed actively to maintain the community composition and structure. Because prairies are dynamic systems that readily and continuously undergo succession, active management is crucial for long-term prairie preservation. In the Pacific Northwest, prairies that are not managed are rapidly overtaken by native and exotic trees and shrubs. Prairie maintenance includes burning, weed control, animal introductions, and further planting (Packard and Mutel 1997). This long-term management requires knowledge of the native community and continued intervention, such as disturbance removal, fire creation, and grazing maintenance (Dunn and Ewing 1997).

Maintaining Functioning Communities

The restoration of prairies is likely to entail the reestablishment of a variety of key ecological processes. Historically in the Pacific Northwest, these included fire and processes that resulted in significant soil disturbance, such as animal burrowing and Native American harvesting of plants (Jackson et al. 1995, Collins and Wallace 1990). Although prairie habitat results from climate and

Case Study: Yellow Island Restoration over the Long Term

The restoration of grasslands is best approached from a long-term perspective. The experience of The Nature Conservancy at its Yellow Island Preserve in the San Juan Islands of Washington illustrates how careful use of small-scale plots and experiments can inform ongoing management decisions and achieve this successfully with limited resources at a small site. The 11-acre (4.5-hectare) island was originally used (and probably regularly burned) by Native Americans to harvest camas and other food plants and was acquired by The Nature Conservancy in 1980. Primary features included several hectares of high-quality grassland with abundant native fescue, camas, and a diverse array of forbs. Several hectares had been overgrown by native shrubs and trees.

In 1981, three plots were established to determine whether and how quickly the grasslands were being encroached on by native shrubs and trees. Once it was determined that invasion by woody plants was rapid and extensive, these and other plots were used to experiment with mechanical, chemical, and fire treatments to arrest and reverse this trend (Figure 3.12). In this way, a series of studies over the last 25 years have focused on critical questions and are used to develop and refine management. The earliest studies demonstrated that despite the abundance and proximity of native vegetation, removal of invasive plants alone was insufficient to establish native species; active planting of native species and subsequent control of invasive grasses and forbs was necessary.

Over time, prescribed burning has emerged as a major tool for maintaining the grasslands. The vegetation plots established in 1981 were again used, together with others installed in different habitats, to evaluate the responses of native and nonnative species to repeated burning. The first burn took place in 1987, with five others since then. Although the responses of different species to each fire were complex, several patterns have recurred after repeated burns that suggest common trends with important management implications regarding the relative dominance of annuals and perennials, forbs and grasses, and natives and nonnatives. Future monitoring will refine management practices to minimize invasive species and to enhance the long-term viability of the native communities.

periodic drought, frequent fires favor this landscape type over forests through a variety of mechanisms. Fires keep trees from invading prairie ecosystems. Fire reduces grass thatch and moss and lichen cover, which releases forbs from suppression. Removing thatch creates suitable microsites for additional seed germination, increased establishment, and therefore possibly greater recruitment. The ash may also fertilize the site. The high-frequency and low-intensity fires set by Native Americans are ideal. Using prescribed burning in prairies entails timing and an understanding of the life histories of the plant community (Emery and Gross 2005).

Disturbance by fire promotes native species diversity and limits the invasion of exotic species because native species often are more adapted to fire. Invasive perennials are the greatest source of restoration failure and therefore are a significant challenge for successful prairie restoration (Perrow and Davy 2002). These invasive species can degrade habitats and increase the community recovery time (Packard and Mutel 1997). For example, introduced Scotch broom reduces habitat quality by competing for resources with native plants and ultimately displacing them (Dunn and Ewing 1997). Scotch broom is an especially problematic invasive plant for restoration because of its high reproductive capacity, by which one plant can produce thousands of seeds. Invasive insects also compete with native insects. In Puget Sound

FIGURE 3.12. Fire is an important restoration tool at Yellow Island Preserve. *(Photo by Peter Dunwiddie)*

prairies, introduced ladybird beetles may compete with native ladybirds and other insect species for food (Dunn and Ewing 1997). Exotic insects damage native plant communities that lack chemical defenses against these introduced species. Bowles et al. (2003) found that native species richness increased in burned sites, and Dunn and Ewing (1997) found that nonnative species declined after a fall burning, which was less damaging to the native plants.

The role of faunal activity in maintaining healthy prairies is less well understood, although small mammal burrowing, grazing, and foraging probably are components of a key ecological process in prairie systems. For example, burrowing by pocket gophers disturbs the soil and is believed to enhance soil conditions for native plant establishment. In contrast, western gray squirrel foraging on young tree species contributes to prairie landscape maintenance by reducing the invasion of forests (Dunn and Ewing 1997). In some cases, small mammals also may reduce introduced species abundance (Perrow and Davy 2002), but the inverse may be true as well because many introduced taxa may flourish in the bare ground created by rodent burrowing.

Soil disturbance by Native American harvesting practices probably played an important role in shaping historic prairie composition and structure. However, how it could—or should—be reintroduced in modern prairie restoration efforts has not been addressed by current restorationists.

Social and Economic Context (Current and Past)

Quantifying the social and economic context of prairie restoration can be difficult. Unlike systems such as old-growth forests, restored prairies do not provide obvious economic benefits to landowners. However, restoration goals that emphasize social and cultural values include natural service conservation, erosion prevention, ecotourism, improved habitat quality, and protection of rare and endangered species. In some cases ecosystem services can be assigned monetary values (Costanza et al. 1997, Ehrenfeld 2000). This field of ecosystem services is just gaining ground, and it will be interesting to see whether these concepts can be applied to prairie restoration. Currently, challenges facing this field include issues of scale, value, and prioritization.

Practices

Strategies and Techniques

A common error among restorationists is to begin a project by defining a suite of techniques. Sometimes restoration projects themselves are visualized as techniques, such as controlling Scotch broom so that native species will flourish. Yet rarely is the delineation of techniques the best place to start. Even when goals are well defined, developing a comprehensive strategy is just as significant as determining the best techniques. Comprehensive strategies build a framework in which intermediate goals are well defined, allowing the use of adaptive management to guide the complete restoration process, not just reach for the primary goal. On the flip side, understanding the full palette of potential techniques is also important because a combination of these will build the strategy. It is this interplay between techniques and goals that gives life to restoration and differentiates restoration from engineering.

Several key components make up an effective comprehensive strategy. These include a site assessment that defines existing conditions of substrate, vegetation, climate, hydrology, and landscape context; an identification of key ecological processes that have historically shaped the community; a clear articulation of the restoration goals; a careful consideration of alternative techniques and approaches; and a realistic assessment of available resources and logistical constraints. The primary purpose of a restoration strategy is to ensure that all of these facets fit together in a seamless and logical fashion so that site and resource constraints are thoroughly considered, that ecological processes move the project toward its goals rather than in another direction, and that the goals themselves are realistic and well thought out. If any of these aspects are overlooked, they typically result in restorations that are slow, resource intensive, and fraught with setbacks and failures.

By beginning with an assessment of a site's potential, based on physical and environmental conditions and ecological processes, restoration goals can be established that are more likely to be ecologically attainable. There are a wide range of valid goals, from establishment of a low-growing vegetation structure to full replication of historic prairie composition, structure, and function. It is easiest to determine goals when they are imposed by an outside regulation or mandate, such as The Nature Conservancy's mandate for biodiversity conservation. But even under such a mandate, the goals must be tempered by a thoughtful assessment of a site's constraints and possibilities.

Starting Point

Prairie restoration sites tend to be either agricultural fields or degraded prairies. Agricultural fields that were historically prairie habitat typically are devoid of native species, but they are abundant in the Willamette Valley. They often have highly altered soil structures and nutrient status and may contain significant weed seed banks. These sites can range from xeric to mesic, with some sites including a range of microhabitats. When this range of conditions is available, they provide opportunities to create a fuller mosaic of prairie habitats. Degraded prairies, more commonly found on dry upland sites ill suited to agriculture, contain a range of native plant species and may support prairie animals as well. Many protected preserves fall into this category also because few pristine sites remain that have not been degraded to some extent.

Several characteristics dictate restoration strategies on these sites, including the extent, distribution, and abundance of both invasive weeds and native prairie plants; the size and configuration of the site; and the landscape context, which may constrain the use of certain techniques. For instance, although the agricultural site is likely to be nearly devoid of native plants and animals, this may allow use of a wider range of techniques such as tilling or broadcast herbicide application, which may be inappropriate where there are established prairie plants and animals. Invasive species usually have been controlled in agricultural fields but may be abundant in pastures and upland degraded prairies. In addition to numerous invasive grasses and forbs, woody species such as Scotch broom are particularly problematic in many Northwest prairies. Even native species, such as cascara and Douglas fir, can impede prairie restoration in some areas. Some invasive species with long-lived seed banks, such as Scotch broom, necessitate up to a decade or more of control and a long-term commitment for a restoration to be successful. Establishing the full suite of prairie plants on an agricultural site typically involves larger acreages, where the cost of plant materials becomes a significant consideration. It may be necessary to modify techniques when rare prairie species are present. The variety of characteristics possible at each site again substantiates the importance of site assessment, goal delineation, and restoration strategy development for each site independently.

Restoring Prairie in Agricultural Fields

Though important throughout the Northwest, the restoration of prairies in former agricultural fields is particularly advanced in the Willamette Valley, with its deep alluvial soils and significant agricultural land use. The restoration of these sites holds the greatest promise for conserving prairies in that region (Figure 3.13). Field restoration projects are also under way on conservation areas in south Puget Sound, in British Columbia, and on Whidbey Island.

The basic strategy for agricultural fields is to prepare the site by reducing or eliminating invasive species and returning key ecological processes, establish a native plant community, and maintain it over time. Each of these steps has multiple alternatives for implementation and combinations of techniques that could prove most effective.

Site Preparation

The basic goal of this step is to control or eradicate habitat-modifying invasive species and to return ecological processes at the correct time. These efforts may need to be altered in sequence within a site because of microhabitat considerations. For instance, if a field has been drained, control of invasives may be easiest after natural hydrology is restored. A field with elevated soil nutrients may need a period of haying or prescribed fire to help reduce nutrient levels to those typical of natural systems.

In agricultural fields that contain few natives, the full spectrum of control techniques can be used

Figure 3.13. Small prairie remnants often occur adjacent to agricultural fields and provide a starting point for many prairie restoration projects in the Willamette Valley. (*Photo by Ed Alverson*)

on invasives, including those that are completely destructive to vegetation. These range from maintenance of the established tilling program to broadcast application of a nonselective herbicide and may also include solarization and prescribed fire. Of course, more selective techniques, such as spot spraying and hand control of invasive plants, can be used if appropriate.

Tilling has several advantages in old fields. It is inexpensive, can be used to treat large acreages, and uses equipment readily available in the agricultural community. The type of tilling used in the project depends on the field soil profile. For instance, areas that have been deeply tilled for years have lost their soil profile in the upper layers and can be tilled without negative effects. This is certainly true for fields that are currently active, but is also true for many old fields, where weed seeds may be concentrated near the soil surface.

In contrast, deep tilling is not as productive when soil profiles are undisturbed or when weed seeds are well dispersed throughout the soil profile. Light harrowing is an effective alternative. Harrowing can also be appropriate in nonagricultural areas if the native components of the community have been eliminated, such as highly disturbed areas where Scotch broom has been established for decades. Shallow harrowing leaves undisturbed seeds that are too deep to germinate.

Tilling or harrowing must be repeated until the top layer is weed free. This may involve tilling every 3 weeks or so while conditions are suitable for germination and growth. Tilling can also be combined with other techniques. For example, early tilling to stimulate germination is followed by application of glyphosate or another suitable herbicide.

Herbicide application may be preferred where soil profiles are intact or wet soils make tilling difficult. On sites where few or no native plants remain, a broad-spectrum herbicide such as glyphosate is the first choice. If sufficient broadleaf native species are present, it may be appropriate to use grass-specific herbicides such as Poast. Although many restoration groups shy away from spraying entire sites with herbicide, it may be an acceptable early step, particularly on sites with a history of chemical use or where short-term application of herbicides will alleviate long-term problems.

Solarization involves covering the site with plastic and allowing solar heat to be trapped under the plastic to kill plants and seeds. Performance improves when soils are tilled and covered while wet. Although solarization is widely used on home gardens, its success in prairie restoration in the Northwest is not great. Two trials on sites in Oregon and one in Washington showed only short-term control of weeds. Solarization on larger sites is constrained by cost and handling of the plastic sheeting, although agricultural equipment can ease handling on larger areas.

Prescribed fire is widely used in site preparation, especially when a clean field is not the short-term goal. The direct and indirect effects of fire are multifaceted and vary with burn conditions. Primary goals include creating open sites for seed germination by the removal of thatch, mosses, and lichens; reducing biomass to increase the effectiveness of herbicide application; and eradicating specific weed species. Fire can also be an important tool in reducing excessive nutrient levels in agricultural soils, especially when combined with haying later in the year. Prescribed fire can be used as a first step, for instance before herbicide application to enhance penetration of the herbicide and to ensure that target plants are growing rapidly when the herbicide is applied. Finally, prescribed fire can be used as the sole site preparation technique where the ultimate goal is not a pristine native community but rather grassland structure and function. Fire can remove woody invaders and return the site to conditions conducive to a wide variety of grassland-dependent animals.

Regardless of the tools chosen to eliminate nonnative species from a restoration site, sufficient resources must be devoted to carry out this stage successfully. Many restoration efforts have foundered in a sea of invasive species that were not adequately treated before natives were planted.

Returning Natural Processes

One advantage to using prescribed fire is that it returns one of the most important natural grassland processes. Hydrologic regimes and soil characteristics work in harmony with fire. Fire, moisture, and soil are the primary abiotic influences that create and maintain grasslands. Reestablishing their balance is significant to the ultimate success of prairie restoration (Figure 3.14).

Prescribed fire is most appropriate in upland sites. Most burns by Native Americans occurred in the fall, although spring burning, especially along the edges of prairies, also occurred. There are multiple direct and indirect effects of prescribed fire on grasslands, with results being dictated by interactions between the quantity, type, arrangement, and continuity of fuels; fuel and atmospheric moisture;

Figure 3.14. Three stages of wet prairie management: *left*, before clearing of invading woody vegetation, *top right*, after burning, and *bottom right*, flowering camas (*Camassia quamash*) the next spring. (*Photos by Ed Alverson*)

wind speed; and ignition pattern. Although general trends can be delineated, the effects of each prescribed fire vary with the unique combination of factors associated with each burn. The dynamic nature of prescribed fire should be taken into consideration when one is generating goals, and plans should be altered as needed.

In the Northwest, late summer and early fall is the typical window for many prescribed burns. Fuels typically are drier in fall, and fires are quicker, hotter, and more complete. In some years, when fall rains hamper ignitions, burning in late winter is an alternative. Although this may seem difficult in the Northwest, there are typically a few weeks in February when temperatures drop and rain subsides, and if morning inversions burn off, prescribed fire is possible. These late winter burns are much cooler and typically slower moving. Burn bans have led some restorationists to institute early summer burns timed just before burn bans are put in place, typically in mid-July. These early summer fires can vary in intensity and speed, resembling winter burns. Although summer burns may have some negative effects on prairie birds and vegetation, they may meet goals for controlling woody invasives and opening thatch and moss layers.

The importance of fire in prairie management has become widely recognized, but this tool often is applied inappropriately. The most common mistake arises from a failure to understand what fire can accomplish at a site. Often, sites that contain few or even no native species are burned, with an expectation that simply returning fire to the system will restore native prairie. Clearly, if sufficient native propagules are not present, no amount of burning (or herbicide application, tilling, or anything else) will bring back a prairie. In other cases, little attention is paid to ensuring that the weather and fuel conditions under which a burn is conducted and the manner in which the burn is carried out actually will achieve the desired results. The eagerness to return fire in any form to a site often blinds practitioners to the fact that their particular burn was conducted under conditions that virtually precluded thatch removal, shrub kill, or other planned objectives.

Though not as important on upland sites, reestablishing the hydrologic regime is crucial for wet prairies. Many fields or portions of fields have been drained or sculpted to increase agricultural production. Specific techniques to restore hydrology vary with original land use patterns, current site conditions, and goals for the restoration. Typical actions include removing drain tiles, recontouring lands, and connecting wet areas. Adjacent riparian areas may also need restoration if they were modified to improve field drainage. Additionally, some microtopography can be added to a leveled field to increase microsite diversity and wet prairie characteristics. Care should be taken to ensure that adequate soil structure is maintained with this technique. In general, where hydrologic conditions are important determinants of site conditions, the biggest cause of restoration failures is inattention to ensuring correct initial hydrologic conditions.

Many agricultural fields have altered soil characteristics that affect the likelihood of success for a restoration. Although the range of potential problems is wide, the most common is elevated soil fertility from crop fertilization or nitrogen fixation from weeds. In either case, there is strong evidence that this condition can lead to increased invasion of pest plants. Although some of the other ecological processes, such as prescribed fire and an enhanced hydrologic regime, can ameliorate soil conditions, a more aggressive program may be needed. This can be as simple as continuing to hay and remove the biomass from the site twice a year for a couple of years. Or a combination of techniques can prove more rapid, such as instituting a prescribed fire in fall or winter, followed by haying the next spring. Some restorationists have experimented with addition of carbon, such as wood shavings or sugar, to the soil. This improves the uptake of nitrogen by soil organisms by increasing the carbon–nitrogen ratio. Results from these attempts have been mixed, with varying responses to levels and type of carbon amended into the soil.

Establishing Native Plant Communities

Establishing a native plant community on large agricultural fields requires efficiency. Techniques that are effective on small portions of habitat, such as plugging diverse native seedlings, are likely to be too costly to apply over a large field. The wide range of site preparation techniques used on agricultural fields allows the restorationist to establish favorable germination conditions. Consequently, most agricultural field restorations involve direct seeding to establish the plant community.

Hand seeding, seed drilling, and hydroseeding have been used successfully in the Northwest. The choice of technique hinges on the amount of seed and equipment available. Broadcast seeding can be as simple as distributing seed by hand, although using a hand-held or tractor-mounted seeder allows more uniform distribution. Hydroseeding spreads seed with mulch. Although broadcasting establishes seed in a heterogeneous pattern, it typically relies on a higher seeding rate. Usually some soil management is needed after seeding to enhance germination. Light harrowing or roller packing can help improve seed contact with the soil, thereby facilitating germination and establishment.

Seed drills provide a more efficient distribution of native seed and improve seed–soil contact for many species. These implements produce furrows and meter seeds into the furrows. The no-till drills make much smaller incursions into the soil and typically can be used with less rigorous site preparation. Seed drills have been used successfully in both well-prepared fields and degraded natural prairie, where preparation is limited. To avoid creating linear rows, the field can be drilled more than once in a crisscross pattern. This will also help ensure that no portions of the field are accidentally missed. Care must be taken to ensure that seeds are planted at their optimal depth, which may vary widely between species.

In the Northwest the infrastructure to commercially raise native prairie seed is limited and still developing. As a consequence, most restorationists must produce or gather their own seed. This can be a significant effort and a major hurdle. As the infrastructure continues to expand, some alternatives can help. Consider planting the site over a period of years. This also has the advantage of spreading plant establishment over multiple years of weather because germination and establishment are strongly affected by soil moisture and precipitation patterns. Establish the primary grass components first, and follow with interseeding and interplantings of forbs and secondary graminoids (grasslike plants). Some restorationists believe this reduces the ultimate diversity of the restoration site by increasing the competition on the forbs from the larger established grasses. Yet this method also may be preferable if there is still a seed bank of invasive broadleaf plants at the site. Even with an established grass layer, broadleaf-specific herbicides can be sprayed over the top of the grasses to control weed seedlings. This option may be a conservative way to jumpstart the establishment of grassland structure before the complete site preparation is performed.

Restoring Degraded Upland Prairies

Prairies along the glacial outwash plains of the South Puget Sound region once extended for miles. Much of this prairie—the upland knolls and hills of the Willamette Valley and the outwash and balds of North Puget Sound and Vancouver Island—has been lost, and what remains has been degraded. Typically, they are invaded by woody species including Douglas fir, Scotch broom, and hawthorn. Pasture grasses may have been added to improve forage for cattle. Overall plant community diversity has been lost through fire suppression and other land uses. Yet many of these sites still maintain native prairie components, including some extremely rare species. This combination of a native component and possible rare species demands a different set of techniques and offers its own set of challenges.

Site Preparation

With the prevalence of woody invasive species on degraded sites, an initial goal often is to return the site to a grassland structure. Invading Douglas fir can be removed using traditional forestry methods. However, woody species such as Scotch broom that have long-lived seed banks present a greater challenge.

Scotch broom is the primary biological threat to most protected prairies in the south Puget Sound region, despite a decade of dedicated efforts. It generally demands a combination of mechanical cutting, hand-pulling, prescribed fire, and chemical control over many years, matched to its density and distribution. In locations with well-established seed banks, it may take 8–10 years of consistent, effective control to reduce the infestation to maintenance level. Older broom dies readily when cut, so mechanical control is a good first technique, followed by a prescribed fire to kill seedlings from the second cohort. Once the density of broom falls, herbicide can be wiped on in the spring or spot sprayed in the fall. Finally, small plants can be hand pulled. Integrating these techniques across a site over time has yielded positive results at several locations.

Pasture grasses pose a similar threat. Tall species can change structural components of the prairie that are critical for some sensitive species, including some ground-nesting birds and rare butterflies. Tall stands of oatgrass become barriers that limit butterfly foraging and egg laying. In contrast to native bunchgrasses, many pasture grasses are rhizomatous, forming dense swards that limit space and locations for prairie forbs, thereby reducing both abundance and diversity.

Techniques for controlling invasive grasses in this region are limited. These grasses seem most susceptible early in the season, just as they get ready to flower. Grazing in this season has been a suggested control method, although experimental trials have not been undertaken. A prescribed fire early in the season could be productive but probably would have negative effects on ground-nesting birds and on native bunchgrasses and other prairie plants. Grass-specific herbicides, such as Poast and Fusilade, have proven effective. This is especially notable because fine-leaved grasses, such as Roemer's fescue, are not affected by these herbicides. However, other native grasses, such as California oatgrass and blue wildrye, are affected, so these species may need to be replaced once pasture grasses are controlled.

Enhancing Native Community Composition

The control of pest species and return of key ecological processes usually are insufficient to restore high-quality prairie. The species diversity may be too low to provide the native propagules to establish high-quality prairie. Reestablishment of native species is so slow and patchy that nonnative species overtake opened sites. In most cases native plants must be actively reestablished.

Many of the direct seeding techniques described for agricultural field restoration are also appropriate for degraded prairies. This is especially true of no-till seed drills and broadcast seeding. Both of these methods have been used effectively after prescribed fire or chemical control of woody species. This type of interseeding into prairie vegetation should be considered part of nearly all pest control efforts.

Although propagating and plugging native plants is expensive and time consuming, it should be considered under some conditions. Plugging Roemer's fescue and a variety of forbs has proven very successful, with extremely high survivorship and growth rates. This can be important in mitigation, when success must be documented over short periods. Plugging also helps establish specific habitat conditions for rare species, such as new small patches or the supplementation of larger areas with selected species. Where reestablishing rare butterflies is a goal, plugging specific host plants can be a quick and effective way to restore small habitat patches.

Techniques for collecting, propagating, and planting plugs of Northwest prairie species are straightforward and well documented. Volunteers can do much of the work (Figure 3.15), although commercial propagation services are becoming more common. It is important to emphasize local genetics and take great care in identifying species during collection and propagation. Recently, wild-collected Roemer's fescue was inadvertently mixed with red fescue used to establish a seed production field, illustrating the need for care at every step of the process.

Maintaining Restored Prairies

It is critical that ecological processes be maintained to ensure the long-term viability of restored prairies. Prescribed fire and the hydrologic regime are two obvious processes that help maintain prairies, and certain biological processes help maintain plant and animal diversity. Fire kills encroaching woody vegetation, including both native and nonnative species. The hydrology of mesic systems precludes the establishment of upland species. The role and importance of many biological processes are less clear in Northwest prairies. Whether or not species-specific pollinators are common in these systems and whether they are present in sufficient numbers to ensure the long-term viability of native populations are unknown. Soil disturbance by pocket gophers and other burrowing animals may be important in creating open, safe sites for plant establishment and microhabitats for invertebrates, but this is largely conjectural.

FIGURE 3.15. Volunteers can assist by collecting seed. (*Photo by Ed Alverson*)

Maintaining these processes at appropriate intervals and scales can be a significant effort and may be particularly difficult to justify if the benefits of the process are unclear or are mixed. For example, prescribed fires typically involve significant permitting and implementation costs. Burning a large percentage of the site all at once can have negative effects on invertebrates. Therefore, smaller parts of the site may need to be burned each year on a 3-year to 7-year rotation. Although some effects of fire can be duplicated by other actions, such as grazing or mowing, the full range of effects is very difficult to replicate. The integration of a prescribed fire program into the management of a prairie should be explored in every case. There are few sites that cannot be burned, even small parcels in urban and suburban settings.

It is essential to continue ongoing surveillance and treatment of new invasive pests, particularly as the numbers of invaders steadily increases. New invaders should be located and treated as quickly as possible. The smaller the infestation, the higher the likelihood that treatment efforts will be successful.

Restoration Challenges and Solutions

Although prairie restoration is proceeding throughout the Pacific Northwest, these efforts face many challenges. Perhaps foremost is the highly altered state of the ecosystems. The large blocks of prairies that once supported diverse populations of organisms and were themselves continually shaped by frequent fires and Native American harvesting have become highly fragmented. Key ecosystem processes have been severely altered or removed altogether. Connections that linked prairies have been severed, most prairies have been lost, and those that remain are small and highly degraded (Figure 3.16). Most sites have lost many of their original species, often replaced by invasive taxa.

The implications of these changes are enormous. Particularly problematic is the lack of sites in which to carry out any restoration at all. The loss of habitat to urban development, forest, and agriculture has been so extensive that there are few sites where any restoration can be contemplated. Those that do exist are mostly small, costly to acquire, and in need of extensive restoration of species and processes. In most cases, practitioners are forced to scale back goals, settling for compromised natural systems. Ecologically significant processes may be diminished in extent or intensity or absent altogether. The use of fire may be precluded at some sites, and disturbance from burrowing animals or Native American harvesting may be absent and difficult to replicate. Species may be missing or impossible to reestablish. Some grassland birds with large habitat area needs may be excluded from smaller sites. Butterflies that depend on large population dynamics in a large landscape to sustain local populations may be impossible to recover. The nutrient status, structure, or suite of fungus–plant associations in prairie soils may be extensively or perhaps irrevocably altered by decades of agricultural use.

But identification of restoration goals is further impeded by another problem not encountered where systems are less degraded. Few reference sites exist, making it unclear what species should be restored to sites, how abundant they should be, or how they should be distributed. Some of these

FIGURE 3.16. In many cases, only tiny fragments of prairies remain, such as this one along Highway 101 near Shelton, Washington. (*Photo by Ed Alverson*)

Case Study: West Eugene Wetlands

Many practitioners and natural resource managers regard the West Eugene Wetlands as one of the finest examples of wet prairie restoration in the Willamette Valley. The project owes its success to a serendipitous convergence of location, circumstance, regulatory drivers, key players, and timing. These factors led to creation of a two-decade project that has conserved and restored wet prairie habitat surrounding Amazon Creek within the city limits of Eugene that today provides urban open space, educational programming, preservation of rare species, and flood management. Despite its unique qualities, a number of the lessons of the West Eugene Wetlands experience are transferable to other locations.

Many people intuitively believe that rural land provides the least degraded habitat and should be the primary focus of public restoration investment. However, in many cases lands on the fringes of urbanizing areas have been held by investors in anticipation of development but not actively altered from their condition at the time of purchase. As a result, they have not been degraded by agricultural use or succumbed to urban development but instead retain a complement of native species. This was the case with many of the parcels in the West Eugene Wetlands (Figure 3.17).

Oregon's land use laws require cities to establish urban growth boundaries to contain sprawl and conserve rural lands. The wetland complex was slated for future development but in the 1980s remained outside Eugene's urban growth boundary. In the late 1980s, inventories identified more than 1,300 acres of wetland, mostly wet prairie. In an effort to comply with both emerging federal wetland regulations and state stormwater regulations, the City of Eugene, working with the Lane Council of Governments, determined that instead of building expensive facilities they would use the existing wetlands as part of their stormwater and flood control system. When they sought federal funding to purchase the land for conservation and restoration, they were told they needed a plan.

The West Eugene Wetlands Plan, developed by an interdepartmental team of city staff dubbed the "Wetheads," recommended protecting 75% of 1,500 acres. The city developed a land acquisition program and formed a mitigation bank. The plan provides a broad vision and strategy embraced by a committed partnership that today includes the U.S. Bureau of Land Management, U.S. Army Corps of Engineers, U.S. Environmental Protection Agency, U.S. Fish and Wildlife Service, The Nature Conservancy, McKenzie River Trust, Willamette Resources and Education Network, and Oregon Youth Conservation Corps (Figure 3.17b). In addition, collaborators include the Oregon Department of Land Conservation and Development, the Native Plant Society, Audubon, the League of Women Voters, west Eugene business organizations, and many others. The plan has been integrated into other Eugene and Springfield metro area plans, provides regulatory certainty for west Eugene developers and landowners, and has been an important tool in securing more than $12 million of federal land and water conservation funds and $5 million from other sources.

The West Eugene Wetlands illustrate the value of a long-term, broad-scale vision and plan. Its federal, state, local, and nonprofit partners collectively present a united lobbying front and contribute a variety of skills and capacity. The plan seeks to achieve some balance between habitat protection, recreation, and urban development through a landscape-scale scheme in which smaller degraded sites can be traded for higher-value contiguous habitat, ultimately conserving a functioning wet prairie ecosystem in an urban setting. (More information can be found at www.eugene-or.gov/portal/server.pt. Follow the links to Parks and Openspace, Natural Resources, and Wetlands Program Info.)

FIGURE 3.17. *(a)* A volunteer counts rare species on a west Eugene wet prairie. *(b)* West Eugene Wetlands partners collaborate on implementing a long-term acquisition and restoration plan. (*City of Eugene and West Eugene Wetland Partners*)

constraints may be overcome as research improves our understanding of the habitat needs of individual taxa and the relationships between species within prairie communities. However, much work is needed to understand how fragmentation may be countered effectively. The most successful approach is likely to be a multitiered conservation strategy, nesting restored areas of different quality within a landscape of varying intactness, functionality, and levels of protection, with each area differing in the level of investment in restoration and management.

The most appropriate approach may be establishing core and buffer areas. Under this strategy the greatest effort is expended to restore key core areas by securing high levels of protection, intensively managing them to restore rare species, eliminating or significantly reducing invasive species, reestablishing native species assemblages and critical ecological processes, and maintaining overall ecological integrity. Buffer areas are managed to retain some ecological functions but permit a greater range of compatible land uses (military training, low-impact recreation). In these areas, management is focused on minimizing particular threats, such as real estate development and selected invasive species. In the surrounding matrix of private land, where development, agricultural, and recreational activities occur, other strategies are used to prevent degradation of core and buffer areas.

Education and private landowner incentives are valuable tools for abating key threats, such as invasive species and inappropriate use of pesticides, while encouraging beneficial actions (e.g., use of native plants, retention of native oaks). The conservation community has focused on preservation and restoration of high-quality core habitats, largely ignoring the functionality of these surrounding matrix lands. Critical research is needed to elucidate the ecological roles these matrix lands play in sustaining the long-term viability of core areas and identify strategic actions that will increase their conservation value. In particular, it is critical to avoid well-meaning but misguided efforts that may create ecological sinks for species of conservation concern and to avoid taking actions that feel good but are meaningless or even counterproductive from a conservation perspective.

Prairies exist in a larger setting of varying land uses, and they also function as part of a larger landscape continuum of upland and wet grasslands, savannas, woodlands, and forests. To survive, some species may need this assemblage of habitats and especially the ecotones where they meet. It has been suggested that some butterflies need a mix of habitats to provide nectar sources at various seasons and in different years, when droughts or burns may render suitable habitat temporarily unsuitable. The importance of ecological patterns of various scales and how they should be managed and restored is a key area for future research.

Given the necessity of working in highly degraded sites, perhaps the most important quality for prairie restorationists in the Pacific Northwest to cultivate is patience. Many restoration efforts focus on quickly installing diverse native plants on a site. Such a compositional objective in itself may be an appropriate short- or intermediate-term goal, but it must be approached with caution. Reestablishing natural processes—moisture conditions in a wet prairie, pocket gopher soil disturbance to encourage native annuals, or fire cycles that open moss- and lichen-free substrates for small-seeded perennials—takes time. Eliminating existing weed seed banks or rebalancing soil nutrient conditions to favor desired natives and reducing vectors of weed reinfestations are strategies that can spell the difference between successful restorations and colossal failures. Shortcutting these steps is the primary reason for failure, often leading to situations that rapidly spiral out of control as costs mount, aggressive invasives proliferate, native species dwindle, and tools for effectively remedying the problems evaporate. Restorationists must be forthright in articulating realistic timeframes for accomplishing these initial steps, both to themselves and to those paying the bills.

Much restoration today takes place in a context often called adaptive management. This is not simply learning from one's mistakes but a process by which managers refine their understanding of an ecosystem and how it functions through a series of carefully designed experiments. This approach poses multiple challenges to restorationists. Wherever possible, they must apply scientific rigor in design, replication, monitoring, and analysis of experimental treatments. Too often, the focus is on

Case Study: Garry Oak Ecosystem Restoration in Canada

By Tim Ennis, The Nature Conservancy of Canada

Garry oak (*Quercus garryana*) ecosystems are among the most endangered in Canada. Only an estimated 1–5% of the 1850s distribution remains in a near-natural state, providing habitat for more than 100 species considered at risk either nationally or provincially. Although habitat fragmentation and isolation threaten the long-term viability of Garry oak–associated species, all habitat remnants have been degraded by exotic species of plants and animals, and many suffer from the effects of fire suppression.

Working toward the recovery of Garry oak ecosystems in Canada is a priority for a wide range of organizations including federal, provincial, regional, and municipal governments, First Nations, nongovernment organizations, academia, and citizens. Coming together under the umbrella of the Garry Oak Ecosystems Recovery Team (GOERT), these partners have collectively drafted an innovative ecosystem-based recovery strategy that also addresses the needs of individual at-risk species. Recovery action plans have been drafted to address more focused challenges, including the restoration of Garry oak habitats. The recovery team has produced numerous extension products including a decision support tool for restoration practitioners, various mapping products, and stewardship manuals for both species at risk and invasive species.

Throughout the range of Garry oak ecosystems in Canada, a growing number of agencies with management authority over habitat remnants are initiating ecosystem restoration programs. The first of these fully integrated restoration programs was The Nature Conservancy of Canada's work at the Cowichan Garry Oak Preserve near the town of Duncan on southeast Vancouver Island. Largely considered the best example of low-elevation, valley-bottom Garry oak habitat in Canada, it provides habitat for several robust populations of plants at risk, including the largest Canadian population of the nationally endangered Howell's triteleia (*Tritelia howellii*). Detailed vegetation, soil, and bird inventories have laid the foundation for a management plan and monitoring system. Restoration activities have included manual and mechanical removal of exotic shrubs, control of invasive native woody species, control of exotic grass species using mechanical methods and fire, seeding of native species into treatment areas, and propagation and outplanting of native grasses and species at risk. Restoration efforts have benefited from a strongly supportive local community, with regular and numerous volunteer events and educational tours for community groups and school children. The Nature Conservancy of Canada has supported the work of several leading researchers in conducting investigations on the site that inform the management practices and restoration techniques used by the Conservancy. The results of detailed vegetation monitoring over time suggest that this program is having a profoundly positive effect in decreasing the overall cover of invasive species, increasing the cover of native species, and increasing the populations of plants at risk.

Building on this success, the Conservancy has acquired some adjacent land formerly used as sheep pasture, which no longer contains native elements other than scattered oak trees. Through the Quamichan Garry Oak Restoration Project, the Conservancy is pioneering efforts in Canada to restore an agricultural field back into a "near-natural" Garry oak ecosystem over time. Clearly, increasing the amount of available habitat on the landscape in strategically located places holds the key to improving the long-term viability of Garry oak–associated species. (See Plate 2 in the color insert for images of Garry oak ecosystem species.)

Case Study: Fort Lewis Wildflower Explosion

Fort Lewis occupies more than 86,000 acres of glacial outwash, extending southeast from Puget Sound between Olympia and Tacoma, Washington. It contains some of the largest and highest-quality natural habitats in the region, including an estimated 11,500 acres of prairie—nearly 90% of the total prairie in the south Puget Sound region. Activities at the base are focused on equipping troops with the skills and techniques to function in a wide variety of situations, from fighting and surviving in high-intensity warfare to maintaining peace and providing humanitarian aid. This training entails frequent exercises on varied terrain, including prairies.

The Central Impact Area is one of these prairies. Described as the most violent piece of land on Fort Lewis, it receives regular bombardments of artillery and explosive ordnance. Most observers would think few prairie plants and animals could survive in this area of nearly constant explosions, yet quite the opposite is true. It contains very high-quality prairie, providing habitat for several rare animals. In one portion of the Impact Area, there are four candidates for listing under the federal Endangered Species Act: streaked horned larks, mazama pocket gophers, mardon skippers, and Taylor's checkerspot butterflies. It is likely that the regular wildfires that burn this prairie, ignited by explosives, help maintain the conditions these species need (Figure 3.18).

Although federal regulations, including the Endangered Species Act and Sikes Act, guide the management of Fort Lewis natural resources, wildlife and military managers on the base have discovered that high-quality natural habitats also serve as high-quality training areas. Restoration activities have included removal of dense Scotch broom thickets that had degraded prairie habitat and created obstacles to military maneuvers. The mosaic of habitats found in restored prairies withstands disturbance and provides some of the most desirable training areas on the Fort Lewis base.

As a consequence, managers have actively restored prairies for more than a decade. The Fort Lewis natural resource staff has teamed with a network of partners to complete a full range of restoration actions including weed control, prescribed fire, transplantation of prairie plants, and enhancements for rare species including prairie butterflies and birds. As a result of these efforts, the grounds of Fort Lewis have become one of the most beautiful displays of prairie wildflowers and wildlife in the least likely location.

getting something done, leaving behind golden opportunities for gathering valuable information if sufficient attention is paid up front to designing replicated experiments, gathering pertinent data, and comparing and analyzing results.

Another key to successful restoration is to avoid searching for the best solution, the "silver bullet" that will ensure success. Rather, one of the tenets of adaptive management is to design experiments that capitalize on serendipitous discoveries, that discern critical processes, relationships, or thresholds in ecosystems, and that provide essential information to continually refine ongoing restoration efforts. Deriving answers to a host of initial questions, adjusting management practices based on these answers, and revisiting and refining restoration projects repeatedly over many years based on a studied, iterative approach is essential if the field of prairie restoration is going to move from a series of isolated, anecdotal efforts to a more mature, hypothesis-driven science.

Restoration goals for a site must be regarded as long term, especially because our current understanding of exactly how to return a site to a healthy,

FIGURE 3.18. Scotch broom faces imminent demise at the Fort Lewis Central Impact Area. (*Photo by Patrick Dunn*)

functioning condition is limited. In an impatient world that demands instant results, keeping sight of a long-term objective can be challenging. Establish interim goals that allow restoration to proceed in a thoughtful, prudent manner. Mistakes are certain to occur, but they can often teach us how systems operate and how successive efforts can be improved. Many efforts fail by trying to attain a finished appearance too quickly, not only disregarding financial constraints but, more importantly, often failing to recognize the time and changes that are needed for species to establish a balance with one another. Erecting successive, short-term goals based on an understanding of successional processes, ecological relationships, and key ecosystem functions can sustain forward momentum in a long-term project. Sequencing restoration steps to avoid getting ahead of available resources and to allow natural processes to affect the composition and structure of the community are an important part of a successful practitioner's approach. For example, depending on the ultimate restoration goals, establishing dominant or key species that sustain critical ecological processes or target conservation species may be an important initial goal. Suppressing ecosystem-shaping invasives such as Scotch broom, establishing a continuous matrix of grasses to carry fires, and installing slow-growing taxa, such as oaks, that may be an integral part of a long-term plan may provide important compositional, structural, or functional short-term goals.

Summary

A key problem plaguing nascent Northwest prairie restoration efforts is the lack of clearly defined goals. Far too often, such goals, if they exist at all, are fuzzy because of a poor understanding of historical reference communities, uncertainties over what a restoration site is capable of supporting, or failure to clearly define the objectives. The consequences of this uncertainty include frustrated funders, unhappy clients, wasted money, an inability to achieve meaningful conservation of biodiversity, and successes that are indistinguishable from failures. By adopting an adaptive, iterative approach with clear interim goals, we can open new doors that circumvent the impediments to realistic, meaningful restoration goals. Rather than struggling to precisely define the composition and structure of a target community from the outset, restorationists should develop a more general long-term vision. Progress toward realizing this vision then proceeds by means of a series of more precise, short-term objectives that are revised and refined as new information about site constraints, possibilities, functionality, and viability is obtained.

Recognizing that restoring functionality and a large complement of native species to a prairie system is likely to take years of patient effort raises a final potential problem. Northwest grasslands are dynamic, successional systems that change constantly as they respond to perturbations. Micromanagement—the continual tweaking at a site to shape it to a particular vision—must be avoided. Perhaps the most important strategy for circumventing this pitfall is to articulate the range of variability in a site that is acceptable, given current understanding of how plants and animals interact with the substrate, topography, and climate. Incorporating such variability into restoration goals and objectives acknowledges a reality that is not captured by snapshot descriptions of grassland composition or structure.

Acknowledgments

We thank Tim Ennis, director of Land Stewardship for the British Columbia Region of The Nature Conservancy of Canada for the case study on Garry oak ecosystem restoration in Canada. Additionally, we are grateful to Janice Miller and Claudine Tobalske for producing maps, Sarah Finney for assistance with photographs, and Kara Shaber for assistance with references.

References

Allison, S. 2004. What do we mean when we talk about ecological restoration? *Ecological Restoration* 22: 281–286.

Bakker, J. P. and F. Berendse. 1999. Constraints in the restoration of ecological diversity in grassland and heathlands communities. *Trends in Ecology and Evolution* 14: 63–67.

Bowles, M. L., M. D. Jones, and J. L. McBride. 2003. Twenty-year changes in burned and unburned sand prairie remnants in northwestern Illinois and implications for management. *American Midland Naturalist* 149: 34–45.

Boyd, R. 1999. Strategies of Indian burning in the Willamette Valley. Pages 94–138 in R. Boyd (ed.), *Indians, Fire, and the Land in the Pacific Northwest.* Oregon State University Press, Corvallis.

Caplow, F. and J. Miller. 2004. *Southwestern Washington Prairies: Using GIS to Find Rare Plant Habitat and Historic Prairies.* Natural Heritage Report 2004–02, Washington Natural Heritage Program, Olympia.

Christy, J. and E. Alverson. In prep. *Historic Vegetation of the Willamette Valley, Oregon, in the 1850s.*

Collins, S. L. and L. L. Wallace. 1990. *Fire in North American Tallgrass Prairies.* University of Oklahoma Press, Norman.

Costanza, R., R. d'Arge, R. de Groot, S. Farber, M. Grasso, B. Hannon, S. Naeem, K. Limburg, J. Paruelo, R. V. O'Neill, R. Raskin, P. Sutton, and M. Van den Belt. 1997. The value of the world's ecosystem services and natural capital. *Nature* 387: 253–260.

Crawford, R. C. and H. Hall. 1997. Changes in the south Puget prairie landscape. Pages 11–15 in P. Dunn and K. Ewing (eds.), *Ecology and Conservation of the South Puget Sound Prairie Landscape.* The Nature Conservancy of Washington, Seattle.

Douglas, D. 1959. *Journal Kept by David Douglas During His Travels in North America, 1823–1827.* Antiquarian Press, New York.

Dunn, P. and K. Ewing. 1997. *Ecology and Conservation of the South Puget Sound Prairie Landscape.* The Nature Conservancy of Washington, Seattle.

Dunwiddie, P. W. 2002. Management and restoration of grasslands on Yellow Island, San Juan Islands, Washington USA. Pages 78–87 in P. J. Burton (ed.), *Garry Oak Ecosystem Restoration: Progress and Prognosis.* Proceedings of the Third Annual Meeting of the B.C. Chapter of the Society of Ecological Restoration, April 27–28 2002, University of Victoria.

Ehrenfeld, J. G. 2000. Defining the limits of restoration: the need for realistic goals. *Restoration Ecology* 8: 2–9.

Emery, S. M. and K. L. Gross. 2005. Effects of timing of prescribed fire on the demography of an invasive plant, spotted knapweed *Centaurea maculosa. Journal of Applied Ecology* 42: 60–69.

Floberg, J., M. Goering, G. Wilhere, C. MacDonald, C. Chappell, C. Rumsey, Z. Ferdana, A. Holt, P. Skidmore, T. Horsman, E. Alverson, C. Tanner, M. Bryer, P. Iachetti, A. Harcombe, B. McDonald, T. Cook, M. Summers, and D. Rolph. 2004. *Willamette Valley–Puget Trough–Georgia Basin Ecoregional Assessment.* The Nature Conservancy, Seattle. Available at conserveonline .org/2004/06/g/WPG_Ecoregional_Assessment.

Garry Oak Ecosystem Recovery Team (GOERT) Web pages www.goert.ca/ecoinfo/important.htm and www .goert.ca/about/strat.htm, Victoria, BC.

Groves, C. R. 2003. *Drafting a Conservation Blueprint: A Practitioner's Guide to Regional Planning for Biodiversity.* Island Press, Washington, DC.

Harker, D., G. Libby, K. Harker, S. Evans, and M. Evans. 1999. *Landscape Restoration Handbook.* Lewis Publishers, Boca Raton, FL.

Hobbs, R. J. and D. A. Norton. 1996. Towards a conceptual framework for restoration ecology. *Restoration Ecology* 4: 93–110.

Jackson, L. L., N. Lopoukhine, and D. Hillyard. 1995. Ecological restoration: a definition and comments. *Restoration Ecology* 3: 71–75.

Jancaitis, J. E. 2001. *Restoration of a Willamette Valley Wet Prairie and Evaluation of Two Management Techniques.* M.S. thesis, University of Oregon, Eugene.

Leopold, E. B. and R. Boyd. 1999. An ecological history of old prairie areas in southwestern Washington. Pages 139–163 in R. Boyd (ed.), *Indians, Fire and the Land in the Pacific Northwest.* Oregon State University Press, Corvallis.

Noss, R. F. and A. Y. Cooperrider. 1994. *Saving Nature's Legacy, Protecting and Restoring Biodiversity*. Island Press, Washington, DC.

Packard, S. and C. R. Mutel. 1997. *The Tallgrass Restoration Handbook: For Prairies, Savannas and Woodlands*. Island Press, Washington, DC.

Pendergrass (Connely), K. L. 1995. *Vegetation Composition and Response to Fire of Native Willamette Valley Wetland Prairies*. M.S. thesis, Oregon State University, Corvallis.

Perrow, M. R. and A. J. Davy. 2002. *Handbook of Ecological Restoration*. Cambridge University Press, Cambridge, UK.

Storm, L. E. 2002. Patterns and processes of indigenous burning: how to read landscape signatures of past human practices. Pages 496–508 in J. R. Stepp, F. S. Wyndham, and R. J. Zarger (eds.), *Ethnobiology and Biocultural Diversity. International Society of Ethnobiology*. University of Georgia Press, Athens.

Streatfield, R. and R. E. Frenkel. 1997. Ecological survey and interpretation of the Willamette Floodplain Research Natural Area, W.L. Finley National Wildlife Refuge, Oregon, USA. *Natural Areas Journal* 17: 354–364.

Turner, N. 1999. "Time to burn": traditional use of fire to enhance resource production by aboriginal peoples in British Columbia." Pages 185–218 in R. Boyd (ed.), *Indians, Fire and the Land in the Pacific Northwest*. Oregon State University Press, Corvallis.

Vickery, P. D. and P. W. Dunwiddie, eds. 1997. *Grasslands of Northeastern North America: Ecology and Conservation of Native and Agricultural Landscapes*. Massachusetts Audubon Society, Lincoln.

Westman, W. E. 1991. Ecological restoration projects: measuring their performance. *Environmental Professional* 13: 207–215.

Wilson, M. V., E. Alverson, D. Clark, R. Hayes, C. Ingersoll, and M. B. Naughton. 1995. The Willamette Valley Natural Areas Network: promoting restoration through science and stewardship. *Restoration and Management Notes* 13: 26–28.

Chapter 4

Oak Woodlands and Savannas

PAUL E. HOSTEN, O. EUGENE HICKMAN, FRANK K. LAKE, FRANK A. LANG, AND DAVID VESELY

Oak woodlands of the Pacific Northwest include several oak species, a wide range of plant communities, and important structural variations. This richness reflects diverse environments influenced by climate, soil, slope, aspect, elevation, and human interaction, including recent changes reflecting fire suppression.

Oregon white oak is found primarily west of the Cascade Mountains and is the only native oak in British Columbia, Washington, and northern Oregon. It is often found in pure stands but also mixes with ponderosa pine, Douglas fir, juniper, maple, and other hardwoods. Oak woodlands in coastal areas usually are limited to moist concave swales and benches, often underlain by clays. White oak is more commonly found in dry interior valleys, topographic rain shadows, foothills, and droughty mountain slopes. Also, two large water bodies, Klamath Lake and the Columbia River Gorge, moderate the climate and allow Oregon white oak to persist east of the Cascades at the lower end of its temperature range (Figure 4.1).

Fossil pollen records from California indicate that substantial changes in oak abundance have occurred over the past 10,000 years (Byme 1991). From 5,000 to 10,000 years B.P., records show an increase in oak abundance at upper elevation ranges, probably caused by climatic warming and changes in fire regimes. More recent change in oak abundance at lower elevations along the coast has been tied to cessation of Native American burning since the mid-nineteenth century (Thilenius 1964, 1968, Boyd 1999).

Native American Management and Use of Oaks

Northwest Native people relied on salmon, roots, berries, game, and acorns as key food sources that shaped their cultures. Acorns were highly valued by native peoples from California to Vancouver Island (McCarthy 1993, Dickson 1946). The three most culturally important oaks (and related trees) were the tanoak (*Lithophragma bulbifera*), black oak (*Quercus kelloggii*), and Oregon white oak (*Q. garryana*). Acorns from Saddler's (*Q. sadleriana*), canyon (*Q. chrysolepis*), and coast live oaks (*Q. agrifolia*) were less favored but were used when crops of more desired species were poor (Pullen 1996).

Native people used cultural practices to favor oak savannas and woodlands over mixed conifer forests (Cole 1977). Oak trees were tended and cared for to produce desired stand conditions (McCarthy 1993). Many oak groves were family or tribally owned.

Several forms of management were used to influence oak location, stand configuration, and tree shape. Fire, pruning, and knocking increased acorn production and promoted favorable characteristics (McCarthy 1993). Fire was used to control

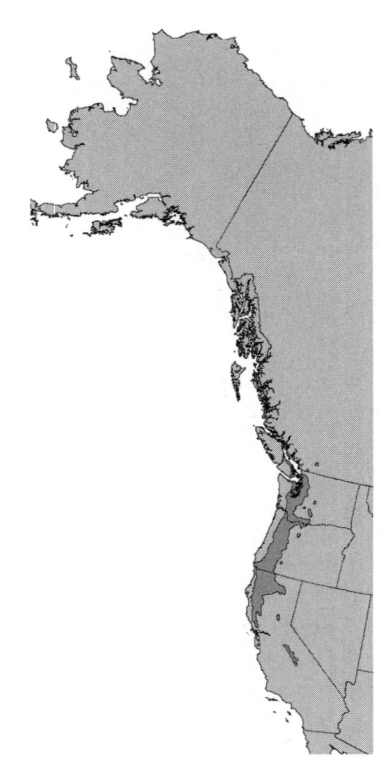

FIGURE 4.1. Distribution of Oregon white oak in the Pacific Northwest. (*Courtesy of U.S. Geological Survey*)

competing vegetation (Boyd 1999). Fires also maintained healthy and productive plant communities (Turner 1999). Prescribed fires also eased collection of acorns by keeping the ground clear and open.

Oak woodlands were also managed to improve harvest of wildlife, including bear, deer, elk, and squirrels, all of which fed on acorns as well as forbs and grasses that sprouted after fire. Woodpeckers and other birds that favor oaks were hunted. Pine nuts, grasses, bulbs and roots (*Allium*, *Brodiaea*, *Calochortus*, *Camassia*, *Chlorogalum*, *Lillium*, *Perideria*), and medicinal plants (*Lomatium* spp., *Wyethia* spp.) were harvested in oak-dominated communities (Dickson 1946).

Management techniques influenced the ecological integrity and productivity of oak communities. Many tribes had cultural practices and rituals that honored oaks and protected their abundance and productivity (McCarthy 1993). The Hupa tribe from the lower Trinity River in California and the Takelma in southwest Oregon had a first acorn ceremony to give thanks (Pullen 1996). The Karuk ritually used fire to pray for abundant acorns and salmon (Kroeber and Gifford 1949). Tribal practices and beliefs helped plan when and how fire would be prescribed and when harvesting began or ended. It is still common practice among many native people in northwestern California to stop acorn harvest after November, thus leaving food for animals.

To have healthy acorns is to have a healthy community. This Native American philosophy linked the quality of environment to the quality of the food it could produce. Even today native people collect acorns each fall. Restoration of oak savannas, woodlands, and mixed oak–conifer forests should consider the historical sociocultural relationship native peoples had and continue to have with oaks as well as the biological diversity of plants and animals associated with management practices.

> Indian have no medicine to put on all places where bug and worm are, so he burn; every year Indian burn. . . . Fire burn up old acorn that fall on ground. Old acorn on ground have lots worm; no burn old acorn, no burn old bark, old leaves, bugs and worms come more every year. . . . Indian burn every year just same, so keep all ground clean, no bark, no dead leaf, no old wood on ground, no old wood on brush, so no bug can stay to eat leaf and no worm can stay to eat berry and acorn. Not much on ground to make hot fire so never hurt big trees where fire burn. (Klamath River Jack, in McCarthy 1993)

The loss of Native American land stewardship has mirrored the loss of healthy oak ecosystems across the region. The influence of Euro-Americans on oak woodlands predates direct settlement impacts by 200 to 300 years, resulting primarily from the spread of diseases that reduced Native American populations and interrupted traditional management practices. Direct European land management further altered plant community dynamics. Loss of frequent fire, overgrazing, and introduction of invasive plants all combined to degrade traditional oak woodlands. The decline of ranching and increased population have triggered further loss of oak woodlands to suburban and rural housing. Altered ecological processes have inhibited the establishment of young trees at some sites, and at other sites an overabundance of saplings creates conditions unfavorable to acorn production or native understory establishment.

Oregon White (Garry) Oak Distribution and Associations

There are many oak-associated plant communities, resulting from a wide variety of ecological and cultural factors. Only the more representative ones are presented in this chapter as a basis for considering restoration.

The northernmost range of Oregon white oak is southwestern British Columbia, along the east coast of Vancouver Island and the Gulf Islands, at Sumas Mountain, and at several locations in the Fraser River Valley on the mainland (Erickson 1996). Oak communities in British Columbia are found where mean annual precipitation varies between 40 and 120 centimeters, with January temperature means between –2.5°C and 2.5°C. To avoid conifer competition, oaks are found mostly in dry sites with shallow, rocky soils and high growing season moisture stress.

In southwest Vancouver Island and the Gulf Islands, a profound rain shadow influences oak locations. A combination of dry summers, shallow soils, excessive drainage, and cultural practices allowed oaks to persist over time (Roemer 1993). A Garry oak, fawn lily, sandwort, and Oregon grape association is found on moist sites with more than 800 millimeters of annual precipitation. A Garry oak–geranium–nonnative annuals association is found where annual precipitation is less than 800 millimeters. A Douglas fir–Garry oak–onion grass community is found below 150 meters elevation in the vicinity of Victoria and the Gulf Islands (Erickson 1996).

Erickson (2002) studied the environmental relationships of twenty-six native plant communities at the northern margin of Garry oak distribution. These communities include those dominated by camas (*Camassia quamash*), camas and fawn lily (*Erythronium oreganum*), camas and shooting star (*Dodecatheon hendersonii*), and camas and buttercup (*Ranunculus occidentalis*). Other communities have moss species as important associates, primarily *Dicranium scoparium* and *Rhacomitrium canescens*. Native grasses such as Roemer's fescue (*Festuca idahoensis*), California brome (*Bromus carinatus*), blue wildrye (*Elymus glaucus*), showy oniongrass (*Melica subulata*), and various forbs are important components of other communities. Wet native Garry oak communities often contain shrubs, such as oceanspray (*Holodiscus discolor*), snowberry (*Symphoricarpos albus*), and nootka rose (*Rosa nutkana*).

On the mainland in Fraser Valley, Fraser Canyon, and Sumas Mountain, small populations of Garry oak occur some distance from the influence of the rain shadow and lower precipitation typical of the Gulf and Vancouver islands. British Columbia Parks established the Yale Garry Oak Ecological Reserve in 1998 to protect the easternmost stand of Garry oak in British Columbia (Figure 4.2).

The Washington Natural Heritage Program (2004) lists Oregon white oak communities according to county. In Island County a snowberry–sedge (*Carex inops*) oak woodland is fairly common. San Juan County has a sedge–camas association. The gravelly soils of Pierce and Thurston counties are known for native prairies and oak woodlands. Local white oak communities include long-stolon sedge–camas, snowberry–long-stolon sedge, snowberry–sword fern, and snowberry–moist forb. The long-stolon sedge–camas community is believed to be similar to presettlement oak savannas. A fifth community, oval-leaf viburnum–poison oak, occurs near the edges of prairies. Oak, Douglas fir, snowberry, and sword fern and oak, Douglas fir,

FIGURE 4.2. Classic Garry oak savanna on Vancouver Island, British Columbia. (*Photo by Ed Alverson*)

snowberry, and sedge communities result from Douglas fir invading former oak woodlands in absence of fire (Chappell and Crawford 1997).

Three important historic Oregon white oak habitats in south Puget Sound include oak savannas and open woodlands, riparian oak woodlands, and wetland oaks (Hanna and Dunn 1996). Oak savannas and open woodlands have virtually disappeared because of land development and lack of fire. Open-crowned single-stem oaks are associated with Roemer's fescue, long-stolon sedge, camas, and shooting star. Riparian oak woodlands typically have an overstory of thin, tall oaks that form a closed canopy, with a dense shrub layer of Oregon ash, bitter cherry, tall Oregon grape, and snowberry and Roemer's fescue. Wetland oak communities are similar to riparian but with a composition that includes Oregon ash, big-leaf maple, vine maple, western hazel, and sword fern.

Along with these historic communities are newer ones, caused by invasion of conifers and alien weeds. These include a closed-canopy mixed oak–conifer woodland with a highly variable shrub and herb layer that includes sword fern, Oregon grape, Scotch broom, and bentgrass. A range oak woodland association is found in areas heavily grazed by livestock and has a variable overstory with Scotch broom and a host of pasture grasses and other weeds underneath. Dense oak woodlands have numerous younger oaks less than 100 years old that form a tightly closed canopy, with scattered shrubs and herbs below, including snowberry and Oregon grape. Clumped oaks often surround an older granny oak. Native species include snowberry, long-stolon sedge, and hairy cats-ear.

In Clark and Cowlitz counties, the Washington Natural Heritage Program (2004) lists the following Oregon white oak communities: Douglas fir–oak–snowberry (*Symphoricarpos albus*); oak–ash–snowberry; oak–poison oak (*Toxicodendron diversiloba*)–blue wildrye (*Elymus glaucus*); and oak–viburnum (*Viburnum ellipticum*)–poison oak. Skamania County, in the Columbia River Gorge, has the following oak communities: Roemer's fescue, snowberry, and viburnum–poison oak. Further east in the Gorge, in Klickitat County, one finds oak–pine, blue wildrye, and Roemer's fescue communities.

On the east side of the Washington Cascades, Oregon white oak is located along streams in western Klickitat, Yakima, and Kittitas counties. There is a community similar to the Oregon white oak–California hazel (*Corylus cornuta*)–common snowberry association found in the Wenatchee National Forest (Lillybridge et al. 1995). It differs primarily in the lack of hazel, bitter cherry, and Rocky mountain maple (*Acer glabrum*). An Oregon white oak–blue wildrye association is found on the Yakama Nation reservation. This has a grass-dominated understory (blue wildrye, elk sedge, and bluegrass) with few shrubs. North of Klickitat, in Kittitas and Yakima counties, are oak–elk sedge and oak–blue wildrye communities.

Four major oak communities have been described in the Willamette Valley (Thilenius 1964, 1968). Oak–hazelnut–sword fern (*Polystichum munitum*) communities include big-leaf maple (*Acer macrophyllum*), Douglas fir, and grand fir (*Abies grandis*) in the canopy. The tall shrub layer includes serviceberry (*Amelanchier alnifolia*), oceanspray, Douglas hawthorn (*Crataegus douglasii*), and introduced mazzard cherry (*Prunus avium*). Sword fern is present in the low shrub layer along with bracken fern, snowberry, poison oak, and several rose and blackberry species. Oak–serviceberry–snowberry communities also include Douglas fir and big-leaf maple in the canopy. Tall shrubs include serviceberry, Indian plum (*Osmaronia cerasiformis*), and mazzard cherry. Snowberry, poison oak, sword fern, and dog and Nootka rose (*Rosa* spp.) are in the low shrub layer. Oak–mazzard cherry–snowberry has a similar tree layer to the others, with a tall shrub layer dominated by mazzard cherry, sometimes forming impenetrable thickets. Oak–poison oak is the most common community in dry habitats, partly as a consequence of grazing and fire suppression.

Southwest Oregon is a unique, ecologically diverse region of the Pacific Northwest and is the natural home for a number of oak species. The most widespread oak in Oregon is Oregon white oak, common in the large warm valleys of the central Umpqua, upper Illinois River, central Rogue River, and lower Applegate River. The driest of these interior valleys are near Medford, where white oak communities persist on a wide variety of environments including droughty foothills, deep clayey plains, loamy riparian bottoms, and dry woodland slopes. Tree size and growth form are wide ranging, depending on the local environment.

Many gentle slopes and terraces of the Cascade foothills have deep, droughty soils that support white oak savanna or woodland and exclude most conifers. Steep south-facing aspects with shallow or gravelly clay loam soils in the Applegate River watershed are too droughty for conifers but allow white oak and bunchgrasses to thrive. At the tops of long forested ridges are natural prairies ringed by scattered oak clumps or bands of white oak woodland. All of the other oak types in southwest Oregon are found mostly in higher-precipitation, forest climatic zones and are discussed later.

Smith (1985) delineated six Oregon white oak plant associations in the Umpqua basin, including oak, poison oak, *Taeniatherum asperum*, and dogtail; oak, poison oak, and *Cynosurus echinatus*; oak, poison oak, and orchard grass; oak, madrone, poison oak, and *Cynosurus echinatus*; oak, ash, sweetbriar rose, and rush; and Douglas fir, oak, poison oak, and sword fern. Fire, agricultural practices, livestock grazing, and the introduction of invasive weeds are strong influences behind present-day communities.

Riegel, Smith, and Franklin (1992) studied oak woodlands in the interior valleys of southwest Oregon, including the Umpqua basin, and described five associations. Oak–California brome is the driest type, often with a sparse shrub layer of poison oak and wedgeleaf ceanothus. Oak–*Cynosurus echinatus* is dominated by Oregon white oak, with a few other tree species present. Annual herbs include *Anthriscus* and soft brome. Oak–Douglas fir–sheep fescue has oak and Douglas fir as codominants. A low, sparse understory of poison oak and white snowberry with sheep fescue and western fescue is also present. Oak–Douglas fir–blue wildrye is the most mesic type, characterized by an open canopy of Oregon white oak and occasional California black oak. Douglas fir is rarely found in the overstory but is consistently present in the reproductive layers. Poison oak provides a patchy cover in understory within a dense matrix of grasses and forbs. Oak–mountain mahogany occurs on ridges and rocky outcrops. Except for poison oak, shrubs are rare. Herbs include wild cucumber, wild carrot, woolly sunflower, and cleavers bedstraw.

Hickman (USDA 1988, 1996) described several mixed conifer–oak communities and seven white oak sites, primarily in Jackson and Josephine counties in warm, dry climatic zones, with less than 35 inches of annual precipitation and at medium to low elevations. These include Droughty Fan, a white oak–bunchgrass type found on open, dry, clayey southerly slopes southeast of Medford. Clayey Hills is a white oak–pine oatgrass community occurring on dry, flat to concave positions on heavy clay soils of east central to southeast Jackson County. Droughty Slopes is a white oak–juniper–Roemer's fescue type found on loamy soils in extreme southeast Jackson County. Loamy Hills has mixed oak–pine–Roemer's fescue and is found on loamy soils in central and southern Jackson and Josephine counties on hills and south slopes. Loamy Slopes includes pine–mixed oak–Roemer's fescue on low to mid-elevations in both counties, occurring on valley plains, hills, and south slopes. Droughty North includes white oak–mountain mahogany–Roemer's fescue on loamy soils and northerly aspects. Deep Loamy Terrace is a pine–mixed hardwood and mixed shrub type with deep loamy soils on valley plains and high terraces where both white and black oak attain very large growth forms.

Atzet et al. (1996) described two Oregon white oak plant associations that occur in southern Douglas County, Jackson County, Josephine County, and parts of Curry County. They are found at lower elevations along valley floors on islands of shallow soils in hot, dry microclimates. Oak–Douglas fir–poison oak has white oak as a dominant, with frequent occurrences of Douglas fir. Poison oak is a common shrub, with mountain sweet-root, hedgehog dogtail, and blue wildrye as common herbs. Oak–*Cynosurus echinatus* is dominated by Oregon white oak with understory trees of white oak and some California black oak. Shrubs include poison oak and buckbrush. Hedgehog dogtail and bur-chervil dominate the herbaceous layer.

Brewer's oak (*Q. garryana* ssp. *breweri*) woodlands, usually found at higher elevation than woodlands formed by Oregon white oak, are characterized by strong sprouting from epicormic buds and form dense clones. Although such woodlands may encompass open patches of herbs, buckbrush, or shrubs from the rose family, taller oak trees retain dominance over time. Photos of Brewer's oak woodlands initially taken in the 1920s and 1960s show little change in the stature and canopy cover by oaks but do show increased abundance of conifers (white fir, Douglas fir, incense cedar, and western juniper).

The Department of Fish and Game (2003) summarizes oak communities in California by providing community names and bibliographic citations. Twelve Oregon white oak communities, twenty-one California black oak communities, and nineteen canyon live oak are listed.

Patterns of Vegetation Change in Oregon White Oak Savanna and Woodlands

Oregon white oak community dynamics are best understood in the Willamette Valley but poorly elsewhere (Atzet and Wheeler 1982, Franklin and Dyrness 1988, Maslovat 2001, MacDougall 2001). Wildlife relationships with woodland habitats are also poorly understood (Block and Morrison 1987, Verner 1987). There are several commonly observed successional changes in white oak woodlands. These include an increase in conifers, the proliferation of a shrub understory, higher oak densities, and an ever-increasing abundance of nonnative annuals and perennials in the herbaceous understory (Barnhart et al. 1996, Reed and Sugihara 1987, Sugihara and Reed 1986, Thilenius 1968, Franklin and Dyrness 1988). The increase in woody cover is generally attributed to the cessation of Native American burning practices, more recent fire suppression (Agee 1993), and the interaction between livestock and fire return interval as a consequence of the reduction in fine fuels. The increase in woody trees and shrubs consists of both native species (Douglas fir, Oregon grape, snowberry, buckbrush, manzanita) and nonnative species (English hawthorn, Scotch broom) (Figure 4.3).

FIGURE 4.3. Looking north toward Ashland, Oregon. *(a)* A 1915 photo shows short-stature Oregon white oak in the foreground, oak-dominated slopes in the left middle ground, and patchy oak woodlands and grassy meadows in the middle right background (*photo by John Gribble*). *(b)* This 2004 retake shows large-stature oaks in the foreground and conifer-dominated slopes in the left middle ground, with the woodland patch in the background remaining largely unchanged (*photo by Paul Hosten*). This illustrates changes common to oak woodlands throughout the Pacific Northwest, including retention of some open oak woodland through edaphic control.

California Black Oak Communities

California black oak or Kellogg oak is an important deciduous hardwood from southern Oregon to California (Figure 4.4). It grows mostly in dry forest communities near and south of Yoncalla, but it is found as far north as Eugene. In the Coast Range and western Siskiyou Mountains, black oak is associated with ponderosa pine and Douglas fir forests in drier climatic zones adjacent to interior valleys. Further east in the dry interior valleys of the eastern Siskiyous and southern Cascades, it is found in on high river terraces and droughty uplands, where it mixes with white oak. It is well represented in both dry and moist mixed conifer forests at low elevations in the western Cascades. At the south end of the Cascade Mountains, black oak is found south of the I-5 freeway summit and east of the Greensprings summit near Ashland in southwest Oregon. Here it occurs with sugar and ponderosa pine and Douglas fir, but it does not extend much beyond the Jackson–Klamath county line or east of the Cascade Mountains.

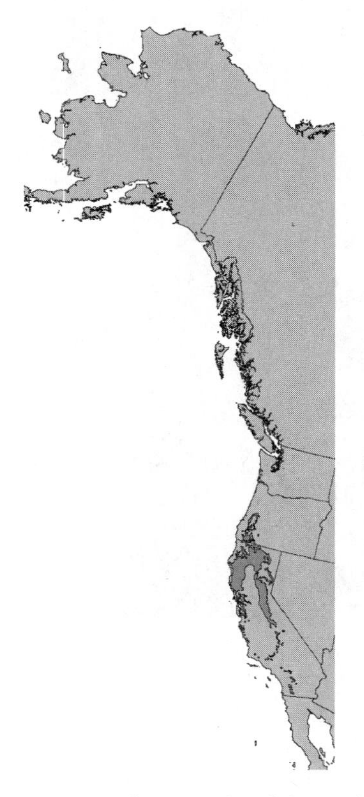

FIGURE 4.4. Distribution of California black oak. (*U.S. Geological Survey*)

California black oak is codominant in some ponderosa pine communities. It also occurs in a number of other plant communities, particularly those dominated by Douglas fir. Ponderosa pine communities have an overstory consisting of ponderosa pine, black oak, white oak, and madrone. Deerbrush, serviceberry, tall Oregongrape, poison oak, mountain mahogany, and snowberry are common shrubs along with fescue, brome, pine bluegrass, blue wildrye, and a number of forbs. The Douglas fir communities have an overstory that includes Douglas fir, madrone, black oak, ponderosa pine, and sometimes sugar pine or incense cedar. Poison oak, tall Oregongrape, deerbrush, oceanspray, baldhip rose, and hairy honeysuckle are common shrubs. Herbs typically include mountain sweetroot, lupine, vetch, woodland tarweed, white hawkweed, slender-tube iris, and bracken fern.

Although stands of black oak in a nonconifer setting may follow some of the same plant community dynamics as those of white oaks, it is the role of black oak as a component of mixed conifer woodlands that stands out. Riparian gallery forests with a black oak, white oak, alder, Oregon ash, and maple overstory have become rare. Generally found on deep alluvial soils, this woodland type has succumbed to house construction, conversion to pasture, and the creation of agricultural fields. In a mixed conifer setting, black oak snags indicate attrition under the development of white fir and Douglas fir canopy as a consequence of fire exclusion. Abundant resprouting of black oak after prescribed fire provides evidence of how well black oak is adapted to frequent low-intensity fire (Kauffman and Martin 1987). Thinning conifer away from black oak is one way to maintain its presence in mixed conifer stands (Tappeiner and McDonald 1979). Black oak can also be rejuvenated by cutting and allowing resprouting.

Open stands of large black oak, ponderosa, and sugar pine are found on river terraces and hills adjacent to historic native village sites. Black oak stands with a California hazel understory were managed with fire to promote basket materials and full-crowned oaks. Fire suppression has resulted in community changes favoring other hardwoods, including madrone, big-leaf maple, and dogwood, as well as Douglas fir and incense cedar (Figure 4.5).

Other Oaks and Related Species

Canyon live oak is part of a broadleaf evergreen woodland that ranges across California and into southwest Oregon south of Canyonville. It is found primarily at low elevations throughout Josephine County, the western Siskiyou Mountains, and the lower Applegate River watershed. Live oak is found in pure stands or with Douglas fir. This woodland type is limited to small areas on very difficult substrates, mainly gravelly soils, sometimes with exposed bedrock, where extreme dryness excludes or greatly limits other trees. Live oak is more common and widespread as an understory shrub or secondary tree under the canopy of dry, low-elevation Douglas fir forests. Other habitats where live oak is found are in colder upper elevation zones, where its growth is a shrub form, and fire-dependent chaparral brushfields. Examples of live oak woodland are abundant on lower slopes of the Rogue River Canyon west of Grants Pass and in the Illinois River Canyon below Cave Junction (Figures 4.6 and 4.7).

Huckleberry oak is found mostly in low to mid-elevation, serpentine soils in higher precipitation zones. It is an understory associate of serpentine-influenced conifer woodland and is also common in shrub stands or brushfields after fire. Huckleberry oak is found in communities associated with

FIGURE 4.5. California black oaks in southwest Oregon occur as large individuals in open monoculture stands, in association with Oregon white oak, or as a component of mixed conifers, particularly along ridges. A short fire return interval can prevent conversion to mixed conifers. (*Photo by Eric Pfaff for the Bureau of Land Management*)

FIGURE 4.6. Distribution of canyon live oak. (*U.S. Geological Survey*)

FIGURE 4.7. Canyon live oaks in southwest Oregon typically grow on shallow, rocky soils with a very low cover by herbaceous species. Woodrat nests are common throughout this plant community type. (*Photo by Eric Pfaff for the Bureau of Land Management*)

different conifers and evergreen hardwoods, including Douglas fir and huckleberry oak; Douglas fir, huckleberry oak, manzanita, and beargrass; and tanoak, white pine, huckleberry oak, and beargrass on serpentine sites (Atzet et al. 1996).

Saddler oak is found only in the cold upper elevation climatic zones where snow is common. It is part of Douglas fir or mixed Douglas fir–white fir forest and is very common in brushfields after stand-replacing fires.

Tanoak is the most abundant hardwood of the western Siskiyous but is not a true oak. Tanoak is widespread from south central Douglas County to the coast and continuing southward through Curry and Josephine counties into California. It is evergreen and associates with Douglas fir in mostly mesic soil temperature regimes. In the Coast Range, it can become a very large tree on all aspects and is generally codominant with Douglas fir. After fires, it sprouts and develops dense woodland, eventually returning to Douglas fir. Further east, in the interior of Josephine County, tanoak nears the eastern boundary of its range, and its growth form is smaller than that of taller trees found to the west. Here, tanoak is found on north aspects on better soils, with Douglas fir as the dominant tree in the canopy. Southern aspects are much drier and have xeric moisture regimes in which tanoak usually is a shrub form growing under Douglas fir. In cold upper elevations of the western Siskiyous, tanoak is abundant but is reduced to a shrub growth form.

Oak Woodland Tree Form, Structure, and Restoration

Restoration of oak woodlands requires understanding not just stand composition and structure but also individual tree form. The most commonly recognized white oak tree form is the statuesque form associated with savannas, widely spaced tree and grass communities dependent on frequent underburns (Thilenius 1968). Tall tree form oaks with interlocking canopies are found on deeper, richer soils, particularly in the Willamette Valley and northward. On dry south-facing slopes in southwest Oregon, oaks may occur in a more shrubby multistemmed form associated with fire-dependent shrubs. These chaparral-like communities persist under less frequent stand-replacing fires. Oak woodlands at higher elevations, colloquially identified as Brewer's oak, form dense coppice stands with or without fire.

The combination of stand composition, stand structure, and tree form aid in identifying past ecological history and understanding conditions consequent to fire exclusion. Land managers interested in restoring oak woodlands normally use thinning, prescribed fire, or a combination of the two. Experience shows that thinning sometimes is used in inappropriate settings. Conditions where thinning may be advantageous include the following:

- Stands where an older cohort of oak trees forms an open overstory, with younger trees filling the gaps, indicate a former oak savanna or open woodland.

- Single-stemmed oaks surrounded by a younger cohort of shrubs indicate a shrub-invaded savanna or open woodland.
- Large circles of oak sprouts coming up through a matrix of seed obligate shrubs also indicate a former savanna that may have experienced a stand replacement fire and subsequent invasion by shrubs.
- Dense, even-aged, single-stemmed (derived from acorns rather than sprouts) oaks may indicate invasion of a former prairie.
- Complete domination of an understory by native shrubs offers opportunities for recreating open spaces for herbaceous communities, assuming research indicates the presence of a historically open situation.
- Nonnative trees and shrubs are obvious targets for complete removal.

Stand structures not amenable to restoration via thinning include the following:

- Small circles of oak sprouts (indicating trees of small stature) coming up through a matrix of seed obligate shrubs indicate natural chaparral with an oak component.
- A cohort of single-stemmed oaks within a matrix of seed obligate shrubs indicates natural chaparral with oaks having returned after a stand replacement fire.
- Many small resprouted oaks at higher elevations in southwestern Oregon may indicate Brewer's oak communities, again typified by stand replacement fire.

These conditions describe areas of at least several tens of acres. Smaller patches of older oaks (of a few acres in extent) within a matrix of younger oaks or shrubs may simply reflect mixed fire regimes. Creating an oak savanna under such conditions is more type conversion than restoration.

More complicated stand structures associated with transitions between prairies, savannas, woodlands, and chaparral are also encountered. For example, large oak trees around the edge of a prairie reflect site-specific fire behavior. Reduced fuels in the ecotone between the prairie and the hinterland may identify a transition in fire behavior. Care should be taken not to transform an entire landscape to oak savanna on the basis of a few large trees growing on the edge of an open area (Figure 4.8).

Wildlife and Oak Ecosystems

Oak woodlands and savannas are used by a wide range of native wildlife. More than 200 vertebrate species are known to use oaks (O'Neil et al. 2001). Of this number, perhaps two dozen reach their greatest abundance in oak-dominated communities. The following nine species represent some of the wildlife most closely associated with oak-dominated plant communities: sharptail snake, southern alligator lizard, western skink, Lewis's woodpecker, western bluebird, acorn woodpecker, white-breasted nuthatch, western gray squirrel, and Columbia white-tailed deer.

Wildlife performs a wide range of ecological functions that shape the plant communities in which they live. For example, acorn caching promotes seed dispersal and is an important determinant of the spatial distribution of oak populations. Scrub jays have a particularly close relationship with western oaks. Jays are known to cache more than 5,000 acorns at individual locations across their territory in a single autumn. Approximately 5% of this amount may never be recovered, thus remaining available to sprout if conditions allow (Carmen 1988). Wild turkey (a new alien introduction in the west), mountain quail, black bears, and dozens of other species feed on acorns in the tree or right where they fall. These species can consume most of an oak's annual seed production without offering much regeneration benefit.

Oaks play host to the most diverse assemblage of leaf-feeding Lepidoptera among North American plants (Hammond and Miller 1998).

FIGURE 4.8. This view at the Cascade–Siskiyou National Monument, near the California–Oregon state line, shows Oregon white oak as a component of chaparral in association with manzanita, buckbrush, and members of the rose family (wild plum, rose, serviceberry, and mountain mahogany). (*Photo by Paul Hosten*)

Foliage-gleaning birds such as warblers and vireos control Lepidoptera populations and limit leaf damage to oaks. Galls formed by tiny wasps also are a part of the native insect community.

Oaks are vulnerable to numerous herbivores. Small mammals such as pocket gophers, ground squirrels, and voles are a major factor in limiting the survival of oak seedlings in California. On more mature trees, new shoots are highly palatable to deer. Even when herbivory does not result in plant mortality, it is thought to cause plants to compensate by directing their energy reserves toward vegetative growth and a corresponding decrease in seed production (Belsky 1986).

Restoring Wildlife Habitat Components

The availability of the following four habitat elements is particularly important in shaping wildlife diversity in oak habitats.

Mature Trees

Mature, large-diameter trees are an important structural element in oak woodlands and savannas. Large trees are preferred as roosts and nest sites by numerous species of birds and mammals. On savannas and agricultural lands, solitary oaks are used as resting sites by red-tailed hawks, bobcats, and other predators. Older oaks support a great diversity of epiphytic plants (lichens and bryophytes) and invertebrates, which offer a corresponding richness of resources available to wildlife communities. For example, deer and squirrels forage on several species of arboreal lichens that are common on oaks (*Lobaria* spp., *Usnea* spp.). The berries of the oak mistletoe (*Phoradendron villosum*) are consumed by cedar waxwings, western bluebirds, and dozens of other species. Several birds such as the brown creeper and white-breasted nuthatch are particularly adapted to gleaning insects from the deeply fissured bark of large oaks.

Acorns

Perhaps the greatest importance of oaks to wildlife is their production of acorns. These have a high caloric content and represent an important food resource in fall and winter, when other forages are becoming scarce. Because annual mast production is highly variable, few species can risk being entirely dependent on acorns. However, good acorn crops can boost survival and reproduction rates, permitting some wildlife populations to attain greater densities than would be possible without this resource. Reduced mast production, as a result of vegetation change consequent to the cessation of Native American burning practices and fire suppression and the introduction of the nonnative turkey, may very well deprive native birds, deer, bear, and other wildlife of an important food source.

Shrubs

Southwestern Oregon chaparral supports an assemblage of wildlife that is regionally unique. Species that have the strongest association with this habitat type include the common kingsnake, western skink, green-tailed towhee, dusky flycatcher, California kangaroo rat, and dusky-footed woodrat. Elsewhere in the region, the presence of shrubs adds another stratum to the vertical structural of oak woodlands, increasing foraging and nesting opportunities for many bird species, particularly neotropical and temperate migrants. Shrubs also provide food and hiding cover for black-tail and Columbia white-tail deer.

Dead Wood

Arthropods that dwell in decaying wood such as carpenter ants, termites, and beetle larvae are primary food for dozens of wildlife species. Ensatina (a salamander), pileated woodpecker, and vagrant shrew are among the dozens of woodland species that forage in dead wood. For some cavity-nesting birds, the availability of large-diameter snags may be the most limiting factor to their populations. Snags and logs are a rare component of oak-dominated communities. In white oak–Douglas fir habitats, there are only about four snags per acre, compared to just under six snags per acre average in conifer-dominated forests. Downed logs are even rarer. The white oak–Douglas fir habitat type typically contains only one third the volume of logs of conifer forests. Perhaps the scarcity of snags and logs isn't surprising considering the frequent fire return intervals that are characteristic of oak-dominated communities. Much of the decaying wood available to wildlife in oak habitats exists in living oak trees (Gumtow-Farrier 1991). Like many other hardwood species, oaks are better able to compartmentalize injuries and insect damage than conifers. Mature oaks often contain a number of large-diameter dead branches and pockets of dead wood in their stems. These dead portions are protected from most wildfires by the same traits that allow oaks to persist on fire-prone landscapes (e.g., thick, corky bark, low amounts of flammable resins).

Ecological Restoration Challenges

The basic challenges of oak woodland–savanna restoration, very similar to those of other ecosystems, are as follows:

- Controlling invasive species, either existing or ready to invade.
- Finding native herbaceous seed sources (or appropriate genotypes).
- Correctly identifying historic conditions (at both stand and landscape levels).
- Crafting appropriate management objectives (changes in site use may preclude restoration of original conditions).
- Developing a clear desired future condition that encompasses natural plant community

dynamics, planning how to get there, and maintaining the restored area in the longer term.

Restoration Practices and Approaches

Oak woodland restoration projects in southern Oregon and northern California often are initiated by a desire to reduce fuel loading and fire hazard. "Restoration" may be only a byproduct of fuel reduction. Changes consequent to fire suppression or nonnative plant invasion are the most likely stimulant for restoration in other areas. The seven-step process described here is recommended as a way to fine-tune objectives once a general restoration goal (i.e., fire hazard mitigation) has been identified. Note that this set of steps is more detailed and cumbersome than would be appropriate for private landowners with more limited objectives and resources. In the latter case, a more flexible process could incorporate some of the ideas and methods described, but with less formality and detail.

Step 1: Restoration should be guided by an *identification of historic conditions* at both stand and landscape levels. This can be accomplished through interpretation of stand structure, examination of historic information (old photos, general land office surveys), and a search for literature pertinent to the area of interest. Resources include local educational institutions, conservation organizations, and historic societies.

Step 2: Site analysis identifies dominant plants and takes note of patterns of vegetation. Direct observations of relationships between plants, wildlife, and the location of weed patches are important aids in refining appropriate restoration objectives and methods. Ideally site analysis includes input from botanists, wildlife biologists, hydrologists, ecologists, fuel specialists, and possibly others. Some things to look for include the following:

- Rare or unique plant communities and specialized organisms. These may form a small but very important part of the diversity of a site or landscape.
- Examples of individual plant, lichen, moss, fungi, and wildlife species that appear dependent on very particular conditions (e.g., an area of dense shrubs within a more open grassland matrix). Maintaining multiple conditions helps ensure the persistence of diversity.
- Patches of invasive weeds and the conditions with which they are associated.
- The general structure and composition of stands as clues to local plant community dynamics.

Step 3: A *landscape context design* implies identification of areas that may warrant different treatments (or no treatment) for a number of reasons. For example, management objectives may identify a need to maintain shrubby areas for nesting birds, whereas another area must be in a grassland condition for a rare plant. The need to leave snags for wildlife may necessitate no-treatment zones for safety reasons. Weeds flourish in areas of recent disturbance, bare soil, and a sparse or nonexistent canopy of native hardwoods. An important landscape design element might be to look for opportunities to leave dense canopies along roads or other areas to interrupt weed dispersal.

Step 4: Build in a *local understanding* of plant–wildlife interactions and vegetation dynamics. In other words, avoid overgeneralizing from study or knowledge of other sites or landscapes.

An example of local understanding is provided in Figure 4.9. This diagram shows Oregon white oak dynamics in shallow soils susceptible to shrub invasion specific to southwest Oregon. Although the depiction of patterns of vegetation change is coarse enough to have some application to other oak woodlands in the region, it should be seen primarily as an example of summarizing local knowledge. Such diagrams enable easy communication of vegetation dynamics that facilitate project planning and communication. Similar models can be

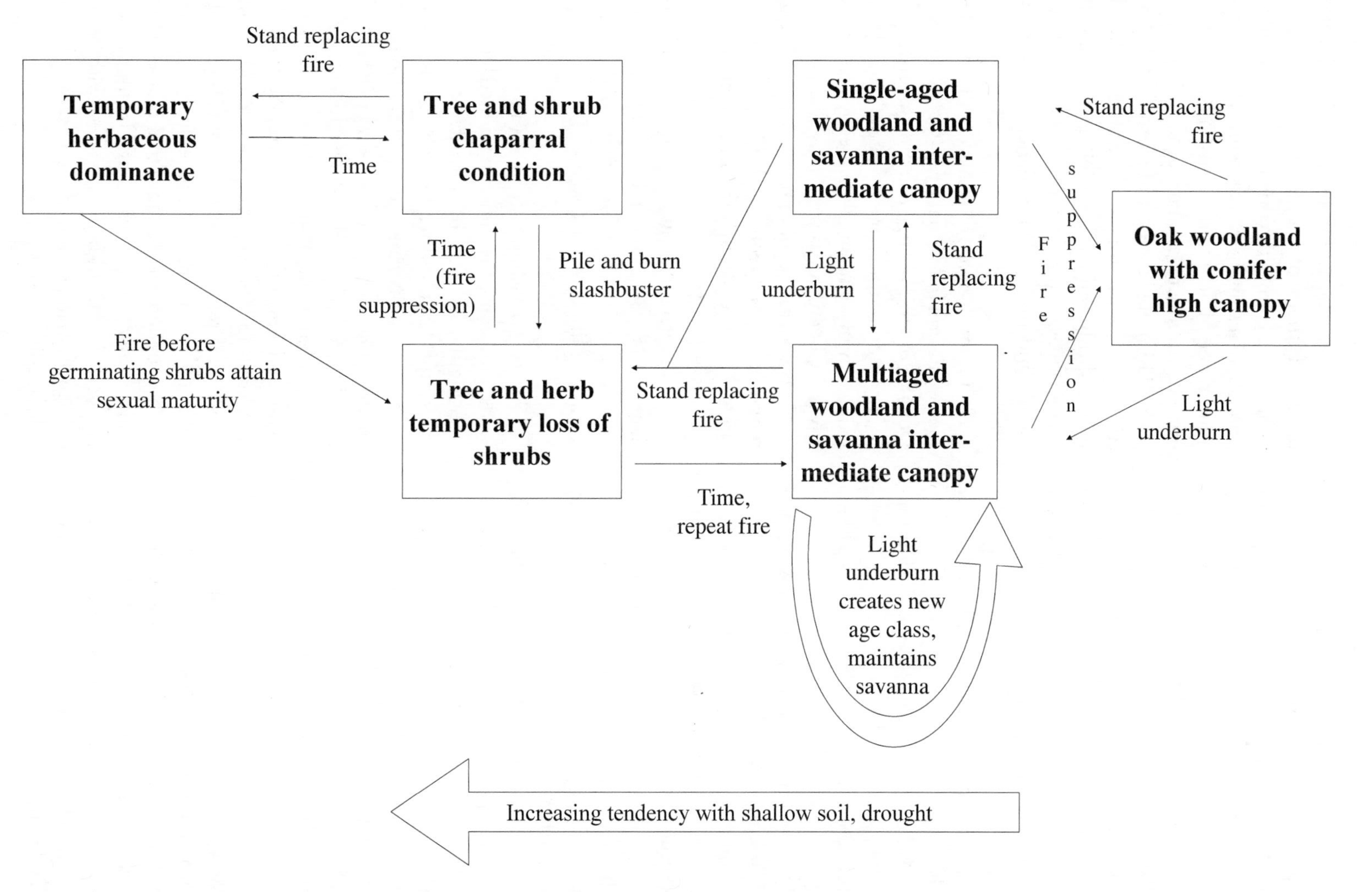

FIGURE 4.9. Basic plant community dynamics for oak woodlands in the Applegate Valley of southwestern Oregon. (*By Paul Hosten*)

created for other local woodland systems. Restoration practitioners are encouraged to study and understand local conditions and to use caution in generalizing from distant oak restoration projects. It would be a shame if *restoration* is added to the existing list of oak woodland afflictions.

Step 5: The improved understanding of the site from the first three steps allows a *fine-tuning of project objectives*. Typical objectives contain some mix of social, ecological, and economic factors, including the following:

- Reducing wildfire hazard and heavy fuel loads
- Creating fire-safe areas (for use during firefighting)
- Accelerating growth of large, fire-resistant trees
- Rejuvenating shrubs to improve browse quality (Thornton 1981, Biswell 1999)
- Maintaining or increasing landscape-level species richness of native plants and wildlife
- Restoring shrubland or woodland life form composition to a more historic condition in each stand while minimizing weed invasion
- Reintroducing fire as a management tool and an ecosystem process
- Restoring diverse conditions across the landscape to ensure the continued existence of all habitats and plant communities
- Creating oak savanna for particular wildlife and plant species (Hagar and Stern 2001)
- Increasing acorn production (Coblentz 1980)
- Maintaining or restoring open conditions for aesthetic values

Step 6: With objectives fine-tuned, the next step is to *develop treatment prescriptions* that focus on target species and size and age classes. Ideally, these are distributed across individual treatment areas, with the larger landscape context in mind. Six basic types of prescriptions are generally used in oak woodlands:

1. Conventional thinning based on ideal spacing and diameter of trees to leave. Though easy to apply and effective at reducing fuels, this prescription often results in stands with an orchard-like appearance. Large reductions in canopy cover often contribute to weed invasion.
2. Thinning based on life form (i.e., trees versus shrubs), with variable amounts of highly flammable shrub removed and retention of less flammable trees. This prescription leaves trees in groups and may retain a higher total vegetation cover.
3. Group selection removes all or selected species around individual trees (e.g., removal of all Douglas fir around large oak individuals).
4. Underburning often is used to selectively reduce shrubs and small-diameter trees and usually is applied in areas with dense canopy cover.
5. Broadcast burning reduces shrub fuels and reinitiates fire-dependent shrublands and grasslands. This is typically used in chaparral-type oak systems.
6. Maintenance burning limits shrub abundance and encroachment. Careful timing of burns promotes low-intensity fire that retains all plant life forms and limits invasive weeds.

Figure 4.10 shows shrub-invaded oak woodlands and how the aforementioned treatments alter them. Although the untreated condition provides bird cover and winter browse, loss of native grasses and forbs can lead to overall biodiversity decline. Oak trees produce fewer acorns when forced to compete with a dense thicket of shrubs. The gnarled, twisted, and hollow oak lunkers are dying and will be lost to stand replacement fire if fuel loads are not reduced. The second condition shows total removal of shrubs and tree thinning. Although this approach reduces fuel loads, it is not the best option for reestablishing native grasses or maintaining neotropical bird habitat. Existing patches of native grass associated with removed oaks may disappear and be replaced by less desirable invasive grasses.

FIGURE 4.10. Oak woodland prescription options illustrate untreated woodlands and alternative prescriptions on vegetation structure. Rigid spacing between trees can be used to attain an oak savanna structure, whereas differential removal of shrubs or thinning relative to tree driplines retains more cover of value to certain wildlife species such as neotropical birds. Each condition favors different plant or wildlife species. (*By Paul Hosten*)

The last three conditions are variations on Prescription 2, with retention of hardwoods in groups and varying levels of shrub removal. Removing shrubs within 10 feet of the dripline offers some fuel reduction while retaining extensive cover and maintaining native grasses and forbs (at least in southwest Oregon). The residual canopy cover is designed to deter weed invasion. The tradeoff is that fuel loading may still be high enough to result in stand replacement fire under extreme conditions.

Step 7: Given that plant communities are ever changing, it is important that a *long-term management plan* be created to ensure that the restoration area is maintained within a set of constraints as defined by vegetation composition and structures. Such long-term goals should encompass the natural temporal variability of the site to enable individual plants and animals to complete their life histories.

Typical Project Implementation Methods and Costs

Manual Clearing

The most versatile method is manual brush cutting, normally using chainsaws, followed by piling and burning of slash. The typical cost is more than $1,200 per acre. It is important to use well-trained crews who can distinguish between species and can improvise. Prescriptions designed to favor rare species are best implemented manually. Hand crews can place burn piles in weedy areas rather than on top of native plants. Postburn seeding then allows the reintroduction of native plants in a competition free area. Manual crews can work in remote areas and on steep slopes.

Machine Clearing

Several machines, such as bulldozers, backhoes, mechanical masticators (also known as slashbusters), and feller-bunchers, can be used in oak woodland restoration. The initial cost is much lower than hand clearing, typically $300–$700 per acre. But these cost savings must be measured against the potential problems of introducing machines to sites, including soil compaction, excessive ground disturbance (and subsequent weed invasion), and smothering or damage to native herbaceous understory. Machines need roads for access and are generally limited to slopes less than 30%.

Because multiple treatments at short return intervals can result in the loss of individual seed obligate species, repeat treatments must be designed to be appropriate for the restoration site. For example, it would be inappropriate to eradicate all buckbrush or manzanita in an oak woodland that is ecotonal with chaparral. Interspersing treated and untreated areas to allow cross-colonization is a good safeguard for maintaining habitat diversity and preventing local extirpation of individual species.

Oak Regeneration

Oak regeneration often is hampered in California woodlands, where weedy annual grasses are abundant. The opposite problem is found in much of the Pacific Northwest, where oak regeneration usually is too prolific. Oak recruitment may be limited in urban areas and around agricultural areas, such as the Willamette Valley, where nonnative grass competition is intense. Oaks can be established by directly planting acorns, but a large proportion will be consumed by wildlife or succumb to plant pathogens. Small mammals may girdle seedlings (Adams et al. 1996). Gophers can devastate seedling establishment (Hibbs and Yoder 1993). Large native and nonnative ungulates prevent the attainment of the tree growth form by browsing tops.

Treating the seedlings or planting area with mycorrhizae facilitates the survival and growth of oak plantings. Their role in the symbiotic partnership is to provide a fine network capable of extending the range of roots. Growth of individual trees and fundamental ecosystem processes depend on mycorrhizal fungi. Decades of experimental work have shown the importance of mycorrhizal symbiosis in restoration, but little of this research has focused on oak ecosystems. Cultivated inocula should be used cautiously to avoid introducing mycorrhiza not native to an area, which could suppress local fungi. An alternative is to use local sources. For example, in a trial that incorporated litter from under Engelmann oak trees into planting spots for new seedlings and acorns, a significant increase in a number of growth variables was reported (Scott and Pratini 1997). It was likely that mycorrhizal fungi from native soil conferred the growth advantage. Berman and Bledsoe (1998) also added soils from valley oak riparian areas to growth media for valley oak seedlings grown in a greenhouse and found that the percentage of mycorrhizal infection and mycorrhizal diversity on the seedlings were increased more by transfer of oak forest and woodland soil than by agricultural field soil.

Recommendations for Acorn Planting

Researchers in California have developed methods for aiding survival of acorns and planted oak saplings (McCreary 1997). If a shallow, impervious soil layer exists, then a hole must be augured to allow roots to penetrate deeper soils and nutrients. As with any tree establishment, competition from grasses and weeds may prevent establishment or stifle growth of saplings. Soil scarification within a 3-foot radius of the planted tree reduces initial competition. However, in areas prone to weed invasion, this facilitates new patches of yellow starthistle, annual grasses, or whatever the local predilection. A more expensive option is to place a weed barrier around the sapling, such as cardboard or synthetic mulch held down by staples. Planting tubes can prevent vole and other rodent impacts. Longer tubes prevent browsing and accelerate vertical growth (McCreary 1997, McCreary and Tecklin 1993). Underground screens can prevent root predation by gophers (Adams and Weitkamp 1992). Here follows a list of recommended steps for planting acorns:

1. Collect acorns (from the local area) in the fall; store dry and cool.
2. Test the germination percentage.

3. If soil in the planting area is hard, loosen to a depth of 0.5 meters.

4. Plant dry acorns, three to five per hole, in late fall and early winter, just before the rains or shortly thereafter and as soon after acorn collection as possible.

5. For inoculum, collect a solid clump of soil from under nearby existing oaks. Use within 4 weeks; store refrigerated.

6. Add the soil clump to the hole, positioned up against the acorn or planted oak roots.

7. Plant acorns near the surface.

8. If planting seedlings, use growing tubes as dictated by local conditions.

Managing Oak Resprouts

The strong sprouting response from latent epicormic buds on most oak species results in multiple sprouts after fire or mechanical damage. Where management favors the tree form, thinning will promote a more rapid attainment of that objective. Repeat removal of resprouts may be needed to kill off oak individuals where between-tree spacing is needed (perhaps to recreate a savanna). Fire should be considered when and where applicable to thin multistem stocks. Herbicide can kill stems effectively, thereby reducing the number of site visits and improving cost effectiveness.

Herbaceous Plant Establishment

A key issue in restoring grasses and forbs in oak woodlands is lack of understanding of plant community associations and successional processes. Choice of species is best made by replicating remnant native plants present in the restoration site. Restored species should include short-lived plants for immediate site occupation and longer-lived plants for longer-term occupation. Borman et al. (1990) found that successful native perennial grasses closely emulate annual grass growth patterns. In annual grass–infested southwest Oregon, perennial grasses included Roemer's fescue.

Small projects can be implemented by using seed collected from the surrounding landscape. Collected seed can be cultivated at nurseries to increase supply. Grass seed application rates depend on local conditions. Ten pounds of grass seed per acre is sufficient where competition from weeds is minimal. Rates should be doubled on very dry sites with no irrigation or in areas where weed competition is likely. Initial examination of species richness at southwest Oregon sites suggests that excess sown seed can suppress native species richness.

The relative ease of converting chaparral to perennial grasslands (Menke 1989) suggests that oak woodlands long under shrub cover respond well to native grass seeding. This has proved to be the case in several grass seed applications in southwestern Oregon. Broadcast seeding across areas pockmarked with pile burn scars shows variable success. Seed on burn piles germinated and established particularly well, whereas seed landing on unburned areas failed to develop. Thick duff may prevent roots from achieving mineral soil contact.

Research in California demonstrates the difficulty of establishing native perennial grasses in the presence of strong competition, particularly nonnative annual grasses. Some shrubs may produce phytotoxic compounds that suppress seed germination. Fire may be needed to reduce woody shrub and tree, litter, and phytotoxic compound abundance.

Other methods of introducing grasses in grasslands, such as planting plugs, may be applicable to open oak woodlands. Stromberg and Kephart (1996) provide a summary and case studies for California grasslands; also see Chapter 3 for best approaches to establishing native grasses and forbs.

Influence of Disturbance Type and Frequency

Plant community members respond differently to stimuli provided by fire and other disturbances.

Among herbaceous plants of fire-dependent plant communities, one might expect herbs that are present both before and after fire, specialized plants restricted to only 1 or 2 years after fire, and yet another group of herbs peaking in abundance several years after a fire before declining to prefire levels. Shrubs can grow from a seed source or can resprout from underground growth points. Managing a stand of vegetation toward a desired plant composition requires an understanding of existing composition and the mode of persistence (whether plants establish from seed or root propagules) in the face of different treatments and disturbances (e.g., fire, mechanical mastication, disking, hand pulling, herbicide application). Treatments are then applied in a manner to favor or disfavor target species.

In the longer term, plants also respond to the frequency of disturbance and the length of the period of rest between disturbances. Seed obligate herbs and shrubs (including invasives) can be reduced in abundance through repeat treatments that prevent seed production and replenishment of the seed bank. Repeat removal of the aboveground portions of resprouters can reduce their vigor and eventually result in their demise. In general, repeat treatments favor plants able to complete their life cycles between treatments. This means that frequent treatments can favor weedy annual grasses and forbs.

Weed Control

Invasive species often are the primary causes of failure in oak restoration projects. The preponderance of nonnative herbs may limit management objectives to establishing a desired structure. In California grasslands with an oak component, the persistence of cheatgrass and medusahead has caused land managers to accept the continued presence of these aliens while preventing the invasion of even less desired ones, such as starthistle, dyers wode, filaree, and knapweed (Melvin et al. 1992). Mesic conditions common over much of the Pacific Northwest provide greater opportunity for the reintroduction of native grasses and forbs than in the drier sites of southern Oregon and northern California.

Disturbance favors weeds. Roads, timber harvest, conversion to pasture, agricultural practices, high use by livestock, and particular soil conditions (e.g., high proportion of shrink–swell clays) may facilitate weed establishment and abundance.

Both native and nonnative trees and shrubs can be invasive. Douglas fir, white fir, manzanita, and buckbrush increase in former oak stands is well documented, and many restoration projects are designed to repel native invaders. Restoration efforts that open the canopy may also increase nonnative invasives, such as Scotch broom, that are restricted under closed canopy conditions (Keeley 2001).

Oak or shrub thinning creates slash that can be scattered over weed patches. Intense heat in slash burns can eliminate weed seed banks and provide a competition-free area for the reintroduction of desired plants. Mechanical contrivances used to reduce fuels (such as slashbusters) may produce deep litter that smothers existing native vegetation. Over time, the native seed bank may be lost, thereby facilitating weed invasion. Additions of carbon to soils are known to affect nutrient dynamics by tying up nitrogen in the microbial biomass. Fire under conditions dry enough to result in intense combustion of debris at the soil surface alters the underlying seed bank, a potential problem in areas that recently had good native seed sources.

Stand- and landscape-level analysis may reveal patterns of weeds across the landscape. Once affinities of weeds with soils, plant communities, and their range of conditions are better understood, precautions can be taken to prevent further invasion. For example, leaving untreated, shady areas adjacent to roads may prevent roadside weeds from spreading into nearby treated areas.

In summary, the basic principles of weed management apply: Be proactive by maintaining healthy communities with an intact balance of native life forms and seed banks, favor management using a natural disturbance regime, minimize soil surface disturbance, quickly eliminate small weed populations as they establish, isolate and contain larger weed populations, wash vehicle undersides and wheels before traveling to a restoration site, use weed-free imported materials, use prescriptions and methods of application that favor desired native plants over weeds, and design a longer-term monitoring and follow-up program to gradually eliminate weed populations or at least reduce them to a level where natives can persist and thrive. See Chapter 16 for more information on invasive plant issues and management strategies.

Using Fire in Oak Restoration

Prescribed fire is commonly used to reduce conifer density in oak woodlands. Young firs are more susceptible to heat from fire than are hardwoods and pines (Biswell 1999). Mid- to late summer is the most natural time for oak woodlands to burn but also the time of highest risk. Land managers may need to wait until later in fall, after some autumnal rains, to conduct a safe burn.

Burning during dry windows in winter or spring must be weighed carefully against possible damage to other resources. Out-of-season fires affect plants differently because of the different intensity of the burn and the plant's response to fire. Areas with accumulated fuels in a matrix of fire-dependent shrubs are more likely to experience stand replacement fire under hot and dry conditions. Fire during the growing season facilitates heat conductance, thereby killing the cambium layer, resulting in at least partial dieback. Out-of-season fires with high soil moisture levels may result in the steaming of plants, seeds, and the soil microbial community. Shrub and hardwood response to season, weather, and moisture indicates that fires characterized as having high duff consumption show high mortality to sprouting shrubs and trees (Kauffman and Martin 1985). In sites dominated by weeds, hot fires may consume all of the duff and alien seed bank, thus providing an opportunity for native seed introduction and occupation of desired plant species.

Spring burning may scorch and kill many weeds (including annual bromes, medusahead, dogtail) and reduce seed set. But unless reseeding is planned, spring burning is useful only where remnant native grasses are present (Keeley 2001). However, burning when perennial grasses are still green, before summer dormancy, will result in the loss of bunchgrass basal area (Menke 1992). In Washington, Tveton and Fonda (1999) found fall fire more effective at promoting native species and communities and favor a fire return interval of 3 to 5 years to maintain fescue prairie and woodland.

Repeat burns favor resprouters over seed obligate shrubs. A threshold is achieved when the return interval is less than the time needed for seed obligate species to achieve sexual maturity. Nonsprouters such as buckbrush are dramatically reduced, whereas yerba santa, poison oak, birchleaf and mountain mahogany, salmonberry, and thimbleberry increase (Figure 4.11).

Although it is a native plant, mistletoe can be considered an invasive of oak woodlands in certain circumstances. Photos near the Cascade–Siskiyou National Monument spanning 80 years indicate that there may be much more mistletoe than was the case historically. White oaks can be overcome by mistletoe within 20 years. Infestations become so dense that oak foliage is greatly reduced. Drought in the 1980s and 1990s may have played a role in weakening trees and making them more susceptible to mistletoe infestation. Fire suppression plays a role by allowing mistletoe to increase in abundance. It is likely that partial or complete dieback and resprouting of oak trees after fire serves as a mechanism for the control of mistletoe. Native

FIGURE 4.11. A prescribed burn in a white oak savanna restoration in the Willamette Valley. (*Photo by Ed Alverson*)

American lore suggests that fire reduces infestation of trees by mosses, lichens, and mistletoe.

Standard practice for mistletoe removal involves cutting the infected branch off below the point of infestation. If hausteria remain visible in the cut surface, then cut again even closer to the tree axis. Placing black sealant on the cut face may prevent undetected hausteria from growing into new mistletoe individuals. Some have sprayed mistletoe with herbicide during winter. Because mistletoe photosynthesizes in winter, they may be susceptible to chemicals while oaks are not.

Another approach is to cut down trees dominated by mistletoe. Because most oaks coppice from the base, removing infected trees might reduce the rate of mistletoe infection while also introducing a younger oak cohort.

Livestock Influence and Management

Many oak woodlands are recovering from livestock impacts that occurred from the mid-1800s through World War I. Livestock impacts can be direct (consider the selection of highly palatable species over less palatable species) or subtle in their longer-term influence on ecological processes such as fire and succession. Livestock have facilitated the spread of weeds, hampered oak regeneration, and altered patterns of plant community change. On the other hand, carefully controlled livestock grazing can reduce palatable weeds and imprint sown seeds for improved germination and establishment. Although nonpalatable weeds may proliferate during periods of livestock use, it is also true that palatable weeds may proliferate upon removal, particularly in areas that have experienced prolonged grazing.

Photo records from the San Joachim experimental station in central California indicate that livestock can play a key role in suppressing regeneration of buckbrush after wildfire in a blue oak plant community (Duncan et al. 1987). Jackson et al. (1998a) found that increased grazing favors oak seedling survival but reduces successful transition to the sapling class.

Research conducted on grasslands of central California provides general livestock management guidelines that probably would also apply to open oak woodlands in the Pacific Northwest. Menke (1992) advocates short-duration high-intensity grazing in the spring to increase the relative abundance of native perennial grasses over annual grasses. Grazing must be timed to reduce annual grass seed production but discontinued before all the soil moisture is depleted. Properly implemented grazing can also reduce bunchgrass decadence and promote vegetative growth in the absence of fire. However, long-term use of livestock without fire may result in the loss of fire-dependent species. Short, intense grazing can be used to reduce weeds if timed to prevent flowering and seed set, but natives with similar phenology also are likely to be affected. A reduction of fine fuels through grazing is thought to lengthen fire return intervals, thereby favoring longer-lived woody species.

In addition to the aforementioned considerations, longer-term grazing strategies in situations with an intact woodland herbaceous component should include a year of complete rest interspersed with periods of grazing. Varying the season of livestock use from year to year ensures that all seasonal suites of species are allowed to complete their life cycle to maintain the seed bank, particularly with short-duration and intense grazing.

Oak Woodland Restoration Cautionary Notes

The conventional view of vegetation change (succession) places excessive attention on the endpoint of change (the climax) and less value on conditions intermediate between disturbance events. Plants and wildlife probably need a full complement of stages of vegetation development across a landscape. Some organisms can be found only in specific conditions along trajectories of vegetation change. Landowners and managers should consider retaining a full range of conditions across large areas. Those working at small scales may lack a landscape perspective and therefore miss the need to retain large-scale natural processes or focus too much on the retention of undisturbed conditions (i.e., protected areas). Such management may be too narrow and serve only a subset of plants and wildlife.

The belief that all oak woodlands were historically open savanna has contributed to misguided management. Many historic oak-dominated communities were woodlands with tree spacing too dense to classify as savanna. Oregon white oak is also a component of southwest Oregon chaparral, where its presence has been mistaken as a clue that the site was a historic savanna. Although Native Americans did burn frequently to clear areas around village sites and to attract wildlife, they may not always have done so at return intervals that resulted in the local extirpation of shrubs. Inappropriate treatment of chaparral is most likely to occur in northwest California and southwest Oregon but may also occur in relict chaparral of the Umpqua Valley. It would be a tragedy to inadvertently lose these relict chaparral in the name of savanna restoration.

Many celebrate the influence of Native American management on species, habitat, and plant

Case Study: Managing Cultural and Natural Elements of the Bald Hills of Redwood National and State Parks

Leonel Arguello, National Park Service

The Bald Hills area of Redwood national and state parks lies between the Klamath River and Redwood Creek in northwest California. The area covers about 4,200 acres (1,700 ha) and includes elevations from 250 to 3,100 feet. The Bald Hills were added to the park in 1978, with additional areas added in 1991. Coastal grasslands (locally called prairies) and oak woodlands generally cover ridgetops, although several prairies extend downslope to Redwood Creek.

Before Euro-American settlement, more than 190 vascular plant species were found in the grasslands, including bunchgrasses consisting of *Danthonia*, *Stipa*, *Melica*, *Poa*, and *Festuca* species. It is likely that Oregon white oak was the most common tree, with dispersed California black oak, California bay, big-leaf maple, tanbark oak, and other less dominant species completing the woodland composition.

During the Native American–European contact period, at least two groups, the Chilula and the Yurok, used the Bald Hills. The Chilula, who spoke an Athapascan language, occupied substantial areas. A number of contemporary Native Americans trace their ancestry to the Chilula, although the group itself is no longer a recognized tribal entity. The Yurok, who spoke an Algic language, resided primarily in villages along the rugged Pacific Coast and lower reaches of the Klamath River (Waterman 1920). Some Yurok traveled seasonally to the Bald Hills to procure resources such as grass seeds, basketry materials, acorns, and large mammals. With the discovery of gold on the Trinity River in the 1850s, miners established a major supply line (the Trinidad Trail) connecting diggings with the Pacific Coast through the Bald Hills. Their interactions led to several bloody skirmishes and the eventual displacement of Native Americans from their ancestral lands.

Permanent Euro-American settlement in the Bald Hills commenced in the late 1860s with the arrival of the Lyons family, who ran sheep and cattle in the area for more than 100 years (Bradley n.d.). Commercial logging of conifers was conducted in the forested portions of the Bald Hills from the 1930s to the 1970s. Postsettlement occupation of the Bald Hills continues to affect the landscape. Altered processes have brought decline in native plant species, radical changes to vegetation, and shifts in wildlife. These changes compelled park managers to take a more active role in managing oak and grassland ecosystems.

Park staff initiated ecological studies in 1978. These investigations identified a number of concerns, including conifer encroachment, exotic species invasion, and erosion and sedimentation from logging and roads. In 1992, the Bald Hills Vegetation Management Plan was developed as a tool to maintain the highest diversity of pre-European plants possible. The plan prescribed a set of strategies to maintain or improve native plant establishment and diversification, prevent conversion of grassland and oak woodlands to coniferous forest, repair or eliminate roads, and preserve the presettlement ethnographic landscape.

Initial efforts focused on restoration of prehistoric fire regimes, invasive weed control, and control of encroaching conifers in the prairies and oak woodlands. By 1998, crews had removed conifers from approximately 2,000 acres of prairie and oak woodlands, resulting in increased native plant diversity, improved wildlife habitat, and preservation of cultural landscapes. Broom was removed from grasslands, allowing prairie plants to reclaim these lands. An aggressive prescribed burn program was initiated to provide ongoing maintenance. About 1,500 acres of prairies and oak woodlands were treated in the first year of full-scale operational burning. The fire program now annually treats 500 to 2,000 acres of oak woodland and prairie, with fire return intervals as short as 3 years in some areas.

Although results generally have been positive, new management directives prompted some changes in restoration work. A general management plan was developed that continued support for restoration strategies in the prairies and oak woodlands. This plan confirmed the Bald Hills as a special zone within the park to be managed for its significant natural and cultural resources. It also provides a stronger foundation for Native Americans to participate in the identification, designation, and management of cultural and ethnographic landscapes and to collect natural materials in conjunction with the maintenance and interpretation of cultural and ethnographic landscapes. A very important addition is that prescribed burning may now be used to promote the quantity and quality of natural materials of value to contemporary Native Americans as long as this is consistent with other resource management goals. What remained to be determined were the identities of those natural materials and the tools to compatibly manage them along with other resources.

Redwood national and state parks contracted with the Yurok Tribe Culture Department to compile information on traditional use of the Bald Hills by Native Americans. Studies included an extensive literature search, field investigations, and interviews with knowledgable elders. These reports underscore the historic importance of Native American management in the Bald Hills. Encompassed in this realization is the need to learn more about the specifics of Indian management practices and the role of contemporary hunting and gathering as tools for both maintaining the Bald Hills as a cultural entity and assisting the local Native American community in carrying on traditional practices.

Case Study: Fuel Reduction and Restoration in Oak Woodlands of the Applegate Valley

The Applegate Valley in southwest Oregon is an area where intensive fuel reduction efforts in oak woodlands have taken place on both private and public lands over the past decade. Experience has shown that compromises must be made between fuel reduction, treatment methods and costs, and restoration of natural systems and processes. Funding for fuel reduction can be harnessed for restoration, but not on every project or in every stand within a project.

Project planning begins at the watershed scale, allowing a combination of objectives that are generally applied at the stand scale. By starting with a large-scale view, the planning process helps shape stand-level objectives to local site conditions, topography, variable land ownership, sensitive habitats, and other concerns. Some projects include plant community mapping to identify rare elements for protection while targeting more common communities for intensive treatment. Other projects focus completely on fuel reduction regardless of other ecological objectives.

Lessons Learned

- Areas with heavy fuels result in high numbers of slash piles per unit area, thus generating enough heat to damage leaves, trees, and shrubs. Burning a subset of piles through the treated area during separate entries helps reduce heat production and concentration.
- Some shrubs remaining after prescription application succumb to heat from pile burns and habitat change. Shrubs left in large clusters help ensure some survival.
- Because buckbrush (*Ceanothus cuneatus*) and whiteleaf manzanita (*Arctostaphylos viscida*) germination from seed is enhanced by heat scarification, stand treatments that create fewer piles result in fewer patches of germinating shrubs, whereas areas with a higher density of piles may experience seed germination across an entire stand. Hand-cut pile-and-burn methods approximate the effects of wildfire more closely than mechanical mastication; they increase the ability of fire-dependent shrubs and herbaceous species to germinate.
- Native herbaceous vegetation can be smothered by slash created during mechanical mastication. This smothering later favors predominance of weedy annual grasses over native perennials.
- Use of a mechanical treatment in chaparral may be inappropriate because low shrub recruitment and altered plant community dynamics do not allow the reestablishment of fire-dependent species.
- Consumption of surface fuels by prescribed fire or wildfire after mechanical treatment may result in high soil temperatures and consequent reduction of desired seed bank.
- Native grass seeding across pile burns and interspaces generally is successful only on burn pile scars.
- The resprouting ability of shrubs depends on the height above ground at which the shrub is cut off. Retention and regrowth of buckbrush for wildlife browse or habitat thus can be governed by lopping height.
- Interspersion of untreated reserve areas with treatments increases habitat heterogeneity and allows possible cross-colonization of plants and wildlife.
- Bird abundance monitoring on small fuel reduction areas shows that shrub-dependent birds are not affected but that edge-dependent birds increase. These include purple finches, which are otherwise in decline.

Support for fuel treatments varies. Local homeowners are mostly pleased about increased safety of their homes and families. Ecologists are divided on whether to use mechanical treatments in disturbance-dependent plant communities rather than relying only on fire. The concern is higher soil surface temperatures that mechanically generated slash may create when fires happen, the threat of spread of weed seed attached to machinery, the fact that mechanical treatments do not truly replicate fire, and the fact that treatments applied out of season (cool season burns) are out of synch with historic fire occurrence.

Ecologists also are concerned that some areas are natural chaparral yet are being treated as former woodlands or savannas. A good compromise for fuel reduction projects is as follows:

- Research the historic condition at the stand and landscape levels.
- Use the most economical treatment methods near homes, where ecological objectives are less crucial.
- Use hand-cut pile-and-burn and prescribed fire strategically on ridges to reduce overall landscape fire hazards but retain a modicum of fire influence.
- Use only natural processes (fire) to treat more remote areas.

Case Study: Ecological Restoration at Bald Hill, Corvallis, Oregon

Greg Fitzpatrick, The Nature Conservancy

The Nature Conservancy, the City of Corvallis Parks Department, and the Greenbelt Land Trust are collaborating to restore Oregon white oak habitat at Bald Hill Park in Corvallis, Oregon. The project was funded in part by the National Fish and Wildlife Foundation and the U.S. Fish and Wildlife Service Partners for Wildlife. The main plant communities that were targets for restoration are oak woodland and oak savanna. For many years oak woodland has been expanding into areas that were historically savanna. These communities are characterized by scattered large, broad-canopied, several-hundred-year-old Oregon white oak trees growing in a dense, young stand of trees and shrubs. Invasive woody plants include young oaks, Douglas firs, English hawthorn, pear, mazzard cherry, and wild apple.

The primary goal was to restore the open savanna community and the understory native grasses and forbs. The main objectives were as follows:

- Reduce or thin the abundance of trees and shrubs.
- Test the effectiveness of goats in clearing dense understory vegetation.
- Use prescribed burns to stimulate native plant growth and to reduce thatch, weed seeds, and invasive plant species.
- Initiate monitoring and research.

Other objectives included testing the collecting and sowing of native seeds in burned or recently cleared areas and providing educational opportunities for the community through volunteer activities and tours.

Methods

Volunteers, county corrections crews, and hired professional crews hand cleared invasive trees and shrubs. Most hardwood stems were treated with a 50% dilution of glyphosate to reduce resprouting. Woody debris was chipped or piled for burning, with larger-diameter material donated as firewood.

Goats were used for 7 days to reduce dense patches of poison oak and blackberry. Temporary fencing and herd dogs were used to confine the goats and protect them from predators. To reduce the chance that goats might bring in unwanted weed seeds from a previous site, they were quarantined and fed weed-free feed for 5–7 days before coming to Bald Hill. Perimeter fire breaks, cutting of exotic blackberries and other shrubs, and limbing of trees were done to prepare the area for prescribed fire. Eighteen photo stations were established and conditions recorded before and after restoration. In addition, a small study was initiated to investigate the effects of burning on the herbaceous plant community (Figure 4.12).

Results

Approximately 3,200 nonnative trees and 650 Oregon white oaks were removed, and 50 Douglas fir trees were either cut or girdled across nearly 17 acres. An informal survey of stumps treated with glyphosate showed that most did not resprout. Crews that worked in areas where goats had grazed reported that exposure to poison oak was less than elsewhere and that it was easier to navigate in the thinned understory. Prescribed burns were successfully carried out in 2001 and 2002. These were probably the first fires to occur at Bald Hill in that past 100 years.

Some Lessons Learned

Mowers and other machinery are more effective than goats for clearing underbrush, except in areas of steep slopes, dense tree spacing, or other factors that inhibit machines. We monitored areas where goats grazed to check for new exotic weeds and made a special point of sowing native grasses into areas where goat feces accumulated to help ensure that native plants usurped the nutrients before nonnatives had a chance. Sown grasses are expected to provide fine fuels to allow future prescribed fire.

Chipping slash was expensive, and burning piles of slash often led to heat-damaged soils. Other methods for managing slash that may reduce these problems include the following:

- Leaving some piles of large and small woody debris unburned as wildlife habitat.
- Leaving more unwanted trees standing, injecting them with herbicide to create snags.
- Placing slash piles on elevated platforms or wet piles of wood chips before burning to reduce the amount of heat reaching the soil.
- Using slashbusters to cut and grind the small to medium size trees in place. This last method may cause wood chips to accumulate to a level where they kill native groundcover or change soil characteristics.

Overall, using machinery to cut and clear invasive trees and shrubs is more cost effective than hand crews. The volume of native seed collected from Bald Hill was insufficient. To help ensure a long-term economical source of seed appropriate for this site, a program to grow native plants for seed harvest has been started with local high schools.

FIGURE 4.12. Illustration of the immediate effects of a prescribed burn at Bald Hill in the Willamette Valley. Note that brush and individual trees are killed, but other trees and groves survive while the understory is opened up. (*Photo by Ed Alverson*)

community richness (Underwood et al. 2003). Others view Native American management as having facilitated the invasion of nonnative weeds after European colonization and associated disturbance patterns (Keeley 2001). Such disparate views result in ambiguity about oak woodland restoration projects.

Summary

We are still at an early learning stage with regard to oak woodland restoration in the Pacific Northwest. Although much good work is being done, from Redwood Park in the south to Vancouver Island in the north, we are still at an experimental stage and consequently ought to proceed with caution and humility. Nevertheless, a few general observations can be stated with some confidence:

- Our oak woodland systems are in dire need of maintenance or restoration. We cannot afford to wait for perfect knowledge to initiate projects.
- Oaks are magnificent trees with high aesthetic and cultural value. This can be used to help generate interest and support for restoration projects. In a real sense, oaks link Native American, rural, and urban communities in ways similar to streams and watersheds.
- Fire reintroduction appears to be an almost crucial element in oak restoration. Without fire, the best one can do is approximate canopy structure. Restoring understory composition depends on burning.
- Oak woodland restoration requires a local understanding of how plant communities react to management, which is likely to vary

through the range of habitats and history of the Pacific Northwest.

- An improved understanding of historic conditions may help define management objectives suitable to local conditions.

Acknowledgments

We thank Lori Valentine for her early reviews of this chapter, and we thank her and Darlene Southworth for their knowledge of mycorrhizal fungi. Work on this chapter was funded in part by the Joint Fire Sciences Council (JFSP project number 3-3-3-36).

References

Adams, T. E., P. B. Sands, and W. H. Weitkamp. 1996. Oak seedling recruitment through artificial regeneration on California rangelands. Pages 1–2 in N. E. West (ed.), *Rangelands in a Sustainable Biosphere*. Proceedings of the 5th International Rangeland Congress, Vol. 1, July 23–28, 1995. Society for Range Management, Denver, CO.

Adams, T. E. and W. H. Weitkamp. 1992. Gophers love oak—to death. *California Agriculture* 46(5): 27–29.

Agee, J. 1993. *Fire Ecology of Pacific Northwest Forests*. Island Press, Covelo, CA.

Atzet, T. and D. L. Wheeler. 1982. *Historical and Ecological Perspectives on Fire Activity in the Klamath Geological Province of the Rogue River and Siskiyou National Forests*. USDA, Forest Service, Pacific Northwest Region, Portland, OR.

Atzet, T., D. E. White, L. A. McCrimmon, P. A. Martinez, P. R. Fong, and V. D. Randall. 1996. *Field Guide to the Forested Plant Associations of Southwest Oregon*. Tech. Paper R6-NR-ECOL-TP-17-96. USDA, Forest Service, Portland, OR.

Barnhart, S. J., J. R. McBride, and P. Warner. 1996. Invasion of northern oak woodlands by *Pseudotsuga menziesii* (Mirb.) Franco in the Sonoma Mountains of California. *Madrono* 43(1): 28–45.

Belsky, J. 1986. Does herbivory benefit plants? A review of the evidence. *American Naturalist* 127: 870–892.

Berman, J. T. and C. S. Bledsoe. 1998. Soil transfers from valley oak (*Quercus lobata* Nee) stands increase ectomycorrhizal diversity and alter root and shoot growth on valley oak seedlings. *Mycorrhiza* 7: 223–235.

Biswell, H. 1999. *Prescribed Burning in California Wildlands Vegetation Management*. University of California Press, Berkeley.

Block, W. M. and M. L. Morrison. 1987. *Conceptual Framework and Ecological Considerations for the Study of Birds in Oak Woodlands*. Gen. Tech. Rep. PSW-100. Pacific Southwest Forest and Range Experiment Station, Forest Service, U.S. Department of Agriculture, Berkeley, CA.

Borman, M. M., W. C. Krueger, and D. E. Johnson. 1990. Growth patterns of perennial grasses in the annual grassland type of southwest Oregon. *Agronomy Journal* 82: 1093–1098.

Boyd, R. B (ed.). 1999. *Indians, Fire, and the Land in the Pacific Northwest*. Oregon State University Press, Corvallis.

Bradley, D. n.d. National Register of Historic Places Registration form, Lyons Ranches Historic District. Draft on file at Redwood National Park, Orick, CA.

Byrne, R., E. Edlund, and S. Mensing. 1991. Holocene changes in distribution and abundance of oaks in California. In *Proceedings of the Symposium on Oak Woodlands and Hardwood Rangeland Management*. Gen. Tech. Rep. PSW-126. USDA Forest Service, Davis, CA.

Carmen, W. J. 1988. *Behavioral Ecology of the California Scrub Jay (*Aphelocoma coerulescens *California), A Non-Cooperative Breeder with Close Cooperative Relatives*. Ph.D. dissertation, University of California Press, Berkeley.

Chappell, C. B., and R. C. Crawford. 1997. Native vegetation of the South Puget Sound prairie landscape. Pages 107–122 in P. Dunn and K. Ewing (eds.), *Ecology and Conservation of the South Puget Sound Prairie Landscape*. Nature Conservancy of Washington, Seattle.

Coblentz, B. E. 1980. Production of Oregon white oak acorns in the Williamette Valley, Oregon. *Wildlife Society Bulletin* 8(4): 348–350.

Cole, D. 1977. Ecosystem dynamics in the coniferous forest of the Willamette Valley, Oregon, U.S.A. *Journal of Biogeography* 4: 181–192.

Department of Fish and Game. 2003. *List of California Terrestrial Natural Communities Recognized by the California Natural Diversity Database*. Department of Fish and Game Wildlife and Habitat Analysis Branch. Available at www.dfg.ca.gov/whdab/pdfs/natcomlist.pdf.

Dickson, E. M. 1946. *Food Plants of Western Oregon Indians*. Master's thesis, Leland Stanford Junior University, Palo Alto, CA.

Duncan, D. A., N. K. McDonald, and S. E. Westfall. 1987. *Long-Term Changes from Different Uses of Foothill Hardwood Ranges*. Gen. Tech. Rep. PSW-100. Pacific Southwest Forest and Range Experiment Station, Forest Service, U.S. Department of Agriculture, Berkeley, CA.

Erickson, W. R. 1996. *Classification and Interpretation of Garry Oak (*Quercus garryana*) Plant Communities and Ecosystems in Southwestern British Columbia*. M.S. thesis, University of Victoria, BC.

Erickson, W. R. 2002. Environmental relationships of native Garry oak (*Quercus garryana*) communities at their northern margin. Pages 179–190 in R. B. Standiford, D. McCreary, and K. L. Purcell (tech. coords.), *Proceedings of the Fifth Symposium on Oak Woodlands: Oaks in California's Changing Landscape*. Gen. Tech. Rep. PSW-GTR-184. USDA Forest Service, San Diego, CA.

Franklin, J. F. and C. T. Dyrness. 1988. *Natural Vegetation of Oregon and Washington*. Oregon State University Press, Corvallis.

Gumtow-Farrior, D. 1991. *Cavity Resources in Oregon White Oak and Douglas-Fir Stands in the Mid-Willamette Valley, Oregon*. M.S. thesis, Oregon State University, Corvallis.

Hagar, J. C. and M. A. Stern. 2001. Avifauna of the Willamette Valley, Oregon. *Northwestern Naturalist* 82: 12–25.

Hammond, P. C. and J. C. Miller. 1998. Comparison of the biodiversity of Lepidoptera within three forested ecosystems. *Annals of the Entomological Society of America* 91: 323–328.

Hanna, I., and P. Dunn. 1996. *Restoration Goals for Oregon White Oak Habitats in the South Puget Sound Region*. Nature Conservancy of Washington, Seattle.

Hibbs, D. E. and B. J. Yoder. 1993. Development of Oregon white oak seedlings. *Northwest Science* 67(1): 30–36.

Jackson, R. D., K. O. Fulgham, and B. Allen-Diaz. 1998a. *Quercus garryana* Hook. (Fagaceae) stand structure in areas with different grazing histories. *Madrono* 45(4): 275–282.

Jackson, R. D., K. O. Fulgham, and B. Allen-Diaz. 1998b. Quercus garryana *Hook. (Fagaceae) with Varying Season, Weather, and Fuel Moisture*. Paper presented at the 8th conference on fire and forest meteorology, Detroit, MI, April 29–May 2, 1985. SAF-AMS.

Kauffman, J. B. and R. E. Martin. 1985. Shrub and hardwood response to prescribed burning with varying season, weather, and fuel moisture. Pages 279–286 in *Proceedings of the 8th Conference on Fire and Forest Meteorology*, April 29–May 2, Detroit, MI. Society of American Foresters, Bethesda, MD.

Kauffman, J. B. and R. E. Martin. 1987. *Effects of Fire and Fire Suppression on Mortality and Mode of Reproduction of California Black Oak (*Quercus kelloggii *Newb.)*. Gen. Tech. Rep. PSW-100. Pacific Southwest Forest and Range Experiment Station, Forest Service, U.S. Department of Agriculture, Berkeley.

Keeley, J. E. 2001. Fire and invasive species in Mediterranean-climate ecosystems of California. Pages 81–92 in K. E. M. Galley and T. P. Wilson (eds.), *Proceedings of the Invasive Species Workshop: The Role of Fire in the Control and Spread of Invasive Species*. Fire Conference 2000: The First National Congress on Fire Ecology, Prevention, and Management. Miscellaneous publication No. 11. Tall Timbers Research Station, Tallahassee, FL.

Kroeber, A. and E. Gifford. 1949. World renewal: a cult system of native northwest California. *Anthropological Records* 13(1): 155, University of California Press, Berkeley.

Lillybridge, T. R., B. L. Kovalchik, C. K. Williams, and B. G. Smith. 1995. *Field Guide for Forested Plant Associations of the Wenatchee National Forest*. Gen. Tech. Rep. PNW-GTR-359. U.S. Department of Agriculture, Forest Service, Pacific Northwest Research Station, Portland, OR.

MacDougall, A. 2001. *Invasive Perennial Grasses in* Quercus garryana *Meadows of Southwestern British Columbia: Prospects for Restoration*. Fifth Symposium on Oak Woodlands: Oaks in California's Changing Landscape, October 22–25, 2001, San Diego, CA.

Maslovat, C. 2001. *Historical Jigsaw Puzzles: Piecing Together the Understory of Garry Oak (*Quercus garryana*) Ecosystems and the Implications for Restoration*. Gen. Tech. Rep. PSW-GTR-184, 2002. USDA Forest Service, San Diego, CA.

McCarthy, H. 1993. Managing oaks and the acorn crop. Pages 231–228 in T. Blackburn and K. Anderson (eds.), *Before the Wilderness: Environmental Management by Native Californians*. Ballena Press, Menlo Park, CA.

McCreary, D. D. 1997. Treeshelters: an alternative for oak regeneration. *Fremontia* 25(1): 26–30.

McCreary, D. D. and J. Tecklin. 1993. Tree shelters accelerate valley oak restoration on grazed rangelands. *Restoration and Management Notes* 11(2): 152–153.

Melvin, R. G., J. R. Brown, and W. J. Clawson. 1992. Application of non-equilibrium ecology to the management of Mediterranean grasslands. *Journal of Range Management* 45: 436–440.

Menke, J. W. 1989. Management controls on productivity. In L. F. Huenneke and H. Mooney (eds.), *Grassland Structure and Function: California Annual Grassland*. Kluwer Academic Publishers, Dordrecht, The Netherlands.

Menke, J. W. 1992. Grazing and fire management for native perennial grass restoration in California grasslands. *Fremontia* 20(2): 22–25.

O'Neil, T. A., D. H. Johnson, C. Barrett, et al. 2001. Matrixes for wildlife–habitat relationships in Oregon and Washington. In T. A. O'Neil and D. H. Johnson (managing directors), *Wildlife–Habitat Relationships in Oregon and Washington*. Oregon State University Press, Corvallis.

Pullen, R. 1996. *Overview of the Environment of Native Inhabitants of Southwestern Oregon, Late Prehistoric Era*. Report prepared for the USDA Forest Service

Siskiyou National Forest and DOI Bureau of Land Management, Medford Resource Area, Medford, OR.

Reed, L. J. and N. G. Sugihara. 1987. *Northern Oak Woodlands: Ecosystem in Jeopardy, or Is It Already Too Late?* Gen. Tech. Rep. PSW-100. Pacific Southwest Forest and Range Experiment Station, Forest Service, U.S. Department of Agriculture, Berkeley, CA.

Riegel, G. M., B. G. Smith, and J. F. Franklin. 1992. Foothill oak woodlands of the interior valleys of southwestern Oregon. *Northwest Science* 66(2): 66–76.

Roemer, H. 1993. Vegetation and ecology of Garry oak woodlands. Pages 19–26 in R. J. Hebda and A. Atikens (eds.), *Garry Oak–Meadow Colloquium: Proceedings*. Garry Oak Meadow Preservation Society, Victoria, BC.

Scott, T. A. and N. L. Pratini. 1997. The effects of native soils on Engelmann oak seedling growth. Pages 657–660 in N. H. Pillsbury, J. Verner, and W. D. Tietje (tech. coords.), *Proceedings—Symposium on Oak Woodlands: Ecology, Management, and Urban Interface Issues*. USDA Forest Service Gen. Tech. Rep. PSW-GTR-160. Pacific Southwest Research Station, Albany, CA.

Smith, W. P. 1985. Plant associations within the interior valleys of the Umpqua River Basin, Oregon. *Journal of Range Management* 38(6): 526–530.

Stromberg, M. R. and P. Kephart. 1996. Restoring native grasses in California old fields. *Restoration and Management Notes* 14(2): 102–111.

Sugihara, N. G. and L. J. Reed. 1986. *Prescribed Fire for Restoration and Maintenance of Bald Hills Oak Woodlands*. Gen. Tech. Rep. PSW-100. Pacific Southwest Forest and Range Experiment Station, Forest Service, U.S. Department of Agriculture, Berkeley, CA.

Tappeiner, J. and P. McDonald. 1979. Preliminary recommendations for managing California black oak in the Sierra Nevada. In T. R. Plumb (tech. coord.), *The Proceedings of the Symposium on the Ecology, Management, and Utilization of California Oaks*. USDA Forest Service. Pacific Southwest Forest and Range Station, Claremont, CA.

Thilenius, J. F. 1964. *Synecology of the White Oak (*Quercus garryana, *Douglas)*. Ph.D. thesis, Oregon State University, Corvallis.

Thilenius, J. F. 1968. The *Quercus garryana* forests of the Willamette Valley, Oregon. *Ecology* 49: 1124–1133.

Thornton, B. 1981. Response of deer to fuel management programs in Glenn and Colusa counties, California. In C. E. Conrad and W. C. Oechel (tech. coords.), *Proceedings of the Symposium on Dynamics and Management of Mediterranean Type Ecosystems*. USDA Forest Service Gen. Tech. Rep. PSW-58. Pacific Southwest Forest and Range Experiment Station, Albany, CA.

Turner, N. 1999. "Time to burn": traditional use of fire to enhance resource production by aboriginal peoples in British Columbia. Pages 185–218 in R. B. Boyd (ed.), *Indians, Fire and the Land in the Pacific Northwest*. Oregon State University Press, Corvallis.

Tveton, R. K. and R. W. Fonda. 1999. Fire effects on prairies and oak woodlands on Fort Lewis, Washington. *Northwest Science* 73(3): 145–158.

Underwood, S., L. Arguello, and N. Siefkin. 2003. Restoring ethnographic landscapes and natural elements in Redwood National Park: the story of the bald hills prairies and oak woodlands. *Ecological Restoration* 21(4): 279–284.

USDA Natural Resource Conservation Service. 1996. *MLRA A5 Range Site Descriptions*. Set of 18 range site descriptions (ecological sites) issued by NRCS on July 1, 1997, for use in fieldwork in S.W. Oregon counties. Portland, OR.

USDA Soil Conservation Service. 1988. *Soil Survey of Jackson County Area, Oregon*. Pages 327–348, vegetative sites by Gene Hickman. Portland, OR.

Verner, J. 1987. *Importance of Hardwood Habitats for Wildlife in California*. Gen. Tech. Rep. PSW-100. Pacific Southwest Forest and Range Experiment Station, Forest Service, U.S. Department of Agriculture, Berkeley.

Waterman, T. 1920. *Yurok Geography*. Trinidad Museum Society, Trinidad, CA, 1993; originally published by University of California Press, 1920.

Additional Reading

Agee, J. 1990. The historic role of fire in Pacific Northwest forests. In J. D. Walstad, S. R. Radosavich, and D. V. Sandberg (eds.), *Natural and Prescribed Fire in the Pacific Northwest Forests*. Oregon State University Press, Corvallis.

Atzet, T. and L. A. McCrimmon. 1990. *Preliminary Plant Associations of the Southern Oregon Cascade Mountain Province*. USDA, Forest Service, Siskiyou National Forest, Grants Pass, OR.

Clark, H. W. 1937. Association types in the north coast ranges of California. *Ecology* 18: 214–230.

Crawford, R. C. 2003. *Riparian Vegetation Classification of the Columbia Basin, Washington*. Natural Heritage Program Report 2003–03. Washington Department of Natural Resources, Olympia.

Dunn, P. 1998. *Prairie Habitat Restoration and Maintenance of Fort Lewis and within the South Puget Sound Prairie Landscape: Final Report and Summary of Findings*. Nature Conservancy of Washington, Seattle.

Eyre, F. H., ed. 1980. *Forest Cover Types of the United States and Canada*. Society of American Foresters, Washington, DC.

Farris, G. 1993. Quality food: the quest for pine nuts in

northern California. Pages 229–240 in T. Blackburn and K. Anderson (eds.), *Before the Wilderness, Environmental Management by Native Californians*. Ballena Press, Menlo Park, CA.

Griffin, J. R. 1977. Oak woodland. Pages 383–415 in M. G. Barbour and J. Major (eds.), *Terrestrial Vegetation of California*. Wiley, New York.

Habeck, J. R. 1961. The original vegetation of the mid-Willamette Valley, Oregon. *Northwest Science* 35: 65–77.

Harrington, C. A., and M. A. Kallas. 2002. *A Bibliography for* Quercus garryana *and Other Geographically Associated and Botanically Related Oaks*. Gen. Tech. Rep. PNW-GTR-554. USDA Forest Service, Portland, OR.

Hastings, M. S., S. Barnhart, and J. R. McBride. 1997. Restoration management of northern oak woodlands. Pages 275–280 in N. H. Pillsbury, J. Verner, and W. D. Tietje (eds.), *Proceedings of the Symposium on Oak Woodlands: Ecology, Management, and Urban Interface Issues*. Gen. Tech. Rep. PSW-GTR-160. USDA Forest Service, Pacific Southwest Research Station, Albany, CA.

Heizer, R. and A. Elasser. 1980. *The Natural World of the California Indians*. California Natural History Guides No. 46. University of California Press, Berkeley.

Hickman, O. E. 1992. *Plant Checklist for Range and Forest Inventories in Southwestern Oregon*. 2nd ed. Rangeland Science Series No. 1. Oregon State University, Corvallis.

Holland, R. F. 1986. *Preliminary Descriptions of the Terrestrial Natural Communities of California*. California Department of Fish and Game, Sacramento.

Holmes, P. M., D. M. Richardson, B. W. Van Wilgen, and C. Gelderblom. 2000. Recovery of South African fynbos vegetation following alien woody plant clearing and fire: implications for restoration. *Austral Ecology* 25: 631–639.

Hull, A. C. 1973. Germination of range plant seeds after long periods of uncontrolled storage. *Journal of Range Management* 26: 198–200.

Jimerson, T. M. and S. K. Carothers. 2002. *Northwest California Oak Woodlands: Environment, Species Composition, and Ecological Status*. Gen. Tech. Rep. PSW-GTR-184. USDA Forest Service, Pacific Southwest Research Station, Albany, CA.

Jimerson, T. M., J. W. Menke, S. K. Carothers, M. P. Murray, V. VanSickle, and K. Heffner-McClellan. 2000. *A Field Guide to the Rangeland Vegetation Types of the Northern Province: Klamath, Mendocino, Shasta–Trinity and Six Rivers National Forest*. R5-ECOL-TP-014. Pacific Southwest Region, USDA Forest Service, San Francisco.

Kagan, J. S., J. A. Christy, M. P. Murray, and J. A. Titus. 2001. *Classification of Native Vegetation of Oregon*. Oregon Natural Heritage Program, Corvallis.

Keeley, J. E. 1992. Recruitment of seedlings and vegetative sprouts in unburned chaparral. *Ecology* 73(4): 1194–1208.

Kimmerer, R. W. and F. K. Lake. 2001. The role of indigenous burning in land management. *Journal of Forestry* 99(11): 36–41.

Klinka, K., H. Qian, J. Pojar, and D. V. Meidinger. 1996. Classification of natural forest communities of coastal British Columbia, Canada. *Vegetation* 125(2): 149–168.

Kroeber, A. and S. Barrett. 1960. Fishing among the Indians of northwestern California. *Anthropological Records* 22: 1–156.

Kuchler, A. W. 1964. *Manual to Accompany the Map of Potential Vegetation of the Conterminous United States*. Special Publication No. 36. American Geographical Society, New York.

Lewis, H. 1993. Patterns of Indian burning in California: ecology and ethnohistory. Pages 55–116 in T. Blackburn and K. Anderson (eds.), *Before the Wilderness: Environmental Management by Native Californians*. Ballena Press, Menlo Park, CA.

Little, E. L. Jr. 1979. *Checklist of United States Trees (Native and Naturalized)*. Agricultural Handbook 541. USDA Forest Service, Washington, DC.

Moerman, D. E. 1998. *Native American Ethnobotany*. Timber Press, Portland, OR.

Munz, P. A. 1973. *A California Flora and Supplement*. University of California Press, Berkeley.

Patterson, R. 1992. Fire in the oaks. *American Forests* 98(11): 32–34, 58–59.

Pavlik, B. M., P. C. Muick, S. Johnson, and M. Popper. 1991. *Oaks of California*. Cachuma Press, Los Olivos, CA.

Peck, M. 1961. *Manual of the Higher Plants of Oregon*. 2nd ed. Binfords and Mort, Portland, OR.

Plumb, T. R. 1979. Response of oaks to fire. In T. R. Plumb (tech. coord.), *The Proceedings of the Symposium on the Ecology, Management, and Utilization of California Oaks*. USDA Forest Service, Pacific Southwest Forest and Range Station, Claremont, CA.

Roberts, R. C. 1987. *Preserving Oak Woodland Species Richness: Suggested Guidelines from Geographical Ecology*. Gen. Tech. Rep. PSW-100. Pacific Southwest Forest and Range Experiment Station, USDA Forest Service, Berkeley, CA.

Sawyer, J. O. and T. Keeler-Wolf. 1995. *A Manual of California Vegetation*. California Native Plant Society, Sacramento.

Sawyer, J. O., D. A. Thornburgh, and J. R. Griffin. 1977.

Mixed evergreen forest. Pages 359–381 in M. G. Barbour and J. Major (eds.), *Terrestrial Vegetation of California*. Wiley, New York.

Schenck, S. and E. Gifford. 1952. Karok ethnobotany. *Anthropological Records* 13(6): 377–392. University of California Press, Berkeley.

Silen, R. R. 1958. *Silvical Characteristics of Oregon White Oak*. USDA Forest Service Pacific Northwest Forest and Range Experiment Station, Portland, OR.

Snyder, S., ed. 2003. *Bear in Mind: The California Grizzly*. Heyday Books, Berkeley, CA.

Stein, W. 1990. *Quercus garryana* Dougl. ex. Hook.: Oregon white oak. Pages 650–660 in R. M. Burns and B. H. Honkala (tech. coords.), *Silvics of North America* Vol. 2, *Hardwoods*. Agricultural Handbook 654. USDA Forest Service, Washington, DC.

Stewart, O. C. 2002. *Forgotten Fires: Native Americans and the Transient Wilderness*. University of Oklahoma Press, Norman.

Storer, T. I. and L. P. Tevis Jr. 1959. *California Grizzly*. University of California Press, Berkeley.

Sugihara, N. G., Reed L. J., and Lenihan J. M. 1987. Vegetation of the bald hills oak woodlands, Redwood National Park, California. *Madrono* 34(3):193–208.

Taylor, R. J., and Boss T. R. 1975. Biosystematics of *Quercus garryana* in relation to its distribution in the state of Washington. *Northwest Science* 49(2):49–57.

Thompson, L. (1916) 1991 *To the American Indian, Reminiscences of a Yurok Woman*. Peter E. Palmquist and Heyday Press, Berkeley.

Thysell, D. R., and Carey A. B. 2001. *Quercus garryana* communities in the Puget Trough, Washington. *Northwest Science* 75(3): 219–235.

Todt, D. and N. Hannon. 2003. *Acorn Economics*. Presentation at the Society of Ethnobiology Conference, Seattle, WA, March 28, 2003. University of Washington, Seattle.

Topik, C., N. M. Halverson, and C. T. High. 1988. *Plant Association and Management Guide for the Ponderosa Pine, Douglas-Fir and Grand Fir Zones, Mt. Hood National Forest*. R6-ECOL-TP-004-88.USDA Forest Service, Pacific Northwest Region, Portland, OR.

Vale, T. R., ed. 2002. *Fire, Native Peoples, and the Natural Landscape*. Island Press, Covelo, CA.

Waring, R. H., and J. F. Franklin. 1979. Evergreen coniferous forests of the Pacific Northwest. *Science* 29(204): 1380–1386.

Washington Natural Heritage Program, Olympia, WA. 2004. www.dnr.wa.gov/nhp/refdesk/communities/index.html.

Watt, B. and A. Merrill. 1963. *Composition of Foods: Raw, Processed, Prepared*. Agriculture Handbook No. 8 (rev.), USDA. U.S. Government Printing Office, Washington, DC.

Withgott, J. 2004. Fighting sudden oak death with fire. *Science* 305(August 20): 1101.

Scientific and Restoration Literature on the Web

British Columbia Garry Oak Ecosystem Recovery Team Literature Review: *Towards a Recovery Strategy for Garry Oak and Associated Ecosystems in Canada: Ecological Assessment and Literature Review*. Available at www.goert.ca/resources/index.html#LitReview.

Campbell, B. H. 2003. *Restoring Rare Native Habitats in the Willamette Valley: A Landowner's Guide for Restoring Oak Woodlands, Wetlands, Prairies, and Bottomland Hardwood and Riparian Forests*. Defenders of Wildlife, West Linn, OR. Available at www.biodiversitypartners.org/pubs/Campbell/01.shtml.

FEIS (Fire Effects Information System). 1999. *Botanical and Ecological Characteristics*. Available at www.fs.fed.us.database/feis/plants.

IHRMP-Oak Regeneration/Restoration Articles 1. Available at danr.ucop.edu/ihrmp/regen.html.

USDA Forest Service Oregon White Oak Bibliography

Harrington, C. A. and M. A. Kallas (compilers) (2002). Gen. Tech. Rep. PNW-GTR-554. U.S. Department of Agriculture, Forest Service, Pacific Northwest Research Station, Portland, OR. Available at www.srs.fs.usda.gov/pubs/viewpub.jsp?index=4822.

Vesely, D. and G. Tucker. 2005. *A Landowner's Guide for Restoring and Managing Oregon White Oak Habitats*. USDI Bureau of Land Management, Salem District, Salem, OR. Available at www.or.blm.gov/salem/html/whatwedo/oak_publication/oak_pub_page.htm.

Conservation and Professional Organizations

Garry Oak Ecosystems Recovery Team (GOERT), British Columbia: www.goert.ca/about/index.html.

Institute for Applied Ecology: www.appliedeco.org.

The Nature Conservancy, www.nature.org.

Oregon Oak Communities Working Group: www.oregonoaks.org.

Chapter 5

Old-Growth Conifer Forests

Jerry F. Franklin, Dean Rae Berg, Andrew B. Carey, and Richard A. Hardt

Old-growth conifer forests of the maritime Pacific Northwest represent ground zero in the economic, social, and philosophical values war that has split communities across the region over the past few decades. The debate over old-growth conservation has changed the way Americans, Canadians, and much of the rest of the world thinks about forests and forestry. "New forestry," "ecological forestry," and "ecosystem management" all owe their origins to research into how Northwest old-growth forests function and what years of timber management have overlooked with regard to sustainability. Emerging management paradigms may offer a fresh opportunity to reconsider forest conservation in the region. As we gain greater understanding of how to adapt forestry to restore old-growth conditions, parties formerly in conflict are beginning to sit down and negotiate mutually acceptable and sustainable solutions. We hope that process will continue and will be furthered by this chapter and volume.

Classic old-growth conifer forests are found west of the Cascade Mountains in Oregon, Washington, and California and west of the coastal mountains of British Columbia and Alaska.

In ecological terms, old-growth forests perform a wide variety of functions, including habitat for a multitude of animals and plants, carbon sequestration, removal of particulates from the air, cooling of the extended environment, clean water, and modulated water flow. People also value old-growth forests for their naturalness, the immensity of the trees, and their unique aesthetic qualities (Franklin 1993a, Carey 1998). Old-growth forests have been the focus of timber harvest in the Northwest for many decades. Natural capital has been converted to economic capital in the form of mills, jobs, income, and timber products that have been used to make homes, packaging, and paper throughout the world. The conflict between the economic, ecological, and aesthetic benefits of old-growth forests has in large part shaped and defined regional environmental controversies for the past several decades. Although logging of old growth has greatly diminished in the U.S. portion of the Northwest, it continues at fairly high levels in British Columbia, although not without controversy.

This chapter does not take a position on how much remaining old growth should be protected from logging. Instead, it addresses the question of whether and how forest land managers could restore old growth from younger stands. It is commonly believed that old-growth forests will reevolve naturally, given enough time and depending on site characteristics and the severity of climate change. The key question this chapter attempts to answer is, "How can younger forests be restored through active management to a state that provides similar functions and services as natural old growth?"

Humans cannot recreate highly diverse old-growth forests because much of their current nature is the result of a climatic and disturbance history we cannot replicate exactly (Carey 1998). At best, we can strive for restored *biological complexity* at multiple scales that will produce forests that develop a structure, composition, and capacity for function similar to that of old growth. Typical young, managed stands have been intentionally simplified to increase fiber productivity, at the expense of many important ecosystem elements. The full extent to which young managed forests and landscapes have been reduced in diversity and ecological function is unknown because the totality of species and their interactions in older natural forests has not been fully described. Attempts at restoration require patience and humility as we continue to build our knowledge of forest components and how they function.

Historical Geography of Old-Growth Forests

Since the time of Euro-American settlement of the Northwest, old-growth forests have been lost as a result of multiple factors, including conversion of forests to farms and towns, the geographic spread of timber harvest, and natural disturbances, such as a 1921 windstorm, multiple large fires in the early to mid-twentieth century, the 1980 eruption of Mt. St. Helens, and several very large recent fires in British Columbia, northern Washington, and southern Oregon. This loss of old growth has far outpaced emergence of "new" old growth and has been accompanied by declines in terrestrial species such as the northern spotted owl and the marbled murrelet (Courtney et al. 2004, Franklin et al. 2002) (Figure 5.1).

The composition, assemblages, and distribution of conifer-dominated forests west of the mountains are traced to about 5,000 years B.P., when a cool, moist climate replaced a warmer and drier one over much of the region. Pollen evidence indicates that western hemlock and cedar were able to move from higher to lower elevations into what had been dry Douglas fir woodlands. From this time forward, but before Euro-American settlement, about half of the total west side conifer forests would have been in an old-growth state at any given time (Agee 1993). The total amount of old growth was highly variable, especially at smaller spatial scales such as fifth-order watersheds. But at the province scale (more than 2 million hectares), the historical amount of old growth is believed to have fluctuated between 25% and 75% of the entire forested landscape (Wimberly et al. 2000).

Currently, only about 10% of the forest in the lower U.S. part of the region is in old growth, most on federal lands (FEMAT 1993). The British Columbia Ministry of Forests reports that of a total of 9.6 million acres of coastal old-growth forest, 2 million (about 20%) has protected status. Environmental organizations report that 80% of the original old growth in large valleys along the "Great Bear Rainforest" (essentially the mainland coast of British Columbia) has been logged (www.savethegreatbear.org). In southeast Alaska, the Tongass National Forest reports that there are 9.4 million acres of "old growth," of which only about half contains commercial-sized trees. Of this amount, less than 500,000 acres has been logged since the 1950s. Under the existing forest plan, 90% of the remaining commercial old-growth forest has protected status (Tongass National Forest Web site, www.fs.fed.us/r10/tongass/). Conservation groups dispute some of these figures and claim that of the original biggest, best, most classic old growth (confined to low-elevation, sheltered valleys), 75% is not protected by law. Additionally, up to 500,000 acres of old growth outside the Tongass (state and tribal lands) has already been logged (www.seacc.org/ForestFacts).

First Nations people lived in and used old-growth forests for many products, particularly the large, seagoing cedar canoes that supported the Northwest salmon culture. The extent to which they affected the distribution of these forests is not fully known. There is evidence that Indians burned

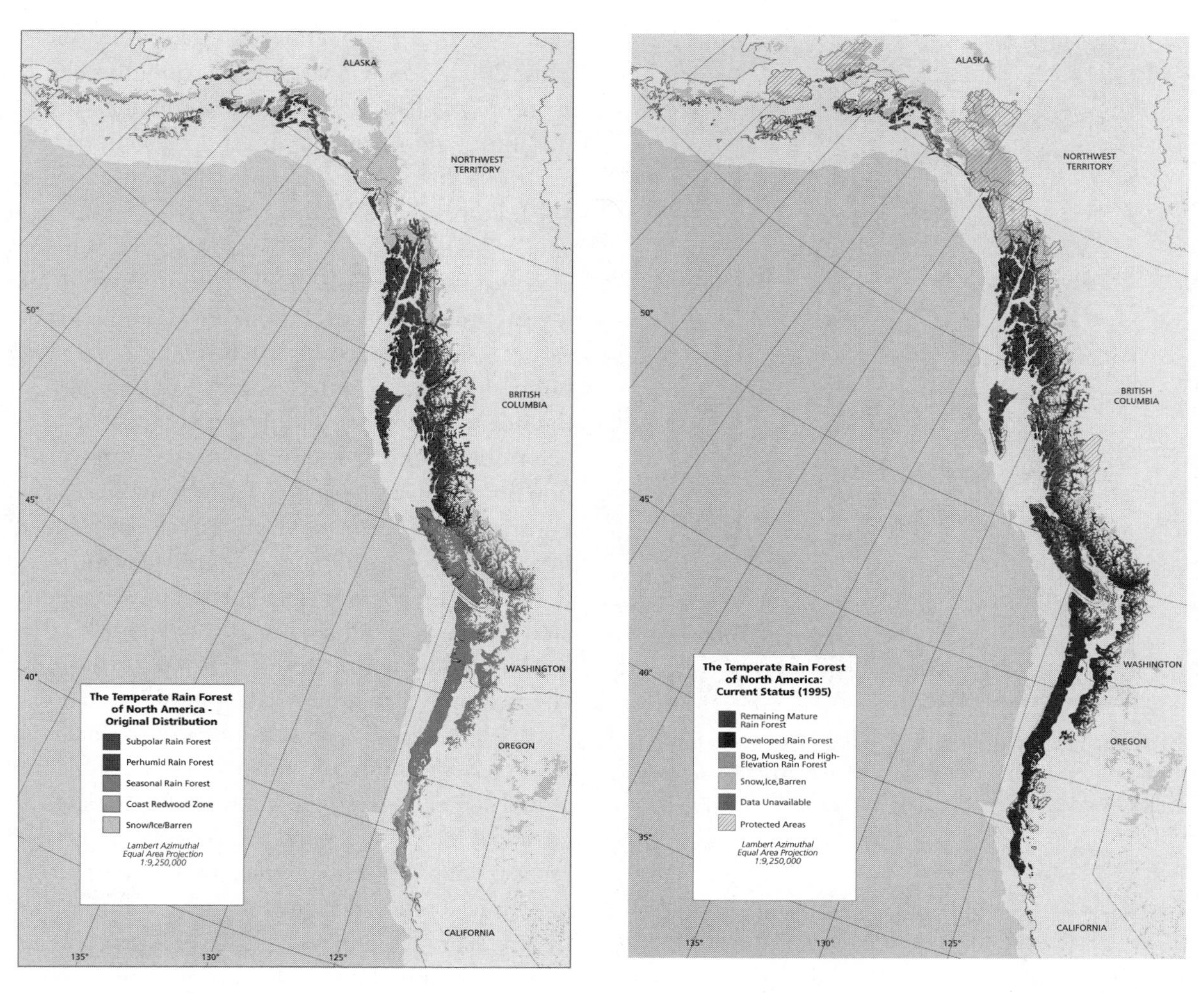

FIGURE 5.1. *(a)* Historic and *(b)* current status of North American coastal temperate rain forest. *(Courtesy of Ecotrust, Rainforests of Home, Island Press)*

grassland clearings in wet conifer forests, right up to the coastline. Fires from the interior valley margins certainly crept into adjacent conifer forests and kept them from encroaching. But it is doubtful that Indian fire played a significant role in the large-scale distribution and structure of old growth (Agee 1993) (Figure 5.2).

Loss of old-growth forest has reduced water and stream habitat quality (Franklin et al. 2002, CSE 1995, Swanson and Berg 1991). Less obvious are the effects in forests at local scales (loss of biotic integrity and decreased abundance of numerous species) and at the landscape level (including inhibition of dispersal and colonization processes and degradation of habitat quality for wide-ranging species) (Oliver and Larson 1990, Carey et al. 1992).

Young managed stands that have replaced old growth are structurally and compositionally different and do not provide many of the same functions. They are usually managed to maximize the efficient production of timber, resulting in simple stand structure, high tree density, and very simple composition, sometimes a true monoculture of

FIGURE 5.2. Old-growth tree near Oregon coast showing fire scars. *(Photo by Jordan Sector, Karnowsky Creek Restoration Project, Siuslaw Watershed Council)*

Douglas fir (Muir et al. 2002, Tappeiner et al. 1997). They often have dense road networks and skid trails, which interrupt natural drainage patterns. Duff layers are thin or absent, as are most understory plants, and soils may be compacted. Interest in shifting management of at least some young stands to move them toward old-growth character and structure has been growing for more than a decade.

Accelerating Restoration of Complexity

Multiple studies on the effects of young forest thinning suggest that forest managers and restoration practitioners can speed the development of old-growth forest structural and compositional characteristics (Oliver et al. 1992, Tappeiner et al. 1997, Carey et al. 1999b, Franklin et al. 1999, 2002, Muir 2002). Old-growth forest restoration can be accomplished by retaining or recreating specific structures and managing forest ecosystem dynamics over a long time horizon (Swanson and Berg 1991, Franklin 1992, Oliver et al. 1992, Franklin et al. 1997, 2002, Carey 2003a, 2003b). Restoration opportunities are enhanced by the inherent biological productivity of regional forests, high timber value, established infrastructure, and extensive public forest ownership. Initially (and perhaps for the next century), we should expect that the character and structure of young forests being intentionally managed toward old growth will be different from those of present old growth in terms of tree size, architecture, and epiphytes. Although development of these attributes can be accelerated, they will take many years to take hold.

The guidelines for restoration put forth in this chapter apply structure and composition to a canvas of landforms, within regional and local climates. Restoration of old-growth forests is not aimed at creating a single stand condition based on an ideal reference ecosystem. Instead, general recovery of biological complexity should drive old-growth restoration design and practice, forest stand by forest stand (Carey 1998, 2003a). This approach allows continued self-organization and development of the ecosystem, in concert with strategic nudges by forest managers in the context of ever-changing natural conditions. In other words, old-growth forest restoration fits very closely to the definition of ecological restoration adopted by the Society for Ecological Restoration International with regard to assisting the recovery of degraded or damaged ecosystems.

Restoration entails taking advantage of previously developed transportation systems, including roads, structured stream crossings, and existing log landings. Old-growth forest restoration often includes related projects, such as road improvements and decommissioning, invasive species removal, elimination or mitigation of aquatic barriers, and placement of in-stream structures. These actions are covered in other chapters of this volume and will not be addressed here, although we acknowledge their importance.

Furthermore, we do not address important concerns about restoring the aesthetic and spiritual qualities of old-growth forests. Does the history of young managed stands prevent them from ever again providing the aesthetic and spiritual values of historic old-growth forests? Does the very act of human intervention spoil nature for all time? This question is dealt with in a general way in Chapter 2 in this book. It may have special significance with regard to old-growth forests because they have been the focus of such heated debate over the years, much more so than has been the case with other regional ecosystems.

The Case for Restoration of Old-Growth Forests

The case for restoration of old-growth forests lies in the need to reestablish important ecological functions now missing or severely diminished in key areas of our region (Swanson and Franklin 1992). Restoring old-growth forests will contribute to the recovery of threatened and endangered species, including the northern spotted owl and marbled murrelet. Additionally, restoration will increase the total amount of habitat for a whole suite of other species associated with old-growth forests (FEMAT 1993). Restored old-growth forests, particularly in riparian areas but also in uplands, will provide essential building blocks for aquatic habitat and will contribute to the improvement of water quality and the recovery of salmon and other species.

We cannot predict with certainty whether or when many young managed stands will develop structural complexity in the absence of intentional management. Left untreated, some young stands may gradually increase in structural complexity if they are partially disturbed by fire, wind, or pathogens. However, it is more likely that the unnatural density and homogeneity of most intensively managed young stands will lead to ecological instability, predisposing them to catastrophic disturbance (Bureau of Land Management [BLM] 2003a). Silviculture can be harnessed to play the role of partial disturbance if designed to accelerate structural complexity of the stand while avoiding a catastrophic disturbance. In many cases young, intensively managed stands (particularly plantations) may have little chance of developing an old-growth structure unless they first experience a catastrophic disturbance and the entire clock of succession is reset to near zero.

Characterization of Old-Growth Forest Ecosystems

Franklin et al. (1981) defined old growth as having numerous old, large trees, more than 250 years old, more than 100 centimeters in diameter, and more than 50 meters tall. They also noted large accumulations of standing dead and fallen trees. Old-growth forests have high complexity that included diverse tree sizes and species, abundant dead trees, high levels of patchiness in overstory and understory cover, high vertical foliage diversity, and a great biological diversity including mosses, lichens, fungi, amphibians, birds, small mammals, arboreal rodents, and bats. Recent definitions have changed emphasis from a focus on structure to include developmental processes and biological complexity (Franklin et al. 2002). Ancient redwoods of the northern California coast, Douglas fir in the Cascade forests, and western hemlock and western redcedar on Vancouver Island centuries to millennia old have the largest accumulation of woody biomass and the most complex forest floors on the planet (Franklin and Van Pelt 2004).

Pacific Northwest old growth includes a heterogeneous set of forests that differ in age, structure, composition, disturbance history, and visual appearance. Physiographic provinces with variations in geomorphology, soils, elevation, topography, distance from the Pacific Ocean, and disturbance regimes produced variety in old growth. Old-growth forests also developed over centuries, experiencing fluctuations in climate that are not likely to recur (Pielou 1991). Quantitative scientific definitions of old growth are important for science but often fail to capture more qualitative and subjective impressions, including the degree of naturalness, wildness, wilderness, and congruence with a preconceived archetypical old growth. Thus, old growth, in a technical sense, often is mistakenly conflated with virgin, native, or unmanaged forest.

U.S. federal foresters have defined old growth as forest stands at least 180–220 years old with moderate to high canopy closure; a multilayered, multispecies canopy dominated by large overstory trees; a high incidence of large trees, some with broken tops and other indications of old and decaying wood (decadence); numerous large snags; and heavy accumulations of wood, including large logs on the ground (FEMAT 1993). However, most federal forest policies focus on late-successional forests, which include both mature and old-growth stands and often are interpreted as stands more than 80 years old (Forest Service and BLM 1994). In practice, a definition of old growth based on stand age is of little value because no amount of restoration can make stands grow older faster.

Structural Features of Old-Growth Forests

Very large conifers, especially Douglas fir, are the most dominant and impressive features of Pacific Northwest old-growth forests. They are the charismatic structural elements that attract so much attention, and rightfully so. Large trees with large branches provide habitat for a diverse community of epiphytic lichens and bryophytes and nesting habitat for arboreal mammals, northern spotted owls, and marbled murrelets (Franklin et al. 1981, 2002, FEMAT 1993, Carey et al. 1997). Large trees are important not only in and of themselves but as the source of large snags and logs. Perhaps ironically, large-diameter trees may be the easiest structural characteristic to develop in young managed stands. Increase tree growing space, and they grow faster. Creating big snags and logs comes only after big trees have grown. Increasing the range of tree diameters and creating shade-tolerant understories can be even more difficult to accomplish.

Diversity of Tree Sizes, Species, and Conditions

It is not only big trees that make old-growth structure but also the wide range of tree sizes that contribute to structural complexity. This diversity is nearly always missing in contemporary plantations. Diversity of tree sizes is needed for multilayered canopies, which may be necessary for important functional attributes, including foraging by northern spotted owls. Biological complexity derives from a wide range of tree sizes, diverse foliage height profiles, and spatial heterogeneity. Spatial structure includes canopy gaps, patches, low shrubs, and herbaceous understory. Depending on the scale, many different types of patches reside in one old-growth ecosystem (Berg and Clement 1992).

Coarse Woody Debris and Snags

Pacific Northwest forests have some of the largest coarse woody debris accumulations of any forests in the world, but coarse woody debris and snag levels are extremely variable in old-growth stands (Spies 1991). Coarse woody debris and snags add to the complexity of the forest structure and have profound effects on food webs, especially fungi, invertebrates, amphibians, and small mammals (Aubry and Hall 1991, Carey 1995). Snags are essential for cavity-nesting species and provide important foraging habitat for many birds and insects associated with old-growth forests (Lundquist and Mariani 1991) (Figure 5.3).

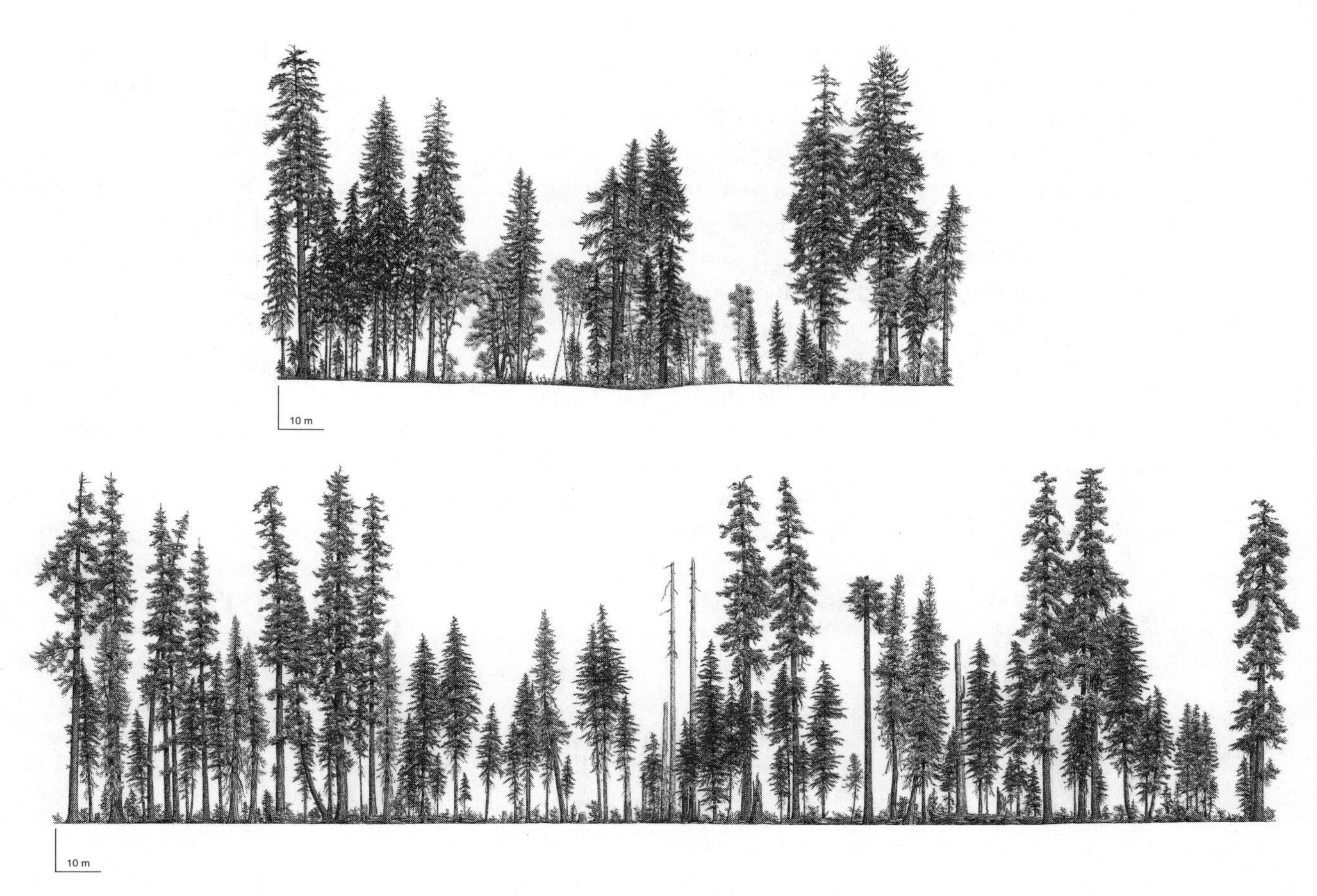

FIGURE 5.3. Two cross-sections showing typical old-growth forest structure at different stages of development. Note canopy gaps, snags, and varying sizes of trees. (*From Franklin and Van Pelt 2004*)

Composition

Old-growth forests are noted for their substantial component of shade-tolerant conifers (particularly western hemlock and western redcedar), which are important contributors to structural attributes such as a range of tree diameters and multilayered canopies (Franklin 1981, Spies and Franklin 1988, 1991). Furthermore, variation in the abundance and size distribution of shade-tolerant conifers may provide much of the variation in structure between old-growth forests (Poage and Tappeiner 2005). Although early forest seral stages may have high overall plant diversity, a broader range of plant habitats can develop in old-growth forests, and these habitats produce a diverse array of vegetation site types that lead to a diversity of invertebrate and vertebrate life.

Function

Important forest functions include biological productivity, capture and cycling of nutrients and water, storage of carbon, and habitat for many organisms. Many of the notable functions of old-growth forest can be attributed to the distinct structural characteristics described earlier. For example, large trees with large branches provide a habitat substrate for lichens that fix atmospheric nitrogen and provide nutrient capture for the forest ecosystem (Neitlich and McCune 1997).

The forest floor is the foundation for the sustainability of forest ecosystems, and in the Northwest old growth provides for the richest small mammal community of any temperate forest (Carey 1995, 2001). The structure, composition, and function of Pacific Northwest forest soils are delineated particularly by the abundance and diversity of ectomycorrhizal fungi (Carey 1995, 2001, North et al. 1997, Smith et al. 2002). Decaying logs and deep organic layers, undisturbed for long periods, may provide important habitat, water storage, and nutrient cycling (Franklin et al. 1981, Aubry and Hall 1991, North et al. 1997, Carey 1995, Olsen 2001).

Approaches to Restoring Structural Complexity

The regional diversity of Pacific Northwest forests, combined with the influence of individual stand history, prevents a single approach to ecological restoration. Strategies are built from the concept of managing forest structural changes along pathways of ecosystem development, with the goal of accelerating development of biological complexity. Before plunging into restoration of an old-growth forest, land managers should apply the following general principles.

Plan at a Landscape Scale

Context usually is very important, so restoration of a particular site or stand should be considered within a larger landscape scale (Spies and Turner 1999). Analysis at this broader scale will allow managers to consider issues such as possible cumulative effects of restoration treatments against long-term restoration gains (Walters and Hollings 1990). Landscape planning includes the choice of rotation age of managed forests, which determines landscape composition and permeability (Tang et al. 1994); watershed analysis, which identifies riparian, mass-wasting, and wetland areas and an aquatic conservation strategy; identification of unique landscape elements, both biological (e.g., grass balds) and physical (talus slopes), that need protection; and transportation system management.

Acknowledge Regional Differences

Differences in regional climate, disturbance regimes, and stand histories dramatically affect stand characteristics, such as the role of shade-tolerant conifers and hardwoods in stand development and typical quantities of snags and coarse woody debris. To the extent possible, restoration strategies should make use of local knowledge on old-growth forests and stand development, rather than more general information, in formulating targets for old-growth restoration.

Develop Multiple Pathways to Reach Multiple Targets

Restoration often centers on thinning designed to develop variability in forest structure. Because variability is a goal, old-growth restoration should not have a single stand condition or treatment as its target (Smith 1986). Existing old-growth forests probably developed along multiple pathways, creating a wide variability in stand conditions. Variability in structure at multiple spatial scales will provide diverse habitats as stands develop. Variability in structure will also maintain future management options in the face of unpredictable disturbance events.

Shift managed stands from *even spacing* of a single species to *variable spacing* of multiple species. This principle is the key to old-growth restoration. A notable distinction between old-growth forests and conventionally managed forests is the even tree spacing characteristic of the latter (Carey et al. 1999a, Franklin et al. 2002, Poage 2000). Thinning of fully stocked young stands must not only reduce stand density but also vary tree spacing. Thinning can also be used to vary tree species, protect cavity trees and legacy trees, and promote understory hardwood development.

Conventional thinning reduces density but retains uniformity of spacing. Local data on existing old-growth stands are particularly important in understanding the role of hardwoods and shade-tolerant conifers in stand development and in assembling stand-level targets for restoration. Reducing stand density and varying tree spacing will allow stand development processes to accomplish many restoration objectives, including crown differentiation, creation of a complex overstory with deep crowns, release of shade-tolerant conifers and hardwoods to create multiple canopy layers, and release of tall deciduous shrubs.

Leave Some Areas Undisturbed

Not every stand need be or should be actively managed to speed the development of old-growth forest conditions. Some stands, especially previously unmanaged stands and those with abundant biological legacies, may already be on a good trajectory to develop old-growth structure. These contrast with those becoming stagnant or beginning to decline in vigor through excessive competition. Additionally, some forest functions may be more related to the time since disturbance than stand age or structure. Habitat quality for some species of plants, lichens, mollusks, and fungi may increase with stand age, independent of stand structure, or may not tolerate new disturbance (Lesher and Henderson 1992, Ruchty et al. 2001, Sillett et al. 2000). Some portion of forests in a landscape should remain undisturbed for periods of time ranging from a decade to a century or more. The landscape context approach noted earlier can help determine which and how many stands should be thinned and which should be left untreated. Short-term impacts and risks must always be weighed against long-term benefits.

Stand-Level Forest Restoration and Variable Retention

One of the difficult conceptual bridges to cross in forest restoration is that logging often is a necessary or desirable part of the job. But the type of logging done is far from what we associate with the thin–clearcut–replant sequence of conventional commercial forestry. Restoration of old-growth forests from younger stands starts with the concept of variable retention. At its core, this concept simply means that the focus is on leaving a complex, nonuniform stand after partial tree removal. (Conventional tree harvest focuses more on what is taken than what is left behind.) Remnants include live trees, standing dead trees, and trees lying on the ground, as well as associated understory vegetation and intact patches of forest floor.

Variable retention systems have three basic attributes: level, type, and distribution of tree retention (Franklin et al. 1997). A first step is to identify critical ecological processes and structures, including biological legacies of the forest that predated the existing one. Enough should be retained to provide a bridge with temporary refuge from the young, simple forest one is starting with to the older, more complex forest that will follow. Variable retention is aimed at rebuilding biological complexity. Restoration is accomplished by focusing on the following approaches.

- At all scales, maintaining reserves, inducing disturbance, steadily shifting the landscape mosaic, anticipating dispersal processes, and generally working toward complex forests (Franklin 1993b).
- Retaining biological legacies (older trees, snags, down wood) as part of the conscious management of forests, whether these result from human actions or natural disturbances (e.g., variable retention, windstorms, fire). Maintaining these legacies takes forethought and planning (Franklin et al. 1997).

- Advancing development of spatial diversity and internal ecosystem dynamics by implementing intentional intermediate to small-scale disturbances that mimic natural ones (Carey 2003a).
- Keeping appropriate levels of decadence in the forest (live trees with decay, snags, and fallen trees) as key components of biological diversity (Carey et al. 1999a).

Figure 5.4 shows how an area previously nearly uniform in forest structure might be made more diverse and put on a path toward greater biological complexity by using variable retention to mimic a small-scale natural disturbance, such as a lightning strike or windstorm. Note how trees left behind are in clusters, clumps, peninsulas, and randomly spaced individuals. The white lines indicate plot boundaries, with different approaches in each area. The upper left is heavily thinned, with perhaps only 15% of the trees left standing. In the upper right, about 30% are left. The two lower quadrants cluster all the remaining trees in a few clumps, an approach loggers favor because it decreases operational costs and increases safety. The one at lower right includes a peninsula of forest attached to the adjacent stand.

Designing the pattern of legacies is important with respect to meeting restoration goals but may also have to be integrated with forest economics (Walters 1986). In addition to the opportunity costs of leaving trees behind, there are also increased operational costs, as noted earlier. Ideally, legacies are selected to maximize habitat value while minimizing operational costs and safety risks (Toumey and Korstian 1937, Polansky et al. 2001). Creating alternative designs can help the planner evaluate varying patterns with respect to the complex interaction of terrain, silviculture, ecological systems, and technical limitations (Berg et al. 1996, 1998). If the existing stand is uniform, then the placement of retained clumps is very flexible, and leave islands can be located to optimize operation efficiency.

Generally, the key decisions include which of the larger, canopy-dominant trees, snags and cavity trees, seedlings, downed wood, and understory

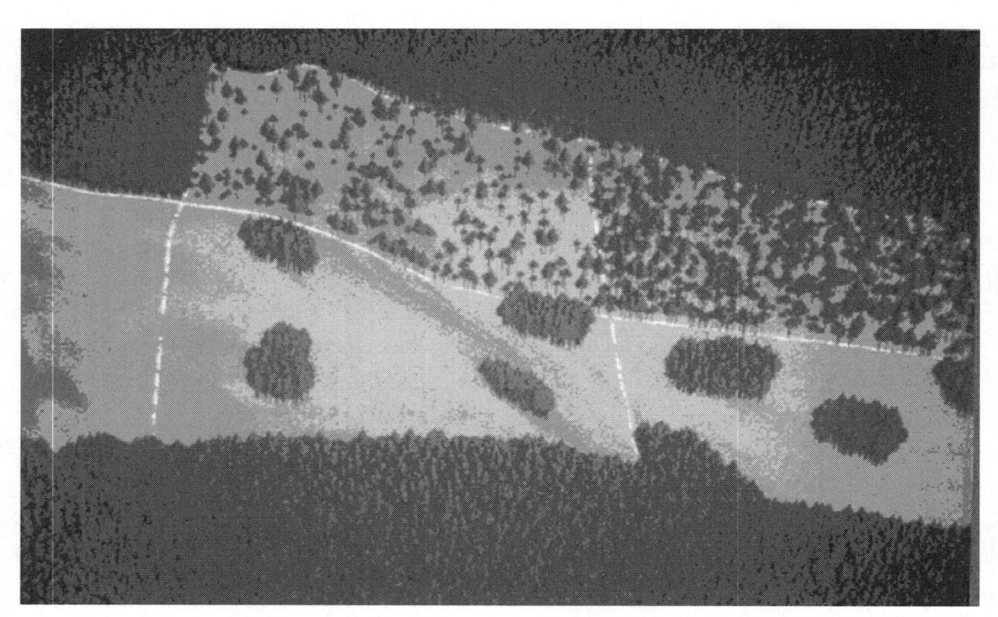

Figure 5.4. Examples of variable retention design demonstrating various levels of dispersed and aggregated retention of trees. In practice a design will call for different retention combinations and levels to maintain small-scale structural diversity over a large area. (*Image by Dean Berg*)

plant clumps to leave undisturbed (Figure 5.5). Existing diversity is an important element to consider (Simpson 1900). For example, if the stand includes only a few hardwoods, the planner may want to retain all of them. As a general rule, it is best to disperse the retained overstory trees and to aggregate reserved clumps around large snags. Clumps have the advantage of maintaining intact patches of an ecosystem (down through the forest floor), with all the biota and maybe with less loss from wind (Scott et al. 1998).

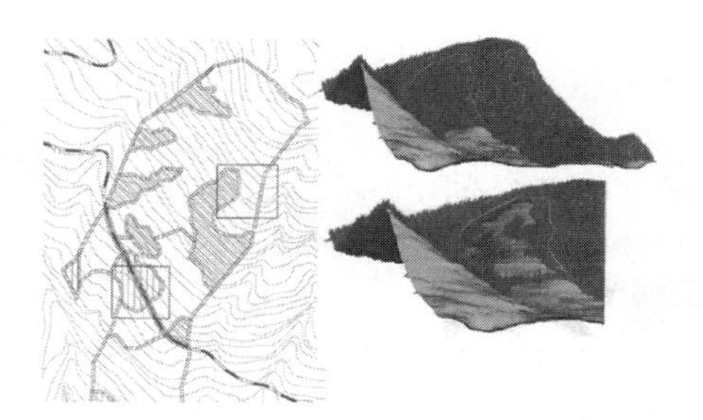

FIGURE 5.5. Plan and perspective views of a timber sale that show retained peninsulas of trees in a harvest unit that tie in with an adjacent unlogged forest. (*Image by Dean Berg*)

Using Digital Information

Choice of logging systems is an important aspect of forest restoration. Ground-based logging, though inexpensive, may disturb low-growing vegetation and soil legacies to a degree that is not appropriate for a restoration project. Digital information helps test design concepts against multiple issues, including aesthetic impacts, operational feasibility, and cost. Aerial views of proposed retained trees and clumps allow comparisons between skyline or ground-based logging systems, location of tower settings, and a number of other variables (Berg et al. 1998). Accurate digital maps that incorporate vegetation cover, topography, existing roads, and other features make planning and design more accurate and faster. Several critical aspects of cable logging systems focus on the interaction of terrain and tree inventory (Twito et al. 1987, McGaughey 1997). Airborne laser imagery (LIDAR) provides stunning details on the landform surface, showing hidden roads, stream centerlines, and landslide remnants, which can be integrated into the design to enhance biological diversity (Reutebuch et al. 2000).

Restoration of a more diverse plant composition can also be accomplished through planting, seeding, or releasing existing plants from competition. Underplanting combined with overstory thinning can restore tree diversity and accelerate canopy layering. Shade-intolerant species such as Douglas fir need canopy gaps of 1–2 hectares to develop successfully. Underplanting with resistant trees in root rot pockets increases resiliency of forests that are heavily thinned. As tree diversity increases, thereby increasing photosynthetic activity, ecosystems allocate additional carbon to seeds and ectomycorrhizal associates.

Thinning and the Economics of Forest Restoration

In typical industrial plantation forestry, after harvest of a native or second growth, stand trees are planted at a density of around 400 per acre. These may be precommercially thinned at 10–15 years of age, leaving perhaps 200–300 trees per acre. At 30 years of age on high-productivity, low-elevation sites, one third of the remaining trees are harvested and sent to the mill. At 35–50 years, all of the 100–200 remaining trees are clearcut, logging slash is cleaned up, and the cycle begins again. At all stages the focus is on maximizing the occupation of space (at the forest floor and in the canopy) by commercially valuable trees, often exclusively Douglas fir. Even in this very intensive approach, forest economics is a dicey proposition for the investor. Long-term land carrying costs can easily exceed profits. There is a continuing risk of drought, fire, catastrophic windstorms, insect outbreaks, theft, and vandalism. Markets can shift quite suddenly. New regulations, such as increased stream buffers, can take significant numbers of trees out of the economic equation for the land owner. New listings of endangered species can also still the chainsaws.

Because it has very different objectives, restoration thinning turns conventional forestry on its head. To the economic forester, decadence is bad, but to the restoration forester it is good. Uniformity is prized by the former and shunned by the latter. Instead of a focus on one or a few tree species, restoration foresters try to make space for many species.

The choice to thin or not to thin revolves around balancing short-term impacts against hoped for long-term benefits, both of which differ from stand to stand and landscape to landscape. Thinning temporarily reduces canopy closure, which changes microclimate and may expose some wildlife to increased predation (Muir et al. 2002, Hagar 1999, Hayes et al. 1997). On the other hand, improving the ability of predators to find food may be part of the point of the restoration action. Although thinning may accelerate development of northern spotted owl nesting habitat, it also temporarily degrades existing dispersal habitat (BLM 2003a, 2003b, Glenn et al. 2004, Anthony et al. 2001).

Thinning may entail some new road construction, which can result in multiple adverse effects, including sedimentation, soil compaction, disruption of hydrology, and new routes for noxious weeds (Jones and Grant 1996, Wemple et al. 1996, Forest Service and BLM 1994, Beschta 1978, Swanson and Dyrness 1975). As a general rule, new restoration access roads should be temporary and put to bed immediately after use is complete unless providing long-term stewardship or recreation access is important. Where roads are intended to remain, they must be designed to allow natural drainage patterns to continue, generally by outsloping. It goes without saying that restoration access roads should not deliver unnatural levels of sediment to streams. In many cases roads should be blocked to prevent unwanted use. Using helicopters for thinning in forest restoration is only rarely economically feasible and often necessitates construction of large landing areas for logs at the nearest road. But if the costs of removal exceed the value of the trees, they can be left as woody debris or scattered as slash. This is common in very young stands where commercial value is low but requires funding to recoup costs and may temporarily increase risks of fire and insect infestation (BLM 2003a, 2003b).

Restoration thinning may reduce the long-term timber volume available for harvest (Zeide 2001). One should expect fewer trees and fewer commercially valuable trees, such as Douglas fir. Some restoration thinning projects increase short-term revenue because more trees are taken than in conventional thinning. Where large, commercially valuable trees are left behind, revenues usually are less than they would have been under a purely commercial operation, but there may be greater economic opportunities for niche products and carbon credits.

Creating Coarse Woody Debris and Snags

Restoration silviculture should always retain existing coarse wood and snags to promote biological complexity (Carey 2000b, Carey et al. 1999a, Berg and Schiess 1996). But most young stands will be deficient in these. Thus restoration treatments may need to be designed to create coarse wood and snags, but how much and from which trees? Studies of small mammals suggest that a 10% cover of coarse woody debris is needed to have healthy forest floor food webs to support small mammal communities (Carey 1995, 2001). Although it is possible to use small logs and snags for some functions, others, such as cavity nesting, need larger wood. To create large snags and down logs one first needs to have large live trees, which by definition may not exist in young stands. Thus initial restoration treatments are limited to creating small logs and snags.

Thinning increases live tree growth rates to create conditions favorable to having large trees that will make future large logs and snags.

Small woody debris created through logging can increase the risk of wildfire and insect infestations. Excessive small fuels can contribute to high fire intensity. Insect populations, most notably that of the Douglas fir bark beetle, increase in response to fresh coarse wood (Ross et al. 2001, Hostetler and Ross 1996). The risk is greater in drier forests. Bark beetle infestations may have restoration value by creating new snags and canopy gaps. But there are too many variables related to bark beetle populations to be able to predict and limit the impacts. For example, bark beetles attracted to an area might damage existing old-growth stands growing nearby, defeating the purpose of the restoration effort.

Where large trees already exist, they can be recruited as snags through artificial wounding, inoculation, or even partial excavation to create cavities and promote decadence to increase habitat quality and diversity (Carey et al. 1999a, Franklin and Forman 1987). Thinning operations incidentally result in some tree wounding. Direct, intentional wounding and creation of snags may be accomplished by a range of techniques, including basal girdling, topping, fungal inoculation, and blasting of tops. Wounding and fungal inoculation are more likely to result in snags with heart rot, particularly important for cavity-nesting species. Basal girdling is far less expensive than other techniques but is unlikely to result in heart rot and may create weakness at the base of the snag, making it likely to snap.

Variations in Old Growth

In planning restorative projects, consideration should be given to the naturally high structural diversity at forests have at multiple scales. The processes and pathways that create old-growth forests differ substantially across the region, depending on elevation, latitude, and distance from the ocean (Franklin et al. 2002, Poage and Tappeiner 2002, Winter 2000, Tappeiner et al. 1997, Spies 1991). Empirical information on structural diversity and recovery targets is essential because local data often are lacking. Local data on existing old-growth stands used as reference sites are particularly important in better understanding the role of hardwoods and shade-tolerant conifers in stand development, understanding typical accumulations of coarse wood and snags, and assembling stand-level targets for restoration. In other words, an old-growth restoration project in a Sitka spruce forest near the coast cannot have exactly the same restoration goals as one in dry Douglas fir in an interior valley or at the 3,000-foot level of the Cascade Mountains. The basic approaches and techniques may remain the same, but the target stand structure has to be different.

Key Challenges in Restoring Old-Growth Coniferous Forest

Old-growth forest restoration takes place in a complex and variable ecological, institutional, and social setting. This variability poses challenges, some unique to old-growth restoration. Although we believe that the potential is very good to restore Pacific Northwest forest ecosystems in ways that continue to produce timber and wealth, restoration must be carefully planned and analyzed to reach this potential.

Land Ownership

Forest ownership patterns complicate planning and execution of restoration, particularly when one considers large-scale landscape patterns and key ecological processes. Many public forests in Oregon and Washington are arranged in a checkerboard with private lands, an unfortunate legacy of historic railroad grants (Figure 5.6). Intensively managed privately owned forests butt up against

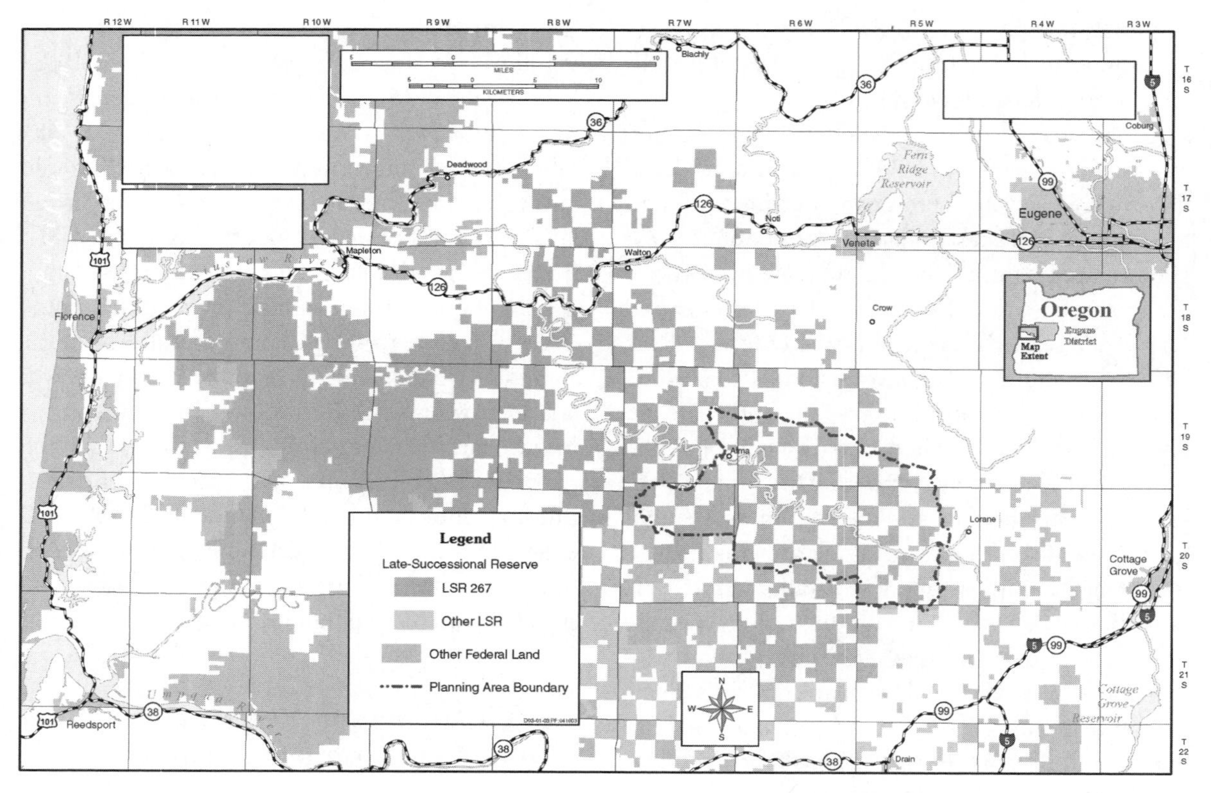

FIGURE 5.6. Eugene area Bureau of Land Management map shows the checkerboard pattern of land ownership that makes forest restoration challenging at the landscape scale. (*Courtesy of the Bureau of Land Management*)

federal forests where restoration is more likely. In some areas, such as the central Washington Cascades and the BLM forests south of Eugene, Oregon, ownership patterns make large-scale restoration problematic if not impossible. In Washington private landowners are participating in restoration planning to a limited extent through multispecies habitat conservation plans under the Timber–Fish agreement. A new, very positive development is the emergence of forest trusts that promote ecological forestry and the growth of the third-party certification movement through the Forest Stewardship Council (FSC). For example, the forestry program at Fort Lewis, Washington, which harvests selectively and with a goal of creating old-growth forest structure on some 40,000 acres near Olympia, has become FSC certified.

Other areas have very large blocks of contiguous public ownership, making landscape-scale old-growth restoration possible. These areas include the Siuslaw National Forest on the central Oregon coast, much of the west slope of the Oregon Cascade Mountains, from the Columbia Gorge south to near Roseburg, the Tillamook and Clatsop state forests in the north Oregon Coast Range, much of the Gifford Pinchot National Forest, north to Mt. Rainier, the north Cascades, and the combined Olympic national forest and parks. In addition, nearly all the potential old-growth forest of British Columbia is in public ownership, although continued heavy economic dependence on old-growth logging makes restoration difficult to initiate there.

Timber Economy

The Forest Service, the BLM, state and local governments, and private landowners own and manage millions of acres of young forests that could be restored to old growth. But there are still limited opportunities to sell small, low-value forest products to recoup restoration costs (Fight and Barbour 2000). Programs to reduce fire risk are a good source of funding in some areas, particularly in the drier forests of southern Oregon. One hopeful note on the horizon is the possible development of forest biomass use for electricity generation. The Warm Springs Indian Nation and other organizations are exploring this concept in central Oregon, which has a huge need for small-diameter tree and brush thinning to relieve fire-starved dryland forests (Preusch 2005). If successful, similar efforts could be made in moist forests west of the mountains. Federal foresters are under a lot of pressure with steadily diminishing budgets, staff, and expertise, all of which limit their ability and inclination to take the initiative on restoration.

In recent years, many mills that could handle large older trees have shut down, so that small to medium-diameter trees now fetch higher prices than large ones. At first glance this may seem a positive development to old-growth conservation advocates because the timber industry now has a diminishing appetite for old trees. Yet it also makes it much harder to convince public and private forest managers to delay harvest and grow older forests as replacements for existing ones. A key argument, higher-value older wood with tight growth rings, has been at least temporarily lost because there are too few mills left to process them.

As was pointed out earlier, managing a young forest with the goal of growing it to old growth has opportunity costs. Trees left behind, turned into snags, or left on the forest floor represent dollar value not realized. Long forest rotations can be quite productive, as demonstrated by models used to plan the Tillamook Forest management in Oregon, but the longer trees stay on the land after reaching economic maturity, the greater the risk to an investor or the school district waiting for trees to become cash and pay teachers' salaries.

The timber economy has driven policy direction and expectations over many years. Towns, jobs, schools, roads, and taxes all have been and to some extent still are dependent on treating forests as commercial enterprises. Revenues from state forests in Oregon and Washington are dedicated to schools. Timber revenues finance more than 50% of the British Columbia provincial government and provide thousands of very high-paying jobs in an otherwise stagnant rural economy. Interestingly, though, the tourist economy has nearly surpassed the timber economy as a revenue and employment generator. If this trend continues it could change old-growth conservation and restoration emphasis.

Overall the institutional challenges of old-growth restoration are perhaps less extreme in the Pacific Northwest than in other regions with substantial federal ownership, such as the interior West and Appalachian Mountains. High growth rates, high timber value, and established infrastructure of roads and markets make restoration thinning more feasible here than elsewhere.

Social and Political Challenges

When the Northwest Forest Plan proposed timber harvest as a restoration tool on federal forests, commentary from environmental groups urged that thinning not be used in late-successional reserves until proven in other areas (FEMAT 1993). Since that time, a broader base of public support has emerged for restoration thinning, even in reserves. Recent restoration thinning projects have been

Case Study: Biological Complexity Restoration at Fort Lewis

In 1991, the Pacific Research Station of the Forest Service, the U.S. Army, and the U.S. Fish and Wildlife Service initiated a landscape-scale experimental restoration program at Fort Lewis, Washington (Carey et al. 1999b). They field tested the use of variable-density thinning to create spatial heterogeneity in 60-year-old Douglas fir–dominated forest. Underplanting was used to restore species diversity. Wounding and inoculating live trees with fungi was a tool for accelerating decadence in the forest canopy. The 60-year-old stands included some that had never been thinned and stands that had been managed with two conventional thins. All had regrown from clearcuts 60 years earlier. Monitoring measured soil organisms, vascular plants, birds, and mammals 3 to 5 years after variable density thinning was conducted, using approaches described previously in this chapter.

Northern spotted owl and other wide-ranging, high-level predators respond to forest conditions from a landscape-scale perspective, so their presence or absence cannot reliably be used to evaluate stand-level prescriptions at small scales (Carey et al. 1992). But a key measure of habitat suitability at the stand level is the presence or absence of owl prey species. Thus abundances of three squirrel species that owls eat were measured and compared with the abundances typically found in complex older forests as reference ecosystems.

A keystone ecosystem complex typical of Pacific Northwest old-growth forests includes northern spotted owl and northern flying squirrel, ectomycorrhizal fungi, and Douglas fir. This ecosystem trio subset was chosen as the target for restoration and analysis (Carey 2000a). The northern flying squirrel is the primary prey of spotted owls, and hypogeous ectomycorrhizal fungal sporocarps (truffles) feed the squirrel. These in turn spread spores and associated microorganisms throughout the forest (Li et al. 1986). The fungi have been shown to help move photosynthetic carbohydrates from trees to the mycorrhizosphere, supporting a vast array of soil organisms (Ingham and Molina 1991). Above ground, the food web expands to include three squirrel species: the northern flying squirrel, Douglas's squirrel, and Townsend's chipmunk. In addition, there are other forest floor small mammals, seeds, fruits, fungal sporocarps, and various trees and shrubs. This complex provides a framework that is both functional and adaptable for use in evaluating forest ecosystem development and change resulting from management actions.

Ecosystem measurements included proportion of total fungi to total bacteria, active fungi to active bacteria, and fungus-feeding nematodes to bacteria-feeding nematodes; biomass of predatory nematodes; biomass and diversity of truffles; diversity of epigeous fungi; and the coverage of fungal mats (Ingham and Molina 1991; Carey et al. 1996).

Soil Food Webs

Both previously thinned and untreated legacy stands had fungus-dominated soils, as evidenced by fungi–bacteria ratios. Variable-density thinning had no effect on total biomass ratios but increased the active biomass ratios to 2.6:1 in both thinned and legacy stands. Truffle standing crop biomass averaged 0.5 kilograms per hectare in both thinned and legacy forest stands. Truffle production was reduced from 18% in control plots to 13% in variably thinned plots.

Understory Plant Communities

Unthinned legacy stands had lower understory diversity than conventionally thinned stands. Community structure varied depending on management history, with thinned stands dominated by aggressive native shrubs and ferns. Thinned stands had greater total understory cover. Three years after variable-density thinning treatments, understory communities recovered, with species richness increasing by 150%. Four invasive nonnative species persisted, with eight natives increasing and seven decreasing in importance.

Small Mammals

Thinned forests had higher biomass of small mammals than unthinned legacy forests. Neither produced mammal communities typical of older natural forests. After variable-density thinning, deer mice, creeping voles, and vagrant shrews all increased in abundance.

Squirrels

Northern flying squirrels were twice as abundant in unthinned legacy as in previously thinned stands. Townsend's chipmunks showed the opposite, preferring the thinned stands. Douglas's squirrels were low in abundance in both. Flying squirrels decreased in abundance after variable-density thinning but recovered to their previous level within 5 years. Chipmunks increased and remained high 5 years afterward.

Wintering Birds

Species richness was higher in thinned than in legacy stands. Richness varied annually but was consistently higher in legacy mosaics than in legacy controls in post–variable density thinned areas after 3 to 5 years. Cavity-excavating birds were present, but low in abundance, in all stands.

Case Study: Upper Siuslaw Late-Successional Reserve Restoration Plan

The Eugene District of the Bureau of Land Management, in cooperation with the U.S. Fish and Wildlife Service, created a landscape-scale restoration program in the Coast Range of Oregon (BLM 2003a, 2003b). This plan encompasses approximately 24,000 acres in the upper Siuslaw River watershed, which totals around 500,000 acres. It takes a comprehensive approach to restoration, integrating three ecosystem components: upland forest stands, riparian and aquatic habitats, and road management. The plan includes commercial and noncommercial thinning, coarse wood and snag recruitment, in-stream restoration, and road decommissioning, all scheduled over a 10-year period.

An environmental impact statement was prepared to analyze effects of different restoration strategies, consider alternatives, and gain feedback from the public. Altogether, six approaches were considered:

- No management (required under the National Environmental Policy Act).
- Restore young forest plantations and repair roads but include no commercial timber harvest.
- Continue the current management approach (conventional thinning).
- Focus on restoring threatened and endangered species.
- Restore forest stand densities rapidly.
- Restore forest structure with multientry and multitrajectory thinning.

The fourth alternative, restoration of threatened and endangered species, became the preferred alternative of the BLM.

Project analysis included computer modeling of forest stand development over a 100-year period in order to compare different restoration approaches. Three characteristics associated with late seral forest structure were chosen (Wykoff et al. 1982):

- The density of large Douglas fir trees
- The density of shade-tolerant conifers (hemlock, cedar)
- Overall variation of tree diameters

Threshold values were developed from local old-growth stand information, using reference sites. According to the model, restoration thinning prescriptions will speed development of complex late-successional forest structure when compared with leaving young plantation stands to grow and develop on their own. Conventional thinning designed to increase marketable tree volume would have little effect on the structural development of stands, although it would make them more ecologically stable. The key tradeoff is between the short-term benefits of maintaining dispersal habitat for owls against developing older, more complex forest structure, much better for owls in the longer term.

Case Study: Monte Carlo Thinning

In this project, the Eugene District of the BLM explored a thinning technique designed to increase the variability of tree spacing using a simple prescription that relies on a simple random number generator for tree selection. Three 1-acre trials were established in a 30-year-old Douglas fir plantation in the Coast Range of Oregon. This stand was initially planted at 600 trees per acre after being clearcut. It was precommercially thinned in 1990, leaving 260 Douglas firs per acre. Density in 2002 in the trial plots was approximately 210 trees per acre, with average stand diameters of 10 inches measured at breast height. Trial plots were dominated by Douglas fir, with a few scattered western hemlock and big-leaf maple.

Prescriptions varied the likelihood of tree selection and the number of trees selected. In all prescriptions, selections were made only among Douglas fir greater than 4 inches and less than 18 inches in diameter. Smaller trees were excluded to avoid trees that were not part of the original planted stand, and larger trees were excluded to retain them as potential future late-successional habitat features (Figure 5.7).

This prescription proved to be quick and easy to implement. In plots as small as 1 acre it resulted in approximately proportional thinning. Variability in tree spacing increased even at a fine scale of around 0.1 acre. Increasing the number of trees selected changed the pattern of spacing and increased variability.

These trials explored a random thinning technique. Implementation of small trials suggests promising topics for more rigorous inquiry. First, mapping pretreatment and posttreatment tree positions in future trials would better quantify the variability in tree spacing resulting from different prescriptions. Second, tracking tree diameter growth rates over time would assess the degree to which variability in tree spacing results in variability in individual tree diameter growth.

enthusiastically supported by environmental groups in the Upper Siuslaw (BLM), Umpqua, and Siuslaw national forests. The Siuslaw has made a clear commitment to old-growth restoration and as a result not only has met commercial harvest targets but has also won praise and support from many environmental organizations.

Some in the timber industry point to the "waste" of wood that results from restoration treatments and express pessimism about future stand conditions (BLM 2003a, 2003b). Wildfire and insect infestation are cited as risks of coarse woody debris accumulations. Loss of access as a result of road decommissioning is sometimes challenged by off-road vehicle users and some local governments.

Ecological Issues

Pacific Northwest forest managers still are at a fairly early stage of designing and implementing old-growth restoration. Nevertheless, old-growth forests are the most thoroughly studied ecosystem in this region. We believe that forest scientists and managers have a very good understanding of old-growth ecology and are knowledgeable enough to initiate restoration. One challenge is the lack a conceptual model for monitoring the effectiveness of restoration treatments, although there are ideas on the table. Because of the brief history of plantation management in the region (barely more than 50 years), we can only speculate about how young

FIGURE 5.7. Photos of the Monte Carlo forest project *(a)* before and *(b)* after restoration thinning. Note the variable spacing of trees and openings. This differs from conventional forest thinning, which would have left uniformly spaced trees behind. *(Photos by Richard A. Hardt)*

managed stands will develop over the long term in the absence of restoration. How can managers be sure that variable thinning will actually speed the development of old-growth characteristics and biological complexity?

Many of the anticipated positive effects of restoration treatments are difficult to prove, particularly when compared with the certainty of short-term impacts. The need to balance long-term restoration gains against short-term impacts requires a careful calculus to determine whether restoration treatments are appropriate. The balance in this choice may be shifted dramatically in some locations by ecological crises, such as declines in northern spotted owl populations or salmon stocks, making the risks of action more difficult to tolerate (Courtney et al. 2004). Clearly there is little value in waiting to begin restoring habitat in the next century if we stand to lose a species in the next few decades. A failure to restore habitat quality over wide areas might cripple recovery efforts. Hard choices must be based on modeling and objective analysis of the expected effects. There can be no certainty because the variables are complex and inexact.

FIGURE 5.8. A classic Northwest old-growth forest used as a reference ecosystem to guide restoration on Mt. Hood National Forest. (*Photo by Pat Greene*)

Summary

This chapter has summarized approaches to restoring biological complexity in young forest stands to accelerate development of characteristics associated with natural old growth. As in other restoration challenges, every stand and every landscape has its own unique history, geography, composition, and attributes. Thus we propose only an approach, not a recipe. Old-growth conifer forests are the signature ecosystem of the Pacific Northwest. They define the biological and social matrix of the entire region. Finding ways to restore "new old growth" is crucial, in part because the old growth we have left is not static. We lose some of it every year. Younger forest plantations may not evolve into old growth on their own, at least not quickly enough to replace these losses. There are many challenges to restoring old growth: institutional, economic, technical, and social. We will learn as we go forward, but go we must if we are to learn at all (Figure 5.8).

Acknowledgments

The ideas presented here are an amalgam of knowledge from foresters and scientists, whose inspiration and sincere concern for the fate of the Pacific Northwest forests prompted this exposition of forestry for the coming centuries. We especially thank professors William Ferrell, Rex Daubenmire, Peter Schiess, David S. Scott, and Chad Oliver of the University of Washington Silviculture Laboratory, Dave DeMoss of the Eugene BLM, and Robert J. McGaughey and the Pacific Northwest Research Station of Seattle for their patience. Also we thank Kathryn Jeanne Young-Berg of Wild@Heart, Regine B. Carey of Holiday Valley Fibers, Phyllis Franklin, and Paola Hardt. The views expressed are solely those of the authors and do not necessarily represent those of BLM or other agencies.

References

Agee, J. A. 1993. *Fire Ecology of the Pacific Northwest*. Island Press, Washington, DC.

Anthony, R. G., M. C. Hansen, K. Swindle, and A. Ellingson. 2001. *Effects of Forest Stand Manipulations on Spotted Owl Home Range and Habitat Use Patterns: A Case Study*. Report to the Oregon Department of Forestry, Salem, Oregon by the Oregon Cooperative Fish and Wildlife Research Unit, Oregon State University, Corvallis.

Aubry, K. B. and P. A. Hall. 1991. Terrestrial amphibian communities in the southern Washington Cascade Range. Pages 327–338 in L. F. Ruggiero, K. B. Aubrey, A. B. Carey, and M. H. Huff (tech. coords.), *Wildlife and Vegetation of Unmanaged Douglas-Fir Forests*. USDA Forest Service Gen. Tech. Rep. PNW-285. Pacific Northwest Research Station, Portland, OR.

Berg, D. R., T. K. Brown, and B. Blessing. 1996. Silvicultural systems design with emphasis on forest canopy. *Northwest Science Special Canopy Symposium Edition* 31–36.

Berg, D. R. and N. C. Clement. 1992. Differences in the diversity of vegetation between mature and old-growth forests in the Cascade Range of the Pacific Northwest, USA. *Northwest Environmental Journal* 8(1): 190–193.

Berg, D. R., S. Hashisaki, and P. Schiess. 1998. Operational timber harvest design with ecological elements of forestry. Pages 23–25 in *Ecosystem Restoration: Turning the Tide. Proceedings 1998 Society of Ecological Restoration Northwest Chapter Symposium*. Society of Ecological Restoration and University of Washington Center for Streamside Studies, Seattle.

Berg, D. R. and P. Schiess. 1996. Setting level design for biological legacies in Clayoquot Sound, British Columbia. In *Clayoquot Symposium CD-ROM*. MacMillan-Bloedel, Ltd., Vancouver, BC.

Beschta, R. 1978. Long-term patterns of sediment production following road construction and logging in the Oregon Coast Range. *Water Resources Research* 14(6): 1011–1016.

Bureau of Land Management. 2003a. *Draft Environmental Impact Statement: Upper Siuslaw Late-Successional Reserve Restoration Plan*. Eugene District Office, Eugene, OR. Available at www.edo.or.blm.gov/planning/lsr/index.htm.

Bureau of Land Management. 2003b. *Environmental Assessment: Upper Umpqua Watershed Plan*. Roseburg District Office, Roseburg, OR. Available at www.or.blm.gov/roseburg/Info/EAs.htm.

Carey, A. B. 1995. Sciurids in Pacific Northwest managed and old-growth forest. *Ecological Applications* 5: 646–661.

Carey, A. B. 1998. Ecological foundations of biodiversity: lessons from natural and managed forests of the Pacific Northwest. *Northwest Science* 72(2): 127–133.

Carey, A. B. 2000a. Ecology of northern flying squirrels: implications for ecosystem management in the Pacific Northwest, USA. Pages 45–67 in R. L. Goldingay and J. S. Scheibe (eds.), *Biology of Gliding Mammals*. Filander Verlag, Fürth, Germany.

Carey, A. B. 2000b. Effects of new forest management strategies on squirrel populations. *Ecological Applications* 10: 248–257.

Carey, A. B. 2001. Experimental manipulation of spatial heterogeneity in Douglas-fir forests: effects on squirrels. *Forest Ecology Management* 152: 13–30.

Carey, A. B. 2003a. Biological complexity and restoration of biodiversity in temperate coniferous forest: inducing spatial heterogeneity with variable-density thinning. *Forestry* 76: 127–136.

Carey, A. B. 2003b. Restoration of landscape function: reserves or active management. *Forestry* 76: 221–230.

Carey, A. B., S. P. Horton, and B. L. Biswell. 1992. Northern spotted owls: influence of prey base and landscape character. *Ecological Monographs* 62: 223–250.

Carey, A. B., J. Kershner, B. Biswell, and L. D. de Toledo. 1999a. Ecological scale and forest development: squirrels, dietary fungi, and vascular plants in managed and unmanaged forests. *Wildlife Monograms* 142: 1–71.

Carey, A. B., D. R. Thysell, and A. W. Brodie. 1999b. *The Forest Ecosystem Study: Background, Rationale, Implementation, Baseline Conditions, and Silvicultural Assessment*. Gen. Tech. Rep. PNW-GTR-457. USDA Forest Service, Portland, OR. Available at www.fs.fed.us/pnw/pubs/gtr_457.pdf.

Carey, A. B., D. R. Thysell, L. Villa, T. Wilson, S. Wilson, J. Trappe, E. Ingham, M. Holmes, and W. Colgan. 1996. Foundations of biodiversity in managed Douglas-fir forests. Pages 68–82 in D. L. Pearson and C. V. Klimas (eds.), *The Role of Restoration in Ecosystem Management*. Society for Ecological Restoration, Madison, WI.

Carey, A. B., T. Wilson, C. C. Maguire, and B. L. Biswell. 1997. Dens of northern flying squirrels in the Pacific Northwest. *Journal of Wildlife Management* 61: 684–699.

Courtney, S. P., J. A. Blakesley, R. E. Bigley, M. L. Cody, J. P. Dumbacher, R. C. Fleischer, A. B. Franklin, J. F. Franklin, R. J. Gutiérrez, J. M. Marzluff, and

L. Sztukowski. 2004. *Scientific Evaluation of the Status of the Northern Spotted Owl*. Sustainable Ecosystems Institute, Portland, OR.

CSE (Centre for the Study of the Environment). 1995. *Status and Future of Salmon in Western Oregon and Northern California*. CSE, Santa Barbara, CA.

FEMAT. 1993. *Forest Ecosystem Management: An Ecological, Economic, and Social Assessment*. Report of the Forest Ecosystem Management Assessment Team. U.S. Government Printing Office, Washington, DC.

Fight, R. and R. Barbour. 2000. The financial challenge of ecosystem management. In *Proceedings of the Management of Fire Maintained Ecosystems Workshop*. Whistler, BC.

Forest Service and Bureau of Land Management. 1994. *Northwest Forest Plan FSEIS: Record of Decision for Amendments to Forest Service and Bureau of Land Management Planning Documents within the Range of the Northern Spotted Owl*. Forest Service and BLM Regional Office, Portland, OR. Available at www.or.blm.gov/ForestPlan/NWFPTitl.htm.

Franklin, J. F. 1992. Scientific basis for new perspectives in forests and streams. Pages 25–72 in R. J. Naiman (ed.), *Watershed Management*. Springer-Verlag, New York.

Franklin, J. F. 1993a. Lessons from old growth. *Journal of Forestry* 91(12): 11–13.

Franklin, J. F. 1993b. Preserving biodiversity: species, ecosystems, or landscapes. *Ecological Applications* 3: 202–205.

Franklin, J. F., D. R. Berg, D. A. Thornburgh, and J. C. Tappeiner. 1997. Alternative silvicultural approaches to timber harvesting: variable retention harvest systems. Pages 111–139 in K. A. Kohm and J. F. Franklin (eds.), *Creating a Forestry for the 21st Century: The Science of Ecosystem Management*. Island Press, Washington, DC.

Franklin, J. F., K. Cromack, W. Denison, A. McKee, C. Maser, J. Sedell, F. Swanson, and G. McKay. 1981. *Ecological Characteristics of Old-Growth Douglas-Fir Forests*. Gen. Tech. Rep. PNW-118. USDA Forest Service, Pacific Northwest Forest and Range Experiment Station, Portland, OR.

Franklin, J. F. and R. T. T. Forman. 1987. Creating landscape patterns by forest cutting: ecological consequences and principles. *Landscape Ecology* 1: 5–18.

Franklin, J. F., L. Norris, D. R. Berg, and G. R. Smith. 1999. The history of DEMO (Demonstration of Ecological Management Operations): an experiment in regeneration harvest of northwestern forest ecosystems. *Northwest Science* Special Issue 73: 3–11.

Franklin, J. F., H. H. Shurgart, and M. E. Harmon. 1987. Tree death as an ecological process. *Biological Science* 37: 550–556.

Franklin, J. F., T. A. Spies, R. Van Pelt, A. B. Carey, D. A. Thornburgh, D. Rae Berg, D. B. Lindenmayer, M. E. Harmon, W. S. Keeton, D. C. Shaw, K. Bible, and J. Chen. 2002. Disturbances and structural development of natural forest ecosystems with silvicultural implications, using Douglas-fir forests as an example. *Forest Ecology and Management* 155: 399–423.

Franklin, J. F. and R. Van Pelt. April/May 2004. Spatial aspects of structural complexity in old-growth forests. *Journal of Forestry* 102(3): 22–28.

Glenn, E. M., M. C. Hansen, and R. G. Anthony. 2004. Spotted owl home-range and habitat use in young forests of western Oregon. *Journal of Wildlife Management* 68(1): 33–50.

Hagar, J. C. 1999. Influence of riparian buffer width on bird assemblages in western Oregon. *Journal of Wildlife Management* 63(2): 484–496.

Hayes, J. P., S. S. Chan, W. H. Emmingham, J. C. Tappeiner II, L. D. Kellogg, and J. D. Bailey. 1997. Wildlife response to thinning in young forests in the Pacific Northwest. *Journal of Forestry* 95(8): 28–33.

Hostetler, B. B. and D. W. Ross. 1996. *Generation of Coarse Woody Debris and Guidelines for Reducing the Risk of Adverse Impacts by Douglas-Fir Bark Beetle*. Unpublished report to the Siuslaw National Forest, on file at Westside Forest Insect & Disease Service Center, Mt. Hood National Forest, Sandy, OR.

Hummel, S., R. Barbour, P. Hessburg, and J. Lemkuhl. *Ecological and Financial Assessment of Late-Successional Reserve Management*. PNW-RN-531. USDA Forest Service, Pacific Northwest Research Station, Portland OR.

Ingham, E. R. and R. Molina. 1991. Interactions among mycorrhizal fungi, rhizosphere organisms, and plants. Pages 169–197 in P. Barbarosa, V. A. Kirsk, and C. G. Jones (eds.), *Microbial Mediation of Plant–Herbivore Interactions*. Wiley, New York.

Jones, J. A. and G. E. Grant. 1996. Peak flow responses to clear-cutting and roads in small and large basins, western Cascades, Oregon. *Water Resources Research* 32(4): 959–974.

Lesher, R. D. and J. A. Henderson. 1992. *Indicator Species of Forested Plant Associations on National Forests of Northwestern Washington*. Field Guide R6-MBS-TP-041-92. USDA Forest Service, Pacific Northwest Region, Portland, OR.

Li, C. Y., C. Maser, Z. Maser, and B. Caldwell. 1986. Role of three rodents in forest nitrogen fixation in western Oregon: another example of mammal–mycorrhizal fungus–tree mutualism. *Great Basin Nature* 46: 411–414.

Lundquist, R. W. and J. M. Mariani. 1991. Nesting habitat and abundance of snag-dependent birds in the southern Washington Cascade Range. Pages 221–240 in L. F. Ruggiero, K. B. Aubrey, A. B. Carey, and M. H. Huff (tech. coords.), *Wildlife and Vegetation of Unmanaged Douglas-Fir Forests*. Gen. Tech. Rep.

PNW-285. USDA Forest Service Pacific Northwest Research Station, Portland, OR.

McGaughey, R. J. 1997. Visualizing forest stand dynamics using the stand visualization system. Pages 248–257 in *Proceedings of the 1997 ACSM/ASPRS Annual Convention and Exposition; April 7–10, 1997*, Vol. 4. American Society for Photogrammetry and Remote Sensing, Seattle, WA and Bethesda, MD.

Muir, P. S., R. L. Mattingly, J. C. Tappeiner II, J. D. Bailey, W. E. Elliott, J. C. Hagar, J. C. Miller, E. B. Peterson, and E. E. Starkey. 2002. *Managing for Biodiversity in Young Douglas-Fir Forests of Western Oregon*. Biological Science Report USGS/BRD/BSR-2002-0006. U.S. Geological Survey, Biological Resources Division, Corvallis.

Neitlich, P. N. and B. McCune. 1997. Hotspots of epiphytic lichen diversity in two young managed forests. *Conservation Biology* 11(1): 172–182.

North, M., J. Trappe, and J. F. Franklin. 1997. Standing crop and animal consumption of fungal sporocarps in Pacific Northwest forests. *Ecology* 78: 1543–1554.

Oliver, C. D., D. R. Berg, D. R. Larsen, and K. L. O'Hara. 1992. Integrating management tools, ecological knowledge, and silviculture. Pages 361–382 in R. J. Naiman (ed.), *Watershed Management: Balancing Sustainability and Environmental Change*. Springer-Verlag, New York.

Oliver, C. D. and B. C. Larson. 1990. *Forest Stand Dynamics*. McGraw-Hill, New York.

Olsen, D. 2001. Ecology and management of montane amphibians of the US Pacific Northwest. *Biota* 21: 55–73.

Pielou, E. C. 1991. *After the Ice Age: The Return of Life to Glaciated North America*. University of Chicago Press, Chicago.

Poage, N. J. 2000. *Structure and Development of Old-Growth Douglas-Fir in Central Western Oregon*. Ph.D. dissertation, Oregon State University, Corvallis.

Poage, N. J. and J. C. Tappeiner II. 2002. Long-term patterns of diameter and basal area growth of old-growth Douglas-fir trees in western Oregon. *Canadian Journal of Forest Research* 32: 1232–1243.

Poage, N. J. and J. C. Tappeiner II. 2005. Tree species and size structure of old-growth Douglas-fir forests in central western Oregon, USA. *Forest Ecology and Management* 204: 329–343.

Polasky, S., J. Camm, and B. Garber-Yonts. 2001. Selecting biological reserves cost effectively: application of terrestrial vertebrate conservation in Oregon. *Land Economics* 77(1): 68–78.

Preusch, M. 2005. Rare alliance favors thinning of forests. *The Sunday Oregonian*, June 26.

Reutebuch, S. E., K. Ahmed, T. Curtis, D. Petermann, M. Wellander, and M. Froslie. 2000. A test of airborne laser mapping under varying forest canopy. In *Proceedings of the ASPRS National Convention*, Washington, DC, May 22–26.

Ross, D. W., B. B. Hostetler, and J. Johansen. 2001. *Douglas-Fir Beetle Response to Artificial Creation of Coarse Woody Debris in the Oregon Coast Range*. Poster presentation. Entomological Society of America, Annual Meeting, San Diego, CA.

Ruchty, A., A. L. Rosso, and B. McCune. 2001. Changes in epiphyte communities as the shrub, *Acer circinatum*, develops and ages. *Bryologist* 104(2): 274–281.

Scott, W., R. Meade, R. Leon, D. Hyink, and R. Miller. 1998. Planting density and tree size relations in coast Douglas-fir. *Canadian Journal of Forestry Research* 28: 74–78.

Sillett, S. C., B. McCune, J. E. Peck, T. R. Rambo, and A. Ruchty. 2000. Dispersal limitations of epiphytic lichens result in species dependent on old-growth forests. *Ecological Applications* 10(3): 789–799.

Simpson, J. 1900. *The New Forestry or the Continental System Adapted to British Woodlands and Game Preservation*. Pawson & Brailsford, Sheffield, England.

Smith, D. M. 1986. *The Practice of Silviculture*. Wiley, New York.

Smith, J., R. Molina, M. Huso, D. Luoma, D. McKay, M. Castellano, T. Lebel, and Y. Valachovic. 2002. Species richness, abundance, and composition of hypogeous and epigeous ectomycorrhizal fungal sporocarps in young, rotation-age, and old-growth stands of Douglas-fir in the Cascade Range of Oregon, USA. *Canadian Journal of Botany* 80: 186–204.

Spies, T. A. 1991. Plant species diversity and occurrence in young, mature, and old-growth Douglas-fir stands in western Oregon and Washington. Pages 111–121 in L. F. Ruggiero, K. B. Aubrey, A. B. Carey, and M. H. Huff (tech. coords.), *Wildlife and Vegetation of Unmanaged Douglas-Fir Forests*. Gen. Tech. Rep. PNW-285. USDA Forest Service Pacific Northwest Research Station, Portland, OR.

Spies, T. A. and J. F. Franklin. 1988. Old growth and forest dynamics in the Douglas-fir region of western Oregon and Washington. *Natural Areas Journal* 8: 190–201.

Spies, T. A. and J. F. Franklin. 1991. The structure of natural young, mature, and old-growth Douglas-fir forests in Oregon and Washington. Pages 91–109 in L. F. Ruggiero, K. B. Aubrey, A. B. Carey, and M. H. Huff (tech. coords.), *Wildlife and Vegetation of Unmanaged Douglas-Fir Forests*. Gen. Tech. Rep. PNW-285. USDA Forest Service Pacific Northwest Research Station, Portland, OR.

Spies, T. and M. Turner. 1999. Dynamic forest mosaics. Pages 95–160 in M. Hunter (ed.), *Maintaining Biodiversity in Forest Ecosystems*. Cambridge University Press, Cambridge, England.

Swanson, F. J. and D. R. Berg. 1991. The ecological roots of new approaches to forestry. *Forest Perspectives* 1(3): 6–8.

Swanson, F. J. and C. T. Dyrness. 1975. Impact of clearcutting and road construction on soil erosion by landslides in the western Cascade Range, Oregon. *Geology* 3: 392–396.

Swanson, F. J. and J. F. Franklin. 1992. New forestry principles from ecosystem analysis of Pacific Northwest forests. *Ecological Applications* 2(3): 262–274.

Tang, S. M., D. R. Berg, J. Greenberg, D. McKenzie, and J. F. Franklin. 1994. *Interactions Between Slope Failure, Economics, and Road Networks on a Small Managed Landscape*. USDA Forest Service Pacific Northwest Research Station, Olympia, WA.

Tappeiner, J. C. II, D. Huffman, D. Marshall, T. A. Spies, and J. D. Bailey. 1997. Density, ages, and growth rates in old-growth and young-growth forests in coastal Oregon. *Canadian Journal of Forest Research* 27: 638–648.

Toumey, J. W. and C. F. Korstian. 1937. *Foundations of Silviculture Upon an Ecological Basis*. Wiley, New York.

Twito, R. H., S. E. Reutebuch, R. J. McGaughey, and C. N. Mann. 1987. *Preliminary Logging Analysis System (PLANS): Overview*. PNW-GTR-199. Pacific Northwest Forest and Range Experiment Station, Portland, OR.

Walters, C. J. 1986. *Adaptive Management of Renewable Resources*. Macmillan, New York.

Walters, C. J. and C. S. Hollings. 1990. Large scale management experiments and learning by doing. *Ecology* 71(6): 2060–2068.

Wemple, B. C., J. A. Jones, and G. E. Grant. 1996. Channel network extension by logging roads in two basins, western Cascades, Oregon. *Water Resources Bulletin* 32(6): 1195–1207.

Wilson, J. S. and C. D. Oliver. 2000. Stability and density management in Douglas-fir plantations. *Canadian Journal of Forest Research* 30: 910–920.

Wilson, S. M. and A. B. Carey. 2000. Legacy retention versus thinning: influences on small mammals. *Northwest Science* 74: 131–145.

Wimberly, M. C., T. A. Spies, C. J. Long, and C. Whitlock. 2000. Simulating historical variability in the amount of old forests in the Oregon Coast Range. *Conservation Biology* 14(1): 167–180.

Winter, L. E. 2000. *Five Centuries of Structural Development in an Old-Growth Douglas-Fir Stand in the Pacific Northwest: A Reconstruction from Tree-Ring Records*. Ph.D. dissertation, University of Washington, Seattle.

Wykoff, W. R., N. L. Crookston, and A. R. Stage. 1982. *User's Guide to the Stand Prognosis Model*. Gen. Tech. Rep. INT-133. USDA Forest Service, Intermountain Forest and Range Experiment Station, Ogden, UT.

Zeide, B. 2001. Thinning and growth. *Journal of Forestry* 99(1): 20–25.

Chapter 6

Riparian Woodlands

DEAN APOSTOL AND DEAN RAE BERG

The focus of this chapter is restoration of riparian forests and woodlands located west of the Cascade and Siskiyou mountains in Oregon and Washington and west of the coastal mountains in British Columbia and southeast Alaska.

Ecological Characterization

To understand riparian woodland restoration, we begin with a description of the variability and distribution of natural riparian plant communities. Many watershed councils and agencies focus on the obvious point that restoration of riparian woodlands involves planting trees along stream banks. But a deeper look reveals that natural riparian plant communities are among the most diverse, dynamic, and complex ecosystems in our region.

Pacific Northwest river systems and associated riparian vegetation took shape after the retreat of the last glacial ice sheet approximately 14,000 years ago. Gradually the land warmed and conifer forests developed as the dominant vegetation type, with deciduous trees occupying a lesser role. Present composition and distribution took root approximately 5,000–7,000 years ago (Benda et al. 1992). As channels migrated across lowland valleys, rich soils and complex forests developed. Braided stream channels with forested islands formed in association with large log jams and floods.

Riparian communities represent highly variable ecotones between terrestrial and aquatic ecosystems (Gregory et al. 1991). The transition zone from riparian to terrestrial is gradual. Forty-three separate riparian plant communities have been classified in the Mt. Hood and Gifford Pinchot national forests in Oregon, varying according to a number of physical factors, including timing, intensity, frequency, and duration of high flows, character of substrate, water table, and landforms (Diaz 1996). The duration of time since the last high flow is a critical variable that determines community composition and distribution. One implication is that riparian restoration should be process oriented in that floods ultimately will influence how plant communities develop (Figure 6.1).

Individual streamside landforms often have their own distinct plant communities. Small landforms such as gravel bars have micro–plant communities in a matrix of more uniform vegetation occurring on a stable river terrace. Intensity of flooding determines succession patterns. A highly variable landscape grain results in variable communities. Pioneer species such as red alder, cottonwood, and horsetail, are more common in riparian areas than in adjacent uplands because of frequent disturbances (Diaz 1996).

Franklin and Dyrness (1973) identified many variations of riparian communities west of the Cascade Mountains. Hardwood communities are typ-

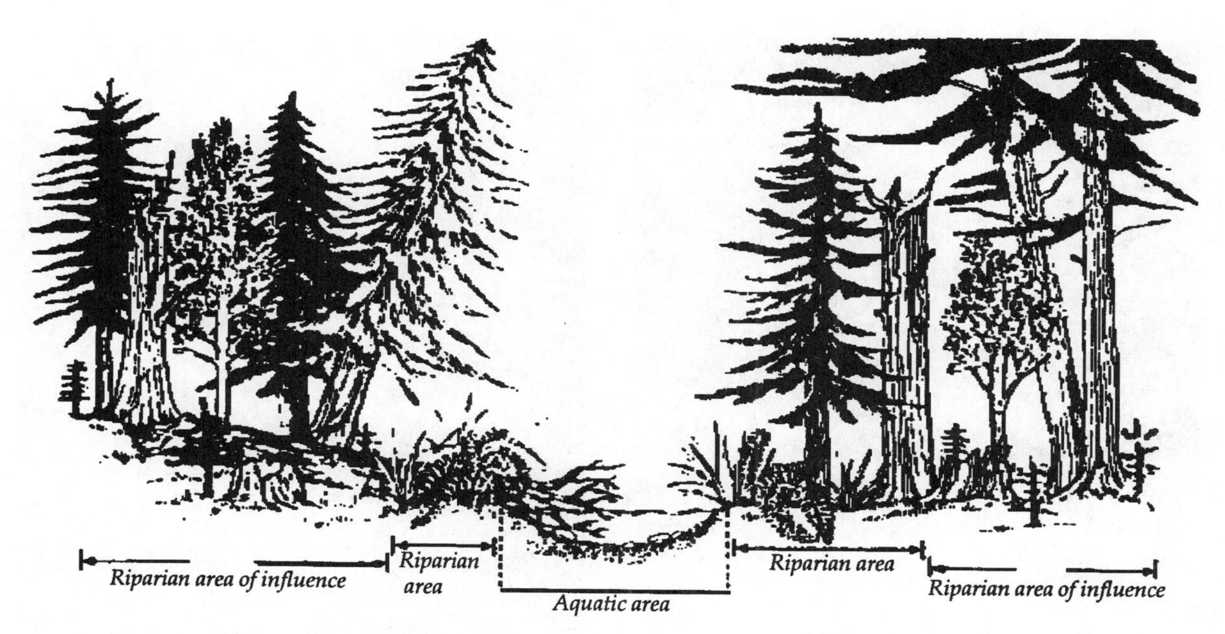

FIGURE 6.1. Idealized riparian vegetation cross-section. (*Courtesy of Washington Department of Fish and Wildlife*)

ical in the lower Willamette Valley and along the Columbia River. These are often dominated by black cottonwood (*Populous trichocarpa*), which forms nearly pure stands on islands and riverbanks. A number of willow (*Salix*) species grow alongside or within cottonwood stands, including rigid (*S. rigida*), red (*S. lasiandra*), river (*S. fluviatilis*), soft-leaved (*S. sessilifolia*), and Scouler's (*S. scouleriana*).

Oregon ash (*Fraxinus latifolia*) forms pure stands on seasonally flooded wetlands along rivers in lower valleys but also occurs in mountain areas (Figure 6.2). Understories of ash stands often are barren where recent silt deposits are found but can also have sedge or dense shrub networks. Big-leaf maple (*Acer macrophyllum*) is common in valley riparian areas, as is red alder. Oregon white oak (*Quercus garryana*) grows alongside ash in some areas, such as the Finley Wildlife Refuge in the Willamette Valley and on Sauvie Island. Ponderosa pine (*Pinus ponderosa*) is a surprising riparian woodland component in a few areas, and California myrtle (*Umbelaria californica*) is a riparian tree in the Umpqua Valley of southern Oregon.

Elevation, latitude, and nearness to the coast all result in conifers gaining dominance in riparian areas (Figure 6.3). Along small or intermittent headwater streams the riparian forest is indistinguishable from the adjacent upland. Sitka spruce (*Picea sitchensis*) is an important species along many coastal streams, as is California redwood (*Sequoia sempervirons*) in the southwestern corner of our region.

Western redcedar (*Thuja plicata*) is considered a keystone riparian species in the Cascade and Coast mountain ranges primarily because of its importance as large, long-lived wood in streams. Redcedar grows best in moist, productive sites several feet above the high water line, usually on deep soils along stable terraces. Much current restoration is aimed at reestablishing redcedar where it has been lost to logging and agriculture.

Riparian communities form an ecological continuum from the headwaters of a stream to the

Figure 6.2. A Willamette Valley gallery forest, dominated by hardwood trees. (*U.S. Fish and Wildlife Service*)

mouth and ultimately to an estuary (Vannote et al. 1980). This allows streams to serve as connecting corridors for vegetation and wildlife across large areas. Although wetlands are found in riparian areas, most communities are not classified as wetlands (Johnson and O'Neil 2001). Wetlands are more common in upper headwaters of mountain streams and in lower, unconfined valleys with frequent flooding and high water tables. Skunk cabbage (*Lysichitum americanum*) wetlands are common along low-gradient coastal streams.

Figure 6.3. An upland conifer riparian forest in Oregon Cascades. Note dominance by mature conifer trees. (*Photo by Dean Apostol*)

Riparian Ecological Functions

Most ecologists focus on the set of ecological functions that riparian vegetation performs as a consequence of its unique landscape position straddling aquatic and terrestrial environments (Triska et al. 1982, Naiman and Sedell 1980). These functions vary spatially and shift over time and can be grouped into five broad categories:

- Mediation of solar energy
- Provision of nutrients
- Filtering of sediment and pollutants
- Provision of large wood
- Provision of terrestrial habitat

Mediation of Solar Energy

A key and obvious function of riparian vegetation is to shade streams and prevent heating of surface water. Conventional restoration wisdom says that more shade is better, yet direct sunlight can have a beneficial role in stream production. For example, there is a consistent increase in density and biomass in open canopy conditions, which can provide food for salmon and trout (Hawkins et al. 1983).

If solar input has positive effects, why should restorationists be concerned about lack of shade?

First, cumulative streamside forest removal may increase stream temperatures to a point where competition from species adapted to warmer conditions can overwhelm more desirable salmon. Second, higher temperatures often result in lower dissolved oxygen levels, also harmful to aquatic wildlife. Third, salmon and trout need cool stream temperatures to survive.

Light naturally increases in streams as channels widen until a point is reached where depth and turbidity result in decreased light penetration relative to overall water volume (Minshall 1978, Naiman and Sedell 1980). Clear, shallow streams in upper watersheds often begin in narrow, confined reaches with heavy shade from adjacent tall conifers. As streams reach lower, unconfined valleys, they widen out, and sunlight easily penetrates to the stream bottom, except in areas of deep, murky water. Light penetration increases again in alluvial fans and estuarine areas, where the water usually is shallow (Vannote et al. 1980, Triska et al. 1982). The overall result is that stream temperatures tend to increase with stream order, related to elevation, ambient air temperature, and stream size. The combination of environmental changes—stream size, light interception, temperature moderation, in-stream photosynthesis, and organic inputs—causes changes in aquatic community structure along the way. In lower reaches, riparian communities are not able to have much of an effect on water temperatures, although this limited effect may still be locally important.

Smaller streams usually are more effectively shaded by riparian vegetation, and it is here that shade can have a large effect on overall stream heating. Shading of small streams is most important in summer. Water temperatures in lower reaches are largely a function of the combined temperatures of smaller streams. Generally the net effect is based on the relationship of temperature and volume but can also be influenced by groundwater and hyporrheic (subsurface) flow, channel conditions, and larger climatic factors (Beschta et al. 1987). In large networks, more than 85% of the total stream length may be in small first- and second-order creeks (Johnson and O'Neil 2001).

Removal of vegetation along upper stream reaches results in increasing stream temperatures primarily as a result of solar energy directly striking the water's surface (Holtby 1988, Murphy et al. 1981). Shade does not lower stream temperatures but rather prevents increase by blocking solar input. Once water temperature has increased, it may gradually moderate to reflect ambient air temperature if downstream riparian vegetation provides sufficient shade. Temperatures in moist coastal streams appear to recover more rapidly than those in more arid ecosystems, and streams at lower elevations recover more rapidly than those at higher elevations (Beschta et al. 1987).

There are a number of unknowns with regard to understanding the role of riparian vegetation in moderating solar energy. We do not know much yet about the relationship between channel orientation and temperature. The geometric relationship between tree height, solar angle, streamside slope, and air temperature probably has a strong influence on stream temperatures, but we do not have a reliable means of measurement. The precise influence of solar angle and canopy density on solar radiation is poorly understood. We lack a landscape-level understanding of how the magnitude of changes in stream temperatures in overall networks adds up. Thus it is difficult to understand exactly where shade is most important in a given watershed and how much overall shade is necessary to maintain a target temperature. However, a general principle is that the focus should be on preserving and restoring shade along upper-watershed, small, perennial streams in order to avoid initial heating.

Provision of Organic Material and Nutrients

Riparian vegetation is a source of organic material to streams and a means of nutrient retention. Deciduous trees provide leaves, branches, and twigs, all of which provide structure, substrate, and nutrients (Bilby and Likens 1980, Bilby 1981, Harmon et al. 1986). Most leaf litter arrives in late summer and fall and is transformed into food by warm water temperatures (Suberkropp 1998). Conifer needles and twigs decompose more slowly, providing a less nutritious but longer-lasting source of energy (Cummins 1974, Bird and Kaushik 1981, Elliott 1999). In headwater streams, riparian vegetation may provide 90% of total organic matter (Triska et al. 1982, Johnson and O'Neil 2001). Herbaceous and shrub leaf litter can also provide significant amounts of organics to streams. Fallen insects from riparian vegetation can contribute significantly to the diet of trout and juvenile salmon. Organic material also enters the stream system after floods, which move material from the land surface to the water.

A key ecological component of healthy streams is large wood, most of which originates in riparian areas. Large wood helps retain leaves and other organics by anchoring debris dams and pools, thus preventing them from being flushed out of the channel (Boling et al. 1975, Sedell et al. 1975, Bilby and Likens 1980, Cederholm and Peterson 1985). Nutrient supplies that accumulate over summer and fall can decrease significantly in winter, particularly in streams that lack debris dams (Bilby and Bisson 1992).

Nitrogen and phosphorus often are limiting factors in productivity for both aquatic and terrestrial species (Bilby and Bisson 1992, Bisson et al. 1982, Naiman et al. 1992). Riparian vegetation produces, accumulates, and incorporates essential elements such as nitrogen, sulfur, and phosphorus. Red alder fixes nitrogen from the atmosphere to the soil and contributes nitrogen to streams through leaf litter.

Filtering of Sediment and Pollutants

Riparian vegetation is important in stabilizing stream banks and filtering sediments. Like solar radiation, all sediment is not bad. In the Oregon Coast Range, for example, debris torrents originating in steep, concave headwalls deliver large amounts of sediment and organic material to streams on a regular basis. But coastal streams have the ability to move, store, and use up this material over time. Charles Dewberry (Ecotrust 2002) coined the term "aquatic digestive process" for the time it takes for the stream system to use up sediment and organics until the next delivery. Rather than being in a constant balance, stream systems are adapted to periodic pulses of large amounts of sediment and organic material, followed by quiet periods in which they can absorb, transport, and adjust. It is when sediment loads become chronic, the timing and magnitude increase, or the stream lacks enough complex structure to handle the sediment that it becomes a long-term problem.

In mountain stream systems, riparian forests growing in concave-shaped headwaters play a role in reducing the frequency of debris torrents and providing an important source of large wood when torrents occur. In forested valleys, riparian vegetation helps filter sediment that originates from logging or roads, although roadside ditches may bypass vegetation and deposit sediment directly into streams. Riparian vegetation moderates how materials enter and move through stream systems by filtering overland sediments, protecting stream banks from erosion, and dissipating stream energy.

In lower-gradient alluvial floodplain areas where agriculture and cities are dominant land uses, excessive nutrients and chemical pollutants attach themselves to sediment and negatively affect aquatic wildlife. Healthy riparian vegetation helps to stabilize sediment and inhibit erosion while also metabolizing nutrients and some chemicals. Rich duff and herbaceous layers help provide avenues for stormwater to infiltrate the ground, where it can be stored and released more slowly to streams, thus lessening velocities and preventing erosion. Riparian plants take nutrients up into their systems, accumulate them, and use them for growth. Riparian woody species are can sequester significant amounts of nutrients (Johnson and O'Neil 2001).

Provision of Large Wood

Some of the trees growing in riparian areas eventually make their way into streams, where they provide key structural elements that create and enhance aquatic habitat (Murphy and Koski 1989, Lienkaemper and Swanson 1987). The many positive roles that large wood plays in riparian and aquatic ecosystems are the reason so much attention is being given to finding ways to regrow not just trees but *large* trees along streams (1–2 meters in diameter). Large wood dissipates erosive energy, stores sediment and nutrients, and forms critical habitat features (Johnson and O'Neil 2001). Diverse wood sizes and shapes (e.g., branches, trunks, root wads) create debris dams and pools, log jams, and bank features that support a range of salmonid life stages. Large wood helps a stream create and maintain pools, riffles, eddies, side channels, meanders, and areas of habitat cover. Together, these features represent habitat complexity and are critical for trout and salmon at all life stages. In addition to creating habitat, large wood controls the way water flows through streams, deposits sediment, dissipates overall water energy, protects stream banks, and stabilizes stream beds by preventing downcutting (Beechie and Sibley 1997). Large wood is clearly important to the geomorphology of stream networks from headwaters to estuaries.

Old-growth forest stands from pre-European settlement times were the original source of large wood in streams, both before and after early logging. Loss of large wood as a consequence of agriculture, urbanization, logging, and stream cleaning have caused changes fluvial action (e.g., bank erosion, debris torrents) and may influence streamside slope failures, floodplains, and terrace development (McKee et al. 1982). Legacy wood, the residual wood that is embedded in the channel from decades or centuries past, is to this day a principal source of large wood in many Northwest stream channels. Second-growth forests generally have not matured to the point where they can deliver wood of sufficient size to moderate stream geomorphology, and continued short-rotation forestry holds little prospect for future production of large wood (Grette 1985, Andrus et al. 1988).

Large wood gradually decays and disintegrates or may be transported out of streams during floods. The persistence, mobility, and aquatic functioning of large wood vary along the length of a river with variations in stream dynamics (Bilby and Likens 1989). Physical abrasion, species characteristics, exposure, and ambient temperature all influence the rate of decay (Harmon et al. 1986, Beechie et al. 2000). Old-growth conifer logs can remain in streams for more than 200 years (Johnson and O'Neil 2001).

Large conifers provide the most desirable structural elements because of their size and slow decomposition rates. Hardwoods tend to be smaller and shorter-lived and decompose more quickly than conifers. Large wood enters streams from random tree fall and through infrequent disturbances (e.g., windstorms, fires, landslides, and stream bank erosion). Once it enters the stream, large wood may immediately begin to affect channel structure or may take some time to settle in and get to work. Montgomery and Buffington (1995) noted that less than one fourth of total in-channel large wood contributes to pool formation. Thus, one should not expect perfect efficiency from the mere presence of large wood in a stream.

An important question about riparian forests is the distance from which large wood found in streams originates. Studies of old-growth forests concluded that more than 70% of large wood in streams originated within 20 meters of the channel, and 11% originated within 1 meter (Murphy and Koski 1989). Of the large wood pieces in mature stands, 83% of the large hardwood originated within 10 meters of the stream channel, compared with 53% of conifer wood. Under natural conditions in mature stands, nearly 50% of large wood may be recruited from within a 20-meter-wide forested strip (McDade et al. 1990). The Northwest Forest Plan uses site potential tree height to determine riparian buffers, in part under the belief that any tall tree in this area could end up at least partly in the stream (FEMAT 1993).

Provision of Terrestrial Habitat

In addition to their importance to aquatic ecosystems, riparian plant communities provide a disproportionate share of habitat for reptiles and amphibians (herpetiles), birds, and mammals. The following information, except where noted, is drawn from *Wildlife Habitat Relationships in Oregon and Washington* (Johnson and O'Neil 2001). This excellent book provides a comprehensive summary of wildlife habitat across a range of Northwest ecosystems. Herpetiles are good indicators of the overall health of riparian and aquatic systems. They are quite sensitive to pollution and are less able to adapt to changes in vegetation cover and habitat. Herpetiles are also important as prey species for birds and mammals, so decline in their abundance can affect other wildlife. Amphibians are long lived and can have important influences on nutrient cycling by tying up nutrients in their bodies. High densities of tadpoles can help regulate the spread of algae and aquatic plants. Ninety-one percent of western Oregon and Washington herpetile species use riparian areas for breeding. Many species also use riparian areas for foraging, winter cover, and migration corridors. The high density of insects and other wildlife commonly found in riparian areas makes them excellent places for herpetiles to forage. Nearly all amphibians need aquatic habitat for at least part of their life cycle. Large wood is important not only in streams but also in the riparian woodlands themselves. Wood on the ground provides habitat for birds, amphibians, reptiles, and small mammals as well as nurse logs.

Riparian habitats provide breeding habitat for more bird species than any other vegetation type in North America. Diversity and abundance of birds are disproportionately high in riparian and wetland ecosystems all across the western United States. All twenty-three waterfowl species that breed in the western United States do so in riparian and wetland habitats, as do all fourteen species of wading birds. Neotropical migrants are critically dependent on riparian vegetation, with between 60–85% breeding primarily in woody, deciduous areas. They also use riparian woodlands at very high rates during migration. Species richness can be as much as fourteen times higher in riparian than in nearby nonriparian habitats. Twenty-nine of forty-seven nongame migratory birds listed as species of concern in the Pacific Northwest breed in riparian habitats. Overall, 266 of 367 bird species in Oregon and Washington are known to use freshwater, riparian, and wetland habitats.

What is it about riparian habitats that appeals to our avian friends? First, high levels of plant and insect productivity diminish competition for food and reduce the time and energy costs associated with finding it. This in turn allows contraction of territories, which has the effect of allowing ecologically similar species to share a given area. Migrating birds use riparian areas as stopover sites in part because it is easy to find food there and replenish energy reserves.

The vegetation diversity and habitat complexity associated with natural riparian areas also provide a wide range of potential nest sites. Cottonwood snags provide cavities for primary and secondary nesters. Stream banks have bare banks and overhanging ledges for burrowing species. Wet meadows and shrubby areas make room for birds that nest in open areas.

Riparian vegetation is likewise disproportionately important for mammals. This is again because high levels of structural diversity, high productivity, connectivity, and high edge to interior ratios provide ample edge habitat. Twenty-seven threatened, endangered, or special interest mammals use riparian habitat in Oregon and Washington. All regional furbearers and most big game animals relay on riparian areas for at least part of their habitat needs. Of the 147 total mammal species in these two states, 95 use riparian areas. Riparian areas have greater mammal species richness than upland areas and provide some habitat for mammals that are not found elsewhere. Mammals use riparian areas for food, shelter, and water and as movement corridors that facilitate dispersal. Fifty percent of mammals that use Northwest riparian areas do so for breeding and feeding. Bats are heavy users because of the abundance of insects. Several vole, mouse, and shrew species are very closely associated with riparian habitats. Predators including minks, river otters, and raccoons are found most often in these areas as well.

Beavers are a keystone riparian species that were nearly eliminated from the landscape by overharvest in the nineteenth century. They are only beginning to recolonize many stream systems in this region. Current beaver populations are still only a fraction of what they were estimated to have been before Euro-American settlement (Lichatowich 1999). We know that when beaver populations become abundant, they alter the structure of riparian areas greatly by creating wetlands, trapping sediment, modifying hydrology, altering nutrient and decomposition cycles, and modifying plant community composition. Recent research shows that juvenile coho salmon favor beaver ponds for summer rearing (Pess et al. 1999). In one small urban watershed in Oregon City, beaver dams probably have saved downstream salmonid habitat from the upstream effects of excessive impervious surface (Apostol and Sinclair 2004). Beaver dams can reduce settlement loads in downstream areas by as much as 90% (Johnson and O'Neil 2001). Up to one half mile of stream channel with adjacent deciduous woodland within 100 meters (willow and alder preferred) may be needed to sustain a beaver colony. Because beavers haul food to their dens and eat only a small portion of what they cut, they end up contributing a large amount of small wood to the aquatic ecosystem. Portions of the stems they cut may embed in banks, take root, and then generate new stands of willow, poplar, aspen, or alder. Willow-dominated habitats, which are common near beaver ponds, have small mammal densities three times higher than in nearby, nonwillow habitats (Johnson and O'Neil 2001).

Historic Changes to Riparian Woodlands

Northwest Indians built their villages along rivers and estuaries, clearing streamside vegetation to make space, build homes, gather firewood, and craft canoes and many other implements. Western redcedar was the tree of life for most people of this region. It was the wood of choice for canoes, paddles, homes, partitions, shelves, carved poles and other art objects, storage chests, bowls, drums, drying racks, masks, musical instruments, and numerous tools. Cedar bark was used for basketry, clothing, and diapers. The tree had strong spiritual significance (Stewart 1984). Most harvest of cedar materials was done in ways that left trees standing. Northwest Indians split planks and gathered bark from standing, live trees. Some trees were cut down to make canoes or to provide larger timbers and boards for homes. Stands of cedar that stood near navigable water were highly prized, carefully guarded, and sustainably managed.

Northwest Indians also used many other riparian trees and plants. Sitka spruce, Pacific yew (*Taxus brevifolia*), and grand fir (*Abies grandis*) were all used for a variety of purposes. Next to cedar, red alder was the most widely used tree for woodworking and was of course used for smoking salmon. Young shoots of cottonwood and willow were used for basketry and sweat lodge construction, among other purposes (Turner 1992).

Indian people of the Northwest widely used fire as a tool for managing vegetation. Interior valleys were burned nearly every year. This practice must have had an impact on riparian vegetation, particularly along small, intermittent streams in uplands.

In some areas, such as northwestern British Columbia, riparian floodplain meadows were burned regularly to maintain rice root bulb production. In the absence of burning, these areas quickly are invaded by cottonwood and willow. On Vancouver Island and in the Bella Coola Valley, areas were burned to maintain and stimulate berry patches (Boyd 2000). It is not clear that these patches were in what we now consider to be riparian zones, but given the local narrow valley profiles it is highly likely.

Clearly, riparian zones were not considered to be no-touch zones, but the total impact Indian people had on riparian areas is unclear. We know that larger streams throughout the Northwest, such as the Willamette, had extensive forested riparian areas at the time of Euro-American settlement (Hulse et al. 2000). The presence of large wood and log jams in streams were widely reported, indicating that despite Indian clearing and burning there was an ample wood supply. One log jam on the North Fork of the Siuslaw River even had old-growth trees growing over the top of it (Scholfield 1853).

Euro-American Settlement

Riparian areas changed beginning with mass trapping of beaver and the rapid population decline of Indian people as a consequence of introduced diseases in the early nineteenth century. These events were followed by clearing for agriculture, dredging for navigation, logging, dam building, and other development actions as the Northwest was settled and domesticated by Euro-Americans. The original extent of Northwest riparian woodlands is unknown. The state of Oregon estimates that 15% of the entire state land area is classified as riparian, or within 100 meters of perennial streams and rivers. If intermittent streams are included, this number increases significantly. We have a 150-year history of modifying streams and riparian areas to "improve" them for production of food and fiber. Modifications include removing natural vegetation, clearing snags, deepening channels, constructing dikes and levees, riprapping banks, and building dams. In urban areas and on some farmland, streams have been placed in pipes, with the land above them completely developed (Oregon Progress Board 2000). Although we were unable to locate any comprehensive study on the full extent of regional riparian loss, a few examples from selected studies illustrate the scope of the problem.

The Willamette River has lost 80% of the riparian vegetation that it had at the time of Euro-American settlement and one half of its stream channel complexity (Hulse et al. 2000) (Figures 6.4 and 6.5). Nonnative plants make up more than half of the total vegetation in the riparian zone of the Willamette River mainstem (Oregon Progress Board 2000). As it courses through the City of Portland, nearly 83% of the total riverbank has been classified as unnatural, meaning that the rest is fill, structure, wall, or riprap. The seminatural banks that remain are disturbed or degraded, with only a small portion having mostly native vegetation (City of Portland 2001).

Studies of riparian vegetation in the Oregon Cascades, Willamette Valley, Coast Range, and Klamath Mountains by various researchers demonstrate that early-succession forest dominates private forestland riparian areas (Oregon Progress Board 2000). Although conifer old growth is found in 20% of the riparian zone in the Cascade Mountains, which are mostly in federal ownership, it covers only 3% in the Oregon Coast Range, which is primarily in private hands. In the 500,000-acre Siuslaw watershed, 36% of the total riparian area (200 feet on each side of streams) is classified as mature forest (more than 80 years old), 38% is either young conifer or hardwoods, and 26% is essentially treeless (Ecotrust 2002). In Johnson Creek watershed, the largest remaining open creek in the Portland area, only 5% of active stream segments have forested riparian vegetation that extends beyond the immediate stream channel edge. Thirty-two

FIGURE 6.4. Computer simulation that compares Willamette River riparian vegetation near Eugene, Oregon in 1851 *(top)* and 1990 *(bottom)*. (*By David Diethelm, courtesy of Oregon State University Press, from the Willamette River Basin Atlas, Hulse et al. 2000*)

percent contain little or no riparian vegetation (Portland–Multnomah Progress Board 2001). In British Columbia, a status report on riparian restoration suggests that it is reasonable to assume that every accessible river valley has been logged and may be in need of at least some restoration (Poulin et al. 2000).

These examples are repeated across our region. Generally, the intensity of development lessens as one moves north, particularly with relation to agriculture and city building. In the north, logging and road construction have been the main causes of riparian disturbance, and the overall impacts are lower than they are in the more highly developed parts of our region. Although logging, mining, dam building, agriculture, transportation, and urban development all have altered or destroyed riparian vegetation throughout the Northwest, the latter three have created more permanent changes. Generally, the timber industry has had stricter standards for riparian protection for a longer period than have agriculture and cities, but of late cities have been improving their standards.

Restoring Riparian Areas

Most riparian restoration is aimed at helping aquatic ecosystems recover, in response to the listing of multiple salmon stocks as threatened or endangered. Restoration projects are taking place in all parts of our region, in many communities and land use types. As with nearly all restoration, the activity is far ahead of the scientific or technical knowledge curve, and success is rarely measured beyond immediate implementation.

Figure 6.5. Historic photo of riverbank hardening along the Willamette River. (*Courtesy of Oregon State University Press, from the Willamette River Basin Atlas, Hulse et al. 2000*)

For the purposes of this chapter, we have grouped riparian restoration into three categories:

- Recreating woody riparian vegetation in agricultural floodplains
- Using riparian silviculture to restore formerly logged areas
- Restoring highly disturbed or modified urban riparian areas

Agricultural Floodplains

Agricultural floodplains were historically a patchwork of forest types and plant communities, with a diverse set of soil and hydrologic conditions. Removal of preexisting forest cover in many cases has been accompanied by installation of tile drainage systems. Many low-gradient floodplains are also protected from flood damage by dikes and levees. Networks of smaller streams and seasonal swales throughout lowland floodplain areas have been deepened, straightened, and in some cases placed in culverts. Other riparian lowland areas, too wet to plow, are generally neglected or used for grazing. In some areas trees with low economic value, such as cottonwood, ash, and alder, have been left to line stream edges. At cultivated edges, nonnative species of shrubs and forbs have invaded fence lines, field margins, roadsides, and unmanaged pastures, often composing the dominant cover (Figure 6.6).

Riparian buffers in agricultural ecosystems for the most part are designed to leave existing native forest strips and to encourage active reforestation of sites heavily colonized by competing, aggressive, and largely nonnative vegetation. Conventional agroforestry systems are an effective way to address initial establishment of native vegetation as a step toward long-term riparian forest development. Agroforestry strategies include the following:

- Site preparation, including fallowing, cover or smother crops, soil-building rotation crops,

FIGURE 6.6. Typical lowland agriculture and remnant riparian vegetation in Oregon's Tualatin Valley. (*Photo by Dean Apostol*)

herbicide treatment, and soil solarization using plastic tarps or sheet mulch
- Use of nursery-grown planting stock or live stakes
- Protection from herbivores (fencing, mesh, chemicals)
- Fertilization and timely irrigation
- Thinning, pruning, and soil and nutrient conservation
- Postinstallation control of invasive species

The abundant resources of riparian areas make establishment of desired native woody vegetation a competitive struggle that entails long-term management. Once established, trees can be given progressively less attention and over time develop into self-sustaining woodland ecosystems.

Agricultural riparian restoration strategies must take into account natural processes, current land uses, and legacies from past land uses. In most cases, riparian restoration on agricultural lands cannot be designed as a full return to a historic condition because the surrounding context is too altered. Generally speaking, restoration should aim at recovery of key natural processes and functions. For example, in most agricultural landscapes, natural channel migration will continue to be constrained. This means that the ability of streams to shape their geomorphology and substrates will also be limited, which in turn means that there are fewer opportunities for natural colonization and development of diversity. But it may be that the critical functions are bank protection, creation of shoreline habitat, and capturing of sediment from overland sources. Restoration can be designed to accomplish these objectives without being held to a higher standard.

Riparian restoration has become a key focus of watershed councils and soil conservation districts throughout the Northwest. Most efforts are voluntary agreements in which willing landowners work with local councils or conservation district staff to plan and implement a grant-funded project. There is usually some negotiation between the landowner and council regarding the width of the area to be protected or restored and the composition of plantings. If the issue is protection of water quality along a small headwaters creek, then providing shade is a key objective. If interception of polluted runoff is the main issue, then grass filter strips bordering a woody shrub zone may be the prescription. The line often blurs between what constitutes restoration, mitigation, or perhaps rehabilitation. If the goal is to help a creek recover its ability to be a suitable home for salmon, then designing a riparian system to improve water quality can be viewed as a component of a creek or watershed restoration rather than riparian ecosystem restoration.

Ideally floodplain riparian restoration should have multiple objectives, including support of a naturally functioning aquatic system and development of forest habitat. Riparian restoration design should have the following components:

- Site analysis, including an assessment of the geomorphology of the stream and its floodplain
- Consideration of natural substrates and processes
- Variable width that fits the land
- A successional approach to planting that identifies beginning, middle, and end points over some time period, possibly with staged interventions (weeding, thinning, underplanting)
- Plant communities that are all or nearly all native and genetically appropriate
- A strategy for responding to natural processes, including floods and herbivory

Site Analysis

A good site analysis is the logical starting point for riparian restoration on agricultural floodplains. Key elements include the nature of existing vegetation and land use, frequency and duration of flooding, evidence of channel migration, depth to water table, site microtopography, and local wildlife. The goal is to understand the dynamics of the site. In its natural state, how would processes sort out riparian plant communities? It helps to prepare a sketch map that captures existing site features and dynamics. Interviews with the landowner or neighbors can be very helpful. Farmers and ranchers usually are quite aware of how frequently a stream floods, how it may have migrated over the years, where the soils stay wet, and what species inhabit an area. It is also helpful to consult with an expert in stream and floodplain geomorphology, assuming the budget does not allow for a complete interdisciplinary team.

Response to Natural Substrates and Processes

This step includes selecting plant communities that fit the nature of the site. For example, gravel bars are likely locations for willows or cottonwoods. Stable, deep, rich soils may be right for cedar, spruce, or other conifers. If a channel is migrating rapidly toward a given area, this might not be the best place to attempt to grow a long-term conifer forest. Areas with high, permanent water tables may best be restored to herbaceous wetland vegetation.

Variable Width and Structure That Fits the Land

Regulation-driven approaches often rely on fixed widths, but natural riparian vegetation nearly always has a variable edge as valleys and channels change shape. In most cases, it is just as easy to design a variable edge as it is one that is rigid. Microtopography clues can be used to establish a meandering line. However, getting carried away with variable width can compromise effectiveness. Riparian vegetation further from a stream is less likely to aid the stream than vegetation that is closer, and because restoration is expensive, a narrower band that restores more stream miles per acre may be a better approach where budgets are very limited.

Successional Planting

Restoration from scratch is a long-term prospect. Growing a mature riparian woodland is for the tortoise, not the hare. It may be that a key goal for a given project is establishment of conifers to eventually contribute large wood to the stream system. Most foresters and aquatic ecologists favor conifers for this purpose because they are longer lived and more stable once in the stream. But cottonwoods or alders may mature and be ready to fall into a stream within only 30 or 40 years, whereas a cedar may take 100 years or more. Thus several generations of cottonwood and alder could grow and fall into streams in the time it takes for one of spruce or cedar. The key is to be clear about long-term intent and start the restoration on a trajectory that will head there but not in too great a hurry. In many coastal streams, where fast-growing hardwoods are abundant but conifers scarce, it makes sense to skip the short-term benefits that a few more acres of hardwoods might provide and go directly to a conifer planting. But in lowland agricultural zones, where all woody structure and summer shade are in short supply, it may be better to start with establishing hardwoods.

Native Plant Communities

Sometimes the argument is made that nonnative trees, which could be ornamental shade or hybrid cottonwoods, are better than native trees, or at least equally effective in providing certain functions (shade, bank stabilization, or nutrient uptake). This may be true, but to gain wider benefits, riparian plantings should be part of the native ecosystem. There are circumstances in which hybrid cottonwood plantations can be integrated into a riparian restoration scheme, perhaps by acting as an additional edge zone between native trees and a plowed field or meadow. The odd ornamental shade tree placed in or at the edge of a restoration planting would not be detrimental. But as a rule, riparian plants should be native and from appropriate genetic stock.

Responding to Natural Processes, Including Floods and Herbivory

Riparian restoration projects ought to be viewed as happening within natural processes, the two most influential of which are flooding and grazing of planted vegetation by wildlife. Successful restoration design takes account of the frequency, severity, and duration of local flooding. But even so, every flood is different and may alter the ecosystem in ways that the designer failed to predict. Over the long term, the designer or property owner may need to come back and make adjustments based on flood-driven changes. For example, the stream may have decided to deliver an uninvited load of fresh gravel on top of newly planted conifers, which were placed in an area that the designer assumed gravel could not reach. What to do? Scrape the gravel away to save the conifers? Or go with the flow and let cottonwoods have that area? Some restorationists are more aggressive, others more flexible. We encourage the latter path.

In the case of native wildlife browsing new plants, we offer a few thoughts. First, one has to consider that a key purpose of restoration is to benefit wildlife. Beavers in particular are such important species in the aquatic ecosystem that they should not be begrudged the odd newly planted tree.

On the other hand, beavers can make short work of an expensive new willow or cottonwood planting, and it is wise to protect at least some or most of the new saplings if the goal is not simply to create a short-term welfare system for aquatic rodents. The key again is to anticipate and be flexible. The beavers may just eat a few trees and move on or may move in, build a den, and eat everything. Girdling of unprotected young willows and cottonwoods by rabbits can also kill a lot of trees, as one of the authors has discovered in his new native willow nursery.

Overgrazing of newly planted riparian areas by moose, elk, or deer can be a more frustrating problem in that these species generally do not do much in return to aid the aquatic ecosystem. Large ungulates can have a profound effect on riparian vegetation if their numbers are artificially high (Johnson and O'Neil 2001). The effects can be very similar to those of overgrazing by cattle. Elk travel in large groups and preferentially use riparian areas, particularly in summer. Various techniques common in reforestation practice should be used to protect plants that are subject to heavy grazing. In some cases, temporary fencing may be the best and only effective solution.

Riparian Silviculture

The second riparian restoration category is riparian silviculture, which is practiced primarily in former or current commercial forest areas. Silviculture is closely linked to the practice of commercial forestry, and restoration ecologists may ask what business commercial forestry has in protected riparian zones. Silviculture is value neutral in the sense that it operates under a presumption that the needs of a particular place, time, and set of objectives are brought into harmony. If the objective is to grow the best forest possible to enhance the aquatic and terrestrial ecosystem, then that falls within the skills and knowledge that a silviculturist is expected to possess.

Much of the practice of riparian silviculture is linked to the new goal of developing large conifers as a key component of the aquatic ecosystem. Decades of logging have left behind riparian woodlands that are hardwood dominated or have only young conifer plantations, but active management of alder- and brush-dominated riparian forests can successfully accelerate mature coniferous riparian forest (Berg 1995).

Riparian silviculture projects should address ecosystem functions that have been lost or degraded. Consequently, a logical first step is to understand what is going on at the point along a stream or river where restoration is being planned. What is the watershed context, and what are the key functional needs to be maintained or enhanced? The goal should be to design prescriptions that will result in a forest structure and composition that support ecological functions. These prescriptions are best linked to a larger watershed- and landscape-level strategy that aims at habitat-forming processes (Montgomery and Buffington 1995). As mentioned earlier, riparian forest communities have varying influences along the river continuum. Therefore, site-scale projects should be designed to enhance the functional needs of the ecosystem at the landscape level (Hollings 1978).

Riparian silviculture ranges from letting a forest develop at its own pace to very aggressive intervention. The degree of intervention should depend on the existing forest stand structure (size, age, density, and composition), present land use and ownership, ecological processes, and acceptable level of risk. As a rule, the younger the stand, the more adaptable it will be to designed changes in its growth and development trajectory (Oliver and Larsen 1990). There is justifiable controversy with regard to the motivation and wisdom of many riparian silvicultural projects. Although there is general agreement that past disturbances have negatively affected riparian forests, strengthened regulatory protections are leading to natural recovery in some areas, and many believe these should simply be left alone. Brush and hardwood trees provide some important riparian functions, including shade, nutrient uptake, food, and habitat. Clearly, in some instances forest managers are wise to allow riparian areas to recover at their own pace. But the one element hardwood areas lack, large conifers, is a critical ecosystem component, and in some places silviculture aimed at restoring this feature is warranted. For example, in the Kennedy Flats watershed on Vancouver Island, British Columbia, most of the riparian trees have been logged right to the stream edge, leaving only small areas of mature conifer (Warttig et al. 2000). In addition, although alders are mostly beneficial to aquatic ecosystems, they are also known to substantially reduce base flows because of their prodigious thirst.

Allowing a Northwest riparian forest to develop on its own in most cases leads to development of mature conifer or a mixed hardwood–conifer structure. However, the end result may be less than optimal. For example, many pure red alder stands naturally thin out as they mature. Eventually, if sufficient seed trees are nearby, conifers may establish in canopy gaps. Even after many years, we would expect less than the full potential conifer stocking. Very dense hardwood thickets often have tenacious understory communities, such as salmonberry (*Rubus spectabalis*). These can be expected to remain patchy, sparsely stocked forests for many decades (Tappenier et al. 1991). Long forestry experience demonstrates that thinning of hardwoods can accelerate conifer dominance by creating needed space and light far earlier than nature would. Ultimately, as the newly established conifers mature, further thinning and natural mortality will result in large wood recruitment to the aquatic and terrestrial ecosystem. Managed succession can greatly accelerate the natural recruitment process for large wood while also creating potential opportunities for financial returns where appropriate, a significant benefit to private landowners (Berg 1997).

Nevertheless, active management of riparian forests to promote large wood development is still new in Pacific Northwest watersheds. As a consequence, there is much to learn. We can draw on numerous sources of information on the yield and dynamics of various tree species (McArdle et al. 1949, Johnson 1955, Nystrom et al. 1984, Long et al. 1988, Curtis et al. 1997). But the process of how large wood gets to streams is still poorly understood. Foresters have made assumptions about the direction trees will naturally fall, but only recently has research been able to measure source distance (McDade et al. 1990).

Riparian woodlands often are wind prone after logging of adjacent areas. Prevailing wind direction and episodic windstorms, such as downbursts, can determine both the timing and future availability of large wood (Grizzel and Wolf 1998). Windthrow is difficult to predict, but light thinning spread over many years can help stands gradually adapt to decreased density.

Thinning

Even after more than 100 years of study, foresters still have many questions on the effects of thinning in Northwest forests (e.g., Brown 1882, Curtis et al. 1997, Zeide 2001). We know that very dense young stands will stop growing for a time, until natural mortality thins them. The historical reason to thin is to concentrate more growth in salable softwoods on fewer, higher-quality stems. But maximizing salable wood is not the goal in riparian restoration. More often, riparian thinning is used to remove less desirable species in order to accelerate the development of those with greater ecological utility. Thinning either releases existing conifers in the understory or makes more growing space for newly planted trees.

Because thinning helps concentrate growth on fewer tree stems, the expected result is large wood more quickly. Thinned trees also can be conscripted to assist in restoration. They can be delivered directly into streams, sometimes located to reinforce existing log jams. In some cases, enough surplus wood may be generated by thinning to allow a landowner to retain a portion for personal use or sale. This return may encourage landowners to try active restoration through thinning. However, economic return should be a secondary product of the restoration effort rather than a main goal.

Some growth simulation models indicate that poorly planned or timed thinning of riparian woodlands can reduce the amount of large wood available to streams (Beechie and Sibley 1997). Thus although thinning is beneficial if done well, there is still reason for restoration foresters to remain humble. Forest thinning for riparian restoration is largely untested, and theory is ahead of the practice (Beechie et al. 2000). Investments of time and resources need to account for risk, largely because of the physically active and dynamic nature of floodplains but also because of a lack of knowledge and experience.

The riparian restoration forester also needs to ensure that enough trees are left to provide shade. The design for a given site should be analytical, considering tree height, solar angle, and stream orientation as well as operational considerations.

Density management diagrams are used to predict growth in simple, even-aged, single-species stands. Most riparian forests are uneven aged, mixed species, and complex. Empirical yield tables (e.g., Meyer 1939, McArdle et al. 1949) can be used to predict growth in natural forests. Some popular models, such as Organon, fail to include information on key riparian trees, such as western redcedar and red alder (Hester et al. 1989).

The natural rate of tree death in riparian forests is not well known. Foresters use simulation models, but estimates often are based on the least robust equations. New tools are being developed to predict large wood recruitment in a variety of management scenarios. Wood recruitment is projected forward in time by taking estimates from growth and yield models and using these numbers as inputs to another model (Kennard et al. 1998, Bragg 2000). This approach is useful in comparing management scenarios from a relative standpoint but is not reliable in predicting how much total wood will get to the stream.

There are two general approaches to thinning riparian trees. In the first, a fairly uniform removal leaves behind a continuous forest cover. In the second, a series of small openings are interspersed in a matrix of forest patches. The decision usually comes down to how the designer wants light to influence the stand. As available light increases, so does the stability of saplings. This is measured by the height-to-diameter ratio (Oliver and Larsen 1990). When available light is at or below 30%, growth results in a ratio that exceeds 100, the trees grow very tall and thin and are easily blown over (Emmingham et al. 2000). An alternative approach is to focus on habitat diversity and the natural resilience and adaptability of multiple species, which allow them to take advantage of diverse ecological niches. Dent and Walsh (1995) found that a series of small openings along a stream section had less impact on temperature than one large opening, even though the total amount cleared was the same. Designers might consider creating a series of small openings, combined with strict preservation of high-quality trees, along with an intermediate level of thinning and planting to diversify the stand.

Planting

Why plant at all? Why not let natural regeneration and succession take place? Previously disturbed riparian forests normally redevelop conifer composition on most Pacific Northwest sites given enough time, reasonable regeneration conditions, and a local seed source (Beach and Halpern 2001). Genetic diversity and long-term site fitness are good reasons to stick with natural regeneration. Avoiding disturbance to naturally recovering systems that may result from planting is another. The main argument in favor of planting is simply that it is a proven way to increase the presence of desired species faster than natural processes, particularly where seed sources are limited.

Variations in planting include dispersed and aggregated patterns. Dispersed planting creates an evenly spaced, uniform stand. One clear advantage is that it eases the work of subsequent brush control. Aggregated, or clustered, planting is advantageous where rapid early growth is needed in order to allow the trees to outcompete brush. An added advantage of aggregated planting is more rapid development of diverse forest structure and composition (Scott et al. 1998).

Given that high water tables and seasonal flooding are characteristics of riparian areas, use of elevated planting sites makes sense if these are available. Along the South Fork of the Hoh River on the Olympic Peninsula, seedlings on downed log substrates far exceeded natural regeneration directly on the forest floor (McKee et al. 1982). Harmon and Franklin (1989) found this to be a regional phenomenon resulting from a number of influences, including pathogens, predation, competition, and hydric soils.

Scott et al. (1998) found the height growth of seedlings planted at about 3,000 stems per hectare to be greater than at all lower densities tested. This is counterintuitive but may have several explanations. One is that the seedlings were able to more completely occupy the space and through root grafting were able to exclude competition for resources. Tight plantings also have the advantage of being frost hardy, a problem in some riparian areas subject to cold air drainage. They also help prevent soil moisture loss during late summer droughts by fully shading the ground. But high-density plantings may need subsequent thinning in order to prevent early stagnation.

Riparian silviculture trials in western Washington are testing planting patterns and overstory retention levels. Stands have been planted at 500 trees per hectare, using both aggregated and dispersed patterns. Aggregates are high-density clumps of nine conifer seedlings spaced 5 feet apart, with the aggregates spaced 40 feet from each other. Dispersed plantings follow the Washington State Forest Practices legal minimum of 4.2-meter spacing (about 500 trees per hectare). Survival, growth rates, and structural development all will be monitored over time.

Hazards to planted stock include animal browse, plant competition, and difficult growing conditions. Competition can be intense. Early brush control is advantageous and can increase seedling survival. The Lummi Indian Nation and University of Washington Center for Streamside Studies found that after intensive site preparation, planted seedlings were able to grow above brush (Wishnie and McClintick 1999). Hot summer droughts can dry out coarse-grained soils and lead to desiccation of seedlings. Past logging may have compacted soils, resulting in slow drainage. Invasion by exotic species can overwhelm planted trees in some areas. Intensive site preparation and invasive species control may temporarily reduce stream bank stabilization, nutrient and carbon sequestration, and wildlife habitat provided by native and nonnative brush species.

In addition to conifers, riparian restoration can include willows, red osier dogwood, or other streamside-adapted species. These are easily grown from live stakes or wattles and can provide bank stability. Streamside willow plantings serve as a final filter for subsurface nutrients, can reduce the force of floodwaters, may help retain organic debris and sediments, and can help attract beavers to an area.

Restoration in Highly Disturbed or Modified Urban Riparian Areas

Given the strong environmental ethic proclaimed by urban inhabitants of the Northwest, it is not surprising to find that a lot of money and effort are being put into restoration of urban streams and riparian areas, in areas where many environmentalists live. But urban riparian restoration presents a set of unique challenges.

As pointed out earlier, riparian vegetation communities are closely linked to the hydrology and fluvial geomorphology of their streams. Yet changes to the hydrology of urban drainage basins are severe and set a series of negative consequences in motion (Booth 1991).

The first issue is that of urban stormwater runoff. According to the Environmental Protection Agency, once a watershed is developed to a point where 10% of it is in impervious surface, water runs off the land quickly and intensively enough to create detrimental changes to stream channels. Rainfall on the west side of the Cascades falls at low intensities over long time periods. This gives plants and the soil an opportunity to hold

Case Study: Riparian Silviculture at Kennedy Flats, Vancouver Island, British Columbia

Kennedy Flats is located in the nearly 13,000-hectare Kennedy Lake watershed on the west coast of Vancouver Island, Clayoquot Sound (Figure 6.7), a UNESCO Biosphere site. The following information is condensed from several published reports on watershed restoration efforts in this area, which have been taking place since 1995 (Warttig et al. 2000). Interestingly, riparian restoration is only one component of the larger watershed work, which has focused on removing old logging roads on unstable ground that were resulting in excessive sedimentation and extensive in-stream habitat enhancement. (This work is further profiled in the watershed restoration chapter.)

Kennedy Flats is classified as a target watershed, making it a high priority for restoration in British Columbia. Until 2002 restoration funding had been generated by a surcharge on each log cut from public lands. Altogether there are nine sub-watersheds within the larger basin. The landscape is characterized as a broad floodplain, with meandering channels, some steep hills, and bedrock knobs. Historically this was a very productive stream network for coho, chum, pink, and sockeye salmon and steelhead, cutthroat, and rainbow trout. Rainfall is high, at 130 inches per year. Streams are low in both gradient and energy (Warttig 2000).

As on much of the west coast of Vancouver Island, logging has been heavy in Kennedy Lake watershed since the 1950s. Problems with salmon habitat began to surface in the 1970s. By 1980, 42% of the watershed had been logged. Heavy slash choked streams and forced water to spread across the valley floor. This killed the few conifers that had been left behind. Poor-quality roads built on unstable upland terrain often failed, resulting in high sediment loads. This problem was exacerbated by undersized culverts, collapsing culverts, and roads built on floodplains. Overall, streams became choked with sediment that filled pools, smothered spawning gravel, and destroyed winter refugia.

Watershed restoration has aimed at reducing sediment loads and getting stream channels back to their natural forms by removing slash and anchoring remnant large wood pieces in key locations. Riparian restoration wisely waited until the upland work was well advanced, to avoid throwing good Loonies after bad.

For restoration inventory purposes, forested riparian areas in western British Columbia have been classified into five categories, or riparian vegetation types (RVTs; Figure 6.8):

- RVT 1: Hardwood brush without conifers
- RVT 2: Overstocked young conifers
- RVT 3: Deciduous forest (usually red alder) over a conifer understory
- RVT 4: Deciduous forest without conifers underneath
- RVT 5: Old-growth or mature conifer or mixed conifer and deciduous forest

Restoration is aimed at influencing stand development toward mature conifer or mixed forests. The focus is on the 30 meters adjacent to large creeks, consistent with streamside protection standards for British Columbia. By early 2003 foresters had treated 81 hectares, most of which are RVT 2 and RVT 3. Province-wide surveys show that these represent the most common riparian condition in coastal watersheds of British Columbia, as they are in Kootowis Creek, the highest-priority sub-watershed at Kennedy Flats. It is too early to tell what beneficial impact riparian restoration has had, but project managers have seen noticeable brush and fern growth in the canopy gaps created by thinning dense conifer stands. Conifer regeneration and growth in treated brush and alder areas have been impressive. The overall watershed restoration impacts have been impressive. Returning coho have steadily increased since restoration began, from less than 1,000 in 1994 to more than 7,000 in 2002. Project managers plan for a total of 533 hectares of riparian restoration, with a cost estimate of $2,800 per hectare on average, or nearly $1.5 million (Canadian) total.

Generally riparian treatments are expected to be one time only. Typically, in RTV 2 young conifers are thinned from an initial density of 1,000 to 5,000 stems per hectare to 200–600. The preferred thinning pattern is clusters and gaps in order to mimic natural tree spacing. Additionally, because formerly hemlock, cedar, and spruce forests were replanted to Douglas fir, efforts are made to favor the first three where possible. In RVT 3 areas, the goal is to release understory conifers by removing some of the alder canopy and treating competing brush. In most cases, some overstory alder are left in place, usually 100–200 trees per hectare. Removed trees are either cut down or girdled and left as habitat.

water, use it, and store it. Much of the water that moves downslope is subsurface. Only when the vegetation and ground are thoroughly saturated (after about 1 inch of rain in a 24-hour period) does water begin to run directly off the surface of the land and into streams. This results from a seasonal rise in water tables (Booth 1991).

In small urban watersheds runoff is influenced by the total amount of rain that falls and the rate at which it falls. As land gets paved over, both the amount and rate of runoff are affected.

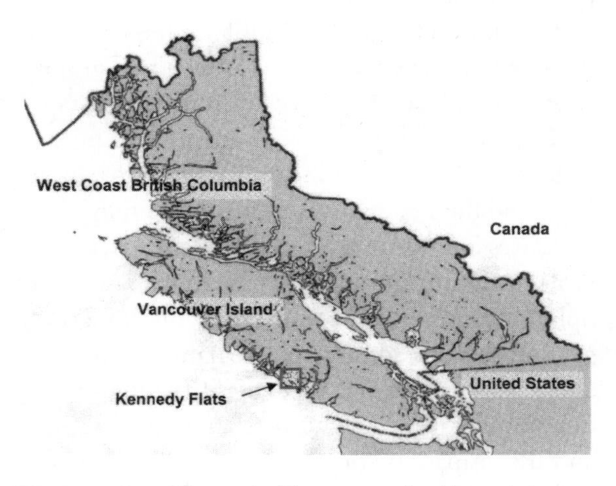

FIGURE 6.7. Kennedy Flats watershed location. (*Courtesy of International Forest Products Ltd.*)

Under natural conditions, very little of the rain that falls reaches streams. Most is held and used by vegetation. Much is held in the soil and stored for long periods. Some reaches the deep groundwater, where it may emerge as stream flow some distance from the source. Water that reaches channels does so slowly because of the high complexity and friction of the surface (Booth 1991).

Ultimately, loss of vegetation, grading, filling, compaction, and paving result in increased peak flows up to five times what they were under natural conditions. These flows produce high levels of channel and bank erosion and floods. Not only do flow rates increase, but the number of peak flows also increases because smaller storms generate high rates of runoff. The traditional engineering response was to deepen and straighten channels in order to convey more water more quickly. This destroys bank side riparian trees and results in further loss of contact between the stream and floodplain, which deprives remnant riparian trees of summer water. Large wood is taken out of streams, and no new large wood is grown or delivered. Lack

RVT	Stand Condition	Function Impaired	Recommended Treatment	Desired Future Condition
1	Brush-dominated, poorly stocked conifer component 	Large wood, shade, bank, and floodplain stability	Improve conifer stocking by planting. Release suppressed trees through competition removal or spot fertilizing. 	
2	Overstocked conifer, >800 stems per ha 	Large wood, forage for wildlife, structural diversity	Thin to 400–600 stems per ha, favoring largest diameter trees; vary densities, creating gaps and clusters. Opportunities for bird and bat nests and wildlife trees. 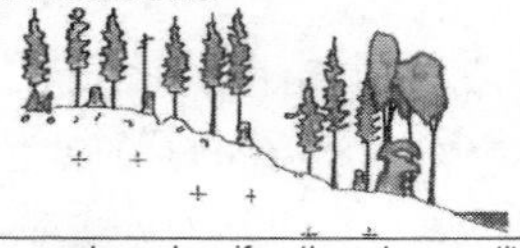	
3	Decicuous forest with a good conifer understory 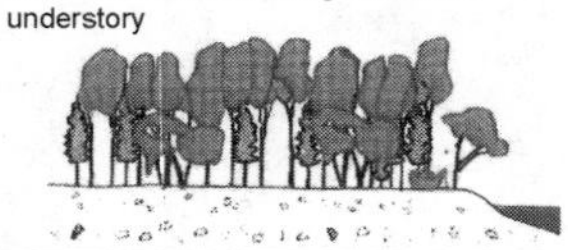	Large wood, bank, and floodplain stability	Release over topped conifers through competition removal or spot fertilizing. 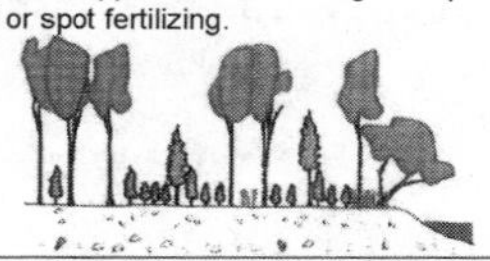	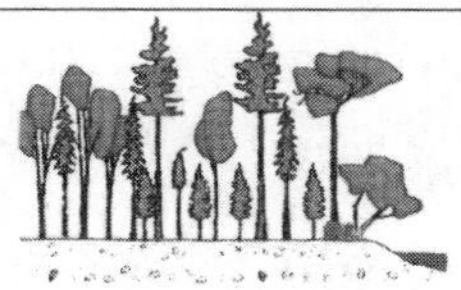
4	Decicuous forest with a poor conifer understory 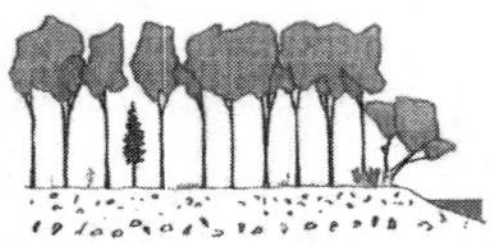	Large wood, structural diversity, bank, and floodplain stability	Improve conifer stocking by planting. Release suppressed trees through competition removal or spot fertilizing. 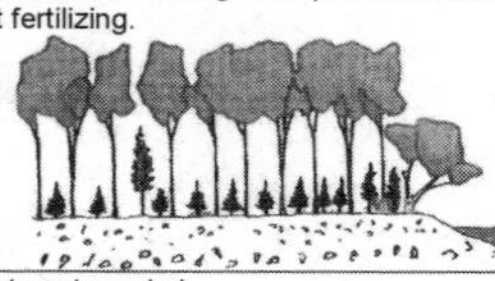	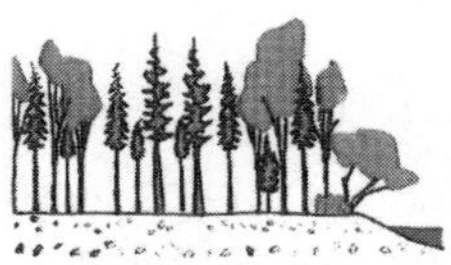
5	Old growth or old second growth (>70 years)		No treatment needed.	N/A

FIGURE 6.8. Riparian vegetation types in British Columbia, including desired conditions and recommended treatments. *(British Columbia Ministry of Forests)*

of wood contributes to channel erosion and simplification. Channel widths and depths increase, and high quantities of sediment fill pools (Booth 1991).

Streams naturally are disrupted by occasional high flows that move materials, wash out log jams, fill pools, and so forth. But in intervening periods low flows allow new wood to get positioned, sediment to rebuild, and the overall ecosystem to renew itself. On average, large-scale channel-changing flows in western Oregon and Washington occur about eight times in a 40-year period. But in an urban watershed, this increases to five such flows per year. As a consequence, urban streams develop uniform beds, have few pools or riffles, have raw and nearly vertical channel banks that are constantly eroding, have no or very little wood, and lack aquatic wildlife. They look like streams that have recently suffered a catastrophic debris torrent, yet they have no time to recover.

Engineers have recently begun to embrace a suite of methods for mitigating urban stormwater flows and their effects on streams. Generally, these measures include detention, bioswales, streamside protection zones, and reductions in the amount of upland impervious surface. The central idea is to get flows back to something like natural conditions and to allow riparian areas to perform their historic functions. Given the severe alteration of streams and their riparian zones in urban areas, how can and should restoration be practiced?

Case Study: Riparian Restoration in Portland, Oregon

The City of Portland, Oregon has taken a leading role in efforts to restore urban riparian systems. This work was first tried on the Columbia Slough, a badly polluted former channel of the Columbia River, located in north Portland. It was initiated by the City Bureau of Environmental Services, the department responsible for stormwater management. The goal of riparian restoration is to improve water quality, particularly temperature. This program has evolved over the years into a larger effort that now includes all of the city's watersheds (A. Curtis, personal communication, 2003).

Portland watershed restoration planners begin by identifying historic riparian vegetation communities recorded in General Land Office Surveys from the 1850s. They also identify reference sites, which are relatively undisturbed riparian areas in and around Portland that serve as templates. They develop a site analysis for a given project area that takes account of past disturbances, soils, land use, and existing vegetation and notes key limiting factors, such as depth of fill and changes in flow regimes. From this, they develop a conceptual planting plan, using native species that are most likely to survive the altered conditions. The planners rely on past experience and do constant monitoring to check on successes and failures.

Weedy species are cut back and often treated with carefully selected herbicides. Portland uses small bare root plants, live stakes, and in some cases wattles where slope stability is a concern. Monitoring shows that survival rates are in the range of 50–75% after 5 years. In areas where beavers are active, efforts are made to protect seedlings with netting or sleeves. Erosion netting and silt fencing are used when appropriate. Contract crews do annual weed control for 5 years. The goal is to get plantings in a "free to grow" condition by that time.

The initial goal of improving water quality has expanded into a more comprehensive set of objectives to improve aquatic and terrestrial habitats. One clear issue is that the city is limited in what it can do to affect upstream conditions, which are mostly agricultural or forestland uses. In the Johnson Creek watershed of southeast Portland, these upstream land uses are generating huge amounts of sediment, thus potentially negating downstream restoration efforts.

Attempts at restoration along the mainstem Willamette River in Portland are typical of the types of challenges faced by urban restoration planners. A recent analysis pointed out that only 17.5% of the entire river within the City of Portland has natural banks, meaning it has not been filled over, riprapped, or engineered (City of Portland 2001). A research project by Oregon Department of Fish and Wildlife surprisingly discovered that anadromous fish use these natural riverbanks as habitat and do not just quickly swim past the city, as had been previously believed. This provides a strong argument to those who are attempting to naturalize the river shoreline, even through downtown Portland. But how does one restore such an altered landscape?

The recently published *Willamette Riverbank Design Notebook* suggests a series of design approaches that can be used to naturalize disturbed sites (Figure 6.9). Generally, these can be classified as bioengineering, or hybrids with traditional engineering. The central goal is to use the notebook to help plan site-specific projects as riverfront areas are redeveloped (City of Portland 2001).

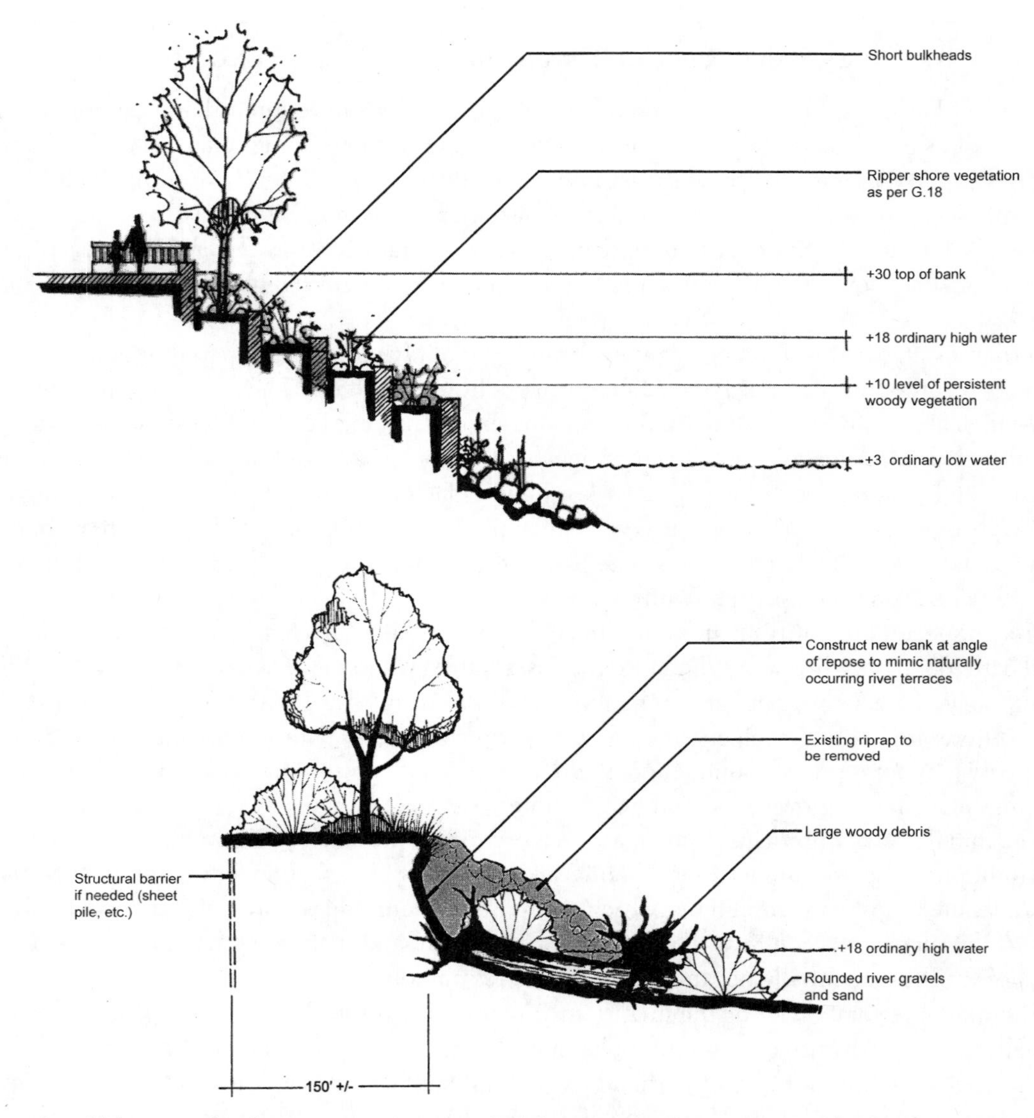

FIGURE 6.9. Urban riverbank riparian restoration details in Portland, Oregon. *(a)* Stepped terraces. *(b)* Naturalistic bank. (*Courtesy of the Portland Development Commission and Greenworks PC, from the Willamette River Restoration Notebook*)

Restoring Riparian Plant Communities in the Northwest

How advanced are the art and science of restoring riparian plant communities in the Northwest? Based on the experience of the past decade, we should have some confidence that woodland riparian communities can be established through good site analysis, preparation, planting techniques, and at least a few years of maintenance. Given the long timeframes needed to monitor results, we have somewhat less confidence with regard to the long-term benefits of riparian silviculture aimed at restoring large wood to streams, although early research results are promising. Forested riparian areas appear to have greater potential for restoration of historic conditions, as compared with agricultural and urban areas, because the degree of landscape alteration is far less. In the latter two, restoration can reestablish some riparian functions, but lack of space, continued disturbance, invasive species, and changed hydrology all conspire to prevent fully realized restoration.

Summary of Riparian Woodland Restoration Practices

Preservation

In our efforts to restore degraded riparian areas, retention of remnant high-quality areas must not be overlooked. This is true at all scales, from large watershed basins down to individual reaches or sites. In our project work we have encountered watershed councils and agencies engaged in heroic efforts to restore woody vegetation in highly disturbed areas at great public expense while ignoring mature patches just upstream or downstream that could be lost because of inadequate regulations. The first rule of riparian restoration is that preservation is cheaper and more effective and must be attended to first. That said, any lowland riparian area is subject to invasive species and should be monitored and treated as needed.

Planting

Various methods for planting denuded riparian areas have proven successful across a range of land uses and sites. As long as planting is embedded in proper site analysis, including an understanding of local stream geomorphology, it stands a good chance of succeeding over time. Site preparation, choice of materials, and plant protection methods all must be chosen on a case-by-case basis.

Invasive Species Control

This is a critical practice in lowland agricultural and urban areas but is less important in most forested areas. Though often daunting, large masses of Himalayan blackberry can be controlled and eventually reduced to manageable levels through a variety of methods (see Chapter 16). Over time, as shade increases, blackberry loses vigor and fades out. In urban areas, species such as English ivy and clematis are much more problematic and warrant longer-term attention. Japanese knotweed (*Polygonum cuspidatum*) is emerging rapidly as a serious riparian invader throughout Northwest riparian areas.

Clearing Native Brush to Make Space for Conifers

This is a proven, effective method for releasing existing conifers or making space to plant. There is a larger question of whether the short-term disruption to habitat is worth the long-term gain. The answer is sometimes yes and sometimes no, depending on ecological context. Along streams where past logging has resulted in many acres of brush, clearing some to help establish conifers can be beneficial, as long as the work is done carefully to minimize aquatic impacts.

Thinning Dense Conifer Plantations

This practice is still unproven in terms of long-term effectiveness, yet it is intuitively sound based on years of experience in forestry. In particular, riparian sites that have been converted to Douglas fir can benefit by thinning to favor naturally regenerated (or newly planted) cedar, hemlock, and spruce. Dense conifer plantations, particularly once they reach a closed canopy stage (20–40 years generally), have very little value to wildlife and may stagnate, thus inhibiting development of the large trees that most aquatic systems lack. Thinning is a restoration practice that should be continued, but land managers must commit themselves to monitoring and reporting results honestly.

Thinning Mature Hardwoods to Release or Plant Conifers

Thinning native hardwoods has proven effective at improving the growth or reestablishment of conifers. The key question is whether the temporary loss of habitat and ecological functions the hardwoods provide is worth the long-term benefits. These projects must be selected carefully, with consideration of the wider landscape context. This is particularly true if sites are along small tributaries in a stream network where high summer temperatures are an issue. Lack of nitrogen or insects may also be a limiting factor and warrant caution. Foresters contemplating treatment of hardwood stands need to pay attention to substrates and stream geomorphology. In many cases, there are good reasons that a site has plenty of hardwoods but few or no conifers.

Techniques

Table 6.1 illustrates some typical riparian restoration practices, along with their importance and effectiveness.

TABLE 6.1.

Riparian restoration techniques, importance, and effectiveness.

Practice	Importance	Effectiveness
Site preparation	Very	Varies
Planting	Varies	High
Plant protection	Varies	Varies
Thinning	Context dependent	Long term
Hardwood overstory removal	Context dependent	Long term
Weed control	Very	High

Summary

Riparian forest restoration integrates native processes and enhancement of natural ecosystem functions within practical constraints (Berg 1995). Armed with effective practices and guided by ecological principles, riparian restoration works with the dynamics of floodplain forests. Full recovery takes many years, even with sound science-based approaches. Multiple land use changes have altered the dynamics of large wood in most Northwest rivers, making full restoration to historical conditions problematic. The significance of riparian forests in promoting the persistence of large wood in rivers is undisputed. Restoring managed riparian forests is important for protecting degraded habitats (terrestrial and aquatic) and maintaining cold, pure water. Practitioners and land managers must recognize that rivers exist in a constantly changing landscape.

References

Andrus, C. W., B. A. Long, and H. A. Froehlich. 1988. Woody debris and its contribution to pool formation in a coastal stream 50 years after logging. *Canadian Journal of Fisheries and Aquatic Sciences* 45: 2080–2086.

Apostol, D. and M. Sinclair. 2004. *A Conservation and Restoration Strategy for Newell Creek Watershed.* Unpublished report, John Inskeep Environmental Learning Center, Oregon City, OR.

Beach, E. W. and C. B. Halpern. 2001. Controls on conifer regeneration in managed riparian forests: effects of seed source, substrate, and vegetation. *Canadian Journal of Forest Research* 31(3): 471–482.

Beechie, T., G. Pess, P. Kennard, R. E. Bilby, and S. Bolton. 2000. Modeling recovery rates and pathways for woody debris recruitment in northwestern Washington streams. *North American Journal of Fisheries Management* 20: 436–452.

Beechie, T. J. and T. H. Sibley. 1997. Relationships between channel characteristics, woody debris, and fish habitat in northwestern Washington streams. *Transactions of the American Fisheries Society* 126: 217–229.

Benda, L., T. J. Beechie, R. C. Wissmar, and A. Johnson. 1992. Morphology and evolution of salmonid habitats in a recently deglaciated river basin, Washington State, USA. *Canadian Journal of Fisheries and Aquatic Sciences* 49: 1246–1256.

Berg, D. R. 1995. Riparian silviculture design and assessment in the Pacific Northwest Cascade Mountains, USA. *Ecological Applications* 5: 87–96.

Berg, D. R. 1997. Active management of riparian habitats. Pages 50–61 in *Wetland and Riparian Restoration, Taking a Broader View*. EPA 910-R-97-007. U.S. EPA Region 10, Seattle.

Beschta, R. L., R. E. Bilby, G. W. Brown, L. B. Holtby, and T. D. Hofstra. 1987. Stream temperature and aquatic habitat: fisheries and forestry interactions. Pages 191–232 in E. O. Salo and T. W. Cundy (eds.), *Streamside Management: Forestry and Fishery Interactions*. College of Forest Resources, University of Washington, Seattle.

Bilby, R. E. 1981. Role of organic debris dams in regulating the export of dissolved and particulate matter from a forested watershed. *Ecology* 62(5): 1234–1243.

Bilby, R. E. and P. A. Bisson. 1992. Allochthonous versus autochthonous organic matter contributions to the trophic support of fish populations in clear-cut and old-growth forested streams. *Canadian Journal of Fisheries and Aquatic Sciences* 49: 540–551.

Bilby, R. E. and G. E. Likens. 1980. Importance of organic debris dams in the structure and function of stream ecosystems. *Ecology* 61(5): 1107–1113.

Bilby, R. E. and G. E. Likens. 1989. Changes in characteristics and function of woody debris with increasing size of streams in western Washington. *Transactions of the American Fisheries Society* 118: 368–378.

Bird, G. A. and N. K. Kaushik. 1981. Coarse particulate organic matter in streams. Pages 41–68 in *Perspectives in Running Water Ecology*. Plenum, New York.

Bisson, P. A., J. L. Nielson, R. A. Palmason, and L. E. Grove. 1982. A system of naming habitat types in small streams, with examples of habitat utilization by salmonids during low streamflow. Pages 62–73 in *Acquisition and Utilization of Aquatic Habitat Inventory Information*. Western Division, American Fisheries Society, Portland, OR.

Boling, R. H., E. R. Goodman, and J. A. V. Sickle. 1975. Toward a model of detritus processing in a woodland stream. *Ecology* 56: 141–151.

Booth, D. 1991. Urbanization and the natural drainage system: impacts, solutions and prognosis. *Northwest Environmental Journal* 7: 93–118.

Boyd, R. 2000. *Indians, Fire, and the Pacific Northwest*. Oregon State University, Corvallis.

Bragg, D. C. 2000. Simulating catastrophic and individualistic large woody debris recruitment for a small riparian stream. *Ecology* 81: 1383–1394.

Brown, J. 1882. *The Forester, or a Practical Treatise on the Planting, Rearing, and General Management of Forest Trees*. 5th ed. Blackwood, Edinburgh.

Cederholm, C. J. and N. P. Peterson. 1985. The retention of coho salmon (*Oncorhynchus kisutch*) carcasses by organic debris in small streams. *Canadian Journal of Fisheries and Aquatic Sciences* 42: 1222–1225.

City of Portland. 2001. *Willamette Riverbank Design Notebook*. Unpublished report.

Cummins, K. W. 1974. Structure and function of stream ecosystems. *BioScience* 24(11): 631–641.

Curtis, R. O., D. D. Marshall, and J. F. Bell. 1997. LOGS: A pioneering example of silvicultural research in coastal Douglas fir. *Journal of Forestry* 95(7): 19–25.

Dent, L. F. and J. B. S. Walsh. 1995. *Effectiveness of Riparian Management Areas and Hardwood Conversions in Maintaining Stream Temperature*. Forest Practices Technical Report No. 3. Oregon Department of Forestry, Salem.

Diaz, N. 1996. *Riparian Plant Communities of Mount Hood and Gifford Pinchot National Forests*. Unpublished report, USDA Forest Service.

Ecotrust. 2002. *Siuslaw River Watershed Assessment*. Unpublished report available online at www.inforain.org/dataresources/.

Elliott, S. R. 1999. *Seston and the Streamside Forest: The Impact of Riparian Disturbance on the Food Quality of Suspended Particles*. Master's thesis, University of Washington, College of Forest Resources, Seattle.

Emmingham, B., S. Chan, and D. Mikowski. 2000. *Silvicultural Practices for Riparian Forests in the Oregon Coast Range*. Res. Contrib. 24. Oregon State University, Corvallis.

FEMAT. 1993. *Forest Ecosystem Management: An Ecological, Economic, and Social Assessment*. Report 1993-793-071 of the Forest Ecosystem Management Assessment Team. U.S. Government Printing Office, Washington, DC.

Franklin, J. and C. T. Dyrness. 1973. *Natural Vegetation of Oregon and Washington*. Gen. Tech. Rep. PNW-8. USDA Forest Service, Portland, OR.

Gregory, S. V., F. J. Swanson, W. A. McKee, and K. W. Cummins. 1991. An ecosystem perspective of riparian zones: focus on links between land and water. *BioScience* 41(8): 540–551.

Grette, G. B. 1985. *The Role of Large Organic Debris in Juvenile Salmonid Rearing Habitat in Small Streams*. Master's thesis, University of Seattle.

Grizzel, J. D. and N. Wolf. 1998. Occurrence of windthrow in forest buffer strips and its effect on small streams in northwest Washington. *Northwest Science* 72: 214–223.

Harmon, M. E. and J. F. Franklin. 1989. Tree seedlings on logs in *Picea sitchensis–Tsuga heterophylla* forests of Oregon and Washington. *Ecology* 70: 48–59.

Harmon, M. E., J. F. Franklin, F. J. Swanson, P. Sollins, S. P. Cline, N. G. Aumen, J. R. Sedell, G. W. Lienkaemper, K. Cromack Jr, and K. W. Cummins. 1986. The ecology of coarse woody debris in temperate ecosystems. Pages 133–302 in A. MacFadyen and E. D. Ford (eds.), *Advances in Ecological Research*. Vol. 15. Academic Press, New York.

Hawkins, C., M. L. Murphy, N. H. Anderson, and M. A. Wilzbach. 1983. Density of fish and salamanders in relation to riparian canopy and physical habitat in streams of the northwestern United States. *Canadian Journal of Fisheries and Aquatic Sciences* 40(8): 1173–1185.

Hester, A. S., D. W. Hann, and D. R. Larsen. 1989. *Organon: southwest Oregon growth and yield model user manual*. College of Forestry, Oregon State University, Corvallis.

Hollings, C. S. 1978. *Adaptive Environmental Assessment and Management*. Wiley, London.

Holtby, L. B. 1988. Effects of logging on stream temperature in Carnation Creek, British Columbia, and associated impacts on the coho salmon (*Oncorhynchus kisutch*). *Canadian Journal of Fisheries and Aquatic Sciences* 45: 502–515.

Hulse, D., S. Gregory, and J. Baker. 2000. *Willamette River Basin Atlas, Trajectories of Environmental and Ecological Change*. Pacific Northwest Ecosystem Research Consortium, Oregon University Press, Corvallis.

Johnson, F. A. 1955. *Volume Tables for Pacific Northwest Trees*. Agriculture Handbook No. 92. USDA Forest Service, Pacific Northwest Range and Experiment Station, Portland, OR.

Johnson, S. H. and T. A. O'Neil. 2001. *Wildlife Habitat Relationships in Oregon and Washington*. Oregon State University Press, Corvallis.

Kennard, P., G. Pess, T. Beechie, R. Bilby, and D. R. Berg. 1998. Riparian-in-a-box: a manager's tool to predict the impacts of riparian management on fish habitat. In *Proceedings of Forest–Fish Conference: Land Management Practices Affecting Aquatic Ecosystems*, May 1–4, 1996, Calgary. Inf. Rep. NOR-X-356. Natural Resources Canada, Canadian Forestry Service, North Forestry Centre, Edmonton, Alberta.

Lichatowich, J. 1999. *Salmon without Rivers: A History of the Pacific Salmon Crisis*. Island Press, Washington, DC.

Lienkaemper, G. W. and F. J. Swanson. 1987. Dynamics of large woody debris in streams in old-growth Douglas-fir forests. *Canadian Journal of Forest Research* 17: 150–156.

Long, J. N., J. B. Mcarter, and S. B. Jack. 1988. A modified density management diagram for coastal Douglas fir. *Western Journal Applied Forestry* 3: 88–89.

McArdle, R. E., W. H. Meyer, and D. Bruce. 1949. *Yield of Douglas Fir in the Pacific Northwest*. USDA Technical Bulletin 201, Washington, DC.

McDade, M. H., F. J. Swanson, W. A. McKee, J. F. Franklin, and J. V. Sickle. 1990. Source distances for coarse woody debris entering small streams in western Oregon and Washington. *Canadian Journal of Forest Research* 20: 326–330.

McKee, W. A., G. LaRoi, and J. F. Franklin. 1982. Structure, composition, and reproductive behavior of terrace forests, South Fork Hoh River, Olympic National Park. Pages 22–29 in E. E. Starkey (ed.), *Ecological Research in the National Parks of the Pacific Northwest*. Oregon State University, Corvallis.

Meyer, W. H. 1939. *Yield of Even Age Stands of Sitka Spruce and Western Hemlock*. Technical Bulletin 544. USDA, Washington, DC.

Minshall, G. W. 1978. Autotrophy in stream systems. *BioScience* 28: 767–771.

Montgomery, D. R. and J. M. Buffington. 1995. Pool spacing in forest channels. *Water Resources Research* 31(4): 1097–1105.

Murphy, M. L., C. P. Hawkins, and N. H. Anderson. 1981. Effects of canopy modification and accumulated sedimentation on stream communities. *Transactions of the American Fisheries Society* 110: 469–478.

Murphy, M. L. and K. V. Koski. 1989. Input and depletion of woody debris in Alaska streams and implications for streamside management. *North American Journal of Fisheries Management* 9: 427–436.

Naiman, R. J., T. J. Beechie, L. E. Benda, D. R. Berg, P. A. Bisson, L. H. MacDonald, M. D. O'Conner, P. L. Olsen, and E. A. Steele. 1992. Fundamentals of ecologically healthy watersheds in the Pacific Northwest coastal ecoregion. Pages 127–188 in R. J. Naiman (ed.), *Watershed Management*. Springer-Verlag, New York.

Naiman, R. J. and J. R. Sedell. 1980. Relationships between metabolic parameters and stream order in Oregon. *Canadian Journal of Fisheries and Aquatic Sciences* 37: 834–847.

Nystrom, M. N., D. S. Bell, and C. D. Oliver. 1984. *Development of Young Growth Western Redcedar Stands*. USFS Research Paper PNW 324. Pacific Northwest Range and Experiment Station, Portland, OR.

Oliver, C. D. and B. C. Larsen. 1990. *Forest Stand Dynamics*. McGraw-Hill, New York.

Oregon Progress Board. 2000. *State of the Environment 2000*. Available at egov.oregon.gov/DAS/OPB/soer2000 index.

Pess, G. R., B. D. Collins, M. Pollack, T. J. Beechie, A. Haas, and S. Grigsby. 1999. *Historic and Current Factors That Limit Coho Salmon Production in the Stillaguamish River Basin, Washington State: Implications for Salmonid Protection and Restoration*. Snohomish County, Everett, WA.

Portland–Multnomah Progress Board. 2001. *Salmon Restoration in an Urban Watershed*. Portland–Multnomah Progress Board, Portland, OR.

Poulin, V. A., C. Harris, and B. Simmons. 2000. *Riparian Restoration in British Columbia: What's Happening Now and What's Needed for the Future*. B.C. Ministry of Forests, Victoria, BC.

Scholfield, N. 1853. *Journal of the Klamath Exploring Expedition, 1850–53*. Unpublished manuscript in Oregon State Library System, Salem.

Scott, W., R. Meade, R. Leon, D. Hyink, and R. Miller. 1998. Planting density and tree size relations in coast Douglas fir. *Canadian Journal of Forestry Research* 28: 74–78.

Sedell, J. R., F. J. Triska, and N. S. Triska. 1975. The processing of conifer and hardwood leaves in two coniferous forest streams: weight loss and associated invertebrates. *Verhandlungen-Internationale Vereinengung für Theorofilche und Angewandte Limnologie* 19: 1617–1627.

Stewart, H. 1984. *Cedar: Tree of Life to the Northwest Coast Indians*. Douglas & McIntyre, Vancouver, BC.

Suberkropp, K. F. 1998. Microorganisms and organic matter decomposition. Pages 373–398 in R. J. Naiman (ed.), *River Ecology and Management: Lessons from the Pacific Coastal Ecoregion*. Springer-Verlag, New York.

Tappenier, J. C., J. Zasada, P. Ryan, and M. Newton. 1991. Salmonberry clonal and population structure in Oregon forests: the basis for persistent cover. *Ecology* 72: 609–618.

Triska, F. J., J. R. Sedell, and S. V. Gregory. 1982. Coniferous forest streams. Pages 292–332 in R. L. Edmonds (ed.), *Analysis of Coniferous Forest Ecosystems in the Western United States*. Dowden, Hutchinson and Ross, Stroudsburg, PA.

Turner, N. J. 1992. *Plants in British Columbia Indian Technology*. Handbook No. 38. Royal British Columbia Museum, Victoria, BC.

Vannote, R. L., G. W. Minshall, K. W. Cummins, J. R. Sedell, and C. E. Cushing. 1980. The river continuum concept. *Canadian Journal of Fisheries and Aquatic Science* 37: 130–137.

Warttig, W. R., D. Clough, and M. Leslie. 2000. *Kennedy Flats Restoration Plan*. Prepared for International Forest Products Ltd., Central West Coast Forest Society, Pacific Rim National Park, Ministry of Forests, Ministry of Environment Lands and Parks, and FRBC.

Wishnie, M. and A. McClintick. 1999. *The Lummi Natural Resources Riparian Zone Restoration Project*. Center for Streamside Studies, University of Washington, Seattle.

Zeide, B. 2001. Thinning and growth. *Journal of Forestry* 99: 20–25.

Chapter 7

Freshwater Wetlands

John van Staveren, Dale Groff, and Jennifer Goodridge

Recognition of the importance of freshwater wetlands has increased in the last few decades. Unfortunately, less than half of the estimated wetland area present in the conterminous United States in the late 1700s remains today (Dahl 1990). This loss has resulted in a reduction of the many ecological functions wetlands provide. These functions include improving water quality, increasing flood control, and providing habitat, particularly for species whose populations are threatened. In the Pacific Northwest, this includes multiple species and stocks of salmon, which use wetlands along rivers and streams as vital nursery habitat (Giannico and Hinch 2003). Freshwater wetlands occur throughout the Pacific Northwest.

Without active wetland restoration, wetland area, function, and diversity will continue to decrease. Successfully restoring wetlands has proven to be difficult, however. Studies show that many restoration projects fail to meet their goals and objectives (Kentula et al. 1992, Johnson and Mock 2000). The most successful restoration projects are those that are carefully planned, where the driving forces of the wetland (e.g., hydrology) are well understood, and where there is long-term stewardship, including ongoing management and monitoring. The Mud Slough wetland mitigation bank in Oregon's Willamette Valley is an example of a successful wetland restoration project in which existing conditions were well understood, the restoration plan was well conceived, and active monitoring and maintenance played an important role.

This chapter focuses on the practice of freshwater wetland restoration as a means of reestablishing wetland area and functions in the Pacific Northwest.

Historical Overview of Wetland Loss and Protection

Freshwater wetlands are transitional between terrestrial and deepwater aquatic ecosystems. They can be inhospitable to humans, make poor farmland unless drained, and provide no viable routes for commerce. Early on, these traits led to a disregard for wetlands and their potential functions. This disregard resulted in widespread acceptance of wetland destruction, particularly their conversion to agricultural uses (Hammer 1992, Mitsch and Gosselink 2000). Because wetland soils accumulate organic matter and nutrients through slow decomposition under saturated soil conditions, they can result in high crop yields, thus providing the incentive to drain wetlands for farmland (Lewis 2001). The conversion of large areas of freshwater wetlands to agricultural uses began in the early nineteenth century in the Pacific Northwest (Taft and Haig 2003).

In Alaska, estimates are that more than 200,000 acres of wetlands have been lost in the last 100 years (Dahl 1990). In the Fraser Lowland and parts of Vancouver Island, British Columbia, estimated losses approach 70% of the original wetland area.

In the ecologically sensitive area of South Okanagan, only 15% of the original wetland area remains (Nowlan and Jeffries 1996). Loss of wetlands is not merely a historic artifact but continues to this day. Losses in the Willamette Valley averaged more than 500 acres per year between 1981 and 1994, with 64% the result of conversion to agriculture (Daggett et al. 1998).

Increased understanding of the connection between declining waterfowl populations and wetland losses in the 1920s and 1930s led to increased public awareness of the value of wetlands (Lewis 2001). This awareness helped to support passage of the Migratory Bird Habitat Stamp Act of 1934, allowing the collection of fees for purchasing large tracts of land for wetland protection (Mitsch and Gosselink 2000).

In the 1970s, researchers began to more publicly state the role of wetlands and articulated the important functions wetlands perform (Novitzki et al. 1997). In 1977, the Federal Clean Water Act was passed and provided a permitting mechanism through Section 404. This was the first legal protection of wetlands in the United States (National Research Council 1995). In 1987, the U.S. Army Corps of Engineers, which administers Section 404, released the *Corps of Engineers Wetland Delineation Manual Technical Report Y-87-1* (Environmental Laboratory 1987), defining wetlands as areas with hydric soils, a dominance of hydrophytic vegetation, and wetland hydrology.

Since the passage of the federal Clean Water Act, some state and local jurisdictions have also enacted wetland regulations. In the Pacific Northwest, Oregon has a state-level wetland regulatory program. Other Pacific Northwest states regulate activities in wetlands through Section 401.

No explicit federal or provincial wetland protection exists in Canada. Federal wetland protection is enacted through such laws as the Fisheries Act, which protects wetlands that provide habitat for fish, and the Canadian Environmental Assessment Act, which requires an assessment of potential environmental effects of development projects. Provincial laws that provide some form of wetland protection include the Water Act, the Wildlife Act, the Land Act, the Waste Management Act, and the Environmental Assessment Act. Most wetland protection decisions are made at the local level by municipalities or regional districts through controls of land use and development decisions.

The process known as wetland mitigation is an integral component of wetland regulation. Wetland mitigation is a federal, state, or local permitting requirement to ensure replacement of wetland loss in area or function through the restoration, creation, and enhancement of wetlands. In a regulatory context, these terms generally are defined as follows:

- Wetland restoration is the reestablishment of wetland hydrology and a hydrophytic plant community to an effectively drained or filled wetland.
- Wetland creation is the establishment of wetland hydrology and a hydrophytic plant community where they did not previously exist.
- Wetland enhancement is the functional improvement of a degraded wetland.

The number of wetland mitigation projects implemented each year has increased with the surge in land development. Private landowners, state and federal agencies, watershed councils, nonprofit organizations, and other groups implement nonregulatory wetland restoration in the Pacific Northwest. They rely on financial assistance or volunteers to achieve restoration objectives. State and federal programs available to fund nonregulatory wetland restoration include the Wetlands Reserve Program (WRP), a voluntary program managed by the U.S. Department of Agriculture Natural Resources Conservation Service (NRCS). This program offers landowners a means to protect, restore, and enhance wetlands on property currently in agricultural production by paying farmers to put farmed land into a conservation easement for wildlife use. Participation usually involves taking land out of production for the duration of the easement, installing native plants, and sometimes implementing hydrologic manipulations. Another federal program stems from the 1989 North American Wetlands Conservation Act, which provides federal cost-share funding to support the North American Waterfowl Management Plan. The purpose of the act is to stimulate public–private partnerships to protect, restore, and manage wetlands for migratory birds and other wildlife.

A variety of groups, including Ducks Unlimited and The Nature Conservancy, manage, preserve, and restore wetlands in the United States and Canada. The Wetlands Conservancy in Oregon performs a similar function at the local level. The restoration efforts of these groups are not subject to regulatory requirements. The resulting restorations usually involve areas that are larger than those involved in compensatory mitigation projects. Because these efforts are not associated with a wetland impact as such, they typically result in a net gain in wetland area and function.

Defining Wetland Restoration

The variety of groups restoring wetlands has resulted in numerous ways of defining wetland restoration. The majority of definitions focus on the repair of anthropogenic actions that have altered or destroyed wetland characteristics (Society of Wetland Scientists 2000). As used in this chapter, *wetland restoration* refers to any act that restores, creates, or enhances wetland characteristics or functions.

Although there is no consensus on the definition of wetland restoration, practitioners do agree somewhat on its goals. The National Academy of Sciences Committee on Restoration of Aquatic Ecosystems states that the goal of restoration, in part, is to emulate a natural, functioning, self-regulating system that is integrated with the ecological landscape in which it occurs (National Research Council 1992). This goal is inherent in the proposed statement of the Society of Wetland Scientists that wetland restoration should produce a persistent, resilient system, one that is not static but has sufficient physical and biological processes intact so that it can respond to human disturbance (Society of Wetland Scientists 2000).

Wetland Function

Wetland function is its capacity to perform hydrologic, geochemical, and biological processes (Azous et al. 1998). Wetland restoration projects generally are able to increase the capacity of the wetland to improve water quality, alter water quantity (e.g., control floodwater), and provide habitat for plants and wildlife (National Research Council 2001). A given wetland generally does not perform all possible functions, nor does any particular wetland perform all functions equally well (Novitzki et al. 1997). It has been widely documented that restoration projects do not uniformly replace lost wetland structure and function (Zedler 1998, Streever 2000, National Research Council 2001). Factors influencing how well a wetland performs these functions include climatic conditions, the location of the wetland in the context of the watershed, the quantity and quality of water entering the wetland, the substrate, and the dominant plant community (Carter 1997).

Water Quality

Increased interest in design standards associated with wetlands constructed solely for improving water quality has led to better understanding of how wetlands need to be restored to provide this function. As a result, restored wetlands can achieve an effective water quality function (National Research Council 2001). Wetlands improve water quality by capturing sediments, transforming nutrients (carbon, nitrogen, and phosphorus), retaining metals, and capturing and assimilating anthropogenic pollutants such as pesticides. A wetland may retain these materials temporarily or serve as a permanent sink if materials are buried in the substrate or released into the atmosphere. For example, nutrients such as nitrogen and phosphorus often are constituents of stormwater runoff. Biogeochemical processes (i.e., nitrification and denitrification) transform nitrogen into forms that can be assimilated by plants or released into the atmosphere. These processes remove up to 90% of the nitrogen entering a wetland (Gilliam 1994). Because phosphorus is attached to sediment particles, the process of sediment deposition can remove phosphorus, depending on the length of time the wetland detains water. Phosphorus retention is among the most important functions of natural and constructed wetlands (Mitsch and Gosselink 2000).

Flood Control

Many wetlands, especially those associated with streams, can temporarily store water during storms. This storage can reduce flood peaks, allowing tributary streams to peak at different times, which distributes stormwater runoff over longer time periods. The ability to store water can reduce downstream erosive forces and allow groundwater recharge. Restoration activities to support the function of flood control entail consideration of the location of a wetland in the context of its watershed. Wetlands that are isolated may be able to store large quantities of water but generally do not affect local or regional runoff because they lack natural outlets.

Habitat

Wetland restoration design often includes components providing habitat for a variety of flora and fauna. In the United States, wetlands provide habitat for an estimated 190 amphibian species, 270 bird species, and more than 5,000 plant species (U.S. Geological Survey 1996). In addition, approximately 26% of the plants and 45% of the animals listed as threatened or endangered either directly or indirectly depend on wetlands for survival (Hammer 1992). The provision of habitat is particularly important in the Pacific Northwest for species that rely on wetlands to complete a portion of their life cycles. Among these are many amphibian species whose populations have declined in recent years (U.S. Geological Survey 2003).

Freshwater Wetlands and Associated Vegetation in the Pacific Northwest

The Pacific Northwest has diverse freshwater wetlands, largely because of local and regional variations in climate and geomorphology. Most of the region experiences a Mediterranean seasonal climate, in which precipitation falls primarily in the winter and spring, and summers tend to be dry. Portions of the region (especially in the south and east) have hotter and drier summers as a rule, which can greatly influence the nature and distribution of wetlands in those areas. Regionally distinct wetlands can be roughly grouped as coastal freshwater, western interior valley, and eastern interior (semiarid). In addition, certain areas contain vernal pools and bogs, two less common systems that are described briefly.

Coastal Freshwater Wetlands

Pacific Northwest coastal areas generally experience less seasonal variation in temperature and higher year-round humidity than interior areas because of the moderating effect of the Pacific Ocean. These climatic conditions are limited largely to areas west of the Coast Ranges in Oregon, Washington, and British Columbia yet extend inland for more than a hundred miles around Puget Sound. Wetlands formed under these conditions are more likely to remain wet most or all of the year and in many cases may be peat-forming, especially in the northernmost portions of the range.

Coastal freshwater wetlands often are located along river estuaries subject to tidal fluctuations;

these wetlands gradually transition to more salt-tolerant marsh communities closer to the ocean or Puget Sound. Other wetlands may be associated with depressions formed by glacial action (especially in Puget Sound area) or in interdunal swales (especially along the Oregon coast). Forested alder and cedar swamps, willow scrub, and herbaceous marsh and bog communities are scattered throughout the coastal regions.

Plants commonly found in coastal wetlands include red alder (*Alnus rubra*), Sitka spruce (*Picea sitchensis*), western redcedar (*Thuja plicata*), willows (*Salix* spp.), salmonberry (*Rubus spectabilis*), bearberry honeysuckle (*Lonicera involucrata*), hardhack spiraea (*Spiraea douglasii*), western crabapple (*Malus fusca*), Labrador tea (*Ledum* spp.), bog blueberry (*Vaccinium uliginosum*), slough sedge (*Carex obnupta*), yellow pond-lily (*Nuphar luteum*), skunk cabbage (*Lysichiton americanum*), and water parsley (*Oenanthe sarmentosa*).

Interior Valley Wetlands, West of Cascade Crest

Palustrine (nontidal) wetlands are found throughout inland areas of British Columbia, Washington, and Oregon, most extensively in the broad valleys between the Coast Ranges and the Cascade Mountains. The wetlands in this region often form where water is discharged from adjacent slopes onto valley floors; these areas may also be subject to seasonal flooding from nearby streams. Depending on current and historic site conditions, wetland plant communities in the region may include forested, scrub–shrub, and emergent types; sometimes all are present as a mosaic in larger wetlands. Because of the low gradients and often fertile soils of these areas, many wetlands were drained over the past century for agriculture and other land development.

In western Oregon, late Pleistocene flood deposits blanketed the lower elevations of the Willamette River Valley with fine clays, producing finely textured soils that drain poorly. Poorly drained soils also are common in the Puget Trough area of western Washington, in some instances formed in highly compacted, unsorted glacial till materials. More recent deposition of finely textured soils from flood events of nearby waterways has also contributed to wetland formation in the region.

Because of the Mediterranean seasonal climate, little rainfall occurs between June and October, and wetlands often are quite dry by the end of summer. As a consequence, some wetland plants have adapted to both significant flooding in winter and drought conditions in summer. Trees and shrubs include black cottonwood (*Populus trichocarpa*), red alder, western redcedar, Oregon ash (*Fraxinus latifolia*), willows, red osier dogwood (*Cornus sericea*), black hawthorn (*Crataegus douglasii*), roses (*Rosa* spp.), and Douglas spiraea. The herbaceous groundcover may include a wide variety of grasslike species and forbs, including sedges (*Carex* spp.), rushes (*Juncus* spp.), reed canarygrass (*Phalaris arundinacea*), tufted hairgrass (*Deschampsia cespitosa*), and many others. The greatest species diversity typically is found in more open, seasonally wet sites (i.e., wet prairies in southwestern Washington and the Willamette Valley) where highly variable microtopography allows plants adapted to early season ponding to grow in close proximity to upland species. As a result, these transitional communities may change in character from dominance by wetland plants in spring to upland plants in late summer.

Semiarid Regions, East of Cascade Crest

Areas east of the Cascade Mountain crest receive significantly less precipitation than most areas west of the crest because of a significant rain shadow effect. Because of the generally higher elevations and a more continental climate east of the Cascades, much of the precipitation falls as snow in the winter. Although naturally occurring wetlands are few and often limited to narrow riparian areas, there are also wetland complexes in the larger intermountain basins. Artificial wetlands resulting from irrigation canal leaks and stream impoundments (reservoirs) also occur at numerous locations in this semiarid landscape. High summer temperatures result in high rates of evapotranspiration, with wetland water levels tending to fluctuate widely with the season.

Common woody species in eastside wetlands include black cottonwood, quaking aspen (*Populus tremuloides*), alder (*Alnus* spp.), willows, red osier dogwood, and currants (*Ribes* spp.). A variety of emergent species (e.g., sedges, rushes) are likely to be present as well. More alkaline areas may include tules (*Scirpus* spp.), sedges, and alkali saltgrass (*Distichlis stricta*). The eastside marshes typically provide important habitat for migratory birds, particularly waterfowl.

Vernal Pool Wetlands

Vernal pool wetlands are a unique type that is nonetheless common east of the Cascades in Washington and Oregon and in southwestern Oregon and northern California. These wetlands generally occur in more arid climates, forming where winter precipitation ponds on a shallow, often cemented soil layer called a duripan. Because surface water infiltrates very slowly through the duripan, increased evapotranspiration as temperatures rise in spring may cause these wetlands to dry up quickly. The pools typically are dry from May through at least October, until the onset of fall rains. Vernal pools may also be found in pool–mound complexes, which may be interconnected by shallow swales. Such pool complexes are uncommon in the region, and their process of formation is still poorly understood. Vegetation in vernal pools typically is herbaceous because the shallow substrate and limited moisture favor early-flowering annuals. Nevertheless, a variety of woody species may colonize the uplands adjacent to the pools.

Because many vernal pools (especially in larger complexes) have been developed for agricultural or industrial uses, plants and animals that rely on this habitat for part or all of their life cycle may be rare. Pool species may include meadowfoam (*Limnanthes* spp.), popcorn flower (*Plagiobothrys* spp.), smooth lasthenia (*Lasthenia glaberrima*), and downingia (*Downingia* spp.). Rare species of fairy shrimp are also present in vernal pools in the southern portions of the range.

Lower-Elevation Bogs

Glacial activity near Puget Sound in Washington and farther north into British Columbia has carved out basins and deposited a lodgment till with poor permeability. Precipitation that falls on these areas may accumulate, forming "kettle pond" wetlands. Because of the high water levels, rates of decomposition are low and eventually result in low-nutrient, acidic conditions in the depressions. The high annual rainfall and cold winters at the higher latitudes also favor perennial cool season grasses and sedges over some of the annual species that occur farther south. The acidic conditions may also be augmented by accumulations of sphagnum moss on the bog surface. Plants include round leaf sundew (*Drosera rotundifolia*), cottongrass (*Eriophorum* spp.), cloudberry (*Rubus chamaemorus*), bog laurel (*Andromeda polifolia*), bog bean (*Menyanthes trifoliate*), and Cusick's sedge (*Carex cusickii*). See Plate 4 in the color insert for wetland plant reference community examples.

Considerations for Restoring Freshwater Wetlands

The wetlands described in this chapter differ in both structure and function. Because of the diversity of wetlands, no single restoration method is appropriate. Even so, a similar process is involved in wetland restoration projects that are effective, regardless of wetland type. This process has several phases, including site investigation and design, design implementation, and monitoring and maintenance. Successful completion of all aspects of wetland restoration is critical to the ultimate success of the restoration project. Restoration projects generally involve similar ecological and technical issues as well. Some of the ecological and technical issues that must be considered in wetland restoration are detailed in the following sections.

Initial tasks in restoration involve establishing the project goals and objectives and collecting sufficient information to adequately characterize the physical and biological influences on the site. The amount of information needed to ensure successful restoration depends on the type of wetland to be restored, the restoration goals and objectives, and the location of the site in a watershed context.

Setting Goals and Objectives

Establishing goals and objectives involves an understanding of what is appropriate and sustainable given the strengths and constraints of the site (Granger et al. 2005). Project goals, such as satisfying regulatory requirements, establish the structure for the design and identify the functions of interest. Project objectives identify components of the goals and potential methods for achieving each component.

One tool particularly useful for setting goals and objectives for wetland restoration projects is the hydrogeomorphic method. This method provides a wetland classification that facilitates the setting of appropriate goals for restoration projects and provides a means for measuring success. Based on a landscape perspective, the method is in wide use for assessing wetland functions in the Pacific Northwest (Brinson 1993). The method takes into account the influence of geomorphic setting, water source, and hydrodynamics on wetland structure and function (Notices 1996). It classifies wetlands on the basis of their ideal functional roles as ascertained through studies of reference wetlands (Whigham 1999). The hydrogeomorphic method describes five major classes of wetlands: riverine, depressional, slope, flats (organic soil and mineral soil), and fringe (estuarine and lacustrine) (Burkhardt 1996). These classes are further subdivided regionally.

Site Location

In the past it was typical for restoration priorities to be made strictly at the site level, without consideration of the project's landscape position (Race and Fonseca 1996). Thus, landscape setting often has not been included in decisions regarding the location and functional goals of wetland mitigation projects. Regulatory agencies often advocate "on-site, in-kind" wetland mitigation projects, thus requiring the wetland mitigation project to take place in close proximity to the wetland impact. Furthermore, economic and practical considerations often play a major role in the choice of a mitigation project location.

The location of a wetland in the watershed can influence its function. Wetland restoration designed to provide an effective water quality function (e.g., to remove nutrients and sediment) must be located in the lower portion of a watershed to ensure sufficient contribution of water from surrounding uplands. Wetland restoration with a goal of increasing the abundance of bird species may be more successful if the site is located in a complex of existing wetlands (Fairbairn and Dinsmore 2001). A wetland designed to provide habitat for amphibians, such as the red-legged frog (*Rana aurora*), must be located in an area with dispersal corridors that allow adults to migrate to forested upland habitats in summer. Recently, restoration practitioners have been taking watershed-based approaches in identifying functional priorities. Restoration projects that have not been bound by permitting requirements have had the flexibility to consider landscape criteria. Therefore, these projects generally have been more successful than compensatory projects in addressing the functional needs and environmental processes of the watershed (Department of Ecology et al. 2004).

Reference Sites

A reference site is an area comparable in structure and function to the targeted restoration site. Such a site provides a model for the restoration project because the climate, soils, hydrology, light, and water quality are substantially the same as those at the restoration site. Although evaluation of a reference site is useful, each aquatic system presents a unique set of conditions. Therefore, restoration practitioners should understand both the similarities and the differences between reference and restoration sites when crafting a project's goals and objectives.

Hydrology

Hydrology must be assessed carefully to ensure that it is sustainable. Ideally, the restoration area should be resilient to climatic changes and watershed changes that alter the hydrologic input (Mitsch and Gosselink 2000, National Research Council 2001). Hydrology can be difficult to characterize accurately. The degree of analysis and data collection needed before a wetland restoration project is designed and implemented depends on the goals, the site's landscape position, and its hydrologic influences.

Restoration projects that reestablish hydrology to a wetland that has been filled or drained by ditching or field tiles may entail little data collection. The hydrologic restoration of these areas may be achieved by filling in drainage ditches, blocking drainage tiles, or removing levees, berms, fill material, or tide gates. Sites such as these need little modification to achieve hydrologic objectives and are generally quite successful.

Information regarding the primary source of hydrology for the site and the pattern of hydrologic input and loss throughout the year is a necessary component of restoration design. Precipitation contributes water to all wetlands; however, the primary source usually is groundwater, surface water, or a combination of the two. Many wetlands receive water from all three sources, depending on the time of year. Characterization of wetland inflows and outflows (i.e., the water budget) includes measures of precipitation, surface water (inflow and outflow), groundwater (inflow and outflow), and evapotranspiration (Carter 1997). The water budget dictates the wetland's hydroperiod, the fluctuation in both surface and subsurface water level.

The hydroperiod defines the hydrologic character of a wetland over a period of time, with seasonal fluctuations the most important. Restoration practitioners can evaluate the relative stability of site hydrology by analyzing the hydroperiod for several years. Hydroperiod data also provide information needed to determine appropriate excavation depths, plant species, and site-specific functions.

Water level fluctuations in wetlands that are fed by surface water or precipitation tend to be greater than they are in wetlands that are groundwater based (Novitzki 1982). In arid areas hydroperiods exhibit pronounced seasonal fluctuations. Such fluctuations occur in vernal pools in eastern Washington and southern Oregon, which receive water primarily from precipitation. The hydroperiods of riverine wetlands along low-order streams in urban areas also can fluctuate dramatically in response to local storms.

In contrast, the hydroperiods of wetlands located along high-order rivers usually do not respond rapidly to storms; instead, they tend to reflect seasonal patterns of precipitation or floods (Mitsch and Gosselink 2000). Exceptions to this pattern include the hydroperiods of wetlands situated along tidally influenced rivers, such as the lower Columbia River and the lower Fraser River in British Columbia, which fluctuate daily. The release of water from dams complicates the hydroperiod. Water levels in these rivers may vary throughout the year and from year to year.

Restoration practitioners can assess the timing, frequency, and period of inundation of riverine restoration sites with data collected from a discharge gauge (Wetlands Research Program 1998). The gauge must be located close to the restoration site and have a period of data collection sufficiently long to adequately characterize the stream's hydrology.

Projects on sites that rely primarily on groundwater are generally the most difficult to characterize and consequently to restore. Groundwater levels are not static; they fluctuate seasonally and with precipitation. Groundwater flow through a wetland is complex and is influenced by a variety of factors, including hydraulic gradients, hydraulic conductivity, soil porosity, and storage coefficients (Wetlands Research Program 1993). Assumptions about the hydrology of these wetlands often are based on observations of existing conditions (e.g., the presence of hydric soils) or on data collected over one season. Two common errors in the design of restorations for groundwater-fed wetlands are the use of redoximorphic (mottling) features in the soil as a determinant of wetland excavation depth and the assumption that the presence of hydric soils is sufficient to characterize the quantity of water available for restoration (Garbish 2002). Field indicators of hydric soils can persist long after the hydrology of a wetland has been altered to the point at which it can no longer support a targeted wetland plant community or satisfy jurisdictional requirements.

Hydrologic assessment of a groundwater-fed restoration site includes information about the amount and timing of rainfall in the area. The hydrologic response of the site to rainfall typically occurs on several time scales. Understanding how the restoration site responds to rainfall is best achieved by comparing daily (preferably hourly) rainfall data for the area with a time series of water table elevations or inundation depths within the site. Water table data collections made over an extended time period are essential for successful restoration of these sites.

Water wells or piezometers often are installed to collect data needed to characterize groundwater depth and movement at a restoration site. A water well is a tube that is slotted along its entire length and usually installed to a depth of 4 feet. Data collected at least weekly during the wet portions of the growing season indicate the depth to the water table (also called the phreatic surface). In contrast, a piezometer is a tube that is open only at its ends (some may have a slit a short distance up from the bottom). Because water flows from a point of high water pressure to a point of low water pressure, it enters the piezometer when it is under pressure greater than atmospheric pressure. Several piezometers placed at different depths provide water level measurements that indicate the direction of groundwater flow across the landscape. In addition to determining the direction of groundwater flow, piezometers are useful for determining whether a basin is a recharge or discharge area and whether shallow groundwater is perched above an impermeable soil layer. The recommended period for collecting data to adequately characterize a restoration site's hydrology and avoid bias from overly wet or dry years is a minimum of 3 years before site excavation (Doyle 1997).

Restoration practitioners should be aware that under certain conditions the depth to a measured water table can change after a site is excavated (Winston 1997). Although this may not happen frequently, it can be a critical factor in the design of groundwater-fed, nonponded, constructed wetlands. This change may result from the perforation of an aquitard or aquiclude within the soil, an increase in evapotranspiration, or a change in how water enters the wetland from the adjacent upland area.

Typically, groundwater flows slowly toward a stream or wetland after rainfall has infiltrated the soil and accumulated as a mound beneath the surface. The elevation of the water table on the slopes of the mound reflects the interaction of lateral groundwater flow with water that percolates through the soil from rainfall. Removing topsoil from the restoration site can reduce upslope groundwater mounding, causing the water table to form a new surface between the elevation of the upslope groundwater source and the discharge point downslope. The new water table surface generally is lower than the pre-excavation water table. The drop can vary from unnoticeable (less than a millimeter) to as much as a meter depending on the geometry of the upslope source, the porosity and hydraulic conductivity of the soils, and the rate of rainfall (Figure 7.1).

Wetland hydrology can be predicted through modeling. The finite difference model (MODFLOW) from the U.S. Geological Survey is a mod-

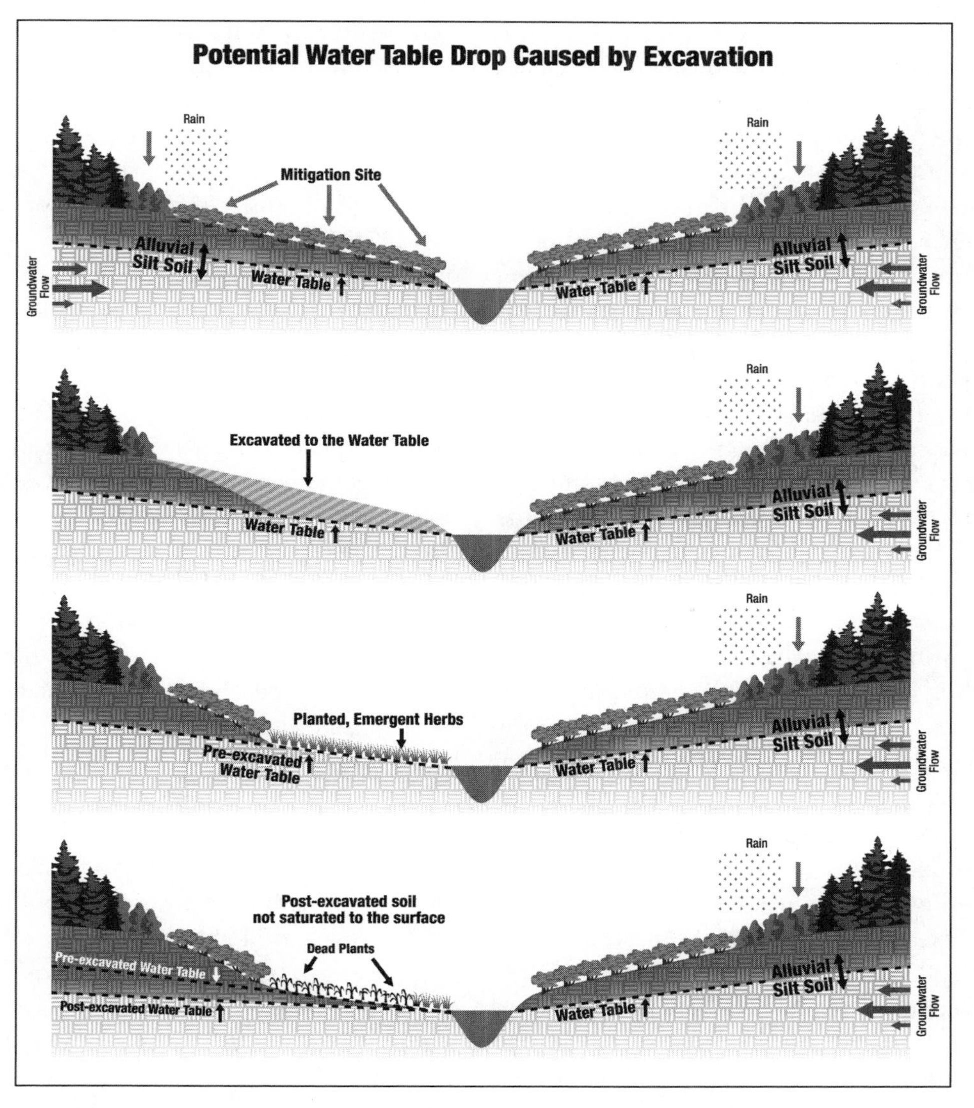

FIGURE 7.1. Timeline illustrating the potential water table drop after the excavation of topsoil from a restoration area. Topsoil removal can alter groundwater mounding beneath the ground surface, causing water tables to be lower than expected and potentially jeopardizing the survival of targeted plant communities. (*Illustration by Sarah Fine*)

ular modeling system well suited for saturated groundwater flow. Other groundwater models, such as FEMWATER, HYDRUS-2D, and SUTRA, describe unsaturated soil hydrology by assuming that the gas phase in the soil is at atmospheric pressure.

Soils

Soils play a critical role in the success of a wetland restoration, particularly in the success of plant growth. Restoration practitioners can anticipate changes in wetland hydrology by assessing the hydraulic properties of soils in and adjacent to the restoration site. One important consideration is the amount of space available for water in the soil (i.e., soil porosity). Because the amount of space is difficult to estimate for fine-grained soils, measurements usually are made in a laboratory by comparison of the mass of saturated soil to the mass of oven-dried soil. Groundwater recharge may be substantial in areas with high soil porosity and thus offset the effects of drought by increasing the height of the water table and augmenting base flows in streams.

An additional property of soils that may affect wetland hydrology is hydraulic conductivity, or the ease with which water moves through the soil. In the Pacific Northwest, many soils vary vertically more than they vary horizontally. Although geomorphic changes may affect the shape of the surface, soil processes generally tend to parallel the ground surface. This pattern of variability must be taken into consideration for restoration sites that are to be excavated in a way that exposes soil horizons at depth.

By affecting hydraulic conductivity, soil texture plays an integral role in the type of wetland that can be restored and the retention and movement of water in the soil. Distinct vertical changes in texture may create conditions that tend to concentrate water in specific horizons, thus exhibiting a set of wetland properties quite different from those in soil with uniform texture.

In addition to affecting hydrology, soils serve two principal functions in wetland restoration projects. First, soils support biological processes. They supply plants with nutrients and provide habitat for mycorrhizae and symbiotic bacteria that facilitate nutrient uptake. Second, soils have physical properties that provide support for plants and functions relating to water retention and transport.

Examination of the biological and physical properties of soils is critical for restoration sites that must be excavated. Excavation often exposes soils that vary widely in organic content, chemistry, and structure from surface layers. In general, the upper 14–24 inches of soil provide the structure, water, and nutrients for wetland plants. Removing these layers alters all of these properties and can have a profound effect on the success of planted vegetation.

Excavation exposes soils low in organic matter content. In fact, studies have shown that the organic component of restoration sites is lower than that of natural wetlands (Kentula et al. 1992, Bischel-Machung et al. 1996). A consequence of low organic matter (i.e., the ability of the soil to retain nutrients and prevent them from leaching beyond the roots) is poor growth of vegetation, reduced habitat and food chain support for invertebrates and fish, and altered nutrient cycling (Schaffer and Ernst 1999). The cation exchange capacity of soils at the depth of the projected excavation and at the surface provides a measure of soil fertility. Because soils tend to vary along elevation gradients, testing at various depths can ensure proper characterization (Reese and Moorhead 1996).

Amending Site Soils

Soils that are not saturated throughout the year may need amendments to buffer the nutrient supply in the subsoil and to introduce sufficient organic matter to start building structure in soil surface horizons. When organic matter is added to soil, it stimulates a sudden increase in microbial growth. For example, E&A Environmental Consultants, Inc. and Adolfson Associates, Inc. (1996) found that rates of survival and growth were increased by more than 20% with the application of a stable, mature compost with high organic and low nitrogen content. Microbes rely on carbon for energy and nitrogen to grow. If nitrogen is not readily available from the organic material, microbes will take it from the soil, thus reducing the amount of nitrogen available to plants.

Among the options for organic amendments are garden compost, animal manures, paper-processing wastes, sewage treatment sludge, mint compost, and mushroom compost. These materials vary greatly in composition and in similarity to naturally occurring soil organic matter. Animal manures that are high in urine content (e.g., chicken manure) tend to be high in urea and ammonium, perhaps too high (in fresh manure) for native plantings adapted to nitrogen-limited conditions. Sewage treatment sludges usually are of uniform nitrogen content but may contain elevated levels of heavy metals or other chemicals. Most available composts have been pasteurized at 140°F for some time to kill pathogens and some weed seeds. Mint compost is pasteurized by the steam distillation process but may contain traces of long-lived pesticides (e.g., clopyralid). Mushroom composts also are high in ionic nitrogen, although high levels of nitrogen salts tend to be short-lived in many environments.

Tilling an amendment into the top of the soil column immediately after excavation provides time for it to become integrated into the soil when the fall rains initiate shrink–swell processes. Temperatures may be high enough in the late summer to further mature the amendment and prevent the sudden chemical shock that can occur when an amendment and plants are installed at the same time.

Plants

Many nonregulatory restoration projects, particularly those that restore large areas, do so without first creating a planting plan. These projects rely on the restoration of hydrologic conditions to facilitate natural revegetation or passive restoration. Projects that are implemented as part of a permitting requirement are almost invariably planted. These mitigation projects provide compensation for a wetland loss and, in general, are monitored for several years. Regulatory agencies expect a well-vegetated wetland at the end of a required monitoring period.

Planted restorations generally have native plants installed throughout the site, including the buffer area. The targeted plant community depends on the project's goals and objectives. A reference wetland can be useful in identifying the appropriate targeted plant community for the restoration site. However, even subtle differences in hydrology or soils between the reference wetland and the restoration site can influence plant establishment.

The growth and succession of wetland plant communities depend on a variety of factors but are most influenced by ecological needs or tolerances of saturated soil conditions, water depth, and frequency and duration of flooding (Niering 1989). Even small changes in the hydroperiod can result in shifts in plant species composition over time. In systems that are influenced by the release of water from dams, the rate of change of soil water pressure between periods of inundation can affect plant establishment (Pacific Habitat Services, Inc. 2003). Soil preparation may help reduce the range of pressure fluctuations. The composition of plants in many wetlands, especially those located in urban areas, is always in flux.

Over time, plant communities respond not only to changes in hydrology but also to factors such as exposure to light. For example, many large restoration projects are established under open sky, and plants are exposed to full sunlight throughout the day. As the plants mature, the light regime in the wetland changes. Species adapted to full sunlight, such as the arroyo willow (*Salix lasiolepis*), lose their competitive advantage to species that are tolerant of shade, such as red osier dogwood. Plant community shifts over time, although difficult to predict, must be considered in assessing the success of a wetland restoration project.

Installing Plants

Because of the dry summers throughout much of the Pacific Northwest, plants should be installed in the late fall and early spring. Planting during this colder and wetter period ensures that plants are dormant when installed and therefore are not subject to the high-transpiration stress of late spring and summer. Even a few weeks of hot, dry weather in the summer can cause moisture stress and wilting.

Woody plants generally are installed as bare root or as container stock. Bare root plants can be installed only when they are dormant, generally after the winter solstice. Although usually smaller than container plants, bare root plants may adapt quickly to the often harsh growing conditions of a restoration site. Container plants may be robust when planted but may grow slowly and not survive as well because of the moderated growing conditions (e.g., raised in growing medium, watered regularly) usually found in a nursery.

Herbaceous plants need to be kept in a moist, shaded condition or in water-filled containers from the time of collection (in the wild or from a nursery) to the time of planting. These plants should be collected in late fall or winter, when they are dormant. The window of time for collecting ripe seeds is short. Depending on the species, collections may be needed at different times of the year to reflect seasonal diversity. Some seeds need special treatment, such as cold hardening or after-ripening, before they will germinate. Although the use of seeds generally is less expensive than the installation of propagated plants, seed-grown plants tend to be more susceptible to predation. In addition, they may be overcome by invasive species, especially in the early stages of germination and while they are young seedlings (Dorner 2004).

The success of wetland restoration projects can be improved by delaying planting for at least a year to observe the wetland's hydroperiod. This delay allows additional consideration of plant species suited to the site's hydrologic conditions or alterations of site hydrology (e.g., changing weir heights or excavating material).

Buffers

Ensuring the presence of an adequately sized and functioning buffer area is an important consideration for wetland restoration. An upland area surrounding a wetland restoration site reduces impacts from adjacent land uses through physical, chemical, and biological processes. In addition, a buffer helps to protect and maintain the functions of the restoration area. The slope, dominant vegetation, and width of the buffer play an important role in the protection and maintenance of these functions. A forested area not only serves as a visual buffer for many species of wildlife using a wetland but also provides the terrestrial habitat needed for a portion of their life stage. For example, the red-legged frog, a species with a declining population throughout the Pacific Northwest, migrates from an aquatic habitat in late spring to adjacent woodlands in summer.

Buffers also help to ensure that the water quality of the restoration site is not impaired. Chemical input from the surrounding landscape can change the characteristics of the restoration site and favor invasive species (Kentula 2002). Regulatory programs often include a requirement for a buffer around wetland restoration projects.

Monitoring and Maintenance

Monitoring and maintenance are important aspects of restoration success. A restoration site will evolve through time, and practitioners or landowners need to monitor sites over long periods to determine whether restoration measures have succeeded (Pickett and Parker 1994). The timeframe that regulatory agencies provide for achieving structure and function, generally 5 years, is not long enough (Parker 1997).

Controlling Invasive Species

Colonization of wetland restoration sites by invasive species is a major problem throughout the Pacific Northwest. Invasive species often exclude native species from the site. For example, a wetland may function effectively to improve water quality and yet be dominated by invasive species (Kadlec and Knight 1996). In the Pacific Northwest, wetlands dominated by large, monotypic stands of reed canarygrass (*Phalaris arundinacea*) remove nitrogen and phosphorus but reduce biodiversity and reduce the quality of wildlife habitat by excluding native species. Although invasive species generally are nonnative, some native species, such as cattail (which colonizes nutrient-rich sites), can be invasive and dominate a plant community.

Invasive plants benefit from our mild climate, particularly west of the Cascades, which provides few periods of freezing or extended drought in many areas. Nonnative species become established under normal weather conditions and are able to survive the periods of extreme weather. In wetlands, species such as reed canarygrass, purple loosestrife (*Lythrum salicaria*), Japanese knotweed (*Polygonum cuspidatum*), and saltcedar (*Tamarix* spp.) can spread both vegetatively and by seed to form dense, monotypic stands with little or no habitat benefit.

Achieving Restoration Success

Wetland restoration efforts, especially those carried out as compensatory mitigation, have met with mixed results (Kentula et al. 1992). According to a recent study by the Washington Department of Ecology, only 29% of the state's mitigation areas reviewed were successful in terms of having been implemented according to plan and having met performance standards (Johnson and Mock 2000). In a related study, Johnson et al. (2002) determined that mitigation projects did not adequately compensate for lost wetland functions, although gains in wetlands that improve water quality and flood control functions were significant. In Oregon, only 24% of compensatory mitigation projects completed between 1995 and 1999 fully complied with required criteria for success (Morrow 2000). Reasons for these failures included human disturbance (e.g., unauthorized maintenance and mowing), lack of sufficient hydrology to achieve required criteria or stated hydrologic goals, infestation by invasive species, and failure to achieve native vegetative cover (e.g., less than 80% of planted species were living at the end of the monitoring period).

Wetland restoration projects that are accomplished through programs that do not involve regulatory permitting requirements, such as the NRCS WRP, often are not subject to the same scrutiny as are wetland mitigation projects. These nonregulatory projects generally are large and are not associated with a wetland loss requiring strict adherence to permit conditions and performance standards. They allow landowners to restore and protect large wetland areas. For example, as of fiscal year 2003, 7,831 projects covering 1,470,998 acres were enrolled in the WRP in North America.

Recent scientific publications indicate that many wetland mitigation projects are failures because they do not meet prescribed standards of success. Yet even an ideal reference site may not meet such standards of success as the common 80% native vegetation cover or a prescribed number of indicator species in a vernal pool. Although wetland creation projects often involve the identification of nearby reference sites to aid in developing proposed conditions for grading and planting, they do not as often include reference sites in developing standards for project success. The Clay Station Wetland Mitigation Bank illustrates the background information review and follow-up monitoring needed for successful wetland creation.

Wetland restoration is young, both as a science and as an art. The first recorded wetland restoration project in the Pacific Northwest took place only in 1978 (Frenkel and Morlan 1990). Today, the methods needed to improve the success of restoration projects continue to be developed, and many existing methods are not uniformly applied. Currently, wetland restoration is occurring on an unprecedented scale in the United States and Canada, and interest in the success of wetland restoration efforts is strong. Restoration to mitigate for wetlands destroyed by development often is a regulatory requirement. Courses and instruction on wetland restoration and a variety of funding opportunities are available for implementing wetland restoration projects. This level of interest in wetland restoration almost assuredly means that freshwater wetland restoration efforts will become increasingly successful in time.

Freshwater Wetland Restoration Case Studies

Case studies involving the restoration of freshwater wetlands in the Pacific Northwest are as varied as the physiography of the region. Certain types of wetlands are especially difficult to restore. These wetlands include bogs, fens, and vernal pools (National Research Council 2001, Kentula 2002), which have formed over the millennia and have very specific soil conditions. Even so, restoration practitioners have achieved some success in these rare ecosystems. The Camosun Bog Restoration Group has restored more than 80% of a small sphagnum-dominated bog in the Pacific Spirit Regional Park, Vancouver, British Columbia, which was undergoing rapid succession toward a western hemlock forest (Baker et al. 2000). In addition, practitioners have achieved partial success in creating vernal pools at the Clay Station Mitigation Bank in Sacramento, California, as reported in the third case study.

Case Study: King County Wetland Mitigation Bank, Washington

Klaus Richter, King County Water and Land Resources Division

The King County Wetland Mitigation Bank, located in the East Lake Sammamish Basin between the cities of Seattle and Issaquah, Washington, serves as a compensatory wetland mitigation site for public agencies and utilities. Constructed in 1996, the mitigation bank includes 0.50 hectare of restored wetland, 2.48 hectares of enhanced wetland, and 1.08 hectares of enhanced upland buffer area.

For decades before restoration, the major use of this site was horse pasture, which resulted in compacted soils. In addition, a portion of the site contained fill material. Tiling, ditching, and runoff inputs from the surrounding urban area created large fluctuations in water levels. These hydrologic alterations also affected the wetland plant community, approximately half of which was dominated by a monoculture of Douglas spiraea. Furthermore, conditions at the site were not favorable for the development of amphibian egg masses.

The project team analyzed the site constraints, including disturbed soils and altered hydrologic regime, before preparing the design for restoration. The design included specific steps for increasing amphibian, small mammal, and bird use of the site and for incorporating adaptive management into site monitoring. Among these steps were the following:

- Removing existing fill to return hydrology to the altered portion of the wetland
- Contouring the shoreline of the ponded area to increase the amount of edge habitat
- Installing thin-stemmed emergent plants along the shoreline to provide amphibian breeding habitat
- Dispersing rock piles throughout the site to provide refuge for amphibians
- Placing large woody debris throughout the site to increase habitat for small mammals and birds, including waterfowl
- Clearing areas of vegetation, creating mounds, and planting trees and shrubs on the mounds to further increase diversity within the spiraea monoculture

The project incorporated an experimental design to enable the team to assess the relative effectiveness of five soil treatments in the wetland and three types of mounds in the monoculture. The soil treatments were 0.15 meters of peat over fill, 0.15 meters of topsoil over fill, 0.30 meters of peat over fill, 0.30 meters of topsoil over fill, and scarified fill. Each soil treatment contained replicate vegetation plots. Materials used in constructing the mounds included native peat soils from the site, spiraea spoils plus native peat, and imported peat. The project team collected data on plant establishment (the dependent variable) and groundwater levels and other independent variables to evaluate the success of the various treatments. Site monitoring provided additional information regarding species-specific mortality, the composition of the recruited community, and the identification of suitable replacement plantings.

During excavation the team learned that the site contained more fill material than expected. Therefore, the team decided to remove only a portion of the fill and to install a weir to increase water levels. The fill removal and weir installation created an additional wetland area at the site, one that supported high water levels. The weir also stabilized water levels, reducing water level fluctuations from 0.82 meters to 0.06 meters per year. The high water levels flooded all but one species of the emergent plantings, water plantain (*Alisma plantago-aquatica*). The stabilized water levels provided habitat for cattail, which became established in a portion of the shallow water area. Several native plant recruits, including *Veronica* sp. and *Eleocharis* sp., became established in the shallow water zone as well. Planted tree and shrub establishment ranged from 50% to 100%, depending on the species and location. In addition, native tree recruits, including willow, black cottonwood, and red alder, became established throughout the wetland.

Monitoring data collected as part of the experimental design indicated that portions of the site containing soil treatments had different water level regimes. Thus, hydrologic regime, rather than soil medium, probably contributed to successful plant establishment. Similarly, monitoring data from the mounds indicated that mound size and depth to groundwater, rather than materials used in constructing mounds, affected plant survival. Project designs that incorporate experimental treatments need to ensure that treatments are replicated throughout the mitigation site if monitoring data are to be useful in testing treatments rather than locations.

The success of this site is based on 6 years of monitoring data. The project team hopes that additional funding will allow longer-term monitoring to assess the later successional development of vegetation communities and wildlife use at this site.

Case Study: Mud Slough, Oregon

Mud Slough is located approximately 145 kilometers southwest of Portland in Oregon's Willamette Valley. This wetland restoration area includes 24.3 hectares that serve as a wetland mitigation bank. Additional wetland restoration activities have occurred on 162 hectares. The project involved a variety of agencies and programs in site design and in funding site acquisition, construction, and management.

The mitigation bank is part of a grass seed farm owned by Mark Knaupp, a duck hunter and farmer with a wildlife degree. The additional restoration activities have taken place primarily on a nearby conservation easement. In 1992, Mark created a shallow 8-hectare pond for waterfowl use on his grass seed farm. The grass seed fields adjacent to the pond were wet and difficult to farm, and the pond led to increased numbers of Canada geese feeding in the fields. In 1996, Mark enrolled nearly 130 hectares in the NRCS WRP. He used the money he received to purchase an adjacent 73 hectares of farmland that was drier and more compatible with grass seed farming.

Constraints and opportunities were identified and used to select and design the restoration site. The selected site is generally flat and contains poorly drained clay soils. It needed little grading to improve site hydrology.

To convert the grass seed crop to a native plant community, the project team scraped the surface, planted an annual cover crop, flooded the area, and sprayed the site with Roundup during the spring and fall of one growing season. These activities allowed grass seed to germinate and grow without producing seed that season, thereby exhausting the seed bank of the crop. The team then planted native wetland grasses without additional tilling, which would have brought grass seed remaining in the soil to the surface for germination. Because the area had been used for grass seed production, both the bank and the easement lacked invasive species (Figure 7.2).

Over the next 5 years, seeded native grasses and native herbs, sedges, and rushes became established at the site. Among the herbs is Nelson's checkermallow (*Sidalcea nelsoniana*), a plant listed as threatened under the federal Endangered Species Act (Lev 2001). The soil seed bank probably served as a source of these native species, previously suppressed through active farm management. Annual flooding from an adjacent slough probably introduced additional seeds to the site. Wildlife monitoring indicated that a variety of bird species, including black-necked stilt, Wilson's phalarope, Virginia rail, bittern, heron, egret, bald eagle, and northern harrier, use the site (Figure 7.3).

Active monitoring and maintenance keep the abundance of nonnative and invasive species at the bank and easement to a minimum. In addition, mowing in the fall maintains the native wetland prairie herbaceous community and increases waterfowl use. Mowing is used as a substitute for the historic burning practices that maintained the prairie, a community that occupies less than 1% of its former range in the Willamette Valley (see Chapter 3). Although wetland enhancement projects typically are less successful than are wetland creation and wetland restoration projects (Johnson et al. 2002), site selection, construction methods, and ongoing site management contributed to the success of this enhancement project (Figure 7.4).

FIGURE 7.2. Wet prairie portion of the Mud Slough dominated by *Downingia* sp. at the Mud Slough wetland mitigation bank near Rickreall in the Willamette Valley, Oregon. (*Photo courtesy of Mark and Debbie Knaupp*)

FIGURE 7.3. Phase 1 of the Mud Slough wetland mitigation bank near Rickreall in the Willamette Valley, Oregon 5 years after initial restoration measures. (*Photo courtesy of Mark and Debbie Knaupp*)

FIGURE 7.4. Wet prairie portion of the Mud Slough wetland mitigation bank near Rickreall in the Willamette Valley, Oregon. (*Photo courtesy of Mark and Debbie Knaupp*)

CASE STUDY: CLAY STATION WETLAND MITIGATION BANK

Peter Balfour, Ecorp Consulting, Inc.

The Clay Station Wetland Mitigation Bank, located in southern Sacramento County, California, is a large-scale wetland complex containing 164 hectares of vernal pools, seasonal marshes, creeks, and surrounding upland areas. This wetland mitigation bank compensates for unavoidable impacts associated with development projects in the area (Ecorp Consulting 2004).

Initial assessments indicated that the site provided wetland habitat for a diverse group of plant and animal species. Furthermore, aerial photographs showed that the site contained extensive vernal pool complexes before its conversion to cropland in 1937. The conversion resulted in changes in the soil profile, hydrologic conditions, and plant community. In contrast to wetlands, where water infiltrates into the soils, vernal pools need a duripan or similarly restrictive soil layer. The hydrologic regime of vernal pools consists primarily of ponding in winter and rapid evapotranspiration in the spring. Soil samples indicated that the clay hardpan layer necessary for the establishment of vernal pools was still present, at a depth ranging from 0.10 to 0.46 meters throughout the site.

Construction of the wetland mitigation bank took place in three phases. Phase I, constructed in 1994, involved of the creation of 9.7 hectares of vernal pools and 5.7 hectares of seasonal marsh habitat in a 68-hectare area. Once Phase I appeared to be successful, the project team began planning and implementing Phases II and III. These two phases involved the creation of 9.3 hectares of vernal pools and 10 hectares of seasonal marsh in a 96-hectare area adjacent to Phase I.

Phase I provided compensation for wetland impacts for six development projects. The 8-hectare area provided compensation for one of these projects, Churchill Downs. Because of the magnitude of impacts associated with the Churchill Downs project, Phase I received extensive review from regulatory agencies, including the U.S. Army Corps of Engineers, the U.S. Environmental Protection Agency, and the U.S. Fish and Wildlife Service. The agencies wanted to ensure that the mitigation area provided adequate compensation.

Hydrology monitoring conducted during the 1995–1997 portion of Phase I documented the establishment of wetland hydrology in the vernal pools and seasonal marsh areas. In 1997, these vernal pools had an average vegetative cover of 89% and an average vernal pool floristic index of 0.88. The ideal vernal pool has an index of 1. Furthermore, invertebrate sampling documented the occurrence of four special status, vernal pool branchiopods. These crustaceans included two federally listed species: vernal pool tadpole shrimp (*Lepidurus packardi*) and vernal pool fairy shrimp (*Brachinecta lynchi*). By 2000, all of the vernal pools in Phase I provided successful vernal pool habitat. In addition, wildlife sampling indicated that bird use increased throughout Phase I (Figure 7.5).

Phases II and III provide mitigation to compensate for potential future impacts on vernal pool habitat. These phases are unique in that the project team based the standards of success and monitoring on comparisons of measures of hydrology, vegetation, and wildlife use with conditions at a reference site. The reference site comprised remaining high-quality vernal pools at Churchill Downs. These pools are part of a wetland preserve, located 19 kilometers from the mitigation bank.

The project team monitored hydrology at both the mitigation and reference sites during Phases II and III and made comparisons that incorporated seasonal and annual variations in weather conditions. For example, hydrology monitoring data collected in 2003 indicated that the constructed vernal pools held water slightly longer than did the reference site pools; however, the percentage inundation in the constructed vernal pools was within the range of that observed in pools at the reference site between February and April 2003. These kinds of comparisons are not possible if success is measured by a predetermined standard (e.g., a requirement that 80% of the pools must be 80% inundated in March of any given year).

According to vegetation monitoring data collected in 2003, 49% of the constructed vernal pools in Phases II and III met all of the standards of success for floristic criteria. In comparison, monitoring data from the reference site indicated that only 68% of the vernal pools met all of these standards of success. The created vernal pools that did not meet the floristic success standards failed because they contained too few indicator species. The reference site pools that did not meet all of the standards also contained too few indicator species. Thus, approximately half of the created vernal pools that did not meet the floristic monitoring requirement for indicator species have a vegetation composition within a range that is normal for the site. The project team continues to monitor vegetation in the created vernal pools to evaluate this aspect of project success. Wildlife monitoring indicated that Phases II and III contained both of the federally listed invertebrate species that occurred in Phase I. These species were not present in samples collected from the reference site.

FIGURE 7.5. *(a)* 1993 black and white aerial photograph of the prerestoration conditions of the Clay Station Wetland Mitigation Bank in southern Sacramento County, California. *(b)* Spring 2003 aerial photograph of the Clay Station Wetland Mitigation Bank in southern Sacramento County, California shows ponding in the created vernal pools after excavation. (*Photos courtesy of Ecorp Consulting*)

Summary

Freshwater wetland restoration in the Pacific Northwest has made great strides in the past few years. Although studies indicate that improvements are still needed, especially for wetland mitigation projects, there is an ever-increasing knowledge base of successful restoration techniques and lessons learned. Numerous private, public, and nonprofit groups conduct wetland restoration projects using a variety of funding sources. U.S. and Canadian wetland regulations protecting and requiring mitigation for wetland loss do not appear to be going away.

The challenge to restoration practitioners is to ensure that wetland restoration plans are well designed, correctly implemented, and monitored and maintained to respond to changes in the environment. Only when wetland restoration practitioners are diligent and their plans successful will the loss of wetlands in the Pacific Northwest be stopped.

REFERENCES

Azous, A. L., M. B. Bowles, and K. O. Richter. 1998. *Reference Standards and Project Performance Standards for the Establishment of Depressional Flow-Through Wetlands in the Puget Lowlands of Western Washington.* King County Department of Development and Environmental Services, Renton, WA.

Baker, N., P. Lilley, T. Sasaki, and H. Williamson. 2000. *Investigation of Options for the Restoration of Camosun Bog, Pacific Spirit Regional Park.* Environmental Studies 400 thesis, University of British Columbia, Vancouver.

Bischel-Machung, L. , R. P. Brooks, S. S. Yates, and K. L. Hoover. 1996. Soil properties of reference wetlands and wetland creation projects in Pennsylvania. *Wetlands* 16: 532–541.

Brinson, M. M. 1993. *A Hydrogeomorphic Classification for Wetlands.* Technical Report WRP-DE 4. U.S. Army Corps of Engineers, Waterways Experiment Station, Vicksburg, MS.

Burkhardt, R. 1996. National action plan to develop the hydrogeomorphic approach for assessing wetland functions. *Federal Register* Notices, August 16.

Carter, V. 1997. *Wetland Hydrology, Water Quality, and Associated Functions.* Water Supply Paper 2425,

National Water Summary on Wetland Resources. U.S. Geological Survey, Washington, DC.

Daggett, S. G., M. E. Boule, J. A. Bernert, J. M. Eilers, E. Blok, D. Peters, and J. Morlan. 1998. *Wetland and Land Use Change in the Willamette Valley, Oregon: 1982 to 1994*. Shapiro and Associates, Inc. Report to the Oregon Department of State Lands, Salem.

Dahl, T. E. 1990. *Wetland Losses in the United States, 1780s to 1980s*. U.S. Fish and Wildlife Service Report to Congress, Washington, DC.

Department of Ecology, U.S. Army Corps of Engineers, Seattle District Environmental Protection Agency Region 10. 2004. *Draft: Guidance on Wetland Mitigation in Washington State, Part 1: Laws, Rules, Policies and Guidance Related to Wetland Mitigation*. 04-06-013a.

Dorner, J. 2004. *An Introduction to Using Native Plants in Restoration Projects*. Plant Conservation Alliance, Bureau of Land Management, U.S. Department of the Interior, U.S. Environmental Protection Agency, University of Washington, Seattle.

Doyle, J. 1997. Site selection for hydrologic modifications to restored or created freshwater wetlands. *Restoration and Reclamation Review* 2.

E&A Environmental Consultants, Inc. and Adolfson Associates, Inc. 1996. *Compost Use in Wetland Restoration Projects*. Report No. CM-96-2. Recycling Technology Assistance Partnership.

Ecorp Consulting, Inc. 2004. *2003 Monitoring Report for Clay Station Mitigation Bank (Sacramento County, California)*.

Environmental Laboratory. 1987. *Corps of Engineers Delineation Manual*. Technical Report Y-87-1. U.S. Army Corps of Engineers Waterways Experiment Station.

Fairbairn, S. E. and J. J. Dinsmore. 2001. Local and landscape-level influences on wetland bird communities of the prairie pothole region of Iowa, USA. *Wetlands* 21: 41–47.

Frenkel, R. E. and J. C. Morlan. 1990. *Restoration of the Salmon River Salt Marshes: Retrospect and Prospect*. U.S. Environmental Protection Agency, Region 10, Corvallis, OR.

Garbish, E. W. 2002. *The Dos and Don'ts of Wetland Construction: Creation, Restoration, and Enhancement*. Environmental Concern, Inc., St. Michaels, MD.

Giannico, G. R. and S. G. Hinch. 2003. The effect of wood and temperature on juvenile coho salmon winter movement, growth, density and survival in side-channels. *River Research Applications* 19: 219–231.

Gilliam, J. W. 1994. Riparian wetlands and water quality. *Journal of Environmental Quality* 23: 896–900.

Granger, T., T. Hruby, A. McMillan, D. Peters, J. Rubey, D. Sheldon, S. Stanley, and E. Stockdale. April 2005. *Wetlands in Washington State*. Volume 2: *Guidance for Protecting and Managing Wetlands*. Publication #05-06-008. Washington State Department of Ecology, Olympia.

Hammer, D. A. 1992. *Creating Freshwater Wetlands*. Lewis Publishers, Chelsea, MI.

Johnson, P. and D. L. Mock. 2000. *Washington State Wetland Mitigation Evaluation Study, Phase I: Compliance*. Publication No. 00-06-016. Washington Department of Ecology, Olympia.

Johnson, P., D. L. Mock, A. McMillan, L. Driscoll, and T. Hruby. 2002. *Washington State Wetland Mitigation Evaluation Study Phase 2: Evaluating Success*. Publication No. 02-06-009. Washington Department of Ecology, Olympia.

Kadlec, R. H. and R. L. Knight. 1996. *Treatment Wetlands*. Lewis Publishers, Boca Raton, FL.

Kentula, M. E. 2002. *Wetland Restoration and Creation. National Water Summary on Wetland Resources*. U.S. Geological Survey Water Supply Paper 2425. Available at water.usgs.gov/nwsum/wsp2425/restoration.html.

Kentula, M. E., J. C. Sifneos, J. W. Good, M. Rylko, and K. Kunz. 1992. Trends and patterns in Section 404 permitting requiring compensatory mitigation in Oregon and Washington. *Environmental Management* 16: 109–119.

Lev, E. 2001. *Heroic Tales of Wetland Restoration*. The Wetlands Conservancy, Tualatin, OR.

Lewis, W. M. Jr. 2001. *Wetlands Explained: Wetland Science, Policy, and Politics in America*. Oxford University Press, New York.

Mitsch, W. J. and J. G. Gosselink. 2000. *Wetlands*. 3rd ed. Van Nostrand Reinhold, New York.

Morrow, S. 2000. Compensatory mitigation study by DSL points to need for improvement. *Wetlands Update* 11: 1.

National Research Council. 1992. *Restoration of Aquatic Ecosystems: Science, Technology and Public Policy*. National Academy Press, Washington, DC.

National Research Council. 1995. *Wetlands: Characteristics and Boundaries*. National Academy Press, Washington, DC.

National Research Council. 2001. *Compensating for Wetland Losses under the Clean Water Act*. National Academy Press, Washington, DC.

Niering, W. A. 1989. Vegetation dynamics in relation to wetland creation. Pages 479–486 in J. A. Kusler and M. E. Kentula (eds.), *Wetland Creation and Restoration. The Status of the Science*. Island Press, Washington, DC.

Notices. August 16, 1996. *Federal Register* 61(160): 42593–42603.

Novitzki, R. P. 1982. *Hydrology of Wisconsin's Wetlands*. Information Circular 40. Wisconsin Geological and Natural History Survey, Madison.

Novitzki, R. P., R. D. Smith, and J. D. Fretwell. 1997. *Restoration, Creation, and Recovery of Wetlands. Wetland Functions Values and Assessment. National Water Summary on Wetland Resources*. United States Geological Survey Water Supply Paper 2425.

Nowlan, L. and B. Jeffries. 1996. *Protecting British Columbia's Wetlands: A Citizen's Guide*. West Coast Environmental Law Research Foundation and British Columbia Wetlands Network, Vancouver.

Pacific Habitat Services, Inc. 2003. *Ross Island Wetland and Riparian Habitat Reclamation Plan*. Pacific Habitat Services, Inc., Portland, OR.

Parker, V. T. 1997. The scale of successional models and restoration objectives. *Restoration Ecology* 5: 301–306.

Pickett, S. T. A. and V. T. Parker. 1994. Avoiding the old pitfalls: opportunities in a new discipline. *Restoration Ecology* 2: 75–79.

Race, M. and S. Fonseca. 1996. Fixing compensatory mitigation: what will it take? *Ecological Applications* 6(1): 94–101.

Reese, R. E. and K. K. Moorhead. 1996. Spatial characteristics of soil properties along an elevational gradient in a Carolina bay wetland. *Soil Science Society of America Journal* 60: 1273–1277.

Schaffer, P. W. and T. L. Ernst. 1999. Distribution of soil organic matter in freshwater emergent/open water wetlands in the Portland, Oregon metropolitan area. *Wetlands* 19: 505–516.

Society of Wetland Scientists (SWS). 2000. *Position Paper on the Definition of Wetland Restoration*. Available at www.sws.org.

Streever, W. J. 2000. *Spartina alterniflora* marshes on dredged material: a critical review of the ongoing debate over success. *Wetlands Ecology and Management* 8: 295–316.

Taft, O. W. and S. M. Haig. 2003. *Historical Wetlands in Oregon's Willamette Valley: Implications for Restoration of Winter Waterbird Habitat*. USGS Forest and Rangeland Ecosystem Science Center, Corvallis, OR.

U.S. Geological Survey. 1996. *National Water Summary on Wetland Resources*. U.S. Department of the Interior, U.S. Geological Survey, Washington, DC.

U.S. Geological Survey. 2003. *The Amphibian Research and Monitoring Initiative in the Pacific Northwest*. FS-020-03. U.S. Department of the Interior, U.S. Geological Survey, Washington, DC.

Wetlands Research Program Technical Note HY-EV-2.2. January 1993. *Wetland Groundwater Processes*. Available at el.erdc.usace.army.mil/elpubs/pdf/hyev2-2.pdf.

Wetlands Research Program Technical Note HY-DE-4.1. January 1998. *Methods to Determine the Hydrology of Potential Wetland Sites*. Available at el.erdc.usace.army .mil/elpubs/pdf/hyde4-1.pdf.

Whigham, D. F. 1999. Ecological issues related to wetland preservation, restoration, creation and assessment. *Science of the Total Environment* 240: 30–41.

Winston, R. B. 1997. Problems associated with reliably designing groundwater-dominated constructed wetlands. *Wetland Journal* 9(1): 21–25.

Zedler, J. B. 1998. Replacing endangered species habitat: the acid test of wetland ecology. Pages 364–379 in P. L Fiedler and P. M. Kareiva (eds.), *Conservation Biology of the Coming Decade*. Chapman and Hall, New York.

Chapter 8

Tidal Wetlands

RALPH J. GARONO, ERIN THOMPSON, AND FRITZI GREVSTAD

The focus of this chapter is the restoration of tidal wetlands along the coasts of Oregon, Washington, and the shores of Puget Sound. Simply put, tidal wetlands are wetlands whose community composition and structure are influenced to some degree by tides. Many of these wetlands are continuously affected by tides, whereas others may be affected only during the highest of high tides (i.e., several times a month). Although tidally influenced habitats occur in both marine and estuarine settings (Cowardin et al. 1979, Dethier 1990, Allee et al. 2000), most of the restoration work in Pacific Northwest tidelands is occurring in an estuarine setting. Estuaries are semienclosed coastal bodies of water that have free connections to the open sea and within which sea water is measurably diluted with freshwater derived from land drainage (Pritchard 1967). This chapter focuses primarily on estuarine tidal wetlands.

Ecological Role of Tidal Wetlands

Tidal wetlands play many roles. They protect coastal areas from storm waves and erosion and may improve water quality by filtering out sediments and sequestering or transforming nutrients (Maurizi and Poillon 1992, Zedler and Callaway 2001). In addition, tidal wetlands provide habitat and prey for coastal organisms and perform many important ecological functions (Table 8.1). People value tidal wetland areas for recreation, aesthetics, food and timber production, and historic and archeological values (Maurizi and Poillon 1992). Attraction to the coast is reflected in the choices people make in where they live. In the United States, according to the National Oceanic and Atmospheric Administration population statistics, coastal areas are home to more than half of the nation's population; in Oregon and Washington, 56% and 78% of the population, respectively, live in coastal areas (Figure 8.1).

There are compelling reasons to be interested in restoring tidal wetlands. There has been a dramatic loss of tidal wetlands in Oregon and Washington. Good (2000) reports that almost 70% of the 30,194-hectare Oregon tidal wetland area was lost between 1870 and 1970. Losses of tidal wetlands in areas around Puget Sound have been estimated to range from 50% to 100% (Lane and Taylor 1996, Tanner et al. 2002). Commensurate with loss of tidal wetland acreage is the loss of wetland functions and interruption of ecological processes. Consequently, many groups have initiated tidal wetland restoration projects. A few of the larger projects are summarized at the end of this chapter.

Definition and Classification of Tidal Wetlands

Each tidal wetland is unique: The types and numbers of plants and animals and the physical environment at any particular tidal wetland probably

TABLE 8.1.

Relationship between physical environment and ecosystem processes in tidal wetlands.

Controlling Factors	Habitat Structure	Habitat Processes	Ecological Functions
Wave energy	Density	Production	Prey production
Currents	Biomass	Sediment flux	Reproduction (forage fish)
Sediment supply	Individual lengths	Nutrient flux	Refuge
Substrate	Diversity	Carbon flux	Carbon sequestration
Slope and depth	Patch size		Nutrient transformation
Light (shading)	Patch shape		Maintenance of biodiversity
Hydrology	Landscape position		Diversity regulation
Pollution and nutrients	Microtopography		
Other disturbance			

Source: After Thom et al. (2005).

don't occur anywhere else. The challenge for tidal wetland restorationists is to look for patterns to emulate in the abundance and distribution of estuarine organisms. These patterns are created and maintained by underlying physical forces and biological interactions. Physical and biological factors operate at spatial scales ranging from those at a particular site to those dependent on the landscape in which the wetland is located. Tidal wetlands, like other ecosystems, change over time as well. Therefore, restorationists must consider both time and space when developing plans.

Classifying a tidal wetland is an important first step to developing a successful restoration plan.

FIGURE 8.1. Wheeler, Oregon marina. Many types of shoreline modifications are associated with our use of coastal areas. (*From Maurizi and Poillon 1992*)

Because the geomorphology of a site is related to the ecological factors that structure tidal wetland communities and physically constrains what is possible in terms of restoration, understanding how geomorphology and biology differ between sites is essential to a successful restoration plan. Therefore, a classification system describing the relationship between a biological community—"a convenient way of dividing the landscape into discrete units that are repeated in space and time" (Bourgeron 1988)—and the physical and biological factors characteristic of that community has predictive power. It follows that restoration targets and the management actions necessary to achieve restoration goals depend implicitly on a good classification system. In this section, several commonly used classification systems are described briefly. For a detailed discussion of ecosystem classification approaches that include marine and estuarine ecosystems see Odum et al. (1974) and Hedgpeth (1957).

There are many ways to classify tidal wetlands. Biologists and ecologists may classify an estuary or its parts based on the dominant plants or animals that inhabit it. Alternatively, geomorphologists may group tidelands that have similar geomorphological settings. Geomorphological schemes are based on the interactions between landforms and changing sea level, between rivers and the land–sea margin, or between geologic formations and erosional patterns over geologic time spans. Although one could group tidal wetlands using plants or animals alone or by geomorphology alone, the most commonly used classification schemes combine biological and geomorphological information.

There are several classification schemes for estuaries and tidal wetlands in use in the Northwest that help to establish restoration targets and select appropriate restoration methods. The *Oregon Estuary Plan Book* (Cortright et al. 1987) groups estuaries into four major categories based on geomorphology: *River-dominated* estuaries (e.g., Columbia River and Rogue River) are heavily influenced by freshwater flow and have small tideland areas. *Drowned river mouth* estuaries (e.g., Coos Bay, Siletz Bay, and Yaquina Bay) are dominated in winter by large amounts of sediment and in the summer by sea water. *Bar-built estuaries* (e.g., Sand Lake and Netarts Bay) typically have very little freshwater inflow and are separated from the marine environment only by narrow sand spits. *Blind estuaries* (e.g., Elk and Sixes rivers) have a sand bar that completely closes off the mouth because of low seasonal river flows.

In Washington, Shipman (2004) is developing a geomorphic typology for the Puget Sound shoreline. Shipman describes a typology as a terminology that synthesizes and organizes the diversity and the variability in the Washington coastline and contrasts this with a classification system. In Shipman's scheme, segments of shoreline can be grouped into the following categories: coastal bluffs, barrier beaches, pocket beaches, rocky shores, lagoons, estuaries, estuarine river deltas, and stream mouths. In both cases Cortright et al.'s (1987) and Shipman's (2004) work establish coarse categories into which tidal wetlands can be grouped. These groupings are based on geomorphological factors that physically constrain development and maintenance of tidal wetlands.

The interactions between the geomorphic setting, wave and current (energy), sediment sources, and depositional environment all determine what type of community will exist in a tidal wetland. Because the distribution of tidal wetland plant and animal communities is closely tied to the distribution of different types of substrates, understanding the factors responsible for patterns in substrate distribution is important to tidal wetland restorationists. Substrates range from exposed bedrock, to boulders and cobbles, to sand and gravel, to fine silts (Figure 8.2; Cowardin et al. 1979; Dethier 1992). Sand flat communities are different from those associated with mud flats, and many tidal marshes are associated with nutrient-rich, fine-textured silts (Alaback and Pojar 1997) deposited by river (terrestrial sources) and longshore currents (marine sources). The close link between substrate types and their associated biological communities allows

FIGURE 8.2. Photo mosaic of intertidal areas in southern Puget Sound shows a range of substrate types. A mud flat with benthic invertebrate burrows *(upper left)*. Footprints on a mud flat demonstrate that sediments can be quite unstable and prone to movement by currents and waves *(upper right)*. Cobble and gravels *(bottom)* serve as points of attachment for estuarine plants and animals. *(Photos by Ralph J. Garono)*

many tideland inventories and regulatory programs to use habitat-based classification systems.

The same factors that structure and maintain tidal wetland communities also constrain restoration possibilities (see Table 8.1). Selecting appropriate restoration targets and necessary restoration actions may be evident through the use of an appropriate classification scheme. For example, restoration actions that depend on the supply of sediments from the upper watershed would not be appropriate for marshes in bar-built or blind estuaries because riverine input to these types of estuaries is minimal. An important first step in developing a restoration plan is linking valued resources (i.e., important plants and animals or wetland functions) with the factors hypothesized to control those resources. Conceptual models are useful tools for examining these relationships. Conceptual models can range from simple diagrams that show the relationship between ecosystem components (see page 195 in Thom et al. 2005) to diagrams showing the types of relationships and even flows and fluxes of materials and energy between ecosystem components (Odum et al. 1974).

Patchwork of Tidal Wetlands in Estuarine Landscapes

Estuaries contain many types of ecological communities in vegetated and unvegetated habitats, both important to Northwest estuarine organisms. In Oregon, the *Oregon Estuary Plan Book* (Cortright et al. 1987) follows the Cowardin et al. (1979) classification of wetlands in the United States. This classification scheme is hierarchical, that is, the various habitat subsystems are grouped into broad, general categories. The Cowardin scheme distinguishes between vegetated and unvegetated areas. Sand and mud are examples of unvegetated, unconsolidated bottom, subtidal habitats. Shrub, fresh marsh, high salt marsh, and low salt marsh are examples of tidal marsh intertidal habitats. There are similar groupings for subtidal habitats. The *Oregon Estuary Plan Book* adds to Cowardin classes by describing the types of substrate or vegetation communities that may occur in each of four estuarine subsystems: marine, bay, riverine, and slough. As in Oregon, marine and estuarine communities in Washington have been classified using the work of Cowardin as a starting point. Dethier (1990, 1992) also considers exposure to wave and water currents as part of her classification hierarchy. Additionally, Dethier's system describes characteristic life forms and species for each community type.

Tidal wetlands exhibit a range of microtopographies. Many of the functions and values of tidal wetlands are directly related to this microtopography. In higher-energy systems tidal wetlands can be dissected by numerous tidal channels. In many natural systems, dendritic tidal channels resemble a stream network when viewed from the air (Figure 8.3). Smaller tidal channels coalesce into larger tidal channels, and the larger channels coalesce into tidal streams that eventually leave the marsh and enter the estuary.

Northwest tidal channel systems are the subject of much research today because of the role they play in the salmonid life cycle. Research from other regions suggests that tidal channels in salt

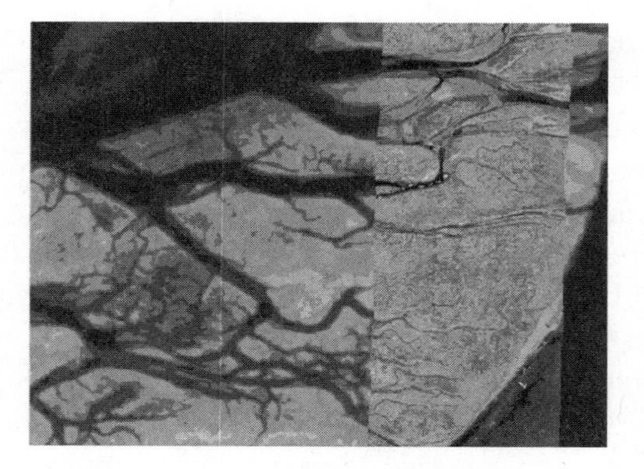

FIGURE 8.3. A network of tidal channels near Russian Island on the Columbia River. The left side shows the tidal channels as seen from a satellite. On the right are tidal channels shown from a special photo system mounted on an airplane. (*Photo by Ralph J. Garono*)

marsh restoration sites do not quickly recover the same functions of natural tidal channel systems mainly because they are designed too simply, have inappropriate hydrologic regimes, have inappropriate elevations or grades, or have incorrect substrate or erosional patterns (see Zedler and Callaway 2001).

The structure of tidal channels is thought to be important to salmonids for several reasons. Dendritic channels create a complexity of varying elevations and patterns in saltwater intrusion and inundation (Figure 8.4). In turn, these can affect rates of primary (plants) and secondary (organisms that eat plants) production, which are the base of the food web exploited by salmonids. That is, different plant species arrange themselves on tidal wetlands according to their tolerance of saltwater and drying, among other factors. Both living and decomposing plants (plant decomposition or organic material processing is an important ecosystem process in estuaries) are used by a variety of invertebrates, which are an important energy source for salmonids.

The dynamism and structural complexity make tidal wetlands some of the most biologically productive areas on Earth (Begon et al. 1986). Tidal wetlands may produce much of the prey used by young migrating salmon. Current thinking among ecologists is that the greater the structural complexity in the tidal marsh, the more material processing and invertebrate production can occur. For example, multilevel plant heights on salt marshes may provide important areas for invertebrates to escape being washed away during high tides (Scatolini and Zedler 1996). Tidal marshes that lack tidal channels or that are colonized by uniform-height, homogeneous vegetation (e.g., *Spartina*) may not produce adequate prey to support

FIGURE 8.4. Sedge-dominated areas occur adjacent to mud flats. Sedge roots bind sediments and prevent them from eroding, resulting in increasing structural complexity. (*Photo by Ralph J. Garono*)

salmonid populations. Or tidal marshes infested with invasive plants (e.g., *Lythrum salicaria*) may support entirely different prey communities than marshes dominated by native vegetation (Garono and Schooler, unpublished data). Invertebrate diversity and abundance often are the most difficult things to reestablish in tidal restoration projects.

In addition to providing food web support, tidal channels are also important places for salmonids to escape predators and high-flow events and to physiologically adapt to ocean conditions. Some species of young salmonids cruise around the tidal channels, not only to find food but also to escape their predators. Deeper tidal channels can provide cool places to escape from warmer water during low tides. Tidal channels also provide low-flow environments for young salmon to escape fast-moving water in the estuary proper, especially in the winter. Finally, the saltwater–freshwater gradient established during the tidal cycle in tidal channels can be very important for the requisite physiological adaptation young salmon undergo before they move out to sea (see Figure 8.3). Understanding the interactions between salmonids and tidal channels is a topic of current research (see Simenstad et al. 2000 for discussion).

Northwest vegetated tidal wetlands often are dominated by aquatic plants, including tufted hairgrass (*Deschampsia caespitosa*), Lyngby's sedge (*Carex lyngbyei*), Pacific silverweed (*Potentilla anserina*), glasswort (*Salicornia virginica*), sea arrowgrass (*Triglochin maritimum*), and salt grass (*Distichlis spicata*) (Figure 8.5). Many subtidal and intertidal areas are dominated by eelgrass (*Zostera*

FIGURE 8.5. A sedge meadow in Skookum Bay, Washington. Although sedge meadows can be quite dense, they often give way to shrub–scrub and forested areas on the landward side and to shorter vegetation and mud flats on the seaward side as elevation changes. (*Photo by Ralph J. Garono*)

marina); however, other intertidal areas may be completely devoid of vegetation. Once established, vegetation affects the structure of tidal wetland communities. Vegetation can directly affect many of the elements listed in Table 8.1, especially rates of sediment deposition and erosion, light availability, carbon and nutrient flux, and prey production. Plants themselves may modify the environment to the extent that a distinctly different plant community replaces the first through the process of succession. For example, Franklin and Dyrness (1988) describe the successional series in tidal wetlands where a rush meadow is replaced by a wet shrub community, which is, in turn, replaced by a pine–spruce forest (Figure 8.6). If this progression of community types is real and predictable, the length of time needed for this series to occur would depend on a number of factors including the landscape setting and the disturbance regime. Understanding how tidal wetland community types change over time and as a function of different types of disturbance is necessary for developing restoration plans.

FIGURE 8.6. A tidal spruce marsh in the Columbia River estuary. Notice the patchwork of trees, shrub–scrub, and herbaceous vegetation. The pattern of different plant types on this island probably contributes many wetland functions. (*Photo by Ralph J. Garono*)

FIGURE 8.7. Purple loosestrife is invading the Columbia River's tidal marshes. In time, many native herbaceous plants are replaced by the woody invader, thereby altering the physical structure of the plant community and the flow of material and energy in the estuarine food web. (*Photo by Ralph J. Garono*)

Invasive Plant Species

Invasive plant species are of particular interest to tidal wetland restorationists. In particular, *Spartina* spp. and purple loosestrife (*Lythrum salicaria*) both occur in tidal wetlands in Washington and Oregon (Figure 8.7). These invasive plants can modify wetland structure and function just as native plants do. However, the aggressive nature of these plants often leads to vast expanses of monotypic stands of vegetation. These and other organisms thereby have the potential to dramatically alter tidal wetland food webs (Garono and Schooler, unpublished data) and other wetland ecological functions. Control of these two species is of concern to wetland restorationists and is the focus of several control efforts (see Coombs et al. 2004).

Where Tidal Wetlands Are Found

The extent of tidal wetlands is not as great in the Northwest as in the Atlantic and Gulf Coast regions (Odum et al. 1974, Simenstad et al. 2000).

In Washington, tidal wetlands are found along the coast, in the Straights of Juan de Fuca, and in Puget Sound. Grays Harbor, Willapa Bay, Skagit Bay, Padilla Bay, and the mouth of the Columbia River exhibit expansive tidal wetlands. Lane and Taylor (1996) estimate that there are 81,746 hectares of estuarine wetlands in Washington.

In Oregon, tidal wetlands are found in estuaries along the coast and the Columbia River. According to Cortright et al. (1987), all of Oregon's major and minor estuaries (approximately 53,000 acres) could fit inside Grays Harbor estuary in Washington. However, this estimate excludes the vast expanses of tidal wetlands at the mouth of the Columbia River. Wetlands along the lower Columbia experience the effects of the tides as far upriver as the Bonneville Dam, a distance of more than 142 river miles (Figure 8.8). The following information sources describe the extent of tidal wetlands in Oregon and Washington: Percy et al. (1974), Thomas (1980, 1983), Cortright et al. (1987), Lane and Taylor (1996), National Oceanic & Atmospheric Administration (NOAA) (1997), Allen (1999), Good (2000), Garono and Robinson (2002), Garono et al. (2004).

FIGURE 8.8. Bonneville Dam is the upper limit of tidal influence on the plants and animals of the lower Columbia River. *(Photo by Ralph J. Garono)*

Disturbances to Tidal Wetlands

Tidal wetlands are dynamic. The previous sections have briefly described some of the factors that create and maintain these ecosystems. Before launching into a discussion of disturbance, it is important to note that disturbances (either natural or induced or exacerbated by humans) to tidal wetlands can be caused by the presence or absence of some factor. For example, the presence of a tide gate can modify patterns in tidal exchange, or the absence of a key predator can alter food web structure. Disturbances can also be more subtle. They can be too much or too little of a factor. For example, some tidal wetlands depend on a continual supply of sediments from terrestrial watershed sources. Land use can dramatically increase sediment supply to a wetland and choke vegetation. Conversely, a road or bridge can reduce sediment supply to a tidal wetland, leading to erosion or subsidence of a tidal wetland. Finally, disturbances can be a shift in the timing of some factor. For example, dams tend to prevent large pulses of water (spates) from flushing tidal wetlands. In any case, tidal wetland restorations should be viewed as whole system phenomena because the estuary depends on the total interaction of all the chemical cycles, water circulations, and species behaviors; the ecological and physical limitations must be considered in planning for wetland restoration or creation (Odum et al. 1974, Kentula 1996).

Disturbances can be grouped into three major categories: physical, chemical, and biological (Maurizi and Poillon 1992). Physical alterations include filling, ditching, draining, excavating, altering water delivery patterns, altering land use, shading, and conducting activities in adjacent areas (Figure 8.9). Chemical alterations include changing nutrient levels or introducing toxic materials. Biological alterations include grazing, introduction of nonnative species, and disruption of natural populations. Knowledge of the magnitude of altered physical, chemical, and biological processes is extremely important to restorationists (see

FIGURE 8.9. Many areas along the lower Columbia River estuary that were once tidally influenced are now diked. (*Photo by Ralph J. Garono*)

Table 8.1); however, this information often is very difficult to acquire for individual sites.

Historic maps and photographs often aid the development of restoration targets. It is important to realize that some ecosystems may never be completely restored (to historic or presettlement conditions) because adjacent ecosystems are no longer connected as they were historically or because the pool of species available to colonize an ecosystem has changed because of extinction or the presence of nonnative or invasive species (see Chapter 2). Comparisons with intact reference sites also are used to establish restoration targets. The assumption is that if the restored site looks like the reference site, the two will function similarly (although this may not be correct; see Simenstad and Thom 1996). Computer models and other studies often are used in planning restoration projects. Fortunately, technological advancements in computer-aided modeling and remote sensing technology give restorationists several new tools with which to design and implement restoration projects. It is clear that disturbances operate at multiple spatial and temporal scales. Restoration projects that are most likely to be successful are those that are implemented in largely undisturbed landscapes that have few site disturbances (Figure 8.10). Restoration projects that are likely to fail are those that have a high degree of disturbance to the site and occur in heavily modified landscapes. Restorationists must evaluate and correct disturbances at both landscape- and site-level scales if restoration projects are to be successful.

Practical Restoration

Ecological restoration has been defined as "intentional activity that initiates or accelerates the recovery of an ecosystem with respect to its health, integrity, and sustainability" (Society for Ecological Restoration [SER] 2004). What does this mean in a practical sense for tidal wetlands? Ecosystems are *systems* of interrelated components that do things. Tidal wetlands are inhabited by plants and animals that interact with each other (predation and competition) and with their environments. These organisms are adapted to and depend on characteristic patterns of disturbance, substrate types, food type and availability, salinity patterns, and temperature regimes (see Table 8.1). Thus, ecosystems can be thought of as places that have characteristic fauna and flora and have characteristic rates, pathways, and intensities of interactions.

Restoration can be thought of as manipulating the types and numbers of plant and animal species or as returning flows and fluxes of materials to

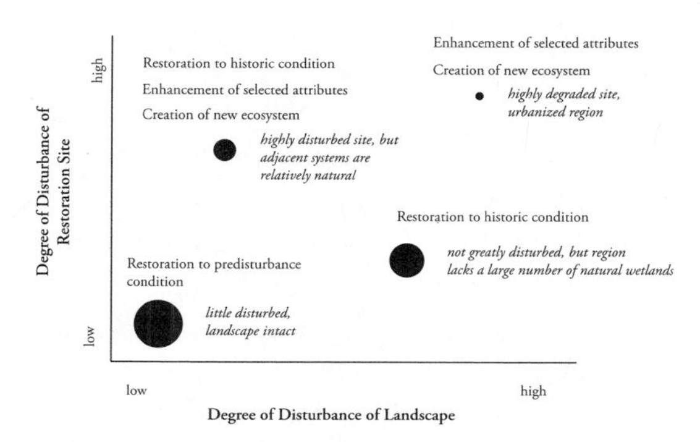

FIGURE 8.10. Restoration potential for wetland sites as a function of site and landscape disturbance. Large circle indicates high potential for success, and small circles indicate low potential. (*From Maurizi and Poillon 1992, as modified by Thom et al. 2005*)

within their normal range of variability of the natural condition. The natural condition is one that is self-maintaining (SER 2004). The normal range of variability representing the natural condition can be measured directly from reference sites, obtained from historic information, modeled, or developed empirically from a series of similar ecosystems.

Restoration of tidal wetlands includes the following steps:

1. *Select the site*: Although tidal restoration projects are initiated for many reasons, sometimes it is necessary to work with an available site (i.e., with a willing landowner) as opposed to an ideal site. At other times sites are selected from a prioritized list, with plans developed as sites or agreements are acquired. This is the case for many basin or watershed planning efforts. Brophy (2005) describes a procedure for mapping and prioritizing estuarine wetlands for restoration and conservation. Her method is well suited to watershed groups interested in developing restoration plans involving tidal wetland parcels for entire estuaries. Dean et al. (2001) have also developed a method for prioritizing restoration sites.

2. *Establish restoration goals*: Many wetland restoration projects fail because goals are not developed or adequately stated. Restoration goals should be as detailed and descriptive as possible and should include both the types of communities and the wetland functions being restored. Adamus (2005) developed a rapid method to assess tidal wetland functions. A conceptual model is useful for describing important estuarine components (e.g., species) and the links between those components and the environmental factors that maintain and support them. Goals should also include rationale and (science-based) supporting evidence for decisions that must be made during the restoration process.

3. *Evaluate ecological constraints*: The restoration plan should identify factors that need to be modified to ensure a successful restoration. Major site and landscape disturbances should be noted at this time. Many restoration projects fail for lack of ecological expertise and inability to recognize important ecosystem connections.

FIGURE 8.11. Capitol Lake, Olympia, Washington. This urban freshwater lake was once an estuary. Studies are being done to evaluate costs and benefits of maintaining this area as freshwater or removing a dam and restoring the estuary. *(Photo by Ralph J. Garono)*

4. *Develop a restoration plan*: This involves sequencing restorative actions, allowing for active revegetation or for passive plant recruitment, and ensuring that the necessary conditions are being met in areas that are ecologically important to the restored wetlands (e.g., upper watershed).

5. *Develop a postrestoration monitoring plan*: This includes identifying thresholds for implementing corrective actions, such as a set of guidelines to be followed if invasive plants move in. Also, provisions should be made to measure wetland functions and biological communities that are expected to develop at the site. Use a conceptual model to assist in developing corrective actions in case restoration goals are not met.

Tidal Wetland Restoration Case Studies

This section presents overviews of three important tidal wetland restoration programs in Northwest estuaries: the Deschutes River estuary restoration project, biocontrol of an invasive plant in Willapa Bay, Washington, and the Lower Columbia River Estuary Partnership (www.lcrep.org).

Case Study: Deschutes River Estuary and Capitol Lake Restoration Study

In 1951, construction of a dam across the southern portion of Budd Inlet (Olympia, Washington) was completed, creating a freshwater lake in what was once a tidal wetland (Figure 8.11). Since the dam was completed, Capitol Lake has been plagued with water quality and sedimentation problems. Consequently, the Capitol Lake Adaptive Management Plan Steering Committee initiated a project to study the feasibility of restoring the estuary.

What is remarkable about this endeavor is that it involves evaluation of several restoration alternatives using a new high-tech tool. Costs and benefits of the various restoration alternatives and a do-nothing alternative are being evaluated, and restoration performance goals will be set before any restoration action is taken. The tool being used is a sophisticated model developed by U.S. Geological Survey (Menlo Park, California) that predicts sediment transport, water velocity, and salinity patterns in the estuary to be restored. The modeling tool uses river gauge, tidal, wind, bathymetry, sediment, and salinity data as input variables. Using the estuary's underwater topography (morphometry), or shape of the estuary basin, the computer model describes patterns of sediment transport, salinity, water velocity, and inundation patterns modeled under different river flow and restoration scenarios. The model predicts where sand and mud flats probably will occur and expected erosional patterns associated with channels and different flow regimes.

An interesting aspect of this study is that the model was used to target specific types of tidal wetland communities from Dethier's (1992) community list for field sampling. Field crews visited nearby estuaries to sample plant and animal communities and measure key environmental variables at these reference estuaries. The field data were then used to develop statistical relationships describing the association of plant and animal communities with the environmental factors that organize those communities. These statistical relationships were then used in the model to predict what communities were likely to occur under each of the potential restoration scenarios. Managers will use this information to evaluate each of the scenarios. In this suite of studies, field data and computer modeling were used to evaluate alternative restoration plans before one brick of the dam was removed. As these types of studies become more common, restoration planning may make better use of scarce funds and optimize restoration potential. Models are useful tools for examining the influence of site-specific and landscape variables on restoration projects.

Case Study: Control of Invasive Plants in Willapa Bay, Washington

Located along the southern coast of Washington State, Willapa Bay is a 25,000-hectare estuary fed by seven rivers and a number of smaller streams. This bar-built estuary is considered one of the country's least spoiled, with minimal development (other than logging) in the surrounding hills and shorelines (Wolf 1993). Willapa Bay provides critical foraging grounds for migrating shorebirds and waterfowl and healthy stocks of chum, coho, and chinook salmon. It is also one of the most productive estuaries for shellfish. In combination with nearby Grays Harbor estuaries, it produces 60% of the oysters in Washington State and 15% in the United States (Hoines 1996, Conway 1991).

Current restoration efforts in Willapa Bay are dominated by the removal of one exotic species. *Spartina alterniflora* is a cordgrass native to the Atlantic and Gulf coasts that was accidentally introduced into Willapa Bay in the late 1800s (Figure 8.12). For 80 or 90 years, the plant spread slowly without raising alarms, but by the 1980s the potential threats of the invasion were realized (Sayce 1988). *Spartina* grows in the intertidal zone, where it acts as an ecological engineer (*sensu* Jones et al. 1994), transforming open mud flats into densely vegetated marsh and eliminating critical habitat for birds, fish, and shellfish. The roots and stems trap sediment, gradually raising the substrate elevation and altering the hydrology (Ranwell 1964, Redfield 1972). This leads to occupation by more terrestrial (salt marsh) plants and animals (Zipperer 1996, Hedge and Kriwoken 2000, O'Connell 2002).

Beginning in the mid-1990s, Washington State began efforts to control *Spartina* using chemical and mechanical methods. The soft mud, intermittent tides, and limited accessibility were challenging initially (Hedge et al. 2003). Lacking experience in what worked and what did not and with scant financial resources relative to the scale of the problem, the plant continued to spread faster than it was controlled. Legal conflicts over the use of chemicals in the estuarine environment also delayed the program (Hedge et al. 2003). In 2003, the Washington State Department of Agriculture estimated that *Spartina* had covered 7,300 hectares of former mud flat and that the area of coverage had been increasing by an average of 17% per year (Murphy 2003).

In 2000, traditional control approaches were supplemented with a classic biological control program. In classic biological control, a natural enemy is introduced into the environment with the goal of establishing a permanent population that will provide long-term control of target weeds or pests (see Chapter 16). Although it is often considered a nontraditional approach, classic biological control has a long history and an excellent track record as a safe, economic, and environmentally sustainable solution for widespread weeds (McFadyen 1998). In the United States, ninety-four natural enemy species have been introduced against thirty-eight weeds (Coombs et al. 2004). Worldwide, more than 350 natural enemies have been used against 133 weeds (Julien and Griffiths 1998).

The planthopper (*Prokelisia marginata*; Figure 8.12) was selected as a promising biocontrol agent because of its narrow host range and its known potency against S. *alterniflora* from Willapa Bay (Daehler and Strong 1997). The *Spartina* biocontrol program is unique in being the first classic biocontrol program applied in an estuarine intertidal environment and the first to target a grass. The insect was introduced into Willapa Bay only after extensive testing indicated a high level of host specificity to invasive *Spartina* (Grevstad et al. 2003). The project was reviewed and approved by the Technical Advisory Group on Biological Control of Weeds and permitted by the Washington Department of Agriculture and the U.S. Department of Agriculture.

After some initial trial and error, several populations of *P. marginata* are well established, growing, and spreading. Local densities of these tiny (2-millimeter) insects have occasionally exceeded 50,000 per square meter. To date, only localized impacts have been seen, including a 50% reduction in biomass in a field cage experiment (Grevstad et al. 2003) and a 90% reduction in seed set in the open field at one site (F. Grevstad, unpublished data). The planthopper population experiences periodic setbacks from winter mortality and spider predation, although summer reproduction typically is good (Grevstad et al. 2004). Overall populations have grown slowly from year to year, and, as is typical for biocontrol programs, the full impact on the plant population will take several years to a decade or more to be fully realized. The screening and introduction of additional biocontrol agents from *Spartina*'s native range would increase the odds of success with this program (Denoth et al. 2002). If effective, biological control will serve as a highly economical and sustainable method of permanent control.

Beginning in 2002, Washington state agencies and the U.S. Fish and Wildlife Service dramatically scaled up their control program and adopted a 6-year plan for eradication. Approximately $2.5 million per year is spent on this program (Murphy 2004). Approval of a new chemical for use in aquatic environments (imazapyr), better machinery for operating in the mud, tidal exposure mapping, and the use of helicopter application of herbicide have all improved the scale and efficacy of control (Patten 2002, Dethier and Hacker 2004, Murphy 2004). However, the spray program is still hindered by variable and unpredictable efficacy, which may reflect the range in variation in the physical conditions in different locations in the bay. Although progress is being made toward reducing the amount of *Spartina*, it is still unclear whether complete eradication will be possible.

FIGURE 8.12. *Spartina* in Willapa Bay, Washington. A native of eastern U.S. tidal marshes, *Spartina* did not historically occur in the Pacific Northwest. Once established, it can quickly transform mud flats into monoculture meadows, displacing native organisms that depend on mud flat habitat (see burrows in Figure 8.2). Planthoppers *(inset)* are being tested as biocontrol agents aimed at stopping the spread of *Spartina*. (*Photos by Ralph J. Garono*)

Case Study: Habitat Mapping in the Lower Columbia Estuary

In 2000, the Lower Columbia River Estuary Partnership initiated a multiphase project to produce a spatial data set describing the current location and distribution of estuarine and tidal freshwater habitat cover types along the lower Columbia River from the river mouth to the Bonneville Dam using a consistent method and data sources (see Plate 4 in the color insert). Spatial data (geographic information system data layers) are extremely useful for planning site-specific restoration actions while maintaining a landscape context. Some of the current and planned projects are listed here.

Two spatial data sets were developed, taking advantage of recent advances in remote sensing technology and desktop computer technology. The first (Garono et al. 2003b) was a broad-brush description of estuarine and tidal freshwater wetland habitat cover classes for the entire tidally influenced study area (about 146 river miles) using Landsat 7 ETM+ satellite imagery (see Figure 8.3). The second produced high-resolution habitat maps using hyperspectral imagery collected in 2000 and 2001 for key focal areas in the larger study area (Garono et al. 2003a). At the time of these studies, no single data set had been produced using a consistent method and uniform scale data describing current estuarine and tidal freshwater floodplain cover types from the Columbia's mouth to the Bonneville Dam. These studies took a limited number of field observations made from tidal wetland communities throughout the estuary and scaled them up by matching plant communities with spectral signatures obtained from the remotely sensed imagery. As prices drop, remote sensing is becoming a more common component of coastal restoration projects.

The Lower Columbia River Estuary Partnership and its cooperators used results from this study to develop indicators of habitat health for target species and populations and biological integrity at the community and ecosystem scales, develop definitions of important salmonid habitat, identify and evaluate potential wetland conservation and restoration sites, track invasive species, and develop an understanding of how estuarine and floodplain habitats have changed over the past 200 years.

Other Columbia River Projects

The Lower Columbia River Estuary Partnership was admitted into the Environmental Protection Agency's National Estuary Program in 1995. The Comprehensive Conservation and Management Plan calls for the restoration of 3,000 acres of tidal wetlands along the lower 46 river miles. Based in part on the remote sensing projects described earlier, the Lower Columbia River Estuary Partnership has several regional projects planned, permitted, or constructed. Site-specific restoration techniques used in these projects include dike or levee breaching; installation, removal, or replacement of culverts; channel enhancement or creation; tide gate removal or replacement; native plantings; and fencing. Tide gate removal is one of the more common ways to restore hydrologic connections and is planned for Grays Bay Area in Wahkiakum County Washington, Young's Bay and Walluski River, and Fort Clatsop. Replacement of existing tide gates with new fish-passable designs is planned for Young's Bay and Skamakowa Creek. Dike or levee breaching is planned at Blind Slough, Grays Harbor, and the Lewis and Clark River in the Young's Bay watershed.

Existing culverts often are too small to allow fish passage, and at Conyers Creek four of these will be replaced with a bridge. Replacing undersized culverts with those of more appropriate size will occur at Fort Columbia, along with Blind Slough and the Genesis and Nelson sites. Native plantings and channel enhancement are planned for several of the partnership's project sites, including Blind Slough, Grays Bay, Young's Bay and Walluski River, the Lewis and Clark River, and Vaughn Creek.

Scappoose Bay and Estuary is located between the towns of Scappoose and St. Helens, Oregon. The area covers 8,960 acres east of Highway 30 on the north and south sides of the bay. The goal of the restoration project is to protect high-quality lowland floodplain wetlands by synthesizing existing information, conducting field studies to assess the wetland condition, completing a risk assessment, developing a picture of reference conditions for particular wetland classes and types, prioritizing sites for protection or restoration, assessing restoration potential of sites, and developing restoration goals and performance standards for each prioritized site. Partners include the Wetlands Conservancy, Columbia Land Trust, The Nature Conservancy, Oregon State Parks, Oregon Trout, the U.S. Geological Survey, and the Columbia River Inter-Tribal Fish Commission. The total cost of the project is $133,455, funded by the Bonneville Power Administration.

The Hogan and Malarkey ranches are also in the tidally influenced bottomlands of Scappoose Bay. This is one of the most complete restoration projects by the partnership at this writing. The restoration goals for this 300-acre area are to restore native plant communities and emergent wetlands through riparian plantings, replace barriers to aquatic species migration with fish-friendly bridges, implement new fencing and grazing practices, and reconnect hydrology. Nearly 200 restored acres are secured and protected by a conservation easement. The Wetlands Conservancy, an important partner with the Lower Columbia River Estuary Partnership, has performed a habitat inventory of the study area to assess potential habitats from the restoration. Other partners include the Scappoose Watershed Council, the City of Scappoose, Columbia County Soil & Water Conservation District, local landowners, Ducks Unlimited, the National Resource Conservation Service, and the Oregon Department of Fish and Wildlife. The total cost was estimated at $214,970, with $100,000 from the Environmental Protection Agency's Watershed Initiative Program and $90,000 from the Bonneville Power Administration.

The Lower Columbia River Estuary Partnership has two other regional projects at Blind Slough and Grays Bay, which are just beginning.

Blind Slough is located near the mouth of the Columbia River, in the Brownsmead area, approximately 17 miles east of Astoria, Oregon. The main goal is to restore tidal connection of approximately 10 miles of slough channels between the Columbia River Estuary and Blind Slough through the replacement or installation of culverts, installation of water control devices, breaching of dikes, and channel enhancement. This project encompasses sloughs located in the Clatsop Diking Improvement Companies 1 and 7, which are in the oligohaline (brackish) zone of the Columbia River Estuary adjacent to Cathlamet Bay. The oligohaline zone is where many salmonid species undergo the physiological transition from freshwater to ocean conditions. Project effectiveness will be determined by improvement in water quality and increased access to spawning and off-channel habitat for salmonid species. This project is a joint effort between the Columbia River Estuary Science Taskforce, Clatsop Diking Improvement Company No. 7, U.S Army Corps of Engineers Portland District, Bonneville Power Administration, U.S. Fish and Wildlife Service, Northwest Power Planning Council, Nicolai Wickiup Watershed Council, North Coast Watershed Association, and Sea Resources Watershed Learning Center. The total project cost is estimated at $550,000, with more than $114,000 raised as of the summer of 2005.

Grays Bay is located between river miles 19 and 23 along the Columbia River in Wahkiakum County, Washington. The project area encompasses the watersheds of three tributaries that empty into the bay: Deep River, the Crooked River, and the Grays River (including Seal Slough). This project will permanently protect 880 acres, restore floodplain connectivity to a minimum of 440 acres of tidal backwater, riparian, and wetland forested habitat, and restore more than 300 acres of potential salmonid-rearing habitat. Restoration will include levee breaches, tide gate removal, ditch filling, swale enhancement, and restoration of native plant communities. All conservation properties will be acquired fee simple and protected in perpetuity for their conservation values by the Columbia Land Trust. The trust will establish a stewardship fund for all long-term maintenance and management as needed. Partners include the Columbia Land Trust, Ducks Unlimited, the Washington State Salmon Recovery Funding Board, the Lower Columbia Fish Recovery Board, the Pacific Coast Joint Venture (U.S. Fish and Wildlife Service North American Wetland Conservation Act program), the U.S. Fish and Wildlife Service Private Stewardship grant program, the Washington Department of Fish and Wildlife, the Wetland Reserve Program, the Charlotte Y. Martin Foundation, and the National Fish and Wildlife Foundation, Lower Columbia River Estuary Partnership, and Columbia River Estuary Science Taskforce. The total project estimate is nearly $4.5 million. More than $500,000 has been raised to date from grants through the Bonneville Power Administration and the Environmental Protection Agency Watershed Initiative Program.

Case Study Snapshots

This section summarizes restoration activities in Oregon and Washington tidal marshes.

Case Study: Padilla Bay Estuarine Research Reserve

This is at the saltwater edge of the large delta of the Skagit River and is approximately 8 miles long by 3 miles across. It is part of the "Salish Sea" that includes Puget Sound, Hood Canal, Georgia Straits, Straits of Juan De Fuca, and waters around Gulf and San Juan islands. The bay is intertidal, flooding at high tide and with almost 8,000 acres of exposed eelgrass mud flats at low tide.

The area is characterized by agricultural, industrial, residential, and recreational land uses, with increasing developmental pressure. The Washington Department of Ecology performs onsite management and administration.

A system-wide monitoring program tracks temperature, salinity, dissolved oxygen concentration, turbidity, pH, water depth, and meteorological data (temperature, humidity, wind velocity and direction, light, and rainfall) and analyzes water samples at four sites each month for various nutrients and chlorophyll. Citizens are involved with water quality monitoring that measures temperature, fecal coliform bacteria, turbidity, depth, and dissolved oxygen data. Data for all National Estuarine Research Reserve monitoring projects are available through the Centralized Data Management Office. Specific titles from the technical reports and reprint series can be requested through the reserve.

There is no specific information on tidal restoration. Invasive plant removal and control are performed through the stewardship project. Two geographic information system (GIS) projects map intertidal habitats of the bay and develop data and information management libraries (www.padillabay.gov).

Case Study: Nehalem Bay

The Nehalem River and Bay are on the northern Oregon Coast, just north of Tillamook. The watershed covers 855 square miles, and the river is 118.5 miles long. Marsh and tide flats cover 1,078 of the 1,800 acres of bay and river waters. Past projects included those funded by Portland State University (PSU), the Oregon Watershed Enhancement Board (OWEB), and the Coast Range Association.

A GIS-based assessment of Nehalem watershed provides baseline data produced by PSU, OWEB, and local watershed councils. Other documents available through the lower council's Web site include a 2000 Action Plan and a report by the Coast Range Association. A Ducks Unlimited report on Alder Creek restoration alternatives and a 1-foot contour survey of the farm are available. The Web site is updated with current newsletters and goals and accomplishments for the year. Funded proposals, planning documents, meeting minutes, budgets, and policies are also available.

Alder Creek is a tributary to Nehalem and has potential to be very productive salmon habitat. Alder Creek restoration began in 2003 with efforts to acquire the Alder Creek Farm, a 55-acre dairy. Restoration plans include 5 acres of tidal salt marsh, 6 acres of freshwater wetland bird habitat, and stream improvement for fish species. Ducks Unlimited has provided several restoration alternatives for the trust to choose from. Initial monitoring data of subsurface water levels and flora and fauna will be beneficial in planning restoration. The Lower Nehalem Community Trust is pursuing the purchase of Alder Creek Farm for $216,000. Grants for acquisition include $50,000 from OWEB, $50,000 from the U.S. Fish and Wildlife Service Bird Habitat Conservation program, $68,000 from the Natural Resources Conservation Service Wetlands Reserves Program Conservation Easement, and more than $28,000 in private donations. A $6,000 OWEB grant for plantings and nearly $16,000 from OWEB restoration planning have also been awarded to the trust.

The Lower Nehalem Watershed Council has completed other projects including large wood placement, riparian enhancement, fish passage enhancement, and solid waste cleanup in the estuary. Nehalem Bay and the Nehalem River Watershed are overseen by the Upper and Lower Nehalem River Watershed Councils (www.nehalemtel.net/~lnwcouncil).

Case Study: Tillamook Bay

The Tillamook Estuary Partnership (TEP) conducts research, education, and restoration for five estuaries and watersheds in Tillamook County, Oregon, covering 1,800 square miles. In 1992 Tillamook Bay Estuary was nominated as one of twenty-eight national estuary projects. The Tillamook Bay & Estuary Feasibility Study was conducted in 2000 with the U.S. Army Corps of Engineers and Tillamook County to examine flood damage reduction and restoration issues in the watershed. Various study documents are available online (usace.co.tillamook.or.us/). GIS data layers of the Tillamook Bay watershed are available through the Tillamook County GIS server.

Tidal marsh restoration is included as an action step in the Comprehensive Conservation and Management Plan, a sixty-two–item action plan that addresses environmental problems in Tillamook Bay. Restoration of high-salt marsh, brackish marsh, and forested wetland ecosystems in the Wilson–Trask Wetlands parcels has begun. Wilson–Trask and two other parcels make up a total of 377 contiguous acres in the river delta of southern Tillamook Bay between Wilson and Trask rivers. Actions began in 1999, and grants for purchase totaled $1,600,000.

Hoquarton Slough links upper Tillamook Bay with downtown Tillamook. The project focuses on 1,500 feet of riparian restoration on the south side of the slough within the city limits to improve water temperature and water quality. Volunteers and students from the Tillamook School District removed invasive plant species and replanted with native plants. Partners include the Tillamook School District, TEP, the Tillamook Chamber of Commerce, the Bureau of Land Management, George Fox University, and community members. This project has received funds from the NOAA Community-Based Habitat Restoration Program and has a budget of $59,000.

Vaughn Creek is a lowland tributary of the Kilchis River in the Tillamook Bay Estuary watershed and drains a 675-acre agricultural watershed. Fish passage was improved by replacement of a partial barrier tide gate at the confluence of Vaughn Creek and Kilchis River. This gave fish access to 1 mile of rearing and spawning habitat. In addition, the project enhanced approximately 1.9 miles of stream-size riparian vegetation to mitigate impacts from livestock grazing and golf course runoff. This project involved the partnership of four private landowners, the watershed council, and several others including TEP, the Oregon Department of Fish and Wildlife, the Tillamook County Creamery Association, the Tillamook County Soil & Water Conservation District, the Tillamook Native Plant Cooperative Riparian Committee, the National Fish and Wildlife Foundation, and the U.S. Fish and Wildlife Service. The total cost was $175,725, with some of the funding provided by the NOAA Community-Based Habitat Restoration Program.

Case Study: South Slough National Estuarine Research Reserve

South Slough covers 5,000 acres of natural area, of which 600 are tidal marsh, mud flat, and open water of the Coos Estuary on the southern Oregon coast near Charleston. Through the South Slough Web site (www.southsloughestuary.org) the public can access miscellaneous reports and publications including workshop proceedings, the Estuaries Features Series, summaries of projects, and a few research publications. No data sets are available.

A large restoration project called the Winchester Tidelands was initiated in 1993, encompassing several smaller restoration projects of various ecosystems in the reserve. A project advisory council includes specialists from universities, nonprofit and community organizations, private consulting firms, and state and federal agencies. Specific groups involved in tidal restoration projects include the U.S. Fish and Wildlife Service, the NOAA, the Environmental Protection Agency, the Oregon Department of Fish and Wildlife, the Oregon Watershed Enhancement Board, the Coos Watershed Association, the Oregon Institute of Marine Biology, Seminole Environmental Inc., David Newton and Associates Inc., Marzet Marine and Estuarine Research, the Boys & Girls Club of Southwestern Oregon, the Oregon Youth Conservation Corps, and South Slough Reserve volunteers.

The Cox, Dalton, and Fredrickson Creek Marsh Restoration is recreating complex tidal channels through dike removal, new channel construction, and channel enhancement via placement of large wood root wads. Projects have been monitored for at least 3 years, with the opportunity to test different restoration methods.

The surface elevation of Kunz Marsh was raised using dredged material to accelerate the restoration of emergent marsh vegetation and tidal channel development. Material was added to test the effectiveness of three elevations (low, medium, and high) on improvement of subsided salt marsh wetland structure and function.

Summary

There are numerous ongoing tidal wetland restoration projects along the coasts of the Pacific Northwest. In a sense, each restoration project is its own experiment. Successful restorations will occur where clear goals are articulated and appropriate methods are selected. Conceptual models will help planners understand links between important ecosystem components. New technology will help evaluate site-specific restoration actions within the constraints imposed by the physical environment. As monitoring data are analyzed, additional restorative actions may be proposed in an adaptive management fashion. (See Plate 5 in the color insert for a landscape perspective of habitat cover classes in the Columbia River Estuary.)

References

Adamus, P. R. 2005. *Rapid Assessment Method for Tidal Wetlands of the Oregon Coast*: Volume 1 of *Hydrogeomorphic (HGM) Guidebook*. Coos Watershed Association, Oregon Department of State Lands, and U.S. Environmental Protection Agency, Salem, OR.

Alaback, P. and J. Pojar. 1997. Vegetation from ridgetop to seashore. Pages 69–87 in P. K. Schoonmaker, B. von Hagen, and E. C. Wolf (eds.), *The Rain Forests of Home*. Island Press, Washington, DC.

Allee, R. J., M. Dethier, D. Brown, L. Deegan, R. G. Ford, T. F. Hourigan, J. Maragos, C. Schoch, K. Sealey, R. Twilley, M. P. Weinstein, and M. Yoklavich. 2000. *Marine and Estuarine Ecosystem and Habitat Classification*. NOAA, Silver Spring, MD.

Allen, T. H. 1999. *Areal Distribution, Change and Restoration Potential of Wetlands within the Lower Columbia River Riparian Zone, 1948–1991*. Oregon State University, Corvallis.

Begon, M., J. L. Harper, and C. R. Townsend. 1986.

Ecology: Individuals, Populations and Communities. Blackwell Scientific Publications, Sunderland, MA.

Bourgeron, P. S. 1988. Advantages and limitations of ecological classification for the protection of ecosystems (in comment). *Conservation Biology* 2(2): 218–220.

Brophy, L. 2005. Estuary assessment chapter (draft). Oregon Watershed Enhancement Board, Salem.

Conway, R. 1991. *The Oyster Industry Economic Impact Study*. Dick Conway & Associates, Seattle.

Coombs, E. M., J. K. Clark, G. L. Piper, and A. F. Cofrancesco, eds. 2004. *Biological Control of Invasive Plants in the United States*. Oregon State University Press, Corvallis.

Cortright, R., J. Weber, and R. Bailey. 1987. *Oregon Estuary Plan Book*. Oregon Department of Land Conservation and Development, Salem.

Cowardin, L. M., V. Carter, and E. T. LaRoe. 1979. *Classification of Wetlands and Deepwater Habitats of the United States*. U.S. Department of the Interior, Fish and Wildlife Service, Office of Biological Services, Washington, DC.

Daehler, C. C. and D. R. Strong 1997. Reduced herbivore resistance in introduced smooth cordgrass (*S. alterniflora*) after a century of herbivore-free growth. *Oecologia* (110): 99–108.

Dean, T., Z. Ferdana, J. While, and C. Tanner. 2001. *Identifying and Prioritizing Sites for Potential Estuarine Habitat Restoration in Puget Sound's Skagit River Delta*. Proceedings of the 2001 Puget Sound Research Conference, Olympia, WA.

Denoth, M., L. Frid, and J. H. Myers. 2002. Multiple agents in biological control: improving the odds? *Biological Control* 24: 20–30.

Dethier, M. N. 1990. *A Marine and Estuarine Habitat Classification System for Washington State*. Natural Heritage Program, Washington Department of Natural Resources, Olympia.

Dethier, M. 1992. Classifying marine and estuarine natural communities: an alternative to the Cowardin system. *Natural Areas Journal* 12(2): 98–100.

Dethier, M. N. and S. D. Hacker. 2004. *Improving Management Practices for Invasive Cordgrass in the Pacific Northwest: A Case Study of* Spartina anglica. Washington Sea Grant Program, Seattle.

Franklin, J. F. and C. T. Dyrness. 1988. *Natural Vegetation of Oregon and Washington*. Oregon State University Press, Corvallis.

Garono, R. J. and R. Robinson. 2002. *Assessment of Estuarine and Nearshore Habitats for Threatened Salmon Stocks in the Hood Canal and Eastern Strait of Juan de Fuca, Washington State: Focal Areas 1–4*. Point No Point Treaty Council, Kingston, WA.

Garono, R. J., R. R. Robinson, and C. A. Simenstad. 2003a. *Estuarine Landcover along the Lower Columbia River Estuary Determined from Compact Airborne Spectrographic Imager (CASI) Imagery*. Lower Columbia River Estuary Partnership, Portland, OR.

Garono, R. J., R. R. Robinson, and C. A. Simenstad. 2003b. *Estuarine Landcover along the Lower Columbia River Estuary Determined from Landsat 7 ETM+ Imagery*. Lower Columbia River Estuary Partnership, Portland, OR.

Garono, R. J., C. A. Simenstad, R. Robinson, and H. Ripley. 2004. Using high spatial resolution hyperspectral imagery to map intertidal habitat structure in Hood Canal, WA (USA). *Canadian Journal of Remote Sensing* 30(1): 54–63.

Good, J. W. 2000. Summary and current status of Oregon's estuarine ecosystems. Pages 33–44 in *Oregon State of the Environment Report*. Oregon Progress Board, Salem.

Grevstad, F. S., D. R. Strong, D. Garcia-Rossi, R. W. Switzer, and M. S. Wecker. 2003. Biological control of *Spartina alterniflora* in Willapa Bay, Washington using the planthopper *Prokelisia marginata*: agent specificity and early results. *Biological Control* 27: 32–42.

Grevstad, F. S., M. S. Wecker, and R. W. Switzer. 2004. Habitat tradeoffs in the summer and winter performance of the planthopper *Prokelisia marginata* introduced against the intertidal grass *Spartina alterniflora* in Willapa Bay, WA. Proceedings of the XI International Symposium on Biological Control of Weeds, CSIRO, Canberra, Australia.

Hedge, P. and L. K. Kriwoken. 2000. Evidence for effects of *Spartina anglica* invasion on benthic macrofauna in Little Swanport Estuary, Tasmania. *Austral Ecology* 25: 150–159.

Hedge, P., L. K. Kriwoken, and K. Patten. 2003. A review of *Spartina* management in Washington State, USA. *Journal of Aquatic Plant Management* 41: 82–90.

Hedgpeth, J. W. 1957. Classification of marine environments. Pages I:17–27 in J. W. Hedgpeth (ed.), *Treatise in Marine Ecology and Paleoecology*. Memoir 67. Geologic Society of America, Washington, DC.

Hoines, L. 1996. *Washington State Fisheries Statistical Report: 1993*. Washington State Department of Fish and Wildlife, Olympia.

Jones, C. G., J. H. Lawton, and M. Shachak. 1994. Organisms as ecosystem engineers. *Oikos* 69: 373–386.

Julien, M. H. and M. W. Griffiths. 1998. *Biological Control of Weeds: A World Catalogue of Agents and Their Target Weeds*. CABI Publishing, Wallingford, UK.

Kentula, M. E., ed. 1996. Wetland restoration and creation. *The National Water Summary on Wetland Resources*. Water-Supply Paper 2425. U.S. Geological Survey, Washington, DC.

Lane, R. C. and W. A. Taylor. 1996. *Washington's Wetland Resources*. U.S. Geological Survey, Tacoma, WA.

Maurizi, S. and F. Poillon, eds. 1992. *Restoration of Aquatic Ecosystems*. National Academy Press, Washington, DC.

McFadyen, R. E. C. 1998. Biological control of weeds. *Annual Review of Entomology* 43: 369–393.

Murphy, K. C. 2003. *Report to the Legislature: Progress of the* Spartina *Eradication and Control Programs*. Washington State Department of Agriculture, Olympia.

Murphy, K. C. 2004. Spartina *Eradication Program 2004 Report*. Washington State Department of Agriculture, Olympia.

National Oceanic & Atmospheric Administration. 1997. *Columbia River Estuary Land Cover Change Project*. NOAA Coastal Services Center, Charleston, SC.

O'Connell, K. A. 2002. *Effects of Invasive Atlantic Smooth-Cordgrass (*Spartina alterniflora*) on Infaunal Macroinvertebrate Communities in Southern Willapa Bay, WA*. Western Washington University, Bellingham.

Odum, H. T., B. J. Copeland, and E. A. McMahan. eds. 1974. *Coastal Ecological Systems of the United States*. The Conservation Foundation, Washington, DC.

Patten, K. 2002. Smooth cordgrass (*Spartina alterniflora*) control with imazapyr. *Weed Tech* 16.

Percy, K. L., D. A. Bella, C. Sutterlin, and P. C. Klingeman. 1974. *Descriptions and Information Sources for Oregon Estuaries*. Sea Grant Program: 294. Oregon State University, Corvallis.

Pritchard, D. W. 1967. What is an estuary?: physical viewpoint. Pages 3–5 in G. H. Lauff (ed.), *Estuaries*. American Association for the Advancement of Science, Washington, DC.

Ranwell, D. S. 1964. *Spartina* marshes in southern England, 2. Rates and seasonal pattern of sediment accretion. *Journal of Ecology* 52: 79–94.

Redfield, A. C. 1972. Development of a New England salt marsh. *Ecological Monographs* 42: 201–237.

Sayce, K. 1988. *Introduced Cordgrass,* Spartina alterniflora *Loisel., in Salt Marshes and Tidelands of Willapa Bay, Washington*. Willapa National Wildlife Refuge, Ilwaca, WA.

Scatolini, S. R. and J. B. Zedler. 1996. Epibenthic invertebrates of natural and constructed marshes of San Diego Bay. *Wetlands* 16(1): 24–37.

SER. 2004. *Society of Ecological Restoration International Primer on Ecological Restoration*. Society of Ecological Restoration, Tucson, AZ.

Shipman, H. 2004. *Developing a Geomorphic Typology for the Puget Sound Shoreline*. Discussion Paper (draft) 36. Washington Department of Ecology.

Simenstad, C. A., G. W. Hood, R. M. Thom, D. A. Levy, and D. L. Bottom. 2000. Landscape structure and scale constraints on restoring estuarine wetlands for Pacific Coast juvenile fishes. Pages 597–630 in M. P. Weinstein and D. A. Kreeger (eds.), *Concepts and Controversies in Tidal Marsh Ecology*. Kluwer Academic Publishers, Boston.

Simenstad, C. A. and R. M. Thom. 1996. Functional equivalency trajectories of the restored Gog-Le-Hi-Te estuarine wetland. *Ecological Applications* 6(1): 38–56.

Tanner, C. D., J. R. Cordell, J. Rubey, and L. Tear. 2002. Restoration of freshwater intertidal habitat functions at Spencer Island, Everett, Washington. *Restoration Ecology* 10(3): 564–576.

Thom, R. M., G. W. Williams, and H. Diefenderfed. 2005. Balancing the need to develop coastal areas with the desire for an ecologically functioning coastal environment: is net ecosystem improvement possible? *Restoration Ecology* 13(1): 193–203.

Thomas, D. W. 1980. *Study of the Intertidal Vegetation of the Columbia River Estuary*. Columbia River Estuary Data Development Program, Astoria, OR.

Thomas, D. W. 1983. *Changes in Columbia River Estuary Habitat Types over the Past Century*. Columbia River Estuary Data Development Program, Columbia River Estuary Task Force, Astoria, OR.

Wolf, E. C. 1993. *A Tidewater Place: Portrait of the Willapa Ecosystem*. Willapa Alliance, Long Beach, WA.

Zedler, J. B. and J. C. Callaway. 2001. Tidal wetland functioning. *Journal of Coastal Research* 27: 38–64.

Zipperer, V. 1996. *Ecological Effects of the Introduced Cordgrass,* Spartina alterniflora, *on the Benthic Community Structure of Willapa Bay, Washington*. University of Washington, Seattle.

Resources Available for Tidal Wetland Restoration

A variety of data and information sources are available to tidal wetland restorationists. The University of Washington, Portland State University, Oregon State University, and the University of Oregon all have strong coastal research programs. Many of these programs have information on their Web sites. In addition, there are consortia of academic, government, and private groups that have a wide range of publicly accessible data. Examples include the Washington Department of Natural Resources shoreline inventory (www.ecy.wa.gov/programs/sea/SMA/st_guide/SMP/index.html), the Coastal Zone Atlas (www.coastalatlas.net/), and the Oregon Coast Geospatial Data Clearing House (buccaneer.geo.orst.edu/).

Chapter 9

Ponderosa Pine and Interior Forests

Stephen F. Arno and Carl E. Fiedler

Successive, widely publicized wildfires have swept through lower-elevation forests of the inland Pacific Northwest and the northern Rockies over the past two decades. Before 1900, many of these areas were dominated by large fire-resistant trees and were visited by frequent low- or mixed-intensity fires. In recent years, hundreds of forest homes and cabins have gone up in flames, from the mountain suburbs of Bend, Oregon to Kelowna, British Columbia. By now anyone mildly interested in Western forests is aware that management and protection policies have failed to account for the historical role of fire and left a legacy of overly dense forests with sickly trees and hazardous fuel levels (Figure 9.1).

Today's forests also commonly lack the diverse and productive grass and shrub communities needed by wildlife. Each year we commit more money, personnel, and technology to fire suppression, yet uncontrollable wildfires continue to threaten forests, rural homes, and treasured recreation areas.

After decades of studying Western forests, the authors observed that magnificent old trees begin to disappear when deprived of frequent fire. Also, when forests of these venerable trees are managed using traditional timber harvesting methods, the features that made them famous ultimately disappear. Even in protected "natural areas" such as parks, wilderness, and primitive areas, big, old fire-resistant trees gradually die and are replaced by thickets of smaller trees. Our experience also confirms that long-lived trees and associated plant communities can be restored through practices that mimic the effects of historical fires. Research studies and practical examples show what is needed to restore interior forests.

This chapter advocates a restoration forestry approach based on historical natural processes. We use the term *restoration forestry* to designate the practice of recreating an approximation of historical structure and ecological processes to tree communities that were once shaped by distinctive patterns of fire. The intent is not to recreate a single, distinct "historical condition" but rather a range of conditions representative of historical ecosystems (Fiedler 2000b). Restoration of tree structure and composition is not only a product, then, but also a key process that initiates and facilitates the broader goal of ecological restoration.

What is the scale of the restoration forestry we advocate? Conditions are well outside their historical range on perhaps 25–30 million acres of fire-prone, lower-elevation forests in the inland Northwest and the northern Rockies. Any strategically located restoration treatments in these forests can produce noticeable benefits and return important

Condensed from Arno and Fiedler, *Mimicking Nature's Fire: Restoring Fire-Prone Forests in the West*, Island Press, 2005.

Figure 9.1. Deforestation resulting from severe wildfire. *(a)* Ponderosa pine forest in central Montana in 1982, before burning. *(b)* Nearly the same viewpoint in 1998, 14 years after the 170,000-acre Hawk Creek wildfire. *(Photos by Carl Fiedler)*

features of historical forests. In some areas, restoration forestry relies on simply guiding natural fires. In others, it entails thinning to produce a more natural forest structure before using prescribed fire. In heavily populated areas it may rely on strategic removal of certain trees and forest fuels. In areas where timber production is not an objective, initial cutting treatments eventually may be fully replaced by fire. This chapter explains the options and considerations involved in planning restoration of lower-elevation inland forests across a range of conditions and ownership objectives. For those interested in more detailed information, we cite publications that elaborate specific topics and refer readers to our restoration forestry book.

How We Got Here

Why did Euro-American settlement and traditional forestry result in deterioration of Western interior forests? Simply put, forestry predates the science of ecology and failed from the beginning to recognize the ecological and cultural importance of fire. Early concepts of Western forestry were developed in moist regions of Europe, where fire was considered entirely a destructive force (Arno and Allison-Bunnell 2002). In contrast, most North American forests (as well as ancient forests in Europe) were shaped over thousands of years by fire (Pyne 1997).

When European-American settlements were expanding across North America in the nineteenth century, fires threatened everything from mining camps to major communities (Pyne 1982). Many fires were started accidentally by railroads, campfires, or settlers clearing land. Although a few visionaries recommended controlled burning of forests to reduce fire threats, forestry leaders, conservationists, and many landowners believed fire should be completely suppressed. In 1908, the U.S. Forest Service advanced a policy to prevent any use of fire as a management tool and over the ensuing decades built a multiagency program to eliminate fire.

By the 1960s, rapidly expanding knowledge in the young science of ecology revealed that fire plays an essential role in natural forests. In the 1970s natural resource agencies changed course and recognized that fire should be used in forest management (Nelson 1979). However, by this time multiple barriers to its use had developed. An influential fire suppression industry had arisen, which depended on continuation of the fire exclusion policy (Arno and Allison-Bunnell 2002).

The public, nurtured on Smokey Bear messages, became convinced that all forest fire was bad.

New laws, including the Clean Air Act, Clean Water Act, and National Environmental Protection Act, created legal impediments to returning fire to forests. New homes and resort developments worth billions of dollars sprang up in increasingly dense, high-hazard forests, creating further resistance to foresters who wanted to thin trees or reintroduce fire. The potential liability land managers face when prescribed or natural fires go awry far outweighs any rewards they might receive for using fire successfully. Government spends vast amounts to suppress fires but resists spending smaller amounts to reduce dangerous fuels.

Besides the emergence of fire suppression as a key management goal, harvesting also contributed to today's forest conditions. In forests historically shaped by frequent fires, conventional logging removed large fire-resistant trees and allowed saplings of shade-tolerant species to develop into thickets.

Roots of Restoration Forestry

The concept of restoring and maintaining more natural conditions in fire-prone Western forests developed slowly, in tandem with the emerging disciplines of forest and fire ecology. In the early 1900s—and much to the chagrin of the U.S. Forest Service—several prominent timberland owners in northern California practiced burning beneath ponderosa pine–mixed conifer forests to reduce fuels and mimic frequent fires of the past (Hoxie 1910, Kitts 1919, Pyne 1982, 2001). In the 1930s, Harold Weaver, a forester with the U.S. Indian Service, began experimenting with fire in ponderosa pine (Weaver 1968). He later penned a remarkable treatise in the *Journal of Forestry*, championing fire as a silvicultural tool in ponderosa pine forests (Weaver 1943). Within a few years Harold Biswell, a forestry professor in California, also began demonstrating the use of fire in ponderosa pine forests (Biswell 1989, Carle 2002). But the voices recognizing fire's ecological value were few, and the deaf ears were many.

The use of silvicultural cutting to restore historical structure in fire-dependent forests was slower to develop. The first restoration research in the West that involved both cutting and burning was initiated in 1984 in the University of Montana's Lubrecht Experimental Forest. Researchers evaluated modified selection cutting and prescribed burning aimed at sustaining uneven-aged ponderosa pine stands, including old trees, while also producing timber.

By the 1990s mounting environmental controversy had dramatically reduced timber harvest on public forests and forced land managers to emphasize environmental protection. The Forest Service adopted a concept called *ecosystem management*, aimed at restoring and maintaining natural communities of trees, other plants, and animals (i.e., biodiversity). Although the theory of ecosystem management has been well publicized, evaluations of projects applying it are rare.

Private, state, and tribal forestlands have also come under pressure to improve ecological stewardship, sometimes linked to programs that certify sustainable forest management. Conservation easements are also increasingly replacing preservation goals with stewardship management plans that include hands-on treatments, including silvicultural cutting designed to restore more natural conditions.

Restoration treatments initially leave the largest, oldest trees and remove small to medium-sized, mostly shade-tolerant ones. Prescribed fire is used to kill most seedlings and saplings, thus mimicking effects of historical fires (Fiedler 2000b). Valuable timber remains behind, some of which may be harvested later to help pay for tending the stand in the future.

Historical Fire Regimes as the Basis for Restoration

Until the 1970s foresters and fire scientists were so absorbed in the battle against fire that they largely overlooked fire's key role as an ecological process. Ponderosa pine forests were known to have survived frequent low-intensity fires in the past, but many foresters assumed that nonlethal fires were a small percentage of the total. August 1973 marked the first time federal forest managers in the inland West chose to allow a lightning fire to burn (Daniels 1991, Moore 1996). Over several weeks, this fire in the Selway–Bitterroot Wilderness (SBW) of northern Idaho grew sporadically under constantly changing weather, eventually encompassing 2,800 acres. Here at last was an opportunity to study an unsuppressed fire interacting with highly variable natural forests in rugged terrain. Inspection of this momentous burn revealed puzzling sights. For example, the fire burned through a dense stand of lodgepole pine saplings, blackening their bases and charring the ground, yet the trees remained alive. This was inconsistent with conventional wisdom that fires in lodgepole kill nearly all of the trees.

In years that followed, several large natural fires in a few other remote areas of the West were allowed to burn for weeks at a time until extinguished by rain and snow. These free-ranging fires killed trees in patchy patterns related to the time of burning and variations in fuels, weather, and topography. In some places all trees were killed, but there were patches that escaped burning entirely and other places where fire spread across the ground without killing overstory trees. In some parts of the burn, fire thinned forests, killing small trees and thin-barked species, while larger thick-barked trees survived. The variation and complexity of natural fires corroborated evidence from studies of older fires. The new findings suggested that historically 75% of Western forests experienced fires that killed some trees while others survived (Arno 1980, 2000, McKelvey et al. 1996, Quigley et al. 1996). Most of the remaining 25% of forests showed evidence of ancient fires that killed nearly all the trees and led to establishment of replacement stands. For centuries fires had shaped the forest, favoring trees and entire communities of organisms that were adapted to burning.

The knowledge gained from studying how fire interacts with forests provides key insights for restoring fire-prone forests. Despite immense variability in how fires burn and in their effects, each forest type in a region tended to burn in similar patterns over long periods of time. In the 1990s, fire scientists devised a classification of three *fire regimes* that encompasses the different patterns of fire occurrence over long time periods: *understory*, *mixed*, and *stand replacement* fire regimes. Together they provide a frame of reference for understanding how trees, other plants, animals, and soil organisms formed dynamic communities that adapted to certain frequencies and intensities of burning (Brown 2000). Understanding fire regimes helps us compare forests of the past and present. The remainder of this chapter focuses on the understory and mixed regimes; see Chapter 5 in this volume for discussion of the stand replacement fire regime.

Understory Fire Regime

Before 1900, the understory fire regime was familiar to both Native Americans and Euro-Americans because it dominated the dry lower-elevation forests that most people lived in, traveled through, and used for basic needs. Fire swept through a given location at intervals ranging from 1 to 30 years. These frequent fires killed few overstory trees but scorched (and thus pruned) low branches and heavily thinned saplings and maintained a grassy understory. This fire regime encompassed around 40% of the total forest in the interior Pacific Northwest (Arno 2000, Paysen et al. 2000, Quigley et al. 1996, Gruell 2001). It was also found in forests and woodlands of western Oregon (Arno 2000) (Table 9.1). These sunny forests are witnessed in century-old photographs, pioneer accounts, records from nineteenth-century land surveys, and ancient stumps of large trees with datable scars from many fires. Fire suppression and logging have dramatically changed most of these forests, but a few remnants are intact and are being restored.

TABLE 9.1.

Pacific Northwest forest types and their historical fire regimes.

Historical Forest Type	Fire Regime Types		
	Understory	Mixed	Stand Replacement
Average fire interval	20–30 yr	30–100 yr	100–400 yr
Subalpine			M
Douglas fir, coastal		M	M
Douglas fir, inland	m	M	
Hemlock–Sitka spruce			M
Lodgepole pine		M	M
Oregon white oak	M		
Ponderosa pine, east	m	M	m
Ponderosa pine, west	M	m	
Quaking aspen	m	M	
Coast redwood	M	m	
White pine–cedar–hemlock	M	M	
Western larch		M	M
Whitebark pine		M	M

Source: Adapted from Arno (2000).

M indicates a major part of the forest type; *m* signifies a smaller representation.

The historical understory fire regime forests of today are commonly dense stands of small conifers. It takes a sleuth to reconstruct the original forest by inventorying old stumps and tree remnants (Arno and Allison-Bunnell 2002). Some of these relicts record dates of frequent fires back 500 years in ponderosa pine and 2,000 years in the ancient giant sequoias of California's Sierra Nevada (Arno 1976, Dieterich 1980, Swetnam 1993). This primeval pattern of frequent burning favored trees that develop fire resistance at an early age, including ponderosa and Jeffrey pine, giant sequoia, western larch, Oregon white oak, and some other oaks in California and the Southwest (see Chapter 4 in this volume). Open stands featured abundant grass, flowering herbs, and low shrubs that readily resprout. Leaf litter, dead wood, and ladder fuels (tall shrubs, small trees, and lower limbs of larger trees) were controlled by frequent burning.

Studies also show that trees regenerated in small, scattered, even-aged groups (Arno et al. 1995, Bonnicksen 2000, Cooper 1960). New trees established in open areas where old trees had been killed by insects, wind, or fire. Often these openings had concentrations of fuel from dead trees. Repeated fires burned these fuel concentrations, eventually creating bare ground favorable for establishment and rapid growth of seedlings. Subsequent fires burned lightly through seedling patches because of a dearth of needle litter, allowing some seedlings to survive—perhaps only one or a few per patch—and develop into a new age class of trees. Historical, open forests regenerated themselves through the centuries in a fine-grained mosaic that was so subtle the forest appeared to be uniform and unchanging. Frequent fires maintained the open conditions essential for developing large, long-lived trees and for regenerating and thinning young, sun-loving offspring.

In the early 1900s, foresters trained in moist regions thought that these open, frequently underburned Western pine forests were understocked. They wanted more saplings to develop and therefore opposed even the light burning favored by some timberland owners because it killed small trees (Weaver 1968). When fires were eliminated, the understory filled up with saplings. Eventually thickets of small trees suppressed the growth of all trees through excessive competition for moisture, nutrients, and sunlight, leaving the forest vulnerable to damage from insects, disease, and severe wildfires. In recent years, huge stand-replacing firestorms have begun to supplant the low-intensity surface fires that characterized the historical understory fire regime.

Today's high-intensity fires also damage soils, creating a heat-induced water-repellent surface that accelerates runoff and erosion (Arno and Allison-Bunnell 2002). The native vegetation, poorly adapted to intense fire, often recovers slowly, while introduced species gain a foothold and prosper (Arno 1999). By about 15 years after a stand replacement wildfire, the dead trees have fallen, creating jackstrawed layers of old trees that can burn white hot. Heavy downed fuels that decompose slowly in a dry environment, coupled with a dense undergrowth of herbs, shrubs, and saplings, may lead to a second fire that is even more damaging to the soil than the first (Arno and Allison-Bunnell 2002, Gray and Franklin 1997).

Mixed Fire Regime

Historically, mixed fire regimes covered about 40% of the forest in broad regions of the West (Quigley et al. 1996). The mixed regime generally occurred in cooler, moister, higher-elevation forests than the understory regime. Fuels dried sufficiently to allow burning for several days or a few weeks each summer, and in any given place fires occurred at intermediate intervals, from about 30 to 100 years (Brown 2000). Individual fires ranged from low-intensity understory burns to stand replacement fires. However, many were of intermediate intensity, killing most fire-susceptible trees—species with thin bark and saplings of all species—but fewer fire-resistant trees.

The mixed fire regime created great diversity in a particular forest and across the landscape. In the aftermath of an individual stand replacement fire, the mixed fire regime could produce a forest mosaic similar to the large-scale patchwork found in stand replacement fire regimes more typical west of the Cascade Mountains. After a series of low-intensity fires, the mixed regime might have a uniform, open forest, as in the understory fire regime; however, the trees would be of similar age, having arisen from the occasional high-intensity fire (Arno et al. 1995, 1997).

In areas where the mixed fire regime had many intermediate-intensity fires, the resulting forest was a quilt of contrasting patches dominated by multiple age classes and species of trees. In moist inland forests, fires often left surviving western larch representing different age classes that established after earlier fires (Table 9.2). Some old white pine and red cedars survived along with occasional lodgepole pines, but hardly any hemlock or subalpine fir remained after burns. Shade-intolerant conifers dominated tree regeneration after fires. Fires also triggered production of a rich assortment of shade-intolerant deciduous trees and fruit-bearing shrubs important for wildlife and people.

Mixed fire regimes created variations even in forests with only a few tree species, such as cool, dry forests in the Rocky Mountains. When fire is suppressed, these forests may grow into dense monocultures of inland Douglas fir with sparse undergrowth. In contrast, the mixed regime fire maintained a patchwork of mountain grassland, aspen groves, young lodgepole pine and Douglas

TABLE 9.2.

Historic and modern stand structures contrasted.

	Historical Stand		Modern Stand (second-growth)	
	Pines	Douglas Firs	Pines	Douglas Firs
0–4 inches	5	2	7	25
4–8 inches	4	2	55	45
8–12 inches	4	2	50	25
12–16 inches	4	2	26	14
16–20 inches	6	1	7	2
20–24 inches	4	1	1	
24–28 inches	3	1		
28–32 inches				
32–36 inches	1			
36–40 inches	1			
Basal per acre	65 square feet		115 square feet	
Trees >8 inches in diameter per acre	30		125	
Average overstory diameter	18		11	

fir, and old Douglas fir with open, grassy understories (Arno and Gruell 1983).

Land managers are attempting to restore mixed fire regime ecosystem mosaics in parts of several large wilderness areas and national parks by allowing some natural fires to burn and applying only limited suppression to others. In many areas, however, fire exclusion is still the rule, and the mixed fire regime has missed one or two historical fire cycles. Fire's extended absence leads to decline and disappearance of historically dominant shade-intolerant trees such as aspen, ponderosa pine, sugar pine, western larch, western white pine, and whitebark pine along with fruit-bearing shrubs and fire-maintained grasses.

Moist and productive areas readily transform into dense forests of shade-tolerant trees when fire is kept out, ultimately favoring higher-intensity stand replacement fires. Large, long-lived intolerant trees are lost to biotic agents, weather events, or stand replacement fire.

Applying Ecological Knowledge

Three ecological issues characterize forests that were historically shaped by the understory fire regime: increased stand density, development of ladder fuels, and replacement of shade-intolerant tree species with shade-tolerant ones. Historical stand characteristics can be simulated by using uneven-aged management approaches that perpetuate trees of different ages while maintaining an open understory. Restoration treatments must be conducted at short intervals (perhaps 15 to 35 years) to limit fuel buildup and promote shade-intolerant species.

Forests in the historical mixed fire regime had a patchwork of different age classes of mostly shade-intolerant trees. These areas are now characterized by increasingly dense stands, loss of patchiness, and proliferation of shade-tolerant species. Restoration is designed to prevent the increase of tolerant trees by killing or removing them in patchy patterns while retaining intolerant trees and encouraging regeneration of the herbs, shrubs, and trees that benefit from fire. Restoration treatments can vary in intensity, pattern, and interval (40–100 years).

Historical Ignitions

Native peoples occupied and traversed the West for more than 10,000 years, shaping historical fire regimes in unquantifiable but probably significant ways (Stewart 2002). Evidence compiled by such disparate sources as archeologists, anthropologists, ethnographers, historians, and ecologists indicates that Native American burning significantly shaped drier forests and grasslands west of the Sierra Nevada and Cascade ranges as far north as southwestern British Columbia (Anderson and Moratto 1996, Boyd 1999, Greenlee and Langenheim 1990, Lewis 1973). There is much evidence to support the conclusion that Native American burning also influenced the drier forests east of the Cascades and in the Rocky Mountains (Barrett and Arno 1982, Gruell 1985, Shinn 1980, Stewart 2002). We can characterize the *effects* of historical fire regimes in the structure of pre-1900 forests and document the *process* that created these forests in terms of frequencies, severities, and patterns of past fires. However, we do not know what forests would have been like without Native American ignitions.

Fuels

Fuels in today's interior Northwest forests differ markedly from historical conditions (Arno 2000, Quigley et al. 1996). Duff accumulations are now so deep that old, fire-resistant trees are killed by prolonged heating even in low-intensity fires (Covington and Moore 1994, Harrington 2000). This is partly because roots grow in the duff, whereas under historical conditions they were insulated in the mineral soil. Buildup of down wood also leads to severe burning. Increased understory thickets allow surface fires to torch up into the main canopy, creating crown fires in areas where they were once rare (Cottrell 2004). The proportion of stand replacement burning in Western forests has approximately doubled from pre-1900 levels (Quigley et al. 1996).

Historical Forest Structure

Studies focused on historical understory and mixed fire regimes show that the structure of contemporary stands contrasts with pre-1900 conditions, with many outside the range of historical variation (Agee 1993, Arno 2000, Morgan et al. 1994). This can have a major effect on a stand's response to fire. An historical stand dominated by large fire-resistant trees growing at wide spacing responds to a summer lightning fire quite differently than modern stands, including both densely regenerated stands that have followed logging and unlogged forests that are filled with understory thickets. Historical stands were much more likely to survive fire than either of these current structures.

In 2003, the Mineral–Primm wildfire in western Montana provided a dramatic example of differential fire behavior in adjacent stands with contrasting structures. Dense postlogging stands of pole-sized trees near Primm Meadow mostly burned as a crown fire. An adjacent old-growth ponderosa pine stand with a conifer understory burned in a mixed severity pattern. But when the fire reached an open, old-growth ponderosa pine stand that homesteaders had maintained for livestock grazing by removing saplings and most dead trees, the wildfire damped down to a light underburn (Arno and Fiedler 2003, unpublished observations).

Political and Economic Issues

Programs that allow return of natural fires have been operating in a few national parks and wilderness areas since the 1970s. However, only in the largest and most remote forest in the western United States, a block of more than 4 million acres in Idaho, has such a program been able to restore some semblance of the historical fire regime. Even in the greater Yellowstone National Park region, restoration efforts have been significantly constrained by the need to protect resorts and other facilities (Keane et al. 2002).

Because of the precedents set by government fire suppression policies since 1908, the public and its elected representatives expect valiant efforts to protect all private property and development from forest fire. Had the government instead established a policy of adapting to fire-prone forests, a tradition of personal responsibility for fire protection might have been fostered among landowners, as is the case today in parts of Australia (Mutch 2001).

Some suggest that natural fire could be allowed to return if suppression funding were shifted toward fireproofing homes and communities. (Fire suppression and rehabilitation costs in the West generally total about $1–2 billion annually.) National Fire Plan funding has been available for a few years for fuel reduction near homes, but little progress has been made because many homeowners are uninterested, unmotivated, or unwilling to modify their combustible roofing, siding, decks, landscaping, and hazardous vegetation. Moreover, crown fires in surrounding untreated forests can loft firebrands up to a mile, so even when fuel reduction is accomplished, vigilance and maintenance are necessary to ensure that dwellings remain fire safe.

The reality is that national forest and park managers do not have the administrative backing to permit natural fires to burn if they might spread to private lands and developments. Most areas of public land are too small, or they include resorts and other developments. They are not sufficiently insular to provide high assurance of containing a fire that might burn for several weeks and be fanned by high winds. Some argue that natural fires should be allowed to develop and then, if necessary, suppressed near area boundaries. However, such action would be considered negligent because experience has proven that rapid initial suppression is critical. Once a fire grows large it becomes costly and problematic to control. Pity the fire manager who did not quench a lightning ignition while it was small and that later cost millions of dollars to control when it threatened—or worse, overran—people, homes, and other developments.

Three decades of attempts by dedicated fire managers to expand natural fire programs confirms that it is not be possible to allow lightning fires to burn on most federal forestlands (Agee 2000, Parsons 2000, Zimmerman and Bunnell 2000). Moreover, even if such a goal could be achieved, it would fall far short of restoring historical fire regimes.

In view of the crippling constraints that prevent reestablishment of understory and mixed fire regimes, the predictable outcome of relying exclusively on "natural" fires for restoration would perpetuate the current situation. Even under let-burn programs, many fires are still extinguished. There isn't enough funding or political will to allow any other outcome. The government's liability for court costs and fire damages alone would be staggering. The undesirable result would be continuation of the shift toward more and more stand replacement burning, an ironic consequence for a policy intended to restore natural conditions.

The sobering reality is that we cannot rely on returning natural fires as the primary means of restoring most Western forests. The Nature Conservancy, Defenders of Wildlife, and the Rocky Mountain Elk Foundation are among leading conservation organizations that recognize the need for prescribed burning and tree thinning to restore Western forests (Brown 2001, Stalling 2003, Wilkinson 2001).

Recognizing that we cannot simply step aside and let nature do the restoration provides one advantage. It shifts responsibility to us to guide ecological processes toward a semblance of historical conditions. Conceding that we have long disrupted a crucial creative force in forests, we must now use methods to mimic the effects of historical fire regimes to sustain our forests.

Restoring Ponderosa Pine Forests

In a few ponderosa pine forests, managers, scientists, and citizens have developed new approaches to address deteriorating conditions. They have patterned restoration on the historical fire regimes that produced forests of beautiful, long-lived trees in semiarid environments.

Lick Creek

In 1906, the Lick Creek area in the Bitterroot Valley of western Montana was a hub of activity, befitting its status as one of the first large national forest timber sales in the country. The forest at Lick Creek was dominated by old-growth ponderosa pine ranging from 200 to more than 400 years old. The classic "yellow pines" were mostly open-grown, with only thirty to forty large trees per acre, the trunks averaging more than 30 feet apart (Arno et al. 1995, Leiberg 1899, Smith and Arno 1999). Old-growth stands were interspersed with scattered individuals and patches of younger pines and saplings. About 10% of the big trees were inland Douglas fir (*Pseudotsuga menziesii* var. *glauca*), most abundant on north-facing slopes and in swales. Early photographs depict large pines with smooth, fire-pruned trunks extending high above the ground and an open, grassy understory, presumably groomed over the centuries by low-intensity fires. Fire scars on living trees and stumps of logged trees show that fires occurred at average intervals of 7 years between 1600 and 1895.

The 1906 sale was meant to initiate a new era of sustainable timber management. The plan was to eventually harvest all trees as they reached financial maturity, defined as the point at which their growth began to slow, at age 100 years or so. The initial harvest involved selective cutting that left 5% to 50% of the original trees for future harvests. Elers Koch (1998) recounts a visit to Lick Creek by Gifford Pinchot, chief of the U.S. Forest Service. Pinchot thought that too many trees were being marked for cutting, so he had the crew go back and make adjustments. Tree markers were instructed to leave thrifty pines with good crown form and to mark all Douglas firs more than 10 inches in diameter. Foresters worried that allowing Douglas fir to increase would favor spread of dwarf mistletoe. Some clumps of small pines were also thinned. Overall, this project yielded more than 37 million board feet of timber from 2,100 acres. Branches and tops from the harvested trees were piled by hand and burned, but fires were no longer allowed to spread as they had in the past.

Over the decades that followed the remaining trees grew vigorously. Within 20 years of the harvest, pine and Douglas fir seedlings and saplings had established throughout the area. Much of the young Douglas fir had regenerated after the last recorded understory fire in the 1890s, giving it a head start over the shade-intolerant pine that managed to regenerate in openings created by logging. By 1946 the timber volume had increased 60%, to about 6,100 board feet per acre.

Between 1953 and 1981 additional harvests removed some of the trees from crowded groups, damaged trees, and those with poor crowns, including slow-growing older trees. Saplings and pole-size trees were thinned and left to decay where they fell. A photographer captured several views of the Lick Creek forest in 1909. The camera positions were permanently marked in the 1920s, and the scenes have been rephotographed every decade thereafter. Interpretations of these photo series and data on tree growth show that the 1907–1911 logging operations created openings, while skidding and pile burning removed litter, duff, and understory vegetation in patches (Gruell et al. 1982). The 1920s and 1930s photos show conifer regeneration and establishment of vigorous Scouler willow and bitterbrush shrubs, whose young twigs provide important forage for deer, elk, and moose on winter range. By the 1950s, young conifers had formed dense thickets (Figure 9.2). These expanded through time, despite scattered thinning. Even though most of the large Douglas fir trees had been removed in the early 1900s, the proportion of young firs kept increasing.

By the 1980s, despite decades of selective cutting and thinning intended to promote pine, forests at Lick Creek became crowded with small trees. They had become increasingly vulnerable to insects, disease, and crown fire and had decreased aesthetic values and wildlife forage production. These undesirable changes spurred the Bitterroot National Forest to evaluate possible management options for the Lick Creek area. A primary goal that emerged from planning meetings was to guide management toward more sustainable forest structure while also improving wildlife habitat, aesthetics, and tree growth.

In 1991 the Forest Service Intermountain Research Station and the University of Montana's School of Forestry agreed to cooperative studies with the Bitterroot National Forest to provide a basis for restoration management at Lick Creek. The collaborators agreed to work toward developing stands with large, medium, and small pines in proportions that would continuously maintain a large-tree component. This would entail adapting traditional shelterwood and selection silvicultural systems because neither was designed to perpetuate large old trees. The modified system would remove excess small, medium, and large trees every 25–30 years while retaining enough trees to perpetuate an uneven-aged, pine-dominated forest. Each cut would induce regeneration of a new age class of ponderosa pine.

The collaborators agreed that fire should be used in tandem with tree harvest. Low-intensity fires would reduce slash, duff, and litter, recycle nutrients, control excess conifer regeneration (especially Douglas fir and grand fir), and stimulate herb and shrub growth. Collaborators were also aware that reintroducing fire after a long period of exclusion might stress remaining trees and invite bark beetle attacks.

Three kinds of stands covered contiguous areas large enough to accommodate the multiple treatment units or "replications" needed for research

FIGURE 9.2. Photo sequence from a camera point at Lick Creek showing change in forest structure after elimination of the understory fire regime and selective harvesting in 1907 or 1908. *(a)* In 1909 the stand is nearly pure ponderosa pine with an open understory. Stumps indicate the few trees that were removed. *(b)* By 1948 a dense Douglas fir understory had developed. (*USDA Forest Service photos*)

experiments. Douglas fir made up less than 20% of the overstory in all stands, but fir seedlings and saplings were abundant in many areas.

The youngest stand was about 70 years old, and a thinning treatment was prescribed to reduce basal area to about 50 square feet per acre, down from the existing 85. A second stand was 80–85 years old and supported about 120 square feet of basal area per acre. A regeneration harvest was prescribed to reduce basal area to about 40–45 square feet per acre. The third stand averaged about 110 square feet per acre and contained pines of many ages, including some very large, old trees. A modified selection cutting was prescribed to reduce basal area to an average of 50 square feet per acre. This stand was marked to retain pines across the full range of diameters, including about one snag per acre. The most vigorous trees were retained in all three stands, and in the two older stands small openings were created to induce a new age class of pine. The ranger district implemented similar treatments in surrounding stands for a total treatment area of about 530 acres, which produced 1.5 million board feet of timber.

Treetops were left where they fell to provide woody debris and recycle nutrients to the soil. To control the amount of slash left in the units, loggers skidded the topped trees with limbs attached to roadside landings, where limbs were removed, piled, and burned. Stands were subdivided into multiple units, including unharvested controls, harvested areas left unburned, harvested areas receiving a "wet burn," "dry burn," spring burn, and fall burn. A "wet burn" was applied early in the spring, when the lower duff and large woody fuels were still moist. The "dry burn" prescription was implemented after significant drying had taken place, several days later in spring. A successful burn would consume some of the fuel while causing little damage to the overstory trees.

Dry burns caused modest but acceptable levels of overstory mortality. All burn treatments killed an average of about 65% of the trees less than 7 inches in diameter, mostly firs. One goal of underburning was to allow most of the smaller pine (averaging 4 inches in diameter) to survive, and about 80% of them did. Overall, the burn treatments spanned and defined a range of conditions that are broadly applicable to similar projects.

Cutting and burning released a flush of nitrogen into the soil. Some was taken up by the undergrowth vegetation, which was then heavily grazed by deer, elk, and moose. Resprouting willow was browsed so heavily that it could not grow vertically.

The experience at Lick Creek demonstrated that cutting treatments and prescribed fire used in tandem can control excessive numbers of saplings, reduce fuel, and recycle nutrients. However, it also exposed challenges in restoring a fire-dependent forest that has missed several natural fire cycles. Heavy fuel accumulations, coupled with low tree vigor and presence of fine roots close to the surface, make trees vulnerable to fire damage. Trees in such conditions are already stressed, and even moderate fire injury may increase their susceptibility to beetle attack. Invasive, nonnative plants, notably spotted knapweed (*Centaurea maculosa*), were already established at Lick Creek, and their coverage increased soon after treatments. Reducing soil disturbance by winter logging could mitigate this problem. The alternative of ignoring the deteriorating conditions and fuel buildups leaves these forests vulnerable to still greater soil disturbance and weed invasion when large wildfires occur.

The coordinated effort at Lick Creek helped gain acceptance within and outside the Forest Service for new management approaches. Most people involved agreed that management based on natural disturbance processes could meet both ecological and social needs in an area used for recreation and timber production.

Lubrecht Experimental Forest

In the 1930s, the proprietor of a guest ranch in Montana's Blackfoot Valley was alarmed by the proposed cutting of beautiful old-growth ponderosa pine that bordered the road leading to the ranch. Even though the land was owned by the powerful Anaconda Copper and Mining Company, she was able to persuade the logging foreman to spare the splendid yellow pines. A few years later the company donated nearly 20,000 acres of forest land, including the scenic road corridor, to the University of Montana (UM). These lands became the UM Lubrecht Experimental Forest, and here on the scenic corridor a restoration project was initiated in 1984. It included a modified form of selection cutting that focused on the density and sizes of trees to leave rather than concentrating on those to be cut.

At the time, a battle was waging between the prevailing forest management philosophy that focused on wood production and an emerging philosophy that emphasized the ecological and amenity values of forests. Expanding scientific knowledge raised questions about the one-size-fits-all application of even-aged treatments, particularly clearcutting. Even-aged management seemed especially contestable in forests historically dominated by ponderosa pine. Fire scar analyses showed low-intensity fires historically occurring at 5- to 30-year intervals in these forests, and age class determinations typically showed trees establishing not only in different decades within a century but also in different centuries. Remarkably, in this natural uneven-aged system the shade-intolerant pine had continued to dominate and prosper despite the presence of more shade-tolerant Douglas fir.

It was revealing to find that the large pines in a stand that looked reasonably alike and were assumed by many to be similar in age were in fact quite different. For example, imagine four trees in an old-growth stand that were found to be 225, 275, 325, and 475 years old in the year 2000. Now turn back the clock to the year 1775. These same trees were a first-year seedling, a 50-year-old pole-size tree, a 100-year-old medium-size tree, and a 250-year-old old-growth tree. This broad age range suggests a low-intensity disturbance regime in which regeneration occurred in some years and not others. Occasional surface fires killed most trees while they were small, but a few survived the repeated fires to reach large size and old age.

Ponderosa pines have deep roots, thick bark, open crowns, large fleshy buds, and long needles arrayed to deflect rising heat—characteristics that allow them to survive a surface fire regime but confer no protection or advantage in crown fires. Ponderosa pine also has heavy seeds that typically disperse within about 150 feet of the parent tree. This attribute reflects a species adapted to regenerating in small openings created by low-intensity fire, pockets of beetle kill, or occasional lightning strikes—not in expansive openings created by stand replacement fires.

A question then arose: Could silvicultural treatments emulate the disturbances that once sustained ponderosa pine forests for centuries, perpetuating uneven-aged stands with large old pines while also yielding timber products? Years earlier fire ecologists such as Weaver (1943) and Biswell et al. (1973) had provided strong conceptual support for uneven-aged management of ponderosa pine based on historical fire regimes.

The Lubrecht project addresses critical components of an uneven-aged restoration prescription, including density, structure, and species composition. Two treatments—a combined improvement and selection cutting to a reserve density of 60 square feet per acre (from about 100 square feet in untreated stands) and the same cutting in combination with prescribed underburning—were compared with a no-treatment control. Selection cutting methods were chosen as most appropriate for accomplishing the study objectives of maintaining some trees of all sizes and regenerating a new age class of ponderosa pine. Improvement cutting removed pole-size and larger Douglas fir to prevent conversion to this more shade-tolerant species, with underburning used to kill Douglas fir seedlings and saplings.

Proponents of traditional timber management criticized this project, contending that shade-intolerant ponderosa pine would not regenerate under a partial overstory, whereas shade-tolerant Douglas fir would regenerate profusely, resulting in conversion to fir. They argued that treatments cost too much and produced too little timber. They also thought the methods were too complex for operational use and that logging and underburning would significantly damage the smaller trees retained in uneven-aged stands. Some detractors also noted that the system was incompatible with the record-keeping methods used by federal agencies.

Experience since 1984 shows that these problems can be reduced or neutralized using well-designed treatments and improved methods of implementation (Becker 1995, Fiedler 1995). Although the treatments implemented at Lubrecht have shortcomings, the critical question is whether these and similar restoration approaches are more appropriate than traditional timber management for sustaining productive ponderosa pine forests that are also well adapted to natural disturbances.

Modifying a European Approach to Meet Western Needs

The kind of selection cutting that appears to best emulate the effects of historical disturbances on forest structure is a greatly modified version of the selection cutting method described in textbooks. Selection cutting has its roots in Europe and typically was applied in stands of shade-tolerant species such as fir and spruce where fire was not prevalent. Individual trees were selected in a given harvest, hence the name "single-tree selection." Stands were reentered every decade or so, with few trees per acre removed in each entry. Little additional light or soil moisture was freed up in the stand, and the shaded conditions that remained after cutting favored regeneration of shade-tolerant species.

Selection cutting in shade-intolerant ponderosa pine stands must be modified. One change is the focus on the trees to be left. This allows the marker to leave the number, species, size, and spatial arrangement of trees that will make the most progress toward a desired stand of the future. Only leave-tree marking can ensure that desired density levels are achieved. Regenerating pine requires that basal area density be no more than 60 square feet per acre (Fiedler et al. 1988). Although ponderosa pine needs only moderate light levels for seedlings to survive, nearly full sunlight is necessary for seedlings to prosper and develop into overstory trees. Similarly, pole- and medium-size trees need low stand densities to achieve large size. Leaving density too high after treatment is the most common mistake in attempts to restore uneven-aged pine forests (Table 9.3).

Finally, old trees need open conditions to maintain modest growth rates and survive several

TABLE 9.3.

Stand structures before and after treatment in the Archibald timber sale, Seeley Lake Ranger District. Data are trees per acre from an inventory conducted by S. Arno and C. Fiedler after the area was thinned and underburned.

Trees Diameter at Breast Height	Trees Retained	Trees Removed or Killed
1–4 inches	dddddd	56 d, 3 lp
4–6 inches	dddd	48 d, 28 lp
6–8 inches		32 d, 4 w, 32 lp
8–10 inches		16 d, 5 w, 12 lp
10–12 inches	w	5 d, 16 lp
12–14 inches	w	8 d
14–16 inches	w	
16–18 inches		
18–20 inches	dd wwwwww	
20–22 inches	w	
22–24 inches	ww	
24–26 inches	dww	
26–28 inches	ww	
28–30 inches	www	
30–32 inches	pww	
32–40 inches	www	
40–48 inches	www	
Basal area/acre (194 sq. feet before treatment)	128 sq. ft.	66 sq. ft.

Each letter represents an individual tree: d = Douglas fir, lp = lodgepole pine, p = ponderosa pine, w = western larch.

hundred years. Some question the importance of maintaining growth in old-growth trees. After all, what difference does it make if a 2-foot-thick old-growth tree grows an inch in diameter or hardly at all over the next 20 years? The answer is that trees manufacture defensive chemicals to help them survive insect and disease attacks, but this process takes lower priority than diameter growth (Waring and Pitman 1985). Low-vigor trees are unable to marshal enough resources to both grow and manufacture adequate chemicals for their defense. Thus large, slow-growing trees among a dense layer of smaller trees are especially vulnerable, as manifested by accelerated mortality of old pine in many areas of the West (Figure 9.3).

Implementing Treatments

The modified selection cutting prescription developed for the Lubrecht study involved leave-tree marking across a full range of tree sizes, from 2 to 36 inches. Tree marking guidelines focused on keeping healthy pines of all sizes with ratios of live crown to total height greater than 35%, not to exceed an average of 60 square feet per acre. Approximately half of the reserve basal area was in trees larger than 16 inches. The prescription specified that no Douglas fir be retained, presuming that it would seed in from surrounding areas.

Lessons Learned

Two decades of monitoring at the Lubrecht Forest restoration project confirm the importance of low densities. Large trees in the two cutting treatments are growing nearly three times faster than those in the control. Four times as many large trees have died in the control as in treated units (Fiedler 2000a). Sapling-, pole-, and medium-size trees are also growing much faster in the treated areas, indicating that large tree recruitment will continue into the future. The perpetuation of uneven-aged

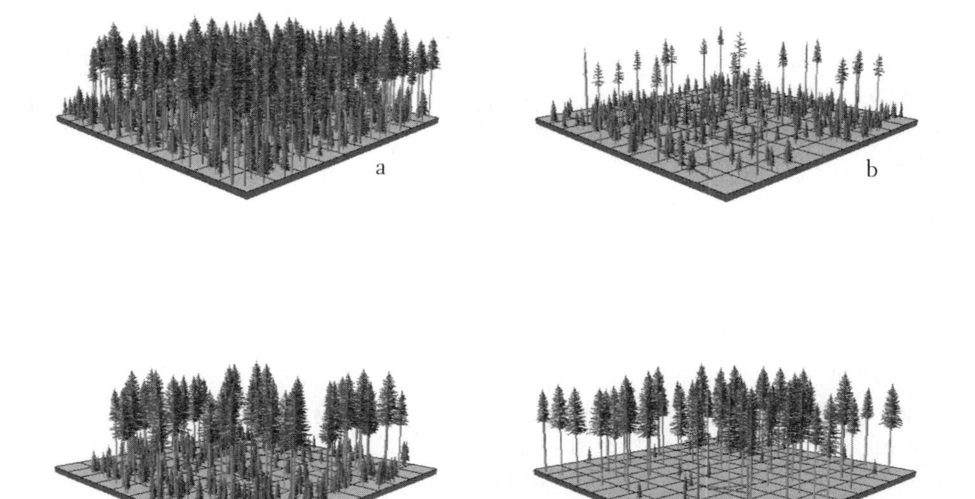

Figure 9.3. Computer visualizations simulating effects of different treatments on stand structure. *(a)* Untreated second-growth stand. *(b)* Result of high-grade logging. *(c)* Logging for timber management. *(d)* Restoration treatment. *(By S. Robertson)*

pine forests will work only if many of the large old trees survive nature's vicissitudes, medium- and pole-size trees develop into large old trees, and pine seedlings and saplings establish and grow well enough to eventually become big trees.

The numbers and kinds of seedlings being recruited differ dramatically between the treated areas and the control. In the two treated areas, ponderosa pine seedlings outnumber Douglas fir approximately 20:1, whereas in the control plots Douglas fir seedlings outnumber pine by more than 2:1 (Fiedler 2000a). Despite the encouraging number and proportion of pine seedlings that have established, height growth is slow and not adequate to ensure development into sapling size in a reasonable time frame. The 60 square feet per acre reserve basal area apparently is too high for adequate development of young trees on this site. The restoration experiments at Lick Creek and long-term monitoring on the nearby Flathead Indian Reservation show that reserve densities of 45–50 square feet per acre provide suitable conditions for pine regeneration and early development on drier sites.

Regardless of ecological benefits, restoration treatments may not be accepted for broader application if people dislike their appearance. The fishbowl setting of the Lubrecht restoration experiment, which borders the scenic Blackfoot Highway, provides a stiff test of aesthetic appeal. Over the 20-year life of this project, Lubrecht Forest director Hank Goetz has received numerous compliments from visitors—and nary a complaint—about its appearance.

Results from the intensive tree-level measurements and observations at Lubrecht Forest can provide guidance for restoration projects in similar forests. Although the 60 square feet per acre reserve basal area evaluated was sufficiently low to spur diameter growth of trees of all sizes, significantly increase survival of large trees, and induce regeneration of ponderosa pine, it was too high to allow the small tree height growth that is critical in perpetuating uneven-aged pine forests.

Another potential problem is the hundreds of Douglas fir seedlings per acre that existed before treatment. In the selection cutting treatment that received underburning, nearly all Douglas fir seedlings were killed (Kalabokidis and Wakimoto 1992). However, in the unburned selection cutting treatment, fir seedlings remained unscathed, and their height growth has exploded. Many of these Douglas firs are now sapling size and will need to be cut at the next harvest entry in 5 or 10 years—a costly, labor-intensive chore—to prevent them from crowding out the pine seedlings that established after treatment. Prescribed underburning is an especially effective means of killing Douglas firs and true firs when they are small, but many landowners are unwilling to risk burning.

Restoring Fire on a Wilderness Landscape

What about restoring understory and mixed fire regime forests where logging is not an option? One such area is Idaho's Clearwater and Salmon River backcountry, which spreads out as an endless succession of rugged ridges and deep canyons. South-facing slopes are clothed with patches of ponderosa pine, grassland, and sagebrush. Cooler, north-facing slopes support dense forests of spruce, hemlock, and fir. Ecosystems in this area were molded over thousands of years by fire. Over most of the last century, this inaccessible country had a history of hard-fought battles to suppress fires. However, it now boasts the most extensive program to restore natural fires in the United States

More than 4 million acres of designated wilderness and roadless backcountry in Idaho and western Montana are being managed to restore the natural role of fire. Further west, the Hells Canyon Wilderness along the Snake River in Idaho and Oregon is also part of this natural fire restoration program. Still other lands in the Clearwater–Salmon region are being considered for inclusion by ten national forests, including the Clearwater, Nez Perce, Idaho

Panhandle, Lolo, Bitterroot, Payette, Salmon–Challis, Boise, Sawtooth, and Wallowa–Whitman.

History

The Great Idaho fire of 1910 engulfed more than 3 million acres, including a huge swath of the Clearwater drainage (Pyne 2001). This holocaust damaged or destroyed several communities and served as the impetus for the U.S. Forest Service's campaign to eliminate forest fire. Nevertheless, fires again swept much of this region in 1919, 1926, 1929, and 1934. Fire suppression relied on punching roads into steep country, carving a network of trails, and establishing dozens of fire lookouts. In 1935, a widely respected forester with 30 years' experience in fire suppression argued that road building and development in the name of fire protection were detrimental to erosion-prone canyons, compromised the aesthetic character of this wild country, and ultimately would fail to eliminate fire (Koch 1935). Koch questioned why it was important in this area to suppress fire, which he viewed as an intrinsic part of the natural environment.

Koch's critique failed to influence fire exclusion advocates in the Forest Service. But a generation later, Bud Moore, a Forest Service fire administrator who had spent much of his life in the Clearwater backcountry, initiated a pilot program to restore natural fires. This was a revolutionary event in the agency that had initiated the national campaign to eliminate fire (Moore 1996). In 1972, under the leadership of Bitterroot National Forest supervisor Orville Daniels, fire scientist Bob Mutch, and others, the natural fire program was launched in one small part of the SBW. In 1979 the program was expanded to cover the entire wilderness, and beginning in 1985 it encompassed 3 million acres, including the Selway–Bitterroot and neighboring Frank Church–River of No Return Wilderness. By the dawn of the twenty-first century, the fire restoration program had expanded to nearby smaller wilderness and backcountry areas.

Evolution of the Fire Policy

Federal fire policy traditionally called for aggressive suppression of *all* fires, with a goal of achieving control by 10 A.M. on the day after a fire was reported. In 1978 a new policy allowed more flexible responses based on risk to natural resources, private property, firefighter safety, and economic efficiency (Benedict et al. 1991). Options included full-blown suppression, limited suppression, and surveillance only. The new policy also opened doors for harnessing "prescribed natural fires," which are lightning fires burning under previously defined conditions.

In the 1980s, fire management plans in the SBW allowed large areas to burn as prescribed natural fires, along with a wildfire confinement strategy that used only limited suppression tactics to keep fires from escaping designated boundaries. Between 1979 and 1990, about 150,000 acres burned in the area that received limited or no suppression—about 12% of the Selway (Brown et al. 1994). The new policies resulted in more fire than had been experienced in many years. The effects included multiple ecological benefits, reduced firefighter risk, and lowered suppression costs (Benedict et al. 1991).

Despite the program's success, even more fire is needed to approximate the historical landscape mosaic (Oppenheimer and Dickinson 2003). Analysis indicates that the annual area burned averaged only about 60% of the historical fire regime (Brown et al. 1994). The largest departures occurred in the low-elevation ponderosa pine and high-elevation whitebark pine ecosystems, where the restoration program was burning only about 40% of historical levels. Also ponderosa pine was experiencing a higher proportion of stand replacement burning than historically occurred.

A new federal forest policy enacted in 1995 eliminated many of the previous limitations to expanding the use of fire and improved the decision process for allowing natural fire to play its ecological role (USDI/USDA 1995). Previously, all fires had been classified as wildfires or prescribed fires, and the designation chosen limited the range of management options available. Under the 1995 policy, all fires not ignited by managers are considered wildland fires. They receive appropriate management response, which includes a broad range of tactical options. The 1995 policy states that fire management should have a foundation based on safety of firefighters and the public, sound risk management, fire ecology, and application of the best available science. Burned area data suggest that the 1995 policy has had a favorable effect in terms of returning more fire to the Clearwater–Salmon region.

Restoring Fire

In the late 1990s, national forests in the Clearwater and Salmon River drainages developed management plans that allow use of lightning fires in areas outside designated wilderness. The Clearwater Fire Management Unit, located north of the SBW, encompasses 515,000 acres where lightning fires can be used for resource benefits (Clearwater National Forest 1999).

Within 2 hours after a fire is discovered, a team of experts either classifies it as wildland fire or designates it for suppression based on threats to the area boundary, life, and property; effects on cultural and natural resources; forecast weather and risk indicators; and the regional and national fire situation. A risk assessment of the fire escaping the designated area is also part of the analysis. The decision considers potential effects of smoke on populated areas, proportion of a watershed allowed to burn in any 10-year period, habitat for endangered species, and protection of structures. Managers may decide to prevent fire spreading along one flank in order to keep it within the designated area, or they may need to protect a historic structure or administrative facility.

The confinement response to wildfires actually returns more fire to the landscape than does prescribed natural fire. In severe wildfire seasons with high levels of firefighting activity, most new fires in the Clearwater and Salmon River backcountry are assigned suppression status. However, many of them are allowed to burn under a confinement strategy except where they pose an immediate threat to structures or private property. Together, prescribed natural and confinement strategy wildfires have returned fire to the land on an impressive scale. Forty-six percent of the 2.3-million-acre Frank Church Wilderness was visited by fire between 1985 and 2000.

The multiple benefits of allowing suppression fires to burn under confinement are illustrated by a comparison of strategies used on two huge fires in the Salmon River Mountains in 2000. Both burned in a mosaic pattern of varying intensities commonly associated with a mixed fire regime. The Clear Creek fire, closer to but not directly threatening communities, was subjected to a $71-million suppression effort, including 200 miles of bulldozed fire lines that ultimately were not needed (Barker 2000). It burned 217,000 acres, including portions of the Frank Church. Meanwhile, the more remote 182,000-acre Wilderness Complex fire was allowed to burn with very little suppression effort at a cost of about $500,000 and little environmental damage.

Despite the impressive extent of burning since about 1980, the fire restoration program has not adequately treated ponderosa pine habitats in the semiarid canyons, which burned at intervals of about 10–25 years (Barrett 1988). As a result of effective suppression, many pine stands have missed several natural fire cycles and exposed old trees to additional physiological stress. The buildup of surface and ladder fuels puts canyon watersheds at risk of uncharacteristic stand replacement fires (Nez Perce National Forest 1999). Such fires are costly and often impossible to suppress, pose safety risks to firefighters, and accelerate soil loss, which degrades water quality and aquatic habitat for threatened populations of salmon, steelhead, and bull trout.

FIGURE 9.4. Postfire mosaic pattern *(foreground)* and snow avalanche community patterns, Selway–Bitterroot Wilderness. *(Photo by Stephen F. Arno)*

National forests are now proposing to conduct spring and fall prescribed fires to reduce fuel accumulations and return a semblance of historical conditions in about 200,000 acres of the Salmon River canyon. After reducing fuels with prescribed fire, managers hope to allow lightning fires to resume their natural role in the spectacular but fragile canyon habitat (Figure 9.4).

Sustaining Success: The Challenge

Armed with officially sanctioned plans and procedures, one might suppose that national forest managers would be able to return fire to an undeveloped region. In reality, the fire restoration program in the Clearwater and Salmon River backcountry still encounters opposition in the 90-year-old federal fire suppression bureaucracy. Most fire restoration occurs during years of extreme wildfire activity when attention is focused on potential damage to communities and dwellings dispersed within or adjacent to this backcountry.

The success of this broad-scale restoration program depends on extraordinary commitment by local fire management personnel, district rangers, and forest supervisors. The natural fire program has perhaps the highest risks and consequences of any operation in the federal land management agencies. Fire is by nature unpredictable, and losing control of a designated natural fire can have devastating professional consequences. If the responsible officials are risk-averse, they will be reluctant no matter how great the potential benefits. During past periods of high wildfire activity, regional officials pressured local managers to suppress new lightning fires and fight ongoing fires in the backcountry. Local managers often resisted and instead allowed fire restoration to continue. Such commitment is crucial to long-term success of the natural fire program. Local managers have a strong sense of mission and shared satisfaction in returning fire despite the difficulties and constraints.

The Wildland–Urban Interface

The largest suburban area affected by the Clearwater and Salmon region's fire program is Montana's Bitterroot Valley, located along the eastern boundary of the SBW. In 1988 three major wildfires spread eastward from the SBW, threatening private residences. After the 1988 fires, highlighted by the firestorms in Yellowstone National Park, forest managers nearly eliminated prescribed natural fires in the easternmost part of the SBW.

The Bitterroot National Forest proposed creating fuel breaks by thinning some of its land between the wilderness and Bitterroot Valley home sites, but local opposition prevented many of these projects. Forest managers have conducted several prescribed burns to reduce fuel buildup, but these operations are continually hampered by complaints about smoke from residents. Smoke is an inseparable part of living in the wildland interface, whether it comes in smaller, more frequent doses from prescribed fire or massive but infrequent doses from wildfire. Until interface residents and air quality regulators accept this reality, broad-scale treatments are unlikely.

Summary

Today restoration forestry is more widely accepted by land managers and is being implemented in small, isolated projects on some national forests and other ownerships. Experience and knowledge have accrued so that we can confidently expand to larger treatment areas and even landscape-scale strategies. Principal constraints on restoration forestry include a dwindling infrastructure (mills) for using trees that are removed, a shortage of people skilled in designing and implementing restoration treatments, and overlapping environmental regulations that favor preservation and fire exclusion strategies (Agee 2002, Arno and Fiedler 2005, Thomas 2002).

Acceptance of restoration forestry is limited by criticism from those who believe that nature knows best, and if there are problems, nature will heal itself. These people believe that even if the alleged forest health problems are real, they result from human interventions, and any activity aimed at restoring more natural conditions is at best misguided thinking that a second wrong will make it right. They fear that restoration forestry is simply a ruse for more tree cutting and forest exploitation, only this time under the pretense of doing good for the forest. Other critics see restoration forestry as another in a long line of obstacles to commercial forest management—a temporary placating of environmental advocates who will never be satisfied anyway.

FIGURE 9.5. Ponderosa pine stand managed using restoration forestry on the E Bar L Ranch. (*Photo by Carl Fiedler*)

These challenges to restoration forestry make it imperative that projects be well designed and clearly communicated, using consistent terminology and based on clear ecological objectives. Restoration activities that entail removing commercially valuable trees might be acceptable to environmental advocates if objectives and treatments are well understood. Likewise, projects that entail removal of only small trees to achieve restoration objectives must be clearly explained to timber advocates who might otherwise perceive such treatments as ineffective (Figure 9.5).

References

Agee, J. K. 1993. *Fire Ecology of Pacific Northwest Forests*. Island Press, Washington, DC.

Agee, J. K. 2002. The fallacy of passive management of Western forest reserves. *Conservation Biology in Practice* 3(1): 18–25.

Anderson, M. K. and M. J. Moratto. 1996. Native American land-use practices and ecological impacts. In *Sierra Nevada Ecosystem Project: Final Report to Congress*. Vol. II. University of California, Centers for Water and Wildland Resources, Davis.

Arno, S. F. 1976. *The Historical Role of Fire on the Bitterroot National Forest*. Research Paper 187. USDA Forest Service, Intermountain Forest and Range Experiment Station, Ogden, UT.

Arno, S. F. 1980. Forest fire history in the northern Rockies. *Journal of Forestry* 78(8): 460–465.

Arno, S. F. and S. Allison-Bunnell. 2002. *Flames in Our Forest: Disaster or Renewal?* Island Press, Washington, DC.

Arno, S. F. 1999. Undergrowth response, Shelterwood Cutting Unit. Pages 36–37 in Gen. Tech. Rep. 23. USDA Forest Service, Rocky Mountain Research Station, Ogden, UT.

Arno, S. F. 2000. Fire regimes in Western forest ecosystems. Pages 97–120 in Gen. Tech. Rep. 42, Vol. 2. USDA Forest Service, Rocky Mountain Research Station, Fort Collins, CO.

Arno, S. F. and C. E. Fiedler. 2005. *Restoring Nature's Fire*. Island Press, Washington, DC.

Arno, S. F. and G. Gruell. 1983. Fire history at the forest–grassland ecotone in southwestern Montana. *Journal of Range Management* 36: 332–336.

Arno, S. F., J. Scott, and M. Hartwell. 1995. *Age-Class Structure of Old Growth Ponderosa Pine/Douglas-Fir*

Stands and Its Relationship to Fire History. Research Paper 481. USDA Forest Service Intermountain Research Station, Ogden, UT.

Arno, S. F., H. Smith, and M. Krebs. 1997. *Old Growth Ponderosa Pine and Western Larch Stand Structures: Influences of Pre-1900 Fires and Fire Exclusion*. Research Paper 495. USDA Forest Service Intermountain Research Station, Ogden, UT.

Barker, R. 2000. Fire officials weigh damage in wake of Clear Creek fire. *Idaho Statesman*, September 22.

Barrett, S. W. 1988. Fire suppression's effects on forest succession within a central Idaho wilderness. *Western Journal of Applied Forestry* 3(3): 76–80.

Barrett, S. W. and S. Arno. 1982. Indian fires as an ecological influence in the northern Rockies. *Journal of Forestry* 80(10): 647–651.

Becker, R. 1995. Operational considerations of implementing uneven-aged management. Pages 67–81 in Miscellaneous Publication 56. University of Montana, Montana Forest Conservation Experiment Station, Missoula.

Benedict, G. W., L. Swan, and R. Belnap. 1991. Evolution and implementation of a fire management program which deals with high-intensity fires on the Payette National Forest in central Idaho. Pages 339–351 in *Proceedings: Tall Timbers Fire Ecology Conference 17*. Tall Timbers Research Station, Tallahassee, FL.

Biswell, H. H. 1989. *Prescribed Burning in California Wildlands Vegetation Management*. University of California Press, Berkeley.

Biswell, H. H., H. Kallander, R. Komarek, R. Vogl, and H. Weaver. 1973. *Ponderosa Fire Management: A Task Force Evaluation of Controlled Burning in Ponderosa Pine Forests of Central Arizona*. Miscellaneous Publication 2. Tall Timbers Research Station, Tallahassee, FL.

Bonnicksen, T. M. 2000. *America's Ancient Forests: From the Ice Age to the Age of Discovery*. Wiley, New York.

Boyd, R., ed. 1999. *Indians, Fire and the Land in the Pacific Northwest*. Oregon State University Press, Corvallis.

Brown, J. K. 2000. Introduction and fire regimes. Pages 1–8 in Gen. Tech. Rep. 42, Vol. 2. USDA Forest Service, Rocky Mountain Research Station, Fort Collins, CO.

Brown, J. K., S. F. Arno, S. W. Barrett, and J. P. Menakis. 1994. Comparing the prescribed natural fire program with presettlement fires in the Selway–Bitterroot Wilderness. *International Journal of Wildland Fire* 4: 157–168.

Brown, R. 2001. *Thinning, Fire, and Forest Restoration*. Defenders of Wildlife, Washington, DC. Available at www.biodiversitypartners.org.

Carle, D. 2002. *Burning Questions: America's Fight with Nature's Fire*. Praeger, Westport, CT.

Clearwater National Forest. 1999. *Clearwater Fire Management Unit: Wildland Fire Use Guidebook*. U.S. Forest Service, Orofino, ID.

Cooper, C. F. 1960. Changes in vegetation, structure, and growth of southwestern pine forests since white settlement. *Ecological Monographs* 30(2): 129–164.

Cottrell, W. H. Jr. 2004. *The Book of Fire*. 2nd ed. Mountain Press, Missoula, MT.

Covington, W. W. and M. M. Moore. 1994. Postsettlement changes in natural fire regimes and forest structure: ecological restoration of old-growth ponderosa pine forests. *Journal of Sustainable Forestry* 2(1/2): 153–182.

Daniels, O. L. 1991. A forest supervisor's perspective on the prescribed natural fire. Pages 361–366 in *Proceedings: Tall Timbers Fire Ecology Conference 17*. Tall Timbers Research Station, Tallahassee, FL.

Dieterich, J. H. 1980. *Chimney Spring Forest Fire History*. Research Paper 220. USDA Forest Service, Rocky Mountain Forest and Range Experiment Station, Fort Collins, CO.

Fiedler, C. E. 1995. The basal area–maximum diameter–q (BDq) approach to regulating uneven-aged stands. Page 94 in Miscellaneous Publication 56. University of Montana, Montana Forest Conservation Experiment Station, Missoula.

Fiedler, C. E. 2000a. Restoration treatments promote growth and reduce mortality of old-growth ponderosa pine (Montana). *Ecological Restoration* 18: 117–119.

Fiedler, C. E. 2000b. Silvicultural treatments. Pages 19–20 in H. Y. Smith (ed.), *The Bitterroot Ecosystem Management Research Project: What We Have Learned*? Rocky Mountain Research Station Proceedings 17. USDA Forest Service, Missoula, MT.

Fiedler, C. F., R. Becker, and S. Haglund. 1988. Preliminary guidelines for uneven-aged silvicultural prescriptions in ponderosa pine. Pages 235–241 in D. M. Baumgartner and J. E. Lotan (comp. and ed.), *Ponderosa Pine: The Species and Its Management*. Washington State University Cooperative Extension, Pullman.

Gray, A. N. and J. Franklin. 1997. Effects of multiple fires on the structure of southwestern Washington forests. *Northwest Science* 71: 174–185.

Greenlee, J. M. and J. Langenheim. 1990. Historic fire regimes and their relation to vegetation patterns in the Monterey Bay area of California. *American Midland Naturalist* 124: 239–253.

Gruell, G. E. 1985. Fire on the early Western landscape: an annotated record of wildland fires 1776–1900. *Northwest Science* 59(2): 97–107.

Gruell, G. E. 2001. *Fire in Sierra Nevada Forests: A Photographic Interpretation of Ecological Change Since 1849*. Mountain Press, Missoula, MT.

Gruell, G. E., W. Schmidt, S. Arno, and W. Reich. 1982. *Seventy Years of Vegetal Change in a Managed Ponderosa Pine Forest in Western Montana: Implications for Resource Management*. Gen. Tech. Rep. 130. USDA Forest Service, Intermountain Research Station, Ogden, UT.

Harrington, M. G. 2000. Fire applications in ecosystem management. Pages 21–22 in Proceedings 17. USDA Forest Service, Rocky Mountain Research Station, Ogden, UT.

Hoxie, G. L. 1910. How fire helps forestry. *Sunset* 34: 145–151.

Kalabokidis, K. D. and R. Wakimoto. 1992. Prescribed burning in uneven-aged stand management of ponderosa pine/Douglas-fir forests. *Journal of Environmental Management* 34: 221–235.

Keane, R. E., K. Ryan, T. Veblen, C. Allen, J. Logan, and B. Hawkes. 2002. *Cascading Effects of Fire Exclusion in Rocky Mountain Ecosystems: A Literature Review*. Gen. Tech. Rep. 91. USDA Forest Service, Rocky Mountain Research Station, Ogden, UT.

Kitts, J. A. 1919. Forest destruction prevented by control of surface fires. *American Forestry* 25: 1264, 1306.

Koch, E. 1935. The passing of the Lolo Trail. *Journal of Forestry* 33(2): 98–104.

Koch, E. 1998. *Forty Years a Forester: 1903–1943*. Mountain Press, Missoula, MT.

Leiberg, J. B. 1899. Bitterroot Forest Reserve. Pages 253–282 in U.S. Geological Survey, *19th Annual Report*, Part V.

Lewis, H. T. 1973. *Patterns of Indian Burning in California: Ecology and Ethnohistory*. Anthropology Paper No. 1. Ballena Press, Ramona, CA.

McKelvey, K. S., C. Skinner, C. Chang, D. Erman, S. Husari, D. Parsons, J. van Wagtendonk, and C. Weatherspoon. 1996. An overview of fire in the Sierra Nevada. Pages 1033–1040 in *Sierra Nevada Ecosystem Project: Final Report to Congress*. Vol. II. Centers for Water and Wildland Resources, University of California, Davis.

Moore, B. 1996. *The Lochsa Story: Land Ethics in the Bitterroot Mountains*. Mountain Press, Missoula, MT.

Morgan, P., G. Aplet, J. Haufler, H. Humphries, C. Hope, M. Moore, and D. Wilson. 1994. Historical range of variability: a useful tool for evaluating ecosystem change. *Journal of Sustainable Forestry* 2: 87–111.

Mutch, R. W. 2001. Practice, poetry, and policy: will we be better prepared for the fires of 2006? *Bugle* 18(2): 61–64.

Nelson, T. C. 1979. Fire management policy in the national forests: a new era. *Journal of Forestry* 77: 723–725.

Nez Perce National Forest. 1999. *Salmon River Canyon Project: Draft Environmental Impact Statement*. U.S. Forest Service, Grangeville, ID.

Oppenheimer, J. and I. Dickinson. 2003. *Fire in Idaho: An Analysis of Fire Policy in Idaho*. Idaho Conservation League, Boise, ID. Available at www.wildidaho.org.

Parsons, D. J. 2000. The challenge of restoring natural fire to wilderness. Pages 276–282 in Proceedings 15, Vol. 5. USDA Forest Service, Rocky Mountain Research Station, Ogden, UT.

Paysen, T. E., R. Ansley, J. Brown, G. Gottfried, S. Haase, M. Harrington, M. Narog, S. Sackett, and R. Wilson. 2000. Pages 121–159 in Gen. Tech. Rep. 42, Vol. 2. USDA Forest Service, Rocky Mountain Research Station, Ogden, UT.

Pyne, S. J. 1982. *Fire in America: A Cultural History of Wildland and Rural Fire*. Princeton University Press, Princeton, NJ.

Pyne, S. J. 1997. *World Fire: The Culture of Fire on Earth*. University of Washington Press, Seattle.

Pyne, S. J. 2001. *Year of the Fires: The Story of the Great Fires of 1910*. Viking Penguin, New York.

Quigley, T. M., R. Haynes, and R. Graham, technical eds. 1996. *Integrated Scientific Assessment for Ecosystem Management in the Interior Columbia Basin*. Gen. Tech. Rep. 382. USDA Forest Service, Pacific Northwest Research Station, Portland, OR.

Shinn, D. A. 1980. Historical perspectives on range burning in the inland Pacific Northwest. *Journal of Range Management* 33: 415–422.

Smith, H. Y. and S. Arno, eds. 1999. *Eighty-Eight Years of Change in a Managed Ponderosa Pine Forest*. Gen. Tech. Rep. 23. USDA Forest Service, Rocky Mountain Research Station, Ogden, UT.

Stalling, D. 2003. The Burnt Fork Ranch: on the cutting edge of stewardship. *Bugle* 20(3): 32–39.

Stewart, O. C. 2002. *Forgotten Fires: Native Americans and the Transient Wilderness*. University of Oklahoma Press, Norman.

Swetnam, T. W. 1993. Fire history and climate change in giant sequoia groves. *Science* 262: 885–889.

Thomas, J. W. 2002. *Dynamic vs. Static Management in a Fire-Influenced Landscape: The Northwest Forest Plan*. Text of presentation at the conference "Fire in Oregon Forests," Oregon Forest Resources Institute, Portland.

USDI/USDA. 1995. *Federal Wildland Fire Management Policy and Program Review*. Final report. National Interagency Fire Center, Boise, ID.

Waring, R. H. and G. Pitman. 1985. Modifying lodgepole pine stands to change susceptibility to mountain pine beetle attack. *Ecology* 66: 889–897.

Weaver, H. 1943. Fire as an ecological and silvicultural factor in the ponderosa pine region of the Pacific Slope. *Journal of Forestry* 41(1): 7–14.

Weaver, H. 1968. Fire and its relationship to ponderosa pine. Pages 127–149 in *Proceedings: Tall Timbers Fire Ecology Conference* 7. USDA Forest Service, Tall Timbers Research Station, Tallahassee, FL.

Wilkinson, T. 2001. Prometheus unbound. *Nature Conservancy* 51(3): 12–20.

Zimmerman, G. T. and D. L. Bunnell. 2000. The federal wildland fire policy: opportunities for wilderness fire management. Pages 288–297 in *Proceedings* 15, Vol. 5. USDA Forest Service, Rocky Mountain Research Station, Ogden, UT.

Chapter 10

Shrub Steppe

Steven O. Link, William H. Mast, and Randal W. Hill

Shrub steppe ecosystems in the Pacific Northwest are found in the long rain shadow of the Cascade Mountains, from British Columbia to northern California and Nevada (Figure 10.1). Shrub steppe vegetation extends beyond the Northwest to Saskatchewan and south to Colorado.

Restoration of the semiarid shrub steppe ecosystem has gained increasing attention over the last 20 years. This is the result of growing recognition of the values that intact shrub steppe ecosystems provide to communities. Soil stabilization may be the highest value of intact shrub steppe (Scott et al. 1998). Healthy shrub steppe ecosystems also moderate wildfire spread, whereas disturbed shrub steppe ecosystems dominated by invasive cheatgrass (*Bromus tectorum*) increase fire frequency and intensity. In addition to increasing risk to lives and property, increased fire causes further loss of big sagebrush (*Artemisia tridentata*), the dominant plant in this ecosystem (Whisenant 1990). Sagebrush is a crucial habitat component for a number of rare birds, including the sage grouse (Rogers et al. 1988, Connelly and Braun 1997). Highly diverse communities dominated by native plant species are more productive and support higher levels of biodiversity (Naeem et al. 1995).

The Columbia Basin has been occupied by people for at least 12,000 years, and Indians used fire to manage vegetation using frequent, low-intensity burns (U.S. Department of Agriculture 1996). The arrival of Europeans to the shrub steppe in the mid-1800s was followed by heavy grazing by cattle and sheep and plowing for wheat (Rogers and Rickard 1988). By the early 1900s, a number of invasive plant species became widespread (Rogers and Rickard 1988). Today only 30% of original grasslands and 70% of shrublands exist, and all the landscape is potential habitat for invasive alien plant species (U.S. Department of Agriculture 1996).

Although it is not possible to restore all shrub steppe lands, it is possible to restore areas that are not likely to be further developed for agriculture and human habitation. This chapter reviews restoration in the shrub steppe, with particular attention to the difficulties of restoration in these semiarid ecosystems.

Description of the Shrub Steppe

The shrub steppe ecoregion is dominated by shrubs and perennial bunchgrasses over about 645,000 square kilometers (250,000 square miles) in North America (Daubenmire 1970, Rickard et al. 1988; Figure 10.1). The Snake and Columbia element of the shrub steppe is lower in elevation than the Wyoming Basin.

Climate and Distribution

At 6,000 years B.P., the shrub steppe occupied almost the same geographic range as it does today. At 18,000 years B.P., it occupied areas that are now

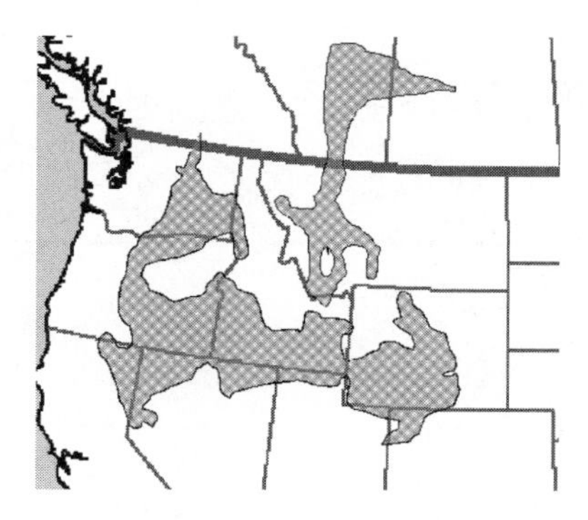

FIGURE 10.1. Map shows general extent of the shrub steppe vegetation zone. *(From Rickard et al. 1988)*

conifer and mixed forests. In the Great Basin, conifers were found in the current domain of the shrub steppe (Thompson and Anderson 2000).

Present yearly average precipitation ranges from about 162 millimeters (6.4 inches) at the Hanford Site in Washington to about 420 millimeters (16.5 inches) at Columbia, Montana, which contains the most productive steppe community in the western United States (Thorp and Hinds 1977, Rickard et al. 1988).

Plant Communities

The dominant natural vegetation of the shrub steppe is sagebrush (*Artemisia* spp.), associated with wheatgrasses (*Pseudoroegneria*), Idaho fescue (*Festuca idahoensis*), and other perennial bunchgrasses (Franklin and Dyrness 1988). Cheatgrass is now dominant in many areas.

A number of plant association zones occur in the Columbia Basin, influenced largely by available moisture. The largest and driest of these is the big sagebrush (*Artemisia tridentata*) and bluebunch wheatgrass (*Pseudoroegneria spicata*) association (Daubenmire 1970). This association is characterized by four layers of vegetation: an overstory composed mostly of big sagebrush up to 2 meters tall, a tall understory of bluebunch wheatgrass, a short understory dominated by Sandberg's bluegrass (*Poa secunda*), and the microbiotic crust composed of algae, lichens, and mosses on the soil surface. The microbiotic crust is a critical component of native grasslands and shrub steppe communities (Link et al. 2000). Perennial and annual herbs are found in the understory layers. Other shrubs include rabbitbrush (*Ericameria* and *Chrysothamnus* spp.), bitterbrush (*Purshia tridentata*), hopsage (*Grayia spinosa*), and three-tip sagebrush (*Artemisia tripartita*). Other bunchgrasses include needle-and-thread (*Hesperostipa comata*), Indian ricegrass (*Achnatherum hymenoides*), Cusick's bluegrass (*Poa cusickii = Poa secunda*), and Idaho fescue (*Festuca idahoensis*).

Other associations, such as big sagebrush–Idaho fescue, bluebunch wheatgrass–Sandberg's bluegrass, and bluebunch wheatgrass–Idaho fescue, occur on moister sites within the big sagebrush–bluebunch wheatgrass association (Daubenmire 1970). The primary large bunchgrasses include needle-and-thread and Indian ricegrass. The dominant shrub in these associations can be either big sagebrush or bitterbrush.

On stony soils or extremely shallow soils over bedrock, various species of buckwheat (*Eriogonum*) and stiff sage (*Artemisia rigida*) dominate the shrub layer, and Sandberg's bluegrass dominates the understory. In the driest areas, associations consist of big sagebrush–Sandberg's bluegrass, hopsage–Sandberg's bluegrass, and winterfat (*Krascheninnikovia lanata*)–Sandberg's bluegrass (Daubenmire 1970). These associations lack large perennial bunchgrasses.

In saline alkaline soils, *Distichlis stricta* and *Leymus cinereus* are the dominant grasses, and *Sarcobatus vermiculatus* is the dominant shrub, with lesser amounts of *A. tridentata* (Daubenmire 1970).

Plant Species

There are numerous native plants to consider for restoration purposes (Table 10.1) and many invasive alien species in need of control (Table 10.2).

TABLE 10.1.

Important native plants of the shrub steppe. They are perennials unless noted.

Family *Species*	Common Name	Life Form	Habitat
Cactaceae			
Opuntia polycantha	Starvation prickly pear	Succulent	Upland
Chenopodiaceae			
Atriplex canescens	Four-wing saltbush	Shrub	Upland
Atriplex confertifolia	Spiny shadscale	Shrub	Upland
Grayia spinosa	Hopsage	Shrub	Upland
Krascheninnikovia lanata	Winterfat	Shrub	Upland
Sarcobatus vermiculatus	Greasewood	Shrub	Riparian
Compositae			
Achillea millifolium	Yarrow	Herb	Upland
Artemisia rigida	Stiff sage	Shrub	Upland
Artemisia tridentata ssp. *tridentata*	Basin big sagebrush	Shrub	Riparian and upland
Artemisia tridentata ssp. *vaseyana*	Mountain big sagebrush	Shrub	Upland
Artemisia tridentata ssp. *wyomingensis*	Wyoming big sagebrush	Shrub	Upland
Artemisia tripartita	Three-tip sage	Shrub	Upland
Balsamorhiza careyana	Carey's balsamroot	Herb	Upland
Balsamorhiza sagittata	Arrowleaf balsamroot	Herb	Upland
Chrysopsis villosa	Hairy goldaster	Herb	Upland
Chrysothamnus viscidiflorus	Green rabbitbrush	Shrub	Upland
Crepis atribarba	Slender hawksbeard	Herb	Upland
Ericameria nauseosa	Gray rabbitbrush	Shrub	Upland
Erigeron filifolius	Threadleaf fleabane	Herb	Upland
Erigeron poliospermus	Cushion fleabane	Herb	Upland
Erigeron pumulus	Shaggy fleabane	Herb	Upland
Gutierrezia sarothrae	Snakeweed	Shrub	Upland
Helianthus cusickii	Cusick's sunflower	Herb	Upland
Machaeranthera canescens	Hoary aster	Biennial herb	Upland
Cruciferae			
Erysimum asperum	Rough wallflower	Herb	Upland
Stanleya tomentosa	Woolly stanleya	Herb	Upland
Thelypodium laciniatum	Thick-leaved thelypodium	Herb	Upland
Graminae			
Achnatherum hymenoides	Indian rice grass	Bunchgrass	Upland
Distichlis stricta	Saltgrass	Rhizomatous	Riparian and upland
Elymus elymoides	Squirreltail	Bunchgrass	Upland
Elymus lanceolatus	Streambank wheatgrass	Rhizomatous	Upland
E. lanceolatus ssp. *lanceolatus*	Bannock	Rhizomatous	Upland
E. lanceolatus ssp. *lanceolatus*	Critana	Rhizomatous	Upland
E. lanceolatus ssp. lanceolatus	Schwendimar	Rhizomatous	Upland
E. lanceolatus ssp. *psammophilus*	Sodar	Rhizomatous	Upland
Elymus wawawaiensis	Snake River wheatgrass, secar	Bunchgrass	Upland
Festuca idahoensis	Idaho fescue	Bunchgrass	Upland
Hesperostipa comata	Needle-and-thread grass	Bunchgrass	Upland
Koeleria cristata	Prairie junegrass	Bunchgrass	Upland
Leymus cinereus	Giant wildrye	Bunchgrass	Riparian and upland
Pascopyrum smithii	Western wheatgrass	Rhizomatous	Upland
Poa secunda	Sandberg's bluegrass	Bunchgrass	Upland
Pseudoroegneria spicata	Bluebunch wheatgrass	Bunchgrass	Upland
Sporobolus cryptandrus	Sand dropseed	Bunchgrass	Upland

(*continued*)

TABLE 10.1. (*continued*)

Important native plants of the shrub steppe. They are perennials unless noted.

Family Species	*Common Name*	*Life Form*	*Habitat*
Laminaceae			
Salvia dorrii	Purple sage	Shrub	Upland
Leguminosae			
Lupinus leucophyllus	Velvet lupine	Herb	Upland
Lupinus sericeus	Silky lupine	Herb	Upland
Petalostemon ornatum	Western prairieclover	Herb	Upland
Psoralea lanceolata	Lanceleaf scurf pea	Herb	Upland
Liliaceae			
Calochortus macrocarpus	Mariposa lily	Herb	Upland
Fritillaria pudica	Yellow bell	Herb	Upland
Triteleia grandiflora var. *grandiflora*	Largeflower triteleia	Herb	Upland
Linaceaea			
Linum lewisii	Lewis flax	Herb	Upland
Loasaceae			
Mentzelia laevicaulis	Blazing star	Herb	Upland
Malvaceae			
Sphaeralcea munroana	Munro's globemallow	Herb	Upland
Onagraceae			
Oenothera pallida	Pale evening primrose	Herb	Upland
Polemoniaceae			
Phlox longifolia	Longleaf phlox	Subshrub	Upland
Phlox speciosa	Showy phlox	Subshrub	Upland
Polygonaceae			
Eriogonum niveum	Snow buckwheat	Herb or shrub	Upland
Eriogonum umbellatum	Sulfur buckwheat	Herb	Upland
Eriogonum sphaerocephalum	Rock buckwheat	Shrub	Upland
Eriogonum thymoides	Thyme buckwheat	Shrub	Upland
Rumex venosus	Winged dock	Herb	Upland
Ranunculaceae			
Delphinium nuttallianum	Upland larkspur	Herb	Upland
Ranunculus glaberrimus	Sagebrush buttercup	Herb	Upland
Rosaceae			
Purshia tridentata	Bitterbrush	Shrub	Upland
Scrophulariaceae			
Penstemon acuminatus	Sand beardtongue	Herb	Upland
Umbelliferae			
Cymopterus terebinthinus	Turpentine springparsley	Herb	Upland
Lomatium macrocarpum	Big seed bisquitroot	Herb	Upland
Lomatium triternatum	Nine-leaf lomatium	Herb	Upland

Sources: Names are from Hitchcock and Cronquist (1976), with the most current names obtained from the PLANTS database (USDA NRCS 2003).

TABLE 10.2.

Common invasive weeds of the shrub steppe.

Family *Species*	Common Name	Life Cycle	Life Form	Habitat
Boraginaceae				
Amsinckia lycopsoides	Fiddleneck	Annual	Herb	Upland
Caryophyllaceae				
Gypsophila paniculata	Baby's breath	Perennial	Herb	Riparian
Holosteum umbellatum	Jagged chickweed	Annual	Herb	Upland
Chenopodiaceae				
Halogeton glomeratus	Halogeton	Annual	Herb	Upland
Kochia scoparia	Kochia	Annual	Herb	Upland
Salsola kali	Russian thistle	Annual	Herb	Upland
Compositae				
Centaurea diffusa	Diffuse knapweed	Annual or biennial	Herb	Upland
Centaurea maculosa	Spotted knapweed	Biennial	Herb	Upland
Centaurea repens	Russian knapweed	Perennial	Herb	Riparian
Centaurea solstitialis	Yellow-star thistle	Annual or biennial	Herb	Upland
Chondrilla juncea	Rush skeletonweed	Perennial	Herb	Upland
Cirsium arvense	Canada thistle	Perennial	Herb	Upland
Cirsium vulgare	Bull thistle	Biennial	Herb	Upland
Lactuca serriola	Prickly lettuce	Annual	Herb	Upland
Onopordum acanthium	Scotch thistle	Biennial	Herb	Upland
Tragopogon dubius	Salsify	Annual	Herb	Upland
Cruciferae				
Cardaria draba	White top	Perennial	Herb	Riparian
Chorispora tenella	Blue mustard	Annual	Herb	Upland
Descurainia sophia	Flixweed tansymustard	Annual or biennial	Herb	Upland
Lepidium perfoliatum	Yellow pepperweed	Annual	Herb	Upland
Sisymbrium altissimum	Tumblemustard	Annual	Herb	Upland
Dipsacaceae				
Dipsacus sylvestris	Teasel	Biennial	Herb	Riparian
Euphorbiaceae				
Euphorbia esula	Leafy spurge	Perennial	Herb	Upland
Geraniaceae				
Erodium cicutarium	Redstem storksbill	Annual	Herb	Upland
Graminae				
Agropyron cristatum	Crested wheatgrass	Perennial	Bunchgrass	Upland
Bromus tectorum	Cheatgrass	Annual	Grass	Upland
Hordeum leporinum	Hare barley	Annual	Grass	Upland
Poa bulbosa	Bulbous bluegrass	Perennial	Bunchgrass	Upland
Secale cereale	Cereal rye	Annual	Grass	Upland
Taeniatherum caput-medusae	Medusahead	Annual	Grass	Upland
Vulpia myuros	Rat-tail fescue	Annual	Grass	Upland
Ranunculaceae				
Ranunculus testiculatus	Hornseed buttercup	Annual	Herb	Upland
Scrophulariaceae				
Linaria dalmatica	Dalmatian toadflax	Perennial	Herb	Upland

Sources: Names are from Hitchcock and Cronquist (1976), Taylor (1990), and Whitson et al. (1992) with the most current names obtained from the PLANTS database (USDA NRCS 2003).

Define Restoration Purpose and Strategy

The best strategy for restoring shrub steppe ecosystems depends on the main functional objectives. For example, a goal for restoration often is to reduce the risk of fire, related primarily to cheatgrass (Whisenant 1990). Returning shrub steppe to a perennial-dominated native plant community can break the cheatgrass–fire cycle. One restoration strategy for reducing the fire risk is to establish strategically placed green strips that can slow or stop the spread of a wildfire (Pellant 1990). Green strips that are 9 meters wide can be restored along roads, where fires often originate. The vegetation planted in a green strip must be fire resistant and capable of surviving occasional burning. Native bunchgrasses serve this purpose very well. Forage kochia (*Kochia prostrata*), though not native, also functions well in green strips and can help reduce the risk of fire to nearby native plant communities (Pellant 1990).

Describe the Site

The topography and soils of a restoration site can strongly influence the plant community. For example, north-facing slopes at Hanford are dominated by bluegrass, whereas south-facing slopes are dominated by cheatgrass (Sauer and Rickard 1979). Elevation, aspect, and slope should be determined so that weed control and native species composition for restoration can be adjusted accordingly.

Soils can have a strong effect on the success of new plantings. Soils in the shrub steppe range in texture size from lithosols, gravels, sand, silt, to clay, with each supporting a different flora. Some plants are found on all substrates, and others are limited to one type. For example, veiny dock (*Rumes venosus*) grows only in sand, whereas cheatgrass seems to grow on most soils. Soil types can be coarsely recognized in the field or can be taken to a soil testing laboratory to define texture (Munshower 1994). Soil information can be found at soils.usda.gov.

It is important to determine the bulk density of the soil. Bulk density can be determined by weighing a known volume of dry soil. Compacted soils lose macropores and become less permeable to water (Munshower 1994). Plants have a harder time establishing in such conditions.

Soil pH should be measured when the original plant community cannot be determined because it can have a strong effect on nutrient availability and species composition. Soil pH near the surface is slightly acidic at Hanford (Link et al. 2000). Basic and alkaline soils favor halophytic species (*Atriplex*, *Sarcobatus*, *Grayia*, and *Distichlis*).

Organic matter tends to be low in shrub steppe soils and very low in subsoils. Organic matter affects water and nutrient availability. It can be determined using a muffle furnace or can be sent to a chemical testing laboratory.

Plant Selection

Creating an appropriate plant list for site restoration is best done by using surveys of nearby undisturbed ecosystems with similar soils and climatic conditions as reference ecosystems. A survey of flora that is on the restoration site also is needed, including both native and nonnative flora. Floristic surveys should be done frequently over the course of a year.

The seed bank also should be analyzed. Past seed bank studies in the shrub steppe can be referred to for methods (Boudell et al. 2002, Hassan and West 1986, Young and Evans 1975). This can take a long time and can slow projects down, but experience shows that seed banks exist in most disturbed areas and may contain desirable species.

Obtaining Plant Materials

There are many suppliers of shrub steppe seed and plants. Some suppliers are listed in Table 10.3 and

TABLE 10.3.

Suppliers of shrub steppe plants.

Supplier	Web Address
Plants of the Wild	www.plantsofthewild.com
Methow Natives	www.methow.com
Rainier Seed	www.rainierseeds.com
Bitterroot Restoration	www.revegetation.com
Fourth Corner Nurseries	fourth-corner-nurseries.com
Wildlands Nursery	wildlands-inc.com

can be contacted on the Web. The native seed network (www.nativeseednetwork.org) connects buyers with sellers. Information about current availability of plants can be found by searching for the scientific or common names of the plants of interest. Current and older names are found at the USDA NRCS (2003) PLANTS Web page (plants .usda.gov).

Seeds and Seedlings

Today, many seed farms produce native shrub steppe plants for restoration purposes. Many producers grow plants for seed production (increase) from the local seed source under contract. This requires that seed be collected from areas very near the restoration site. Seed collection and processing techniques for many shrub steppe species can be found in Young and Young (1986).

It is likely that using seed from plants of the local area will improve restoration success. Linhart (1995) suggests collecting seeds for herbaceous species no further than 100 meters and woody species no further than 1 kilometer from the site. Genetic similarity will be narrower in self-pollinated than in cross-pollinated species (Jones and Johnson 1998). It should be remembered that most commercial sagebrush seed is harvested from less than the full variation of genetic potential because seeds are repeatedly harvested from favorable sites and varieties (Monsen 2002). Most sagebrush seed ripens in the winter and can be collected over a 1-month window (Monsen 2002).

Species exhibit genetic variation associated with varying climatic conditions. For example, bluegrass species from high elevations can grow faster at low temperatures than those from low elevations (Körner and Woodward 1987). Such differences can accumulate and cause an ecotype to fail if planted in climate conditions significantly different from that of its parent plants. Sandberg's bluegrass transplanted 533 meters higher than their original range grew poorly after 5 years, suggesting that low-elevation ecotypes may not survive in colder conditions (Link et al. 2003b). Species can show significant genetic differences within short distances associated with steep environmental gradients.

When a species is rare, there are advantages to propagating from hand-collected seed. The likelihood of success increases if the plant has had an opportunity to grow to a seedling stage before planting. Some perennial species simply do not germinate and establish well in the field unless conditions are optimal. Spiny hopsage (*Grayia spinosa*) seedlings are rarely observed in eastern Oregon and southern Idaho, needing very wet years and lack of competition to become established (Shaw and Haferkamp 1990).

When it is not possible to collect local seed, it can often be purchased from commercial providers (Table 10.3). It is important that the source is known and that the species occurs or occurred in the restoration area. The closer the seed source to the site, the better. When purchasing such seed it is important to recognize that many chromosome races or varieties exist within a species. Restoration success is more likely if the proper variety is chosen. There are many shrub steppe species with recognized varieties, including Sandberg's bluegrass, giant wildrye, green rabbitbrush, and big sagebrush (Jones and Johnson 1998).

Raising and planting seedlings is more expensive than direct seeding but can be more successful. Seedlings are recommended if the species is rare or few seeds are available. They can be used as bareroot stock, container-grown, or salvaged plants. Shrubs such as big sagebrush, gray rabbitbrush, and bitterbrush are commonly planted as seedlings or tubelings (Link et al. 1995, Munshower 1994).

Small, salvaged plants can be transplanted more successfully than larger, older plants. In special construction circumstances it is possible to remove the topsoil and biological material before construction and then place the material on disturbed areas when construction is finished.

Mycorrhizae

Endomycorrhizae are common in semiarid undisturbed ecosystems, although few mycorrhizal plants are found in severely disturbed soils (Reeves et al. 1979). Absence of mycorrhizae severely reduces the establishment of many species (Allen and Allen 1988, Allen 1991). Gray rabbitbrush (*Ericameria nauseosa*) inoculated with endomycorrhizae had better growth and survivorship than those without after planting on coal mine spoils (Moorman and Reeves 1979). Allen and Allen (1988) found that mycorrhizae can regulate succession by improving competitive ability.

Site Preparation

Geomorphic Stability

The stability of planting surfaces must be considered before restoration is started. Many soils, especially sands, are subject to wind erosion, particularly after fires. Stabilizing the surface before planting may be necessary. Blowing sands remove seeds and scour seedlings, which can be blown right out of the soil. Seed usually is applied before or with mulch. Mulches also can be blown away unless stabilized by crimping and tackifiers (Munshower 1994). Crimping uses a disk, wheel, or punch to push part of straw mulch into the soil. This allows part of the straw to stand up, behaving like stubble to reduce wind speed at the surface and thus erosion (Munshower 1994). Hydromulching combines water, mulch, often seed, and a gluelike binding agent to bind the material to the soil surface. Erosion control blankets or mats are a type of mulch used to control erosion on slopes. Seed can be incorporated into the mat or placed under the mat. Mats usually are pinned to the surface with large staples to keep them from blowing away (Munshower 1994).

Blowing sand can also bury plants. In areas that are vulnerable to deposition, snow fences can be used to reduce sand accumulation. Seeds of plants that can germinate from deep depths, such as Indian ricegrass (*A. hymenoides*) and needle-and-thread grass (*H. comata*) are more likely to survive deposition. Rhizomatous plants sometimes can escape accumulating sands by growing away from them.

Soil Ripping and Gouging

Soils at some restoration sites are heavily compacted. Where this is the case, they can be ripped to reduce bulk density, thus improving long-term success of restoration projects (Montalvo et al. 2002). A variety of soil preparation tools and techniques are described in Munshower (1994). In the semiarid shrub steppe, soils can be prepared to concentrate water in local areas, which will aid in plant establishment. Gouged depressions (10–20 centimeters deep, 25–40 centimeters wide, up to 90 centimeters long) have been used successfully in mine land restoration because they reduce runoff and concentrate water in the gouge. Seed usually is then broadcast in the roughened terrain.

Fertilizer

Fertilizer amendment of shrub steppe soils usually promotes weedy annual growth but has little beneficial effect on seeded perennials. Nitrogen and phosphorus fertilizer should be used only where soils have very low organic content such as subsoils, mine spoils, and moving sands. If soil organic matter exceeds 2%, nitrogen should not be added (Munshower 1994). When low-nitrogen organic amendments such as straw are added to a low organic soil, nitrogen must be added so that the ratio of carbon to nitrogen is kept between 12:1 and 20:1 (Munshower 1994).

Irrigation

Irrigation is an expensive addition to any restoration project and should be avoided if possible. It is always better to plant when water is available to give new plantings the best opportunity to establish. Generally, the best time to plant is in the fall, after rains have begun and added enough moisture to maintain new seedlings through a potentially dry winter. When planting is scheduled to take advantage of seasonal rains, plants can establish roots and survive the next summer.

Seeding is best done in fall and winter to mimic the natural history of shrub steppe plants. For example, sagebrush species drop seed in the winter and often are protected by snow cover. Most other shrub steppe species drop their seed in the spring through the fall. Seeding in spring is less likely to be successful without supplemental water.

Water can be added using agricultural techniques such as sprinklers or drip irrigation. Water can also be added by placing a tube near the base of planted seedlings, then watering through the tube. This technique keeps water away from weeds near the surface. A condensation trap can be created around a seedling to direct evaporated water to the seedling. Organic polymer gels have also been used to provide water to seedlings, but they can dehydrate a seedling in very dry soils (Munshower 1994).

Weed Management

Disturbed ground often has a strong component of invasive weeds. The shrub steppe unfortunately has many invasive alien species to consider and control. There are many reviews and books on the topic of steppe weeds and weed control (DiTomaso 2000, Gaines and Swan 1972, Sheley and Petroff 1999, Taylor 1990, Whitson et al. 1992, Zimdahl 1999).

Strategies to control weeds, covered in more detail in Chapter 16, include hand pulling, hoeing, mowing, fire, mulching, competition, fertility management, biological control, and chemical control (Zimdahl 1999). The usefulness of a particular strategy depends primarily on the size of the area. Hand pulling and hoeing annuals can work in small areas, but over large areas biological or chemical control may be the only cost-effective strategy.

Extremely hot wildfires can reduce cover of cheatgrass by destroying most of the seed bank, but the benefits often are transient. Prescribed fires often are cooler than a midsummer wildfire and are not very effective in reducing the seed bank of cheatgrass. But prescribed fire has been used to remove vegetation and litter as a first step to make herbicide application more efficient and spatially consistent. Link et al. (2003a) found that prescribed fire applied in the fall had no effect on cheatgrass.

Mulching has been used to reduce weed competition in steppe riparian zones (Link and Bower 2004). It also retains water, especially helpful in the semiarid shrub steppe. Wide mulch resulted in higher survivorship of ponderosa pine (*Pinus ponderosa*) and snowberry (*Symphocarpus alba*) than narrow mulch in riparian restoration efforts along the Touchet River in eastern Washington (Link and Bower 2004).

Glyphosate (Roundup) and imazapic (Plateau) are herbicides commonly used to control invasive species to prepare a site for restoration. Glyphosate is a broad-spectrum, nonselective contact herbicide. At high enough concentration, it will kill anything that is actively growing. It should be applied with surfactant and fertilizer to increase growth and effectiveness. Its half-life ranges from 32 to 40 days. Imazapic is used as a preemergent and postemergent herbicide. It needs soil contact to act as a preemergent and can be active for up to a few years. After fire, the suggested application rate is 2 to 6 ounces per acre depending on expected precipitation. Application should be made in the fall before emergence. Application in the spring is not advised.

Serious Weeds of the Shrub Steppe

Diffuse Knapweed (Centaurea diffusa) *and Spotted Knapweed* (Centaurea maculosa)

Diffuse knapweed is an annual that is found throughout the shrub steppe. Control can be done by hand pulling of plants if enough of the taproot is extracted to prevent regeneration (Roche and Roche 1999). Pulling needs to be repeated for a few years and therefore is possible only in small areas.

The herbicides picloram, clopyralid, and 2,4-D are effective ways to control knapweed if applied when the plant is at the rosette stage (Roche and Roche 1999). Sheep grazing when the knapweed is green but everything else is brown is also effective.

Establishing competitive bunchgrasses can reduce diffuse knapweed and probably is the best strategy for long-term control. An integrated approach that includes herbicide, grazing, and seeding of bunchgrasses can be very effective. There is also some hope for eventual biological control (Roche and Roche 1999).

Spotted knapweed also occurs throughout the shrub steppe but is more common where annual precipitation is above 200 millimeters. Control techniques are similar to those for diffuse knapweed (Sheley et al. 1999b).

Yellow Starthistle (Centaurea solstitialis)

Yellow starthistle is an annual weed common in the western half of the shrub steppe. It can be a dominant weed where precipitation is greater than 12 inches. Starthistle can be controlled using picloram, clopyralid, dicamba, glyphosate, and 2,4-D herbicides. It can be hand pulled in small areas. Grazing can provide control, but it has to be done before spines form around the flowerhead. It can also be controlled by planting of competitive grasses (Sheley et al. 1999c).

Russian Knapweed (Centaurea repens)

Russian knapweed is a perennial, so control strategies differ from those for the other two knapweeds discussed earlier. It occurs only where it can get roots down to ground water and is commonly found near wetlands. It grows by underground-creeping roots that form clonal monocultures. Roots can be up to 7 meters deep.

Russian knapweed is difficult to control, although Whitson (1999) found that application of clopyralid and 2,4-D three times, followed by seeding with Sodar wheatgrass, resulted in suppression. There is hope that an integrated pest management approach using biological control and competitive grasses may provide a solution.

Rush Skeletonweed (Chondrilla juncea)

Rush skeletonweed is a perennial found in the western half of the shrub steppe. This plant has roots at least 2.4 meters deep and can spread from underground runners.

Controlling skeleton weed is difficult, although an integrated weed management program can reduce populations. Effective strategies include competitive plantings, sheep grazing, biological control agents, and herbicides (picloram, 2,4-D, clopyralid, dicamba) (Sheley et al. 1999a).

Tumblemustard (Sisymbrium altissimum)

Tumblemustard is susceptible to broadleaf herbicides including 2,4-D, MCPA, bromoxynil, atrazine, and chlorsulfon (Adams and Swan 1988,

Eckert 1974, Kidder et al. 1988, Swensen et al. 1986). Phenoxy herbicides such as 2,4-D and MCPA provide the best control (90–99%) (Adams and Swan 1988, Kidder et al. 1988, Swensen et al. 1986).

Cheatgrass (Bromus tectorum)

Cheatgrass is an annual grass native to Eurasia. It has spread throughout the shrub steppe and is considered to be the biological driver of the much increased fire frequency in the area (Whisenant 1990). It becomes more competitive with increasing aridity (Mosley et al. 1999).

Cheatgrass control is effective only when combined with treatments that establish perennial species (Harris and Goebel 1976, Klemmedson and Smith 1964, Mosley et al. 1999). In areas where a significant component of perennials is already present, chemicals can control cheatgrass if applied for 2 to 5 years consecutively (Mosley et al. 1999). Paraquat and glyphosate can be applied in the spring after the plants have reached the two- to three-leaf stage and until seedheads begin to emerge. Application rates should be just enough to kill cheatgrass (6.4 ounces active ingredient per acre) yet not damage the perennials, which are killed at application rates of 9.6 ounces active ingredient per acre (Mosley et al. 1999).

Two years of prescribed grazing in the spring can significantly reduce cheatgrass cover (Mosley et al. 1999). In heavily infested areas, prescribed fire in the fall, grazing or herbicide application in the spring, followed by seed application with a drill or broadcasting combined with animal trampling can control cheatgrass (Link et al. 2003a, (Mosley et al. 1999). Seeding can also be delayed to the next fall or spring in a chemical fallow approach (Mosley et al. 1999).

Preemergent and early postemergent herbicides (sulfometuron, Plateau) also can control cheatgrass. Sulfometuron can be applied in the fall or spring, with perennials seeded 1 year later, but it can damage Sandberg's bluegrass (Mosley et al. 1999). Plateau can also be applied in the fall, with perennials seeded in the spring. One year after applying Plateau at 8 ounces per acre after a prescribe fire, Link et al. (2003a) did not observe damage to Sandberg's bluegrass.

In high precipitation parts of the shrub steppe, it is possible to establish competitive bunchgrasses without using herbicides. After a cheatgrass-infested field was rototilled to a depth of 8 centimeters in August, wildrye was seeded the next May. After 3 years, cheatgrass had been reduced by 85% (Whitson and Koch 1998). In drier areas, Daubenmire (1975) noted that sand dropseed can maintain itself in areas with dense cheatgrass cover and that seedlings are more successful than those of bluebunch wheatgrass. It may be possible to broadcast dropseed into cheatgrass fields without first using herbicides. Squirreltail (*Elymus elymoides*) has also been noted to be competitive with cheatgrass.

Medusahead (Taeniatherum caput-medusae)

Medusahead is an annual grass native to Eurasia, now found in the western half of the shrub steppe in areas of high precipitation. Control is similar to that for cheatgrass except that grazing is not practical because of poor forage quality (Miller et al. 1999).

Planting Seeds

Proper seeding rates are very important to restoration success. Too little will not produce an adequate density to compete with weeds, and too much can cause the desired species to compete with themselves and can lead to failure.

The seed rate is the number of seeds placed in a unit area of soil. It can be expressed as the number or mass of seeds per acre or hectare. Confusion reigns when seeding rates are described for species mixtures based on mass because seed size varies widely. The number of seeds per unit mass can be obtained by searching the literature or by asking the seed provider. An estimate can be made by counting and weighing thirty seeds randomly selected from clean seed. It is important to purchase the amount of seed based on the percentage of pure live seed. Pure live seed labeling is regulated by a state seed-certifying agency.

A seed rate of twenty pure live seeds per square foot is considered the minimum on drill-seeded applications, aiming for a 50% germination success. The seed rate can be increased or decreased depending on competition and other environmental stresses. Less seed may be applied to a north-facing slope because there is less water stress than on a south-facing slope.

Each species has differing needs for successful germination (Young and Young 1986). Generally, small seeds should be placed near the soil surface, and larger seeds can be buried more deeply (Montalvo et al. 2002). This is not a firm rule because the small seeds of Sandberg's bluegrass can be placed to a depth of 25 millimeters (Evans et al. 1977), whereas the larger seeds of bluebunch wheatgrass are optimally planted at a depth of 6 millimeters (Plummer 1943, McLendon and Redente 1997). Some seeds need exposure to light. Seeding strategies thus depend on the species-specific needs.

Big sagebrush seed planted in the fall will germinate from midwinter to early spring and does best with a protective snow cover (Monsen 2002). Control of annual and perennial weed competition is necessary for establishment (Monsen 2002).

Drill seeding is successful in rangelands where there are no tall plants, the land is flat, and there are few rocks. The seeder has a box to hold the seed and commonly has ten disks that open a furrow into which seed is dropped. The furrow is then closed with another wheel or a chain. This technique works best for large seeds such as wheatgrasses and legumes (Munshower 1994). A carrier such as rice hulls can be used to keep small and large seeds well mixed.

Broadcast seeding can use any technique that disperses seed onto the surface. Aerial seeding is a broadcast technique used on rough ground. Broadcasting works better than drilling for very small seeds. Often a chain and cultipacker is used to cover the seed with soil (Munshower 1994). Aerial seeding of sagebrush in late fall and midwinter after a wildfire is successful after the area is chained to cover the seed. Without chaining, aerial seeding is only about 10% effective (Monsen 2002).

Hydroseeding is a form of broadcasting in which seed is dispersed in a water-based mixture of mulch, tackifier, and fertilizer. This technique works best with small seeds (Montalvo et al. 2002). Often the soil–seed contact is better when seed is dispersed without the mulch and tackifier, which can be applied after the seed is dispersed (Munshower 1994). Hydroseeding is used on steep slopes or where common agricultural equipment cannot be used (Roberts and Bradshaw 1985). On sand slopes, the use of long-fibered flexible materials for mulch enhances establishment (Roberts and Bradshaw 1985).

A compression or compact type seeder is advised for sagebrush (Monsen 2002). A special sagebrush seeder has been successful in arid, southern Idaho (Boltz 1994). Sagebrush seed should be planted no deeper than 6 millimeters below the soil surface and should be seeded at rates between 0.11 and 0.22 kilograms per hectare (Monsen 2002). The Dixon imprinter creates an depression in the soil in which dropped seeds can germinate in safe microsites (Montalvo et al. 2002).

Planting Seedlings and Plant Parts

Seedlings should be planted immediately after acquisition and should never be allowed to sit in the sun. Seedlings can be stored in a 32°F to 35°F cooler for several weeks but should be examined every week to see whether fungal growth exists on the stems or whether the plants have broken dormancy.

Competition can be reduced by planting into weed cloth, scalping the surface, or using herbicide to kill nearby weeds. When digging a hole it is important not to compact the soil around the edge of the hole. This can restrict root growth beyond the hole. A fertilizer tablet can be placed in the hole and about 2 inches from roots. Make sure the seedling is upright and the soil is firmly packed but not compacted around the seedling. The rooting media should be covered by about a half inch of soil.

Rhizomes can be used to establish species such as lanceleaf scurf pea in sandy areas. Rhizomes need to have at least one lateral bud and should be planted in the fall just before winter rains begin and placed at least 2.5 centimeters deep (McLendon and Redente 1997).

Herbivore Protection

Protection against herbivory is important in the first few years after planting (McLendon and Redente 1997). Grazing must be eliminated or controlled to allow establishment of seedlings and cuttings in riparian areas. Red and Pacific willow (*Salix lasiandra*) were well established after 4 years where cattle grazing was controlled in one eastern Oregon site (Shaw 1992). A variety of herbivore protection screens and tubes are available to protect shrub and tree seedlings.

Monitoring

Monitoring should determine whether a restoration has met objectives for species richness, density, frequency, and cover. Richness is the number of species in a specified area, usually measured in sample quadrats. An adequate sample size is determined using a species–area curve. The modified Whittaker plot technique captures a better sample than do transect techniques (Stohlgren et al. 1998). Density is the number of individual plants of a particular species in a known area. Some plants are clonal and spread below the surface, with many shoots appearing above ground. Counting shoots of such plants gives a shoot but not plant density. Frequency is the percentage of plots in which a species occurs. Species frequency gives an assessment of how common and widespread a species is in a restoration area. Percentage cover is the amount of ground covered by plants. Percentage cover is a measure of the importance or significance of a species in a community. Cover is estimated using line transects, point intercepts (Bonham 1989), and quadrats. Total foliar and soil cover can be estimated using the modified Whittaker plot technique and the Parker, large quadrat, and Daubenmire transect approaches (Stohlgren et al. 1998).

Case Study: Hanford Prototype Barrier

The Hanford Nuclear Site in south-central Washington may at first seem a strange place for sagebrush steppe restoration. But much of the affected area lends itself to restoration experimentation, in part because of the nearly pristine steppe context and in part because of the large sums of money available. One difficult issue has been the contamination of groundwater by nuclear wastes. Water draining through these wastes can enter the groundwater or surface water systems. One mitigation measure being tried is the construction of earthen berms, or barriers to intercept water before it gets to waste stockpiles (Ward et al. 1997). One test barrier was established in 1994, with plants installed to minimize erosion and maximize loss of water from the surface soils (Link et al. 1998).

The upper surface of the barrier is constructed with 2 meters of fine soils over coarser material. Soils in the upper layers are silt loams (Gee 1987, Hajek 1966). These soils were excavated from below the surface to minimize the invasive alien seed bank. The upper test area of the barrier is a little more than half an acre (Link et al. 1995).

The barrier surface and surrounding disturbed area was planted in the fall of 1994. Restoration work was done separately for establishment of perennial shrubs and perennial grasses. Perennial shrub seeds were collected, and seedlings were grown in a nursery and then transplanted to the site. Seeds of big sagebrush and gray rabbitbrush were collected from local populations in December 1993. The entire inflorescence of sagebrush and the fruits of rabbitbrush were harvested and stored in plastic bags in the field. This material was transported to a laboratory and dried. It was stored in the dark at room temperature until shipped to a nursery. The seed was cleaned later that spring and sown in early May. Seedlings were grown in tubes (Gee et al. 1994). On November 7, planting was initiated and completed by the next day. Twenty-seven hundred holes were drilled and two seedlings placed in each hole. In all, 1,350 rabbitbrush and 4,050 sagebrush were planted.

Perennial grasses were hydroseeded onto the barrier surface and surrounding slopes. The mix included seeds, fertilizer, mulch, and a tacking agent. The seed mixture included Sandberg's bluegrass (34 kilograms per hectare), streambank wheatgrass (5.6 kilograms per hectare), Indian rice grass (22 kilograms per hectare), Sherman big bluegrass (*Poa ampla*, 11 kilograms per hectare), needle-and-thread grass (*H. comata*, 5.6 kilograms per hectare), bluebunch wheatgrass (14 kilograms per hectare), and squirreltail (3.4 kilograms per hectare). Most of the perennial grasses originated from off-site sources. Fertilizer was applied as 67 kilograms per hectare (60 pounds per acre) of total nitrogen, 67 kilograms per hectare of available phosphoric acid (P_2O_5), and 67 kilograms per hectare of soluble potash (K_2O) in solution. Mulch was applied as 2,240 kilograms per hectare of Eco-Fibre 100% virgin wood fiber. Degradable glue was added to the mulch as a tackifier at 67 kilograms per hectare. Hydroseeding was done in early November in a slurry form. The material was mixed with water, using power augers in a large tank on a truck, then dispersed under pressure from large hoses onto the ground.

In the first season after planting Russian thistle covered nearly 100% of the surface. But it was nearly eliminated once the native perennials became dominant. The number of annuals has varied from twelve to sixteen, and the number of perennials increased from eleven to nineteen by 1997. In 1995, 55% of the species were annuals; in contrast, only 46% of the species were annuals in 1997.

After 3 years about 98% of the sagebrush shrubs had survived, but only 39% of the rabbitbrush did so. A significant number of new sagebrush seedlings became established in the third year. Sandberg's bluegrass was the most successful grass, and squirreltail did not establish.

Plants on the surface have been successful at eliminating wind and water erosion and appear to have prevented water from accumulating in the soil (Ward et al. 1997).

Case Study: Reducing Unnatural Fuels in the Shrub Steppe

The widespread presence of cheatgrass has caused unnatural and severe fires to threaten much of the sagebrush steppe ecosystem. A project funded by the Joint Fire Science Program from 2002 to 2005 was initiated to develop strategies for returning steppe lands in the Columbia Basin and the Intermountain West to a highly diverse assemblage of native species that bring fire risk back to natural levels.

Objectives include finding a minimum concentration of herbicides that will shift the competitive balance to native species and away from cheatgrass, determining the effect of prescribed burns on the competitive balance, and determining the effect of seeding native bunchgrasses on the competitive balance. The intent is also to determine a fuel management method that least affects native species and is least expensive.

The establishment of wheatgrass was believed to be the best solution for long-term reduction of cheatgrass and thus reduction of fire risk. Wheatgrasses are competitive where soils and precipitation are advantageous. In areas near the study site that were seeded with bluebunch wheatgrass in 1986, little cheatgrass was present in 2004 (Figure 10.2).

Experiments were initiated in 2002 (Figure 10.3). Ninety split plots were established to test the effects of two herbicides, with five concentrations of each, followed by drill seeding of two bunchgrasses. Percentage cover of each species, litter, soil, and soil cryptogams were measured in each plot. Plots were first burned, and the herbicide Plateau was applied at 2 ounces per acre in fall 2002 (Figures 10.4 and 10.5). Snake River wheatgrass (*Elymus wawawaiensis*) and Sherman big bluegrass were seeded the next February, and Roundup was applied in March. The prescribed fire had no effect on cheatgrass (*B. tectorum*) cover and significantly increased the cover of tumblemustard (*S. altissimum*).

Plateau, applied at 1 ounce per acre, did not reduce cheatgrass cover but nearly eliminated tumblemustard (Figure 10.6). This may have contributed to better establishment of wheatgrass after Plateau application than when Roundup was applied at the same rate (Figure 10.7). Cheatgrass cover was reduced at the highest rate of Plateau and with increasing rates of Roundup. After fire, tumblemustard control may be more important than cheatgrass control for establishment of Snake River wheatgrass. Based on 1 year's results, Plateau may be better than Roundup for establishment of Snake River wheatgrass.

Figure 10.2. Shrub steppe site 18 years after restoration with Snake River wheatgrass (*Elymus wawawaiensis*). There was nearly complete control of cheatgrass. The center area was test burned the previous fall.

Figure 10.3. Experimental area before treatments showing a shrub steppe community with about 50% cheatgrass cover.

FIGURE 10.4. Plateau applied at 2 ounces per acre resulted in nearly complete control of tumblemustard but no effect on cheatgrass.

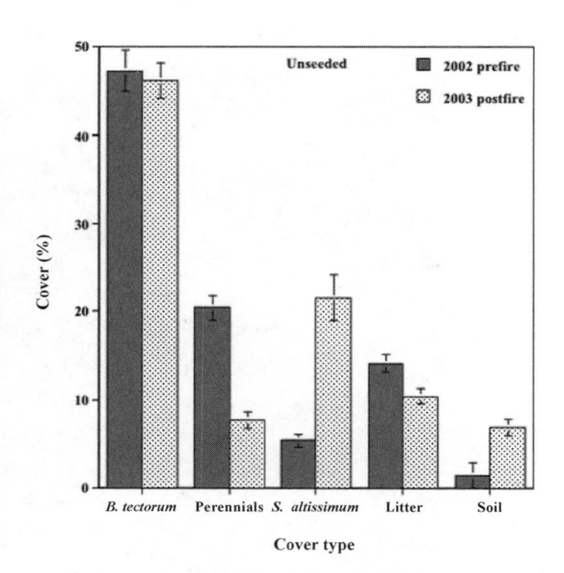

FIGURE 10.5. Effect of prescribed fire on cover elements at the Columbia National Wildlife Refuge. Error bars are one standard error of the mean ($n = 9$).

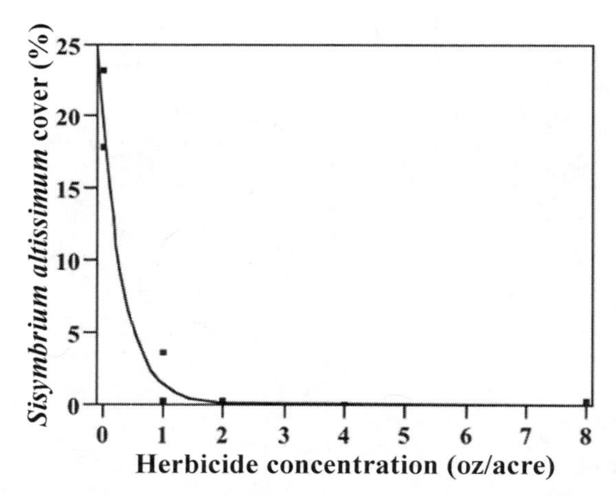

FIGURE 10.6. Relationship between Plateau herbicide concentration and tumblemustard cover.

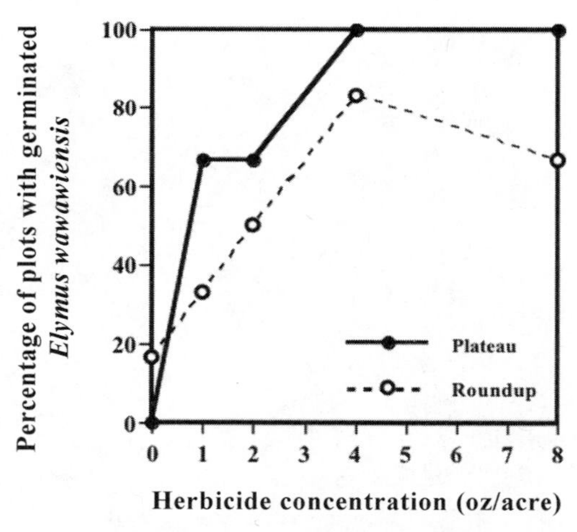

FIGURE 10.7. Percentage of plots with Snake River wheatgrass as a function of herbicide and herbicide concentration (Plateau $n = 3$, Roundup $n = 6$).

Case Study: Restoration of Upland Habitats at Columbia National Wildlife Refuge

Over the last 20 years, several attempts have been made to restore native perennial vegetation to disturbed dry upland sites at Columbia National Wildlife Refuge. The highest priorities for restoration were areas burned by wildfire or otherwise disturbed that had few or no shrubs remaining. A perennial grass was introduced, with selective herbicides to control broadleaf weedy invaders before herbaceous and shrub species were seeded or allowed to replenish naturally. Cheatgrass was suppressed using herbicides. Two restoration projects that may improve dry upland restoration techniques will be profiled.

The first area was a grassland that had been farmed and irrigated as recently as the 1970s. This 40-acre field was dominated by cheatgrass but also had sparse crested wheatgrass (*Agropyron cristatum*), evidence of a conversion from agriculture to perennial grass. The restoration plan included use of a glyphosate spray in mid-November 1996 followed by dormant season seeding of bluebunch wheatgrass, streambank wheatgrass, giant wildrye (*L. cinereus*), and Sandberg's bluegrass. The field was sprayed with 8 ounces of Roundup per acre, with surfactant and ammonium sulphate. Most of the field was sprayed in mid-November, but snowfall (which persisted until late February) prevented treatment of the last 5 acres until late winter. When the snow melted, it was noted that the single day of glyphosate contact before snowfall was enough to kill the cheatgrass that had germinated. The area was drill-seeded in early March, then sprayed with a low rate of glyphosate a week later after extensive germination of cheatgrass but before any drilled seed had emerged. High winter moisture and late cheatgrass control provided excellent establishment. Additional cheatgrass and tumblemustard germinated after the last glyphosate treatment, but favorable soil moisture conditions were sufficient for establishment.

The second area (about 200 acres) had burned in a wildfire in mid-July 1997 and had also burned more than 15 years earlier. This area was dominated by cheatgrass, with scattered patches of Sandberg's bluegrass and bluebunch wheatgrass and little shrub cover. The restoration plan included a late fall glyphosate treatment followed by dormant season drill seeding of Idaho fescue (*F. idahoensis*), bluebunch wheatgrass, and Sherman big bluegrass. Sherman big bluegrass is larger than native Sandberg's bluegrass but with presumed native populations within 25 miles of the site. Additionally, giant wildrye was broadcast seeded on lower areas with deeper soils before drill seeding. Drill seeding was done the same week using 4 pounds of Sherman big bluegrass, 2 pounds of bluebunch wheatgrass, and 1 pound of Idaho fescue per acre. Roundup was sprayed from a fixed-wing aircraft in November. Another wildfire area was treated in a similar manner, but instead of drill seeding the grass seed was applied by aircraft at a 50% higher rate. This area had about 25 of 300 acres harrowed after seeding to improve soil contact, but a majority of the area was too rocky for equipment.

Results were dramatically different between the sites. Aerial seeding resulted in very poor germination, whereas drill seeding got excellent results. By summer, it appeared that all of the new seedlings had died because of extreme heat and dry conditions. In 2002, it was noted that much of the grass, especially big bluegrass, had survived and was doing well, although cheatgrass continued to compete.

Since those projects in 1996–1997, there have been other experimental treatments at the refuge to determine timing and rate using both glyphosate and imazapic (Plateau). Both products are effective in controlling cheatgrass, but timing is very important for optimizing control and minimizing impacts to native species. Both are most effective when litter is reduced after tillage, heavy grazing, wildfire, or prescribed burning.

As with the first example, multiple glyphosate applications can be successful, but application after seeded grasses have emerged is counterproductive. Four to 9 ounces per acre applied from November to late March suppressed cheatgrass, but the higher rates and later applications had increasing impact on Sandberg's bluegrass, which greens up after fall rains or after snowmelt. Nine ounces applied in late March browned out and weakened the bluegrass but did not kill it. An 8-ounce rate applied in mid-May caused noticeable injury to established bunchgrasses and was not effective in controlling cheatgrass.

Imazapic is a very effective contact herbicide when used with a surfactant. It is most effective when applied as a preemergent. Several native perennial grasses tolerate up to 12 ounces per acre if it is applied during dormancy, but new seedlings are more susceptible as the application period approaches germination. Imazapic controls cheatgrass and several weedy mustards. The 1996–1997 restoration may have been more successful if imazapic was used instead of glyphosate. Unfortunately, snow cover precluded treatment.

The combination of imazapic with glyphosate appears to produce a synergistic effect but one that warrants further investigation. During the winter of 2002–2003 cheatgrass was treated by ground application (6 ounces per acre glyphosate) in late December and by helicopter application in mid-February (4 ounces per acre imazapic, 5 ounces per acre glyphosate, 3 ounces per acre imazapic with 5 ounces per acre glyphosate). The December application showed no injury to Sandberg's bluegrass, but all of the February treatments produced some browning, with the combination showing the most injury. The stage of bluegrass probably was the reason because December had top growth but was not growing rapidly, whereas in February growth was faster with warmer weather. When early spring application is the only alternative, greatly reduced rates may still suppress competitive weeds enough to allow seeded bunchgrasses to establish.

Case Study: Canoe Ridge

An inadvertent road clearing project on Canoe Ridge in Benton County, Washington, near the Columbia River, caused disturbance on land managed by the Bureau of Land Management (BLM). As mitigation, the disturbed area was to be restored to a condition similar to that of nearby undisturbed areas. Restoration efforts included assessment of an undisturbed reference area to arrive at restoration goals, planting, vegetation management, and monitoring of the plant populations.

The BLM property is in a big sagebrush needle-and-thread grass habitat type. Soils are a sandy loam and are weakly acidic in the upper 10 centimeters of the soil profile (Daubenmire 1970). The habitat type is dominated by big sagebrush, needle-and-thread grass, and Sandberg's bluegrass. Streambank wheatgrass was also common in the study area, although it is not noted by Daubenmire.

The density of major species in the reference area provided the planning goal for the restoration effort. The road clearing is 0.55 miles long. A plot size (5 meters by 9.3 meters) was chosen to accommodate a large number of the common plants and to cover the width of the disturbed area. In each of twenty-five randomly located plots, each large bunchgrass and shrub was counted in addition to each inflorescence of rhizomatous grasses. Burned stems of sagebrush were found in the undisturbed area, and small clusters of live sage were noted away from the study area. The density of the clusters was not determined.

The mean density of species in the undisturbed BLM property in September 1999 ranged from 3.77 plants per square meter for needle-and-thread grass to 0.00086 plants per square meter for streambank wheatgrass. The overall density of mature bunchgrasses was more than 15,000 plants per acre, and the density of all shrubs was more than 600 plants per acre. Specific restoration goals were a survival of at least 800 shrubs and at least 7,500 native grass plants acre.

Restoration included hydromulching, sowing grass with a drill seeder, planting nursery-grown shrub seedlings, and transplanting plants from an adjacent area. Streambank wheatgrass, needle-and-thread grass, Indian rice grass, and Secar bluebunch wheatgrass were seeded. Big sagebrush seed was collected within 2 miles of the restoration site and grown along with gray rabbitbrush to produce tubelings that were later transplanted. Needle-and-thread grass was also transplanted from adjoining areas. Maintenance activities (weed control and watering) occurred in February for 2 years.

In February 2000 it was observed that cheatgrass was competing with the native grass plantings. Roundup was sprayed at a rate of 3 ounces per acre. It was believed that cheatgrass control would outweigh any loss of native species that might occur at this low application rate. The site was again visited in April to water the plants. At this time, soil moisture was about 10 centimeters (4 inches) below the soil surface. Using previously installed watering tubes, each planted shrub received approximately 1 quart of water. After watering, all tubes were removed and the holes filled with soil.

Monitoring was initiated in November to monitor the success of the restoration effort. Five randomly selected plots were observed for density of grasses and shrubs and two plots for survivorship of the transplanted bunchgrasses. Bunchgrass seedlings and shrubs were counted in addition to each shoot of a volunteer, thickspike wheatgrass (*Agropyron dasytachyum*). In two plots, live and dead transplanted bunchgrasses were counted to determine percentage survival. Mean density data are expressed as the number of plants per acre. Individuals of big sagebrush and gray rabbitbrush were summed to compute shrub density.

The density of native grasses (live transplanted bunchgrasses + germinations from seeding + volunteers) was nearly 20,000 plants per acre. The density of shrubs was only 417 plants per acre, less than the minimum target of 800 shrubs per acre. As a result, more sagebrush shrubs were planted in February. A count of live and dead transplanted bunchgrasses on two sample plots yielded a mean density of 3,885 per acre. The density of live transplanted bunchgrasses was 304 per acre. The survivorship of transplanted bunchgrasses was measured at only 8.2%. Seeding bunchgrasses appears much more successful than transplanting mature bunchgrasses under the conditions of this test.

In March, the same five plots were reexamined. At this time a species list was compiled for plants in the study plots. In each quadrat, bunchgrass seedlings, mature bunchgrasses, shrubs, and herbaceous broadleaf native plants were counted as well as each shoot of thickspike wheatgrass. The density of native grasses (live transplanted bunchgrasses + germinations from seeding + volunteers) was nearly 50,000 plants per acre. This density is much greater than the minimum required. It was expected that the number of surviving grasses would decline over time to a level closer to that of the undisturbed property. The density of shrubs was 975 plants per acre, which satisfied the minimum required. The density of broadleaf herbaceous native plants was more than 5,000 plants per acre.

Twenty-one native and three alien species were identified in the undisturbed BLM property in September 1999. In March 2001 nineteen native species and three aliens were recognized on the restored road surface.

Sample Budget

A well-considered budget estimate can spell the difference between success and failure in restoration. Restoration is similar to agriculture in that it suffers from the vagaries of the weather. It is important that budgets accommodate factors beyond the control of restoration implementers. We present an example budget for restoration in the shrub steppe for 1 acre. Prices are only estimates and will vary depending on local circumstances and the size of the restoration effort. Unit prices will go down with increasing area being restored (Table 10.4).

Restoration Challenges and Research Gaps

Use of herbicides near water is restricted, which makes management of invasive species and establishment of native species more difficult. Strategies must rely on integrated pest management practices, with more emphasis on biological control and the use of weed cloth to promote establishment of planted seedlings. Reducing available soil nutrients can restrict growth of annual weeds, allowing perennials to establish more effectively (Cione et al. 2002).

Restoration in highly disturbed soils and areas dominated by annual grasses can be improved if soil microbiotic crust species can also be restored. Very little work has been done to restore soil microbiotic crusts in practice, although it is believed to be possible (Evans and Johansen 1999). A slurry of microbiotic crust was applied to burned areas, with successful establishment of cyanobacteria and lichens within months (St. Clair et al. 1986). Dry microbiotic crust that has been broken up for application enhanced lichen cover and diversity in disturbed areas. Microbiotic crusts may enhance germination, establishment, and growth by providing cracks for favorable seed environments and nitrogen from nitrogen-fixing algae and lichens

TABLE 10.4.

Estimated costs for restoration of 1 acre of shrub steppe.

Task	Unit Price ($)	Items/Acre	$/Acre
Planning	70/hectare		560
Site description	70/hectare		1,680
Monitoring	70/hectare		3,360
Herbicide application			
Ground	Government rate		10
Ground	Private		100
Fixed wing	Government rate		7.5
Helicopter	Government rate		30
Planting			
Drill seeding	Government rate		100
Drill seeding	Private		600
Broadcast	Government rate		50
Harrow	Government rate		31.25
Cultipacking	Government rate		33
Hydroseeding	Private		800
Shrub seedling planting	3/plant	400	1,200
Bunchgrass salvage and transplanting	1.5/plant	400	600
Materials			
Weed cloth (6-foot-wide roll)	0.04/square foot		1,742
Roundup	25/gallon		
Ammonium sulfate	1/acre		
Surfactant	1/acre		
Roundup + ammonium + surfactant		8 ounces	3.6
Plateau	300/gallon	8 ounces	18.75
Herbivore protection	2/plant	400	800
Seed			
Elymus wawawaiensis	1/pound	5 pounds drilled	5
Achnatherum hymenoides	10/pound	5 pounds drilled	50
Elymus elymoides	45/pound	5 pounds drilled	225
Hesperostipa comata	80/pound	5 pounds drilled	400
Poa secunda	4.25/pound	5 pounds drilled	21.25
Seedlings			
Artemisia tridentata	0.83	400	332
Grayia spinosa	4.25	100	425
Chrysothamnus viscidiflorus	8.95	50	448
Ericameria nauseosa	0.6	400	240
Sphaeralcea munroana	0.6	100	60
Purshia tridentata	1.4	200	280

(Evans and Johansen 1999). See Plate 6 in the color insert for native shrub steppe species.

Acknowledgments

Suggestions by Dean Apostol and Susan Nelson significantly improved the manuscript for this chapter. The Joint Fire Science Program supported some of this work.

References

Adams, E. B. and D. G. Swan. 1988. Broadleaf weed control in Conservation Reserve Program (CRP) grass plantings. Western Society of Weed Science. *Research Progress Reports*, p. 367.

Allen, E. B. and M. F. Allen. 1988. Facilitation of succession by non-mycotrophic colonizer *Salsola kali* (Chenopodiaceae) on a harsh site: effects of mycorrhizal fungi. *American Journal of Botany* 75: 257–266.

Allen, M. F. 1991. *The Ecology of Mycorrhizae*. Cambridge University Press, New York.

Boltz, M. 1994. Factors influencing postfire sagebrush regeneration in south-central Idaho. Pages 281–290 in S. B. Monsen and S. G. Kitchen (comps.), *Proceedings: Ecology and Management of Annual Rangelands*. Gen. Tech. Rep. INT-GTR-313. USDA Forest Service, Intermountain Research Station, Ogden, UT.

Bonham, C. D. 1989. *Measurements for Terrestrial Vegetation*. Wiley, New York.

Boudell, J., S. O. Link, and J. R. Johanssen. 2002. Effect of soil microtopography on seed bank distribution in the shrub-steppe. *Western North American Naturalist* 62: 14–24.

Cione, N. K., P. E. Padgett, and E. B. Allen. 2002. Restoration of a native shrubland impacted by exotic grasses, frequent fire, and nitrogen disruption in southern California. *Restoration Ecology* 10: 376–384.

Connelly, J. W. and C. E. Braun. 1997. Long-term changes in sage grouse *Centrocerus urophasianus* populations in western North America. *Wildlife Biology* 3: 229–234.

Daubenmire, R. 1970. *Steppe Vegetation of Washington*. Washington State University, Pullman.

Daubenmire, R. 1975. Plant succession on abandoned fields, and fire influences, in a steppe area in southeastern Washington. *Northwest Science* 49: 36–48.

DiTomaso, J. M. 2000. Invasive weeds in rangelands: species, impacts, and management. *Weed Science* 48: 255–265.

Eckert, R. E. Jr. 1974. Atrazine residue and seedling establishment in furrows. *Journal of Range Management* 27: 55–56.

Evans, R. A., J. A. Young, and B. A. Roundy. 1977. Seedbed requirements for germination of Sandburg bluegrass. *Agronomy Journal* 69: 817–820.

Evans, R. D. and J. R. Johansen. 1999. Microbiotic crusts and ecosystem processes. *Critical Reviews in Plant Sciences* 18: 183–225.

Franklin, J. F. and C. T. Dyrness. 1988. *Natural Vegetation of Oregon and Washington*. Oregon State University Press, Corvallis.

Gaines, X. M. and D. G. Swan. 1972. *Weeds of Eastern Washington*. Camp-Na-Bor-Lee Association, Inc., Davenport, WA.

Gee, G. W. 1987. *Recharge at the Hanford Site: Status Report*. Pacific Northwest Laboratory, PNL-6403, Richland, WA.

Gee, G. W., H. D. Freeman, W. H. Walters, M. W. Ligotke, M. D. Campbell, A. L. Ward, S. O. Link, S. K. Smith, B. G. Gilmore, and R. A. Romine. 1994. *Hanford Prototype Surface Barrier Status Report: FY 1994*. Pacific Northwest Laboratory, PNL-10275, Richland, WA.

Hajek, B. F. 1966. *Soil Survey, Hanford Project in Benton County, Washington*. Pacific Northwest Laboratory, BNWL-243, Richland, WA.

Harris, G. A. and C. J. Goebel. 1976. *Factors of Plant Competition in Seeding Pacific Northwest Bunchgrass Ranges*. Washington State University, Pullman.

Hassan, M. A. and N. E. West. 1986. Dynamics of soil seed pools in burned and unburned sagebrush semi-deserts. *Ecology* 67: 269–272.

Hitchcock, C. L. and A. Cronquist. 1976. *Flora of the Pacific Northwest*. University of Washington Press, Seattle.

Jones, T. A. and D. A. Johnson. 1998. Integrating genetic concepts into planning rangeland seedings. *Journal of Range Management* 51: 594–606.

Kidder, D. W., I. C. Hopkins, and D. P. Drummond. 1988. Evaluation of bromoxynil, sulfonyl–urea tank mixes in winter wheat. Western Society of Weed Science. *Research Progress Reports*, pp. 343–344.

Klemmedson, J. O. and J. G. Smith. 1964. Cheatgrass (*Bromus tectorum* L.). *The Botanical Review* 30: 226–262.

Körner, C. and F. I. Woodward. 1987. The dynamics of leaf extension in plants with diverse altitudinal ranges. II. Field studies in *Poa* species between 600 and 3200 m altitude. *Oecologia* 72: 279–283.

Linhart, Y. B. 1995. Restoration, revegetation, and the importance of genetic and evolutionary perspectives. *Proceedings Wildland Shrub and Arid Land Restoration Symposium,* October 19–21, 1993, Las Vegas, NV. Gen. Tech. Rep. 315. USDA Forest Service Intermountain Research Station.

Link, S. O. and A. Bower. 2004. Effect of mulch type and size on soil water and survivorship in a semi-arid riparian ecosystem. *Northwest Science* 78: 333–337.

Link, S. O., T. M. Degerman, J. S. Lewinsohn, S. S. Simmons, W. H. Mast, A. L. Ward, and G. W. Gee. 1998. Revegetation of a landfill cover. Pages 18–26 in S. Link and D. Rumsey (eds.), *Native Plants as Minor Crops Conference*, May 5–7, 1997, Washington State University Cooperative Extension, Seattle.

Link, S. O., R. W. Hill, and E. M. Hagen. 2003a. *Management of Fuel Loading in the Shrub-Steppe*. Washington State Weed Association, Kennewick, WA.

Link, S. O., B. D. Ryan, J. L. Downs, L. L. Cadwell, J. Soll, M. A. Hawke, and J. Ponzetti. 2000. Lichens and mosses on shrub-steppe soils in southeastern Washington. *Northwest Science* 74: 50–56.

Link, S. O., J. L. Smith, J. J. Halvorson, and J. H. Bolton. 2003b. Effect of climate change on a perennial bunchgrass and soil carbon and nitrogen pools in a semi-arid shrub-steppe ecosystem. *Global Change Biology* 9: 1097–1105.

Link, S. O., N. R. Wing, and G. W. Gee. 1995. The development of permanent isolation barriers for buried wastes in cool deserts: Hanford, Washington. *Journal of Arid Land Studies* 4: 215–224.

McLendon, T. and E. F. Redente. 1997. In C. J. Kemp (ed.), *Revegetation Manual for the Environmental*

Restoration Contractor. BHI-00971. Bechtel Hanford, Inc., Richland, WA.

Miller, H., D. Clausnitzer, and M. M. Borman. 1999. Medusahead. In R. L. Sheley and J. K. Petroff (eds.), *Biology and Management of Noxious Rangeland Weeds*. Oregon State University Press, Corvallis.

Monsen, S. B. 2002. Ecotypic variability, seed features, and seedbed requirements of big sagebrush. Pages 12–13 in *Restoration and Management of Sagebrush/Grass Communities Workshop*. Elko, NV. www.rangenet.org/trader/2002_Elko_sagebrush_conf.pdf.

Montalvo, A. M., P. A. McMillan, and E. B. Allen. 2002. The relative importance of seeding method, soil ripping, and soil variables on seeding success. *Restoration Ecology* 10: 52–67.

Moorman, T. and F. B. Reeves. 1979. The role of endomycorrhizae in revegetation practices in the semi-arid West. II. A bioassay to determine the effect of land disturbance on endomycorrhizal populations. *American Journal of Botany* 66: 14–18.

Mosley, J. C., S. C. Bunting, and M. E. Manoukian. 1999. Cheatgrass. In R. L. Sheley and J. K. Petroff (eds.), *Biology and Management of Noxious Rangeland Weeds*. Oregon State University Press, Corvallis.

Munshower, F. E. 1994. *Practical Handbook of Disturbed Land Revegetation*. Lewis Publishers, Boca Raton, FL.

Naeem, S., L. J. Thompson, S. P. Lawler, J. H. Lawton, and R. M. Woodfin. 1995. Empirical evidence that declining species diversity may alter the performance of terrestrial ecosystems. *Philosophical Transactions of the Royal Society of London* B 347: 249–262.

Pellant, M. 1990. The cheatgrass–wildfire cycle: are there any solutions? *Proceeding: Symposium on Cheatgrass Invasion, Shrub Die-Off, and Other Aspects of Shrub Biology and Management*. Gen. Tech. Rep. INT-GTR-276. USDA Forest Service Intermountain Research Station, Ogden, UT.

Plummer, A. P. 1943. The germination and early seedling development of twelve range grasses. *Journal of the American Society of Agronomy* 35: 19–34.

Reeves, F. B., D. Wagner, T. Moorman, and J. Kiel. 1979. The role of ectomycorrhizae in revegetation practices in the semi-arid West. I. A comparison of mycorrhizae in severely disturbed vs. natural environments. *American Journal of Botany* 66: 6–13.

Rickard, W. H., L. E. Rogers, B. E. Vaughan, and S. F. Liebetrau. 1988. *Shrub-Steppe: Balance and Change in a Semi-Arid Terrestrial Ecosystem*. Elsevier, Amsterdam.

Roberts, R. D. and A. D. Bradshaw. 1985. The development of a hydraulic seeding technique for unstable sand slopes. II. Field evaluation. *Journal of Applied Ecology* 22: 979–994.

Roche, J. and C. T. Roche. 1999. Diffuse knapweed. In R. L. Sheley and J. K. Petroff (eds.), *Biology and Management of Noxious Rangeland Weeds*. Oregon State University Press, Corvallis.

Rogers, L. E., R. E. Fitzner, L. L. Cadwell, and B. E. Vaughan. 1988. Terrestrial animal habitats and population responses. Pages 181–256 in W. H. Rickard, L. E. Rogers, B. E. Vaughan, and S. F. Liebetrau (eds.), *Shrub-Steppe: Balance and Change in a Semi-Arid Terrestrial Ecosystem*. Elsevier, Amsterdam.

Rogers, L. E. and W. H. Rickard. 1988. Introduction: shrub-steppe lands. Pages 1–12 in W. H. Rickard, L. E. Rogers, B. E. Vaughan, and S. F. Liebetrau (eds.), *Shrub-Steppe: Balance and Change in a Semi-Arid Terrestrial Ecosystem*. Elsevier, Amsterdam.

Sauer, R. H. and W. H. Rickard. 1979. Vegetation on steep slopes in the shrub-steppe region of southcentral Washington. *Northwest Science* 53: 5–11.

Scott, M. J., G. R. Bilyard, S. O. Link, C. A. Ulibarri, H. E. Westerdahl, P. E. Ricci, and H. E. Seely. 1998. Valuation of ecological resources and functions. *Environmental Management* 22: 49–68.

Shaw, N. L. 1992. Recruitment and growth of Pacific willow and sandbar willow seedlings in response to season and intensity of cattle grazing. *Proceedings: Symposium on Ecology and Management of Riparian Shrub Communities, Sun Valley, ID*. Gen. Tech. Rep. INT-284. USDA Forest Service Intermountain Research Station, Ogden, UT.

Shaw, N. L. and M. R. Haferkamp. 1990. Field establishment of spiny hopsage. Pages 193–199 in *Proceedings: Symposium on Cheatgrass Invasion, Shrub Die-Off, and Other Aspects of Shrub Biology and Management*. Gen. Tech. Rep. INT-276. USDA Forest Service Intermountain Research Station, Ogden, UT.

Sheley, R. L., J. M. Hudak, and R. T. Grubb. 1999a. Rush skeletonweed. In R. L. Sheley and J. K. Petroff (eds.), *Biology and Management of Noxious Rangeland Weeds*. Oregon State University Press, Corvallis.

Sheley, R. L., J. S. Jacobs, and M. L. Carpinelli. 1999b. Spotted knapweed. In R. L. Sheley and J. K. Petroff (eds.), *Biology and Management of Noxious Rangeland Weeds*. Oregon State University Press, Corvallis.

Sheley, R. L., L. L. Larson, and J. S. Jacobs. 1999c. Yellow starthistle. In R. L. Sheley and J. K. Petroff (eds.), *Biology and Management of Noxious Rangeland Weeds*. Oregon State University Press, Corvallis.

Sheley, R. L. and J. K. Petroff. 1999. *Biology and Management of Noxious Rangeland Weeds*. Oregon State University Press, Corvallis.

St. Clair, L. L., J. R. Johansen, and R. L. Webb. 1986. Rapid stabilization of fire disturbed sites using a soil crust slurry: inoculation studies. *Reclamation Revegetation Research* 4: 261–269.

Stohlgren, T. J., K. A. Bull, and Y. Otsuki. 1998. Comparison of rangeland vegetation sampling techniques in the central grasslands. *Journal of Range Management* 51: 164–172.

Swensen, J. B., D. C. Thill, and R. C. Callihan. 1986. Broadleaf weed control in spring barley at Potlatch, Idaho. Western Society of Weed Science. *Research Progress Reports*, pp. 191–193.

Taylor, R. J. 1990. *Northwest Weeds*. Mountain Press Publishing Company, Missoula, MT.

Thompson, R. S. and K. H. Anderson. 2000. Biomes of western North America at 18,000, 6000 and 0^{14}C yr BP reconstructed from pollen and packrat midden data. *Journal of Biogeography* 27: 555–584.

Thorp, J. M. and W. T. Hinds. 1977. *Microclimates of the Arid Lands Ecology Reserve, 1968–1975*. BNWL-SA-6231. Battelle Pacific Northwest Laboratories, Richland, WA.

U.S. Department of Agriculture. 1996. Status of the interior Columbia Basin: summary of scientific findings. Page 144. Gen. Tech. Rep. PNW-GTR-385. U.S. Department of Agriculture, Forest Service, Pacific Northwest Research Station; U.S. Department of the Interior, Bureau of Land Management, Portland, OR.

USDA NRCS. 2003. The PLANTS Database, Version 3.5 (plants.usda.gov). National Plant Data Center, Baton Rouge, LA.

Ward, A. L., G. W. Gee, and S. O. Link. 1997. *Hanford Prototype-Barrier Status Report: FY 1997*. PNNL-11789. Pacific Northwest National Laboratory, Richland, WA.

Whisenant, S. G. 1990. Changing fire frequencies on Idaho's Snake River plains: ecological and management implication. *USDA Forest Service Intermountain Research Station General Technical Report* INT-276: 4–10.

Whitson, T. D. 1999. Russian knapweed. In R. L. Sheley and J. K. Petroff (eds.), *Biology and Management of Noxious Rangeland Weeds*. Oregon State University Press, Corvallis.

Whitson, T. D., L. C. Burrill, S. A. Dewey, D. W. Cudney, B. E. Nelson, R. D. Lee, and R. Parker. 1992. *Weeds of the West*. The Western Society of Weed Science, Newark, CA.

Whitson, T. D. and D. W. Koch. 1998. Control of downy brome (*Bromus tectorum*) with herbicides and perennial grass competition. *Weed Technology* 12: 391–396.

Young, J. A. and R. A. Evans. 1975. Germinability of seed reserves in a big sagebrush community. *Weed Science* 23: 358–364.

Young, J. A. and C. G. Young. 1986. *Collecting, Processing and Germinating Seeds of Wildland Plants*. Timber Press, Portland, OR.

Zimdahl, R. L. 1999. *Fundamentals of Weed Science*. Academic Press, San Diego, CA.

Chapter 11

Mountains

Regina M. Rochefort, Laurie L. Kurth, Tara W. Carolin, Jon L. Riedel, Robert R. Mierendorf, Kimberly Frappier, and David L. Steensen

This chapter concentrates on subalpine parklands and alpine meadows of southern British Columbia, Washington, Oregon, and western Montana. These areas lie on the flanks of several mountain ranges including the Olympics, the Cascades of Oregon and Washington, and the Coast Mountains in British Columbia (Figures 11.1 and 11.2).

Ecological Characterization

Subalpine parklands and alpine meadows cover mountainous landscapes above the forest zone (Franklin and Dyrness 1987). Subalpine parkland is the broadest portion and is characterized by a mosaic of individual trees, tree clumps, and meadows (Henderson 1974, Franklin and Dyrness 1987). The lower boundary often is called forest line, which is the upper limit of closed contiguous forests (Douglas 1970, Rochefort et al. 1994). In this chapter, the upper limit of the subalpine parkland will be defined as the tree line, or the highest elevation at which erect trees grow. The alpine zone extends above the tree line and encompasses alpine meadows, krummholz, and areas of permanent snow and ice.

In the Pacific Northwest, the subalpine parkland is a broad ecotone, often 300 to 500 meters wide, characterized by scattered trees or tree clumps interspersed with herbaceous meadows (Franklin and Dyrness 1987). The forest line that forms its lower boundary varies with both latitude and aspect and generally lowers in elevation about 110 meters per degree of latitude (Franklin and Dyrness 1987). Typical forest line elevations range from 2,130 meters at Mt. McLoughlin in southern Oregon to 1,400 meters on Mt. Baker in northern Washington. Although depth and duration of snowpack are the predominant factors that maintain mountain meadows, their establishment may have been initiated by glacier recession or as a result of fires (Henderson 1974). Occasional fires in the closed canopy forest may also prevent the forest line from advancing upslope (Franklin and Dyrness 1987).

High-elevation herbaceous meadows are quite diverse spatially, with plant community patterns reflecting combinations of local slope, aspect, topography, and elevation. Physical characteristics of a site may change over very short distances, and although these differences may seem minor, they may result in steep gradients in soil moisture, temperature, and length of growing season (Billings and Bliss 1959, Canaday and Fonda 1974, Douglas 1970, Henderson 1974, Ingersoll 1991). Plants in upper mountain zones are well adapted to short growing seasons, low summer air and soil temperatures, high interannual variability in climate, and intense ultraviolet radiation. Perennial plants of short stature often dominate plant communities. The few annuals that do grow in this zone must be able to germinate, flower, and set seed within just

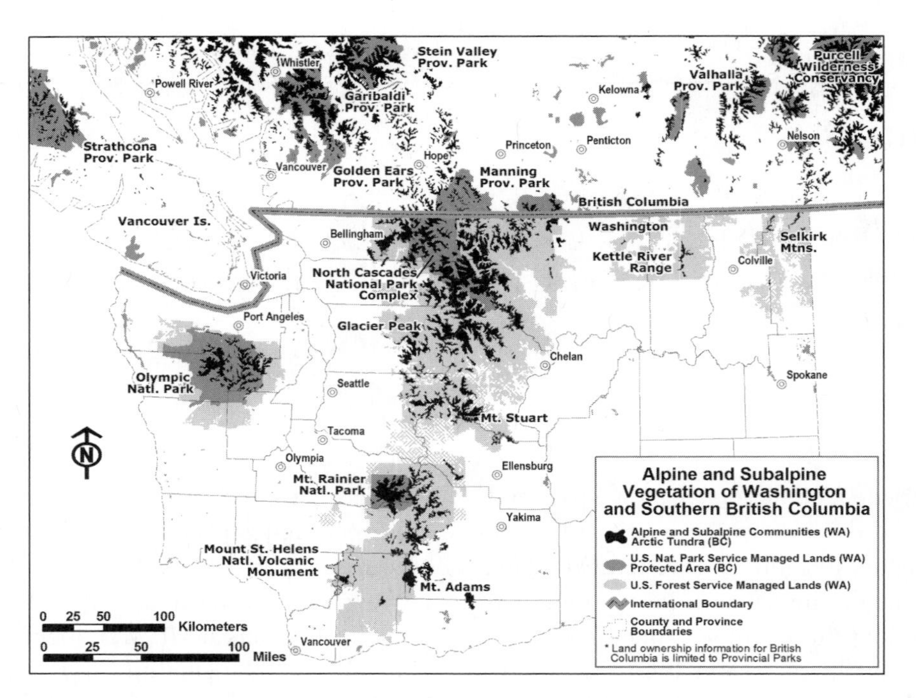

FIGURE 11.1. Map of alpine and subalpine areas in Washington and southern British Columbia. *(Compiled from Eng 2003, Washington Department of Fish and Wildlife 1999)*

a few weeks. Perennial plants often have high root/shoot ratios and have the ability to spread vegetatively to compensate for variable climatic conditions that may not favor sexual reproduction each year (Billings and Mooney 1968, Billings 1973). Nevertheless, plant populations often maintain high genetic variability and show surprisingly small-scale variation between populations in response to climatic gradients or random events (Linhart and Gehring 2003).

The mountains of the Northwest experience a maritime climate that includes long winters with moderate temperatures and abundant snowfall followed by short, cool summers with low precipitation. Thus, nearly all the moisture available for plant growth comes from melting snowpacks, which are greatly influenced by topography and elevation. Windswept alpine ridges often have little or no snow cover during the winter, exposing plants to desiccating winds, abrasion from snow crystals, and frozen soils. Although this results in longer snow-free periods, growing seasons are still short because of low soil and air temperatures. In addition, droughty, poorly developed soils often hold less soil moisture than protected sites with higher organic matter. Convexities or areas on the lee side of ridges or rock outcrops often have late-lying snowbanks with higher soil moisture but shorter growing seasons. Surface topography influences not only the distribution of plant communities but also the height of individual plants. Plant stems that protrude from snow cover during harsh winter winds are exposed to ice abrasion and desiccation and often die back. Restorationists must consider the mesotopographic gradients on adjacent, undisturbed areas and use these areas as references when reconstructing substrates on damaged sites and selecting target plant communities.

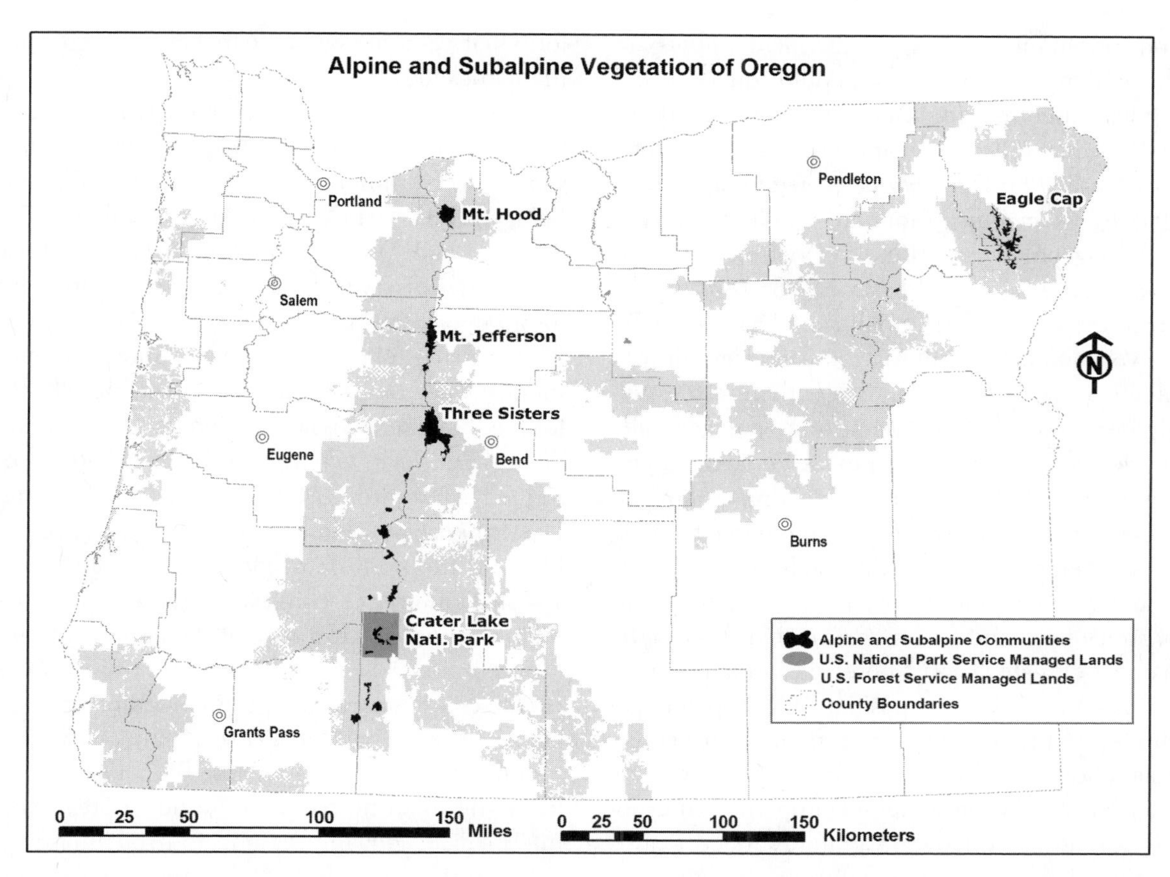

FIGURE 11.2. Map of alpine and subalpine areas in Oregon. (*Compiled from Kagan and Caicco 1992, Oregon Department of Forestry 2003, USGS 2003*)

Plant Communities

Herbaceous meadow vegetation of the subalpine zone can be divided into five broad vegetation types (Douglas 1970, Henderson 1974, Franklin and Dyrness 1987). The first are heath–shrub communities dominated by one or two ericaceous shrubs, including red mountain heath (*Phyllodoce empetriformis*), cream mountain heath (*P. glanduliflora*), western cassiope (*Cassiope mertensiana*), and blueleaf huckleberry (*Vaccinium deliciosum*). The second group includes lush herbaceous vegetation dominated by tall perennials including Sitka valerian (*Valeriana sitchensis*), broadleaf lupine (*Lupinus latifolius*), and false hellebore (*Veratrum viride*). Third are low herbaceous vegetation communities dominated by fanleaf cinquefoil (*Potentilla flabellifolia*) and wooly everlasting (*Antennaria lanata*), often with lesser amounts of black sedge (*Carex nigricans*). Fourth are wet sedge types occupying topographic depressions, dominated by black sedge, showy sedge (*Carex spectablis*), alpine aster (*Aster alpigenus*), and wooly everlasting (*Antennaria lanat*). Last are dry grass communities found on well-drained sites more common on eastside mountains, dominated by green fescue (*Festuca viridula*), Idaho fescue (*Festuca idahoensis*), and broadleaf lupine.

Although many tree species are found in the subalpine zone, subalpine fir (*Abies lasiocarpa*) and mountain hemlock (*Tsuga mertensiana*) are the predominant species. Depending on local conditions, other trees common to this zone are Alaska yellow cedar (*Chamaecyparis nootkatensis*), whitebark pine (*Pinus albicaulis*), Engelmann spruce (*Picea engelmannii*), mountain larch (*Larix lyallii*), lodgepole pine (*Pinus contorta*), silver fir (*Abies amablis*), and Shasta red fir (*Abies magnifica* var. *shastensis*) (Franklin and Dyrness 1987).

Vegetation generally is more sparsely distributed across the alpine landscape than in the subalpine parkland. Growing seasons are generally shorter, summer temperatures are cooler, and precipitation is lower. Stronger winds blow snow off the ridges, exposing vegetation and soils to freezing winter temperatures. Alpine soils generally are drier than subalpine soils because they contain less organic matter and are less mature. Evidence of the harsher climate is easily observed in plant growth forms and species composition. Often plants are shorter and more compact (e.g., cushion plants), adaptations that facilitate survival in cool, windy environments where air temperatures decrease rapidly with distance above the ground surface. Leaves often are evergreen, are narrow, and have hairs or waxy surfaces to provide protection from desiccating winds or ultraviolet rays. Observation of these characteristics is important in planning restoration projects that involve transplants from other field sites or from greenhouses. For example, at Mt. Rainier National Park, mountain and pink heather plants were grown from cuttings for 4 years in a greenhouse. When the plants were transplanted one fall, they were noticeably taller than the adjacent heather plants in undisturbed communities. The next summer, most of aboveground stems of the transplants were dead, but fortunately new buds were sprouting from the root masses. The aboveground stems had been taller than the winter snowpack, even though they were only 6–8 inches high, compared with the existing 2- to 4-inch heathers. In this case, it would have been better to prune the top growth of these plants while they were in the greenhouse to promote more lateral growth and a plant form that was more similar to that of plants growing in the meadows.

Alpine plant communities of the Cascades and Olympics have been classified into five groups: snowbed, krummholz, heath–dwarf shrub, fellfield, and talus (Van Vechten 1960, Hamann 1972, Zwinger and Willard 1972, Douglas and Bliss 1977, Edwards 1980). Snowbed communities are found in areas closest to snowbanks and therefore have the shortest growing seasons. Snowbed community types range from the alpine buckwheat–tolmie saxifrage–luetkea (*Eriogonum pyrolaefolium–Saxifraga tolmiei–Luetkea pectinata*) described by Edwards (1980) in Mt. Rainier National Park, to six types dominated by various sedges or herbs in the north Cascades (Douglas and Bliss 1977). Krummholz consists of stunted trees that often are molded into a prostrate hedge in which individuals are indistinguishable. Krummholz mats can be composed of any species but often are subalpine fir, whitebark pine, Engelmann spruce, or mountain larch (Douglas and Bliss 1977). The heath–dwarf shrub types are found in areas with intermediate growing seasons and soil moistures between those of snowbeds and rocky ridge communities. These are dominated by a variety of shrubs, depending on local conditions. Pink, yellow, and mountain heathers and western cassiope generally are found on moister, more protected sites than willows (*Salix cascadensis, S. nivalis*), crowberry (*Empetrum nigrum*), or kinnikinnick (*Arctostaphylus uva-ursi*) communities. Fellfield and talus are found on the sites with the longest snow-free seasons and the least summer soil moisture. Fellfields are communities in which vegetation patches are interspersed with rocks or small pebbles. Quite a few have been described, ranging from the sparse vegetation of the creamy mountain heath ridge community (Edwards 1980) to the more continuous sedge turf communities dominated by Brewer and showy sedge (*Carex breweri, C. spectablis, Pedicularis contorta*) or grass species such as purple reedgrass (*Calamagrostis purpurascens*) and timber danthonia (*Danthonia interme-*

dia). In these communities, size and distribution of rock cover are extremely important factors influencing conditions for plant germination and establishment. Soils in alpine zones are subject to continuous movement, known as cryopedogenic action, caused by the freezing and thawing of soil water that may result in downhill movement of saturated soils, rock sorting and patterned ground (e.g., rock stripes, sorted or unsorted nets; see Washburn 1956 for a more complete description), or needle ice. Across all these sites, small changes in topography, aspect, and rock cover represent major changes in habitat. Therefore it is important to note conditions in disturbed and reference communities to design a successful restoration project.

Soils

Restoration of soils in damaged alpine and subalpine ecosystems presents challenges for several reasons. First, bedrock geology, parent soil material, and the age of landforms vary widely. Active geologic processes that affect soil formation include volcanism, glaciation, hillslope, fluvial, nival, eolian, and mass wasting. These processes are constantly creating new parent material, and stabilizing soils on steep, still active landforms can be very difficult. In the more recent geologic past, the activity of continental and alpine glaciers and meltwater has left a young landscape in much of the high-elevation region of southern British Columbia, Washington, Idaho, and Montana. Most high mountain landforms are less than 15,000 years old, and many are less than 1,000 years old. The recent retreat of glaciers since the late nineteenth century has left large tracts of terrain that have only recently been colonized by plants. Soil parent material generally is coarse-textured in mountain provinces, although fine-grained soil accumulates in and near small tarns, ridgetops, flat benches, and cirque basins from deposition of loess and volcanic ashes.

Second, climate variability and short growing seasons make restoration of soils difficult at high elevations. In the Pacific Northwest, climate variability is high in space and time, with pronounced summer and interdecadal droughts. Dry periods can lead to difficult moisture conditions late in growing seasons on south-facing slopes. Strong regional, local, and microclimate gradients result from distance from the Pacific moisture source, latitude, elevation, and aspect. In all regional mountain ranges, there are wetter west-slope and drier east-slope conditions. Large annual snowfall levels in the maritime climate of the west slope of the Washington–Oregon Cascades and British Columbia–Alaska Coast Mountains depresses snowline and results in a wide distribution of alpine and subalpine vegetation (Price 1981). Heavy snowfall may also contribute increased toxins (e.g., airborne mercury) and nutrients to soils. Drier and colder conditions prevail in the more continental climate of eastern slopes and more interior mountain ranges, such as the Blues and Bitterroots. The resulting rise in snowline causes a generally narrower band of alpine and subalpine vegetation, a transition that can occur within 75 kilometers. Gradients north to south also are fairly strong, as illustrated by the wider distribution of glaciers and alpine and subalpine zones farther north in the Cascade and Coast ranges (see Figures 11.1 and 11.2). At upper elevations and on shaded northerly aspects, higher moisture conditions and more frigid air temperatures have strong effects on soil formation. Wind also plays an important role in soil formation through desiccation and redistribution of snow. In general, desiccation is greatest on west- and southwest-facing slopes where snow melts off and evaporation is high. Loess, or wind-deposited sediment, can be an important parent material and a source of pollutants and nutrients.

Ecological restoration in mountain zones must begin with a good understanding of the distribution of soil types across the landscape and an understanding of soil-forming factors. A significant challenge is that soils have not been mapped in many subalpine and alpine areas of the Northwest. Although the National Park Service is beginning to develop geographic information system–based soil models for national parks in Washington, information is generally lacking for the Mt. Baker Snoqualmie National Forest on the west slope of the Cascades and in many other areas.

Alternative methods can substitute for the absence of preexisting soil maps. Ecological units and bio-geoclimate zones have been mapped at 1:100,000 scale for much of the United States and Canada (Ecological Stratification Working Group 1996, McNab and Avers 1994). Soil–geology–vegetation links described in these classification schemes can be useful restoration guides. Establishing these relationships is an important aspect of the soil mapping effort in U.S. national parks. Perhaps the most rigorous approach is to use a nearby undisturbed site as a reference.

Although no single soil taxonomy has been developed for North America, the Canadian system of soil taxonomy is closely related to that of the United States (Expert Committee on Soil Survey 1987). Both are hierarchical, and the taxa are defined on the basis of measurable soil properties. Perhaps the main difference between the two systems is that all horizons to the surface may be diagnostic in Canada, whereas horizons below the depth of plowing are emphasized in the U.S. system. This may be because 90% of Canada is not arable.

Great soil orders are the coarsest taxonomic classification in the U.S. soil mapping scheme (Buol et al. 1980). In the subalpine and alpine zones of the Pacific Northwest they include andisols, entisols, spodosols, mollisols, alfisols, and inceptisols. Andisols are particularly important because they form in volcanic ash deposits, which are widespread in mountains of the Northwest. The presence of ash imparts a specific texture and chemistry critical to many plants. Spodosols are well-developed soils found on older, more stable parts of mountain landscapes. They form beneath acidic duff layers, particularly mountain hemlock, and are easily recognized by their reddish and gray subsurface horizons. Mollisols are saturated soils that form in wet sedge– and grass-dominated meadows and seasonally dry tarns (small mountain lakes). They are identified based on higher pH and cation saturation. Alfisols form beneath deciduous vegetation, particularly hardwood-dominated communities. Like mollisols, they have high cation saturation and a higher pH than spodosols. Entisols form in recent glacial, mass wasting, or stream deposits. They are characterized by weak soil horizon development. Development of an ochric horizon can be difficult to distinguish from geologic material in entisols. Inceptisols have better-developed soil horizons than entisols and generally occur on slightly older landforms.

These great soil groups are further differentiated based on temperature and moisture regimes, chemistry, aspect, vegetation, and other factors. Soil temperature regimes found in the alpine and subalpine zones generally are quite cold and are classified as cryic, meaning a range of 0–8°C. Although patterned ground often is seen at higher elevations, warm summers prevent permafrost soil conditions.

Moisture regimes vary widely with aspect and microclimate, and mountains in the Northwest exhibit a full range of conditions, from very dry to saturated. This variability in temperature and moisture, along with other factors such as topography, results in the development of very complex soil patterns in mountain landscapes.

Soil Impacts

Recreational activities, grazing, and resort development in high-elevation areas can result in easily visible loss of soil and vegetation cover, but also there are less obvious changes, including soil physical, chemical, and biotic factors. Physical factors include organic matter content, soil structure, bulk density, soil texture, and soil porosity. Chemical soil factors include total nutrients, available nutrients, pH, conductivity, cation exchange capacity, and organic matter content. Soil biota includes macroinvertebrates, microbial function, and mycorrhizae. Although changes occur at all elevations, high-elevation soils are more vulnerable to lasting damage because of harsh climatic conditions and the fact that they are less mature. Nearly all projects at high elevations must address the potential for soil compaction and erosion. If postproject soil erosion is still occurring, this may affect prioritization of ecological restoration projects. Soil erosion potential is a function of soil texture, aspect, slope, rainfall, snowmelt, vegetation cover, and duration of exposure. Vegetation cover reduces the rate at which water flows downhill, thus protecting against sheet erosion. As cover is decreased, water flow can accelerate and pick up more surface particles, resulting in increased erosion. Recreational trails that are built straight up or down a landform create a high potential for continued erosion, which increases with length and gradient. As water encounters soil, the process of soil erosion is a function of two factors: detachment and transport (Ripple 1989). Both are related to soil texture, or the percentage of clay, silt, and sand. In order for soils to erode, water must have enough energy to detach particles from the soil matrix and transport them down the slope. Clay soils have high cohesion rates that resist detachment, but once they get moving they are easily transported. Sandy soils are easily infiltrated by water and often are too large to be easily transported. Generally, soil erodibility decreases with a decrease in silt fraction (Wischmeier and Smith 1978). In Mt. Rainier National Park, subalpine restoration projects were rated and ranked for restoration using a local adaptation of the Universal Soil Loss Equation so that impacts most susceptible to continued erosion would become the highest priority for restoration (Rochefort 1989). Erosion potential may determine the need to include an erosion mat such as excelsior (and justify the expense). Restoration of social trails that run perpendicular to a steep hillslope may necessitate excelsior or straw mats to reduce erosion, but a flat campsite on the same soils may not need mats for erosion control, although they may be used to help moderate surface physical conditions.

Geographic Extent and Past Uses of Pacific Northwest Alpine and Subalpine Lands

We often think of the Northwest as a mountainous landscape, but true subalpine and alpine ecosystems make up only a small portion of the region (see Figures 11.1 and 11.2). Altogether, less than 0.5% of Oregon and 5% of Washington lie above the continuous forest line. Nearly all of these lands are managed by public agencies, with very few areas under private ownership (Table 11.1). Because of the inaccessibility of high mountains, steep terrain, and lack of merchantable timber, most were not converted to other land uses with Euro-American settlement. However, changing climates have shaped these landscapes over the last

TABLE 11.1.

Ownership of alpine and subalpine areas of Oregon and Washington.

Ownership	Oregon Hectares (%)[a]	Washington Hectares (%)[a]
Bureau of Indian Affairs		25,868 (3.0)
Indian Reservation	5,656 (6.0)	
Bureau of Land Management		323 (<0.1)
U.S. Fish and Wildlife Service		1,148 (0.1)
National Park Service	5,907 (6.3)	235,721 (27.7)
U.S. Forest Service	81,982 (87.6)	562,519 (66.2)
State lands		13,901 (1.6)
Private lands	51(<0.1)	10,170 (1.2)
Total subalpine and alpine	93,596	849,650
Permanent snow and ice	4,109 (4.4)	55,113 (6.5)
Total area in state	25,093,466 (0.4)[b]	17,486,061 (4.9)[b]

[a]Percentage of total alpine or subalpine area, within the state, under this management.

[b]Percentage of state that is subalpine or alpine.

10,000 years, and there is evidence that Native Americans also altered them.

Widespread archeological evidence suggests that Native Americans have continuously used Northwest mountain landscapes for most of the last 8,000–10,000 years, or since glacial retreat. However, the archeology of high-elevation landscapes is poorly known, and reliable radiometric dates are rare. The three oldest radiocarbon dates recovered from subalpine sites in Washington are approximately 6,250 years B.P. in the Goat Rocks Wilderness of the Washington Cascades (McClure 1989), 5,000 years B.P. in the Olympic Mountains (Bergland 1984), and 4,470 B.P. in the North Cascades (Mierendorf 1999). Surveys reveal extensive use of tundra areas by Native Americans for hunting and subsistence-related activities. In addition to the references cited here, other high-elevation archeological evidence is reported from the interior of southern British Columbia (Pokotylo and Froese 1983), the Cascades (Fulkerson 1988, Vivian 1989, Franck 2000, 2003, Reimer 2000), the Rocky Mountains of southern British Columbia (Choquette 1981), and western Montana (Reeves 2003).

Excavation of archeological sites has revealed remains of mountain goats, cooking hearths, and chipped stone debris, indicating a hunting tradition (Burtchard 1998, Mierendorf 1999, 2004). In the north Cascades, subalpine and alpine areas were the source of two types of stone tools used for more than 4,500 years (Mierendorf 2004). In the Coast Range of southern British Columbia, obsidian from Mt. Garibaldi has been recovered from dated contexts spanning the entire Holocene (Reimer 2003). Peeled cedar trees located in forests close to subalpine areas indicate the collection of these materials to make baskets for huckleberry gathering (Burtchard 1998). Oral histories relate the use of fire by Native Americans to improve hunting grounds and growth of huckleberries or other plant foods. In one study at Mt. Rainier, evidence may indicate that Native Americans occupied high-elevation sites during warmer climatic periods when fire frequency was higher (Nickels 2002). Further research is needed to determine whether Native Americans effectively shaped vegetation patterns of high-elevation landscapes and where these changes may have occurred.

Climate Change and Shifting Mountain Ecosystems

Palaeoecological studies indicate that the early to mid-Holocene (9,000 to 5,000 years B.P.) was 1–3°C warmer and drier than today. Tree lines in southern British Columbia at that time were 60–130 meters higher than they are today (Whitlock 1992, Clague and Mathewes 1989, Clague et al. 1992). Other studies in British Columbia, Alberta, and northern California document additional periods in which tree lines went up or down by as much as 100 meters in response to shifting climates (Rochefort et al. 1994). Most general circulation models predict that annual temperatures will increase during the next century in response to increased levels of greenhouse gases, and with these climate changes high-elevation ecosystems can also be expected to shift again (Rochefort et al. 1994).

Tree establishment has been documented in many subalpine meadows of North America after the close of the Little Ice Age (ca. 1850). Increases have been documented in summers with above-average temperatures (Rochefort et al. 1994), after termination of grazing (Dunwiddie 1977, Vale 1981), and after fire (which could be related to both climate and human activity (Agee and Smith 1984, Little et al. 1994). Additionally, studies by Brink (1959) and Gavin and Brubaker (1999) have documented changes in herbaceous species in response to climate changes in subalpine areas.

Need for Restoration in Alpine and Subalpine Zones

Recreation, mining, grazing, and introduced invasive species are the primary activities that result in the need for restoration in subalpine and alpine areas. Impacts can range from small, discrete campsites with minimal plant or soil loss to very large areas that have been completely reshaped to accommodate mining, campgrounds, highways, parking lots, ski runs, or other development. Increased demand for mountain recreation in the Northwest is growing along with urban populations. The population of Washington alone is expected to grow from 5.9 million in 2000 to 7.5 million in 2025. At current growth rates, counties in the Seattle metropolitan area (King, Pierce, Kitsap, and Snohomish) will grow from 3.19 million to 5 million in the next 25 years (Campbell 1996).

Global climate change in itself may not become a primary cause of mountain ecosystem degradation, but its predicted effects will influence the results of restoration programs and perhaps identification of reference communities or evaluations of program effectiveness.

Livestock grazing has been associated with Northwest ecosystems since the early 1700s, when Indians acquired horses from Spanish missionaries. Lewis and Clark saw at least 700 horses in a Shoshone village in western Idaho and thousands more in the adjacent hills (Galbraith and Anderson 1971). Although very few bison made their way into the Pacific Northwest, domestic cattle and sheep have been an important part of the Euro-American history and economy of this region. Cattle were moved from Vancouver, British Columbia to Neah Bay, Washington as early as 1792. By 1825, cattle were becoming an economic factor in the development of some areas of Oregon and Washington, and by 1844 sheep were introduced to Oregon. During this period of expansion, stock were grazed on open range because most areas were still owned by Native Americans. After the Homestead Act of 1863 and subsequent grants of land to railroads, settlers began fencing off lands, and open ranges became limited. By the mid-1880s many arid rangelands were showing signs of overgrazing, and private ownership of lands had changed the way cattlemen managed their stock (Galbraith and Anderson 1971). In 1897, forest reserves were established, and the federal government began to control grazing by requiring permits on public lands.

World Wars I and II placed increasing demands on land for cattle and sheep grazing in order to support the war effort. These demands were felt on all federal lands, not only those administered by the Forest Service. In 1918, the National Park Service received a ruling by the U.S. Food Administration that sheep grazing was incompatible with the mission of National Parks because of the damage it caused to native plants and animals and its interference with visitors (Catton 1996). Unfortunately, the superintendent of Mt. Rainier National Park had already issued a permit for cattle grazing (200 to 500 head) in the Cowlitz Divide area. Although late snow prohibited stockmen from bringing in cattle, the permit was reissued in 1918 along with a second permit for 500 head of cattle in Yakima Park. The Yakima Park permit was reissued in 1919 and 1920, even after substantial damage was noticeable to vegetation and extensive soil erosion observed (Catton 1996). When requests for grazing reemerged during World War II, the National Park Service conducted a system-wide survey that concluded that damages caused by grazing far outweighed the very limited food supply benefits (Catton 1996). Although the National Park Service was able to draw a line against grazing, this was not the case on national forestlands. Because the permit process was not initiated and enforced until after widespread grazing, many areas were severely degraded already. Grazing damage typically includes shifts in species cover from native plant communities to exotic grasses, soil compaction, and soil erosion. Development of soil terraces or soil pedestals is common in heavily grazed areas. In severely damaged areas, restoration must consider repair of both topography and plant communities. An important consideration in revegetation of these areas is whether aggressive noxious weeds are present, even in small quantities. Grazing may have promoted dominance of exotic perennial grasses, but removal of grazing may result in an increase in other species, so revegetation plans must carefully identify which will need to be removed or replaced.

Recreation use or development may also introduce nonnative species to alpine and subalpine ecosystems that can outcompete native plants or cause mortality. Introduced plants such as clover and Scotch broom have been documented on subalpine restoration sites in Mt. Rainier National Park for up to 20 years after restoration. There are also impacts from introduced insects or diseases on mountain plant communities. Whitebark pine is a keystone species of high-elevation ecosystems in western North America. It is an early successional species that establishes both on harsh ridges and in subalpine areas after fire. Today, its long-term survival is uncertain because of the introduction of a Eurasian blister rust (*Cronartium ribicola*) to North America in 1910 (Hoff 1992, Keane and Arno 1993, Campbell and Antos 2000, Kendall and Keane 2001, Zeglen 2002). Surveys in Washington, Oregon, and British Columbia have documented blister rust in about 90% of all stands examined. Mortality is around 20–30%, and average blister rust infection rates range from 13% to 46% (Campbell and Antos 2000, Goheen et al. 2002, Zeglen 2002, Murray and Rasmussen 2003, Shoal and Aubry 2004, Rochefort and Courbois in prep.). Mountain pine beetles and changes in fire regimes may also influence mortality in whitebark pine stands. Restoration or conservation plans to ensure long-term survival of whitebark pine are important for many high-elevation areas in the Pacific Northwest.

Restoring Mountain Ecosystems

Restoration of high-elevation, mountain ecosystems follows the same basic steps as restoration in the other ecosystems covered elsewhere in this book, including site assessment, project planning, implementation, site protection, and monitoring. The specific methods used on a given project reflect not only the goals for the restoration site but also environmental conditions, including severity and extent of the impacts. This section provides a synopsis of each step of restoration as it applies to high-elevation areas and then illustrates application of these steps by discussing three general categories of restoration: small sites with minimal soil loss or substrate alteration (e.g., social trails or campsites), larger sites where geomorphic restoration is necessary, and restoration of a single species.

Site Assessment

Site evaluation is the first step in any ecological restoration process and sets the stage for project goals, objectives, and monitoring. Site evaluation includes location, ecological characterization, quantitative description of impacts, and description of special considerations such as endangered species, habitat, current recreation use, and cultural resources. Ecological characterization should include defining the project boundaries, noting elevation, slope steepness, aspect, soil type, and vegetation communities. Description of impacts should include size and shape of impacts, vegetation damage, soil erosion, and any modifications to streams or water sources. Identifying the origin of the impacts and how management of the area may influence either development or maintenance of the site can influence the success of future restoration. Special concerns in the project area that will influence the ultimate success of the program may include rare species, cultural or historic areas of significance, and scenic vistas or other attractions that influence how people will continue to use the area. It is helpful to produce a map of the site illustrating the location of impacts, specific areas of concern, vegetation, water resources, and site access. In subalpine or alpine areas small-scale changes in topography can result in large differences in soil moisture availability, growing season length, and plant communities. These variations should inform the selection of revegetation methods, timing, and costs. For this reason, it may be helpful to produce a larger-scale map noting plant community changes along with the impacts, late snowmelt areas, ponds, streams, or areas of active erosion. For example, if a social trail traverses a heather community and a lush herbaceous meadow, the percentage of the impact in each plant community type will have a significant influence on the method of revegetation, so it will be important to record this and make it part of a restoration strategy.

Project Planning: Restoration Goals and Objectives

Setting broad goals for a site regarding plant communities, topography, and use is key to a successful project. Broad goals establish a foundation on which to create more detailed objectives. Examples could include the following:

- Restoring original contours and hydrologic functions to a subalpine area that had earlier been regraded as a drive-in campground
- Full reestablishment of a subalpine meadow
- Revegetating bare ground with subalpine plants but leaving the imprint of previous roadbeds or altered contours

With broad goals in hand, project planning can proceed toward a conceptual plan and then a more detailed implementation plan. Sometimes a conceptual plan is developed to pursue funding or build support for a project. Once this support is gained and resources allocated, an implementation plan based on more detailed objectives can be developed. Several factors may influence both broad goals and detailed objectives, which may need refinement as the project moves forward. Examples of some factors that may influence objectives and goals are as follows:

- *Site location*: Is it in a designated wilderness, remote backcountry, or front country near a visitor center? Does the zone designation prohibit camping or other activities?
- *Sensitive resources*: Are there sensitive plant or animal species or habitats, archeological resources, or other cultural resources in the project area?
- *Protection of genetic integrity*: Is it important to protect naturally evolving levels of local genetic diversity or just to ensure that plant stock used in revegetation has been collected from similar environmental conditions (e.g., in seed zones)?
- *Surrounding matrix*: Is the restoration site located in a pristine landscape that provides intact reference communities and potential donor plants or propagation sources? Or, conversely, was the affected plant community locally unique, so that plant materials for revegetation will have to come from a more distant area?
- *Number, size, and severity of impacts*: Are the impacts small, discrete areas, such as campsites or social trails? Or are they large scale, such as a mine or former ski run that is now to be restored to a natural landscape but not necessarily the one that predated the disturbance?
- *Future use of the area*: If the area to be restored is at the edge of a road or on a ski run, long-term goals may include limiting plant height, and occasional mowing or other landscape management may be needed.
- *Cost of restoration and available funding*: Funding sources, labor, and overall costs of restoration may necessitate refinement of initial goals and objectives. Generally, costs and funding availability will influence primarily the methods and the timeframe of a project, but in cases of severe impacts, they may also influence broad goals.

These variables will help determine the extent to which a project aims for complete restoration or a preexisting site ecosystem, a native but not historic system, or mere stabilization that prevents further degradation.

Project Implementation: Site Preparation

Site preparation involves all the work needed to prepare a site for revegetation, including scarification, stabilization, filling, and recontouring. The site assessment should note the extent to which vegetation cover has been lost, the degree of soil compaction, whether soil loss has occurred, and whether erosion is ongoing. If the impacts were of low intensity and short duration, soil may be compacted but still on site. In this case, scarification to a depth of 6–12 inches (15–31 centimeters) may be sufficient to increase soil aeration and permeability and facilitate successful revegetation. The objective of scarification is to break up hardened soil into small soil particles. Scarification may be accomplished in two steps, first breaking the soil into large chunks with a Pulaski, pick, or shovel and then further breaking it up by hand into fine particles.

Soil compaction or loss often results in sites that are below grade with respect to surrounding areas. These may need to be filled and probably stabilized. Sites that have lost less than 2 inches (5 centimeters) of soil often can be filled to grade without additional stabilization. In long, narrow areas, such as social trails, adding silt bars often is necessary. Silt bars are similar to water bars on developed trails, but they differ in two important respects. First, silt bars are built below the surface. Second, they are placed parallel to the contours of the site rather than at an angle. Silt bars can be composed of wood or rocks (Figure 11.3). Once silt bars are installed, the site can be filled to match surrounding grades (Rochefort 1990). If the restoration site is extremely large, as in the case of a road removal, a former mine, or other major development, recontouring of the site with heavy equipment may be needed, and stabilization may be accomplished through reshaping of the ground surface.

FIGURE 11.3. Erosion control bars being installed at Mt. Rainier. (*Photo by National Park Service*)

Adding soil means finding suitable sources of materials, and this can be challenging in alpine environments. The best soil sources will be close to the project site, but in many areas this is not possible because no new excavations are occurring. If the project involves removing a road, material sometimes can be pulled up from the fill slope to completely restore the original contours or at least soften the slopes and mask the original road cut. If soil loss is minimal and limited to the upper organic layer, it might be possible to incorporate imported peat into a scarified soil bed. In the case of old fire rings, it may be necessary to remove burned wood and ash, then scarify the site and add soil or organic matter. Peat is lightweight and easy to transport but is not a renewable resource, so its use should be kept to a minimum. If soil loss has been extreme, areas more than 12 inches (30 centimeters) below the desired grade can first be filled with rocks or gravel, with soil added on top. In this instance, rocks and gravel should be well tamped so that soil will fill only the spaces between the larger particles. In a project at Mt. Rainier National Park, a 3-foot (1-meter) deep social trail was filled first with rocks from a talus slope, then gravel, and finally soil placed on the upper 12 inches (15 centimeters). This site was near a late-lying snowbank, and one restoration objective was to restore sheet flow of snowmelt. However, because the soil had not settled between the larger rocks, there were hollow spaces that channeled water, apparently creating a subsurface stream. Much of the gravel was pushed up and deposited on the surface, creating an unusual soil horizon.

Imported soil should be as similar as possible in physical and chemical characteristics to the native soil. It should be steam sterilized to kill seeds of species exotic to the site.

Once a site has been filled to grade, the surface is reshaped. The filled areas should mimic the contours of the surrounding area. Imagine the flow patterns from local snowmelt. Perhaps the area should include a simulated stream bank or be designed to sheet flow. It is important to be certain that the soil surface is not too flat. Pristine subalpine and alpine areas have quite a bit of microtopography diversity, with many small depressions and convexities, each providing unique conditions for plant germination and survival. Rocks partially embedded in the soil surface provide protection from wind, provide shade and conserve moisture, and provide obstacles to slow or disperse surface water flow. In alpine or young subalpine areas (more recently released from glaciers or snowfields), surfaces often are covered with a "desert pavement" of small rocks or pebbles. This surface helps modify soil temperatures and protect young seedlings from wind and desiccation (Edwards 1980).

Revegetation

Revegetation often is the most time-consuming component of mountain restoration, often occurring over many years. Revegetation methods include seeding, natural seed rain from adjacent areas, transplants from adjacent areas, greenhouse propagation and outplanting, and layering. Size of the disturbed area, distance to undisturbed plant communities, and project goals all help determine revegetation methods. Some important considerations are as follows:

- *Size of the impact*: The larger the disturbed area, the more likely it is that active revegetation techniques will be needed.
- *Intensity of impact*: This includes depth of compaction or erosion and loss of soil organic matter. Loss of soil may mean loss of natural seed banks or an inhospitable substrate for natural seed rain to do the job. Seed banks in subalpine areas normally are confined to surface soil layers and reflect local communities. Most subalpine species have only short distance dispersal capability (Weidman 1984). Highly eroded sites or sites where the organic layer of soil has been lost

probably will not have a seed bank that could provide new germination, even in scarified soils. Zabinski et al. (2000) found that seed density in seed banks declined exponentially with distance from established vegetation, and seed density increased with depth of organic matter. Therefore, the greater the impact, the more active planting is needed.

- *Adjacent plant communities*: Does the adjacent plant community contain species that might easily colonize the disturbed area through seed or vegetative growth? Lush herbaceous communities often contain lupines that produce quite a few large seeds that can successfully colonize bare soil. In contrast, mature heather communities contain few species that would be expected to volunteer or spread into adjacent disturbed sites. Environmental conditions and plant communities change over very short distances in high-elevation areas. Therefore, a detailed survey of plant communities or soil moisture levels may serve as a basis for a revegetation plan and methods.
- *Selection of seed and plant materials*: Direct seeding has been used with success on some projects, but seedling survival is higher on mesic sites, adjacent to rocks or plants that provide shade, associated with mulches or erosion mats, or in small depressions on the soil surface (Ramsay 2004, Juelson 2001, Moritsch and Muir 1993). Plants can be first grown in a greenhouse and transplanted to field sites (Rochefort and Gibbons 1992).
- *Protection of genetic integrity*: Collection of plants or seed from similar environmental conditions (slope, elevation, aspect) will increase survival, but it is best to collect materials within a very short distance from the site. Many high-elevation species are perennial, and seed production may not occur on an annual basis. High levels of variation within and between populations and the presence of ecotypes are evidence that sexual reproduction is an important process in mountain plant communities (Billings 1973, Gehring and Linhart 1992, Linhart and Wise 1997, Rochefort and Peterson 2001, Linhart and Gehring 2003). Linhart and Wise (1997) state that differentiation of herbs, in complex terrain, can be significant over distances of less than 100 meters. Although movement of tree seeds to sites 2–3 kilometers away may be reasonable, movement of herbaceous plants may need to be more conservative, perhaps on the order of hundreds of meters. Environmental factors such as elevation (within bands of 1–200 meters), precipitation, soil moisture, soil type, and aspect all may influence patterns of genetic diversity.
- *Genetic diversity*: Outcrossing species have high genetic diversity, but it helps to spread collection over several days and to collect from different microsites. Small depressions may contain individuals that are more adapted to wetter conditions. Greenhouse studies at Mt. Rainier National Park documented the possibility of genetic variation between seeds collected on different dates. Seeds of *Aster alpigenus* were collected on several dates in Paradise Meadows (approximately 5,500 feet elevation). When the seeds were treated and grown using the same methods, seeds from different seed lots grew at different rates. Because there is always uncertainty with respect to microsite conditions at restored sites, it is best to use plant material from close to the site while maintaining a high degree of within- and between-population diversity. When propagating plants from cuttings, collect them from many plants rather than just those with high growth rates. Where individual plants are difficult to distinguish (e.g., clonal species such as heathers), collect several cuttings (based on rooting success) per plant and then move about 1 meter to another plant to collect more. Cuttings should be taken from a minimum of twenty plants in a given collection.

A seed increase program may be useful in large areas or where protection of genetic integrity is important. Selection can occur in seed collection, field propagation, and location of the seed increase program. Plants that survive in seed increase fields must do well in an agricultural situation (i.e., fungicides and harvesting equipment), which are certainly different from their natural environment. When impacts are large, protection of genetic integrity may be downplayed in order to allow for economies of scale where adherence to a more rigorous standard would compromise the potential success of the project. In areas where there are no adjacent native communities, seeds may be collected from a different watershed. In such cases, finding areas with similar environmental conditions is the key. For instance, in many wilderness areas in the national park system, the goals include restoration of a functioning ecosystem, maintenance of genetic integrity, and continuation of overnight recreational use. Often, when impacts are small, such as individual campsites or social trails, goals are to restore the area to plant communities that probably would cover the site in the absence of impacts. If protection of genetic integrity is a high priority, this may limit the revegetation methods. Seed rain or vegetative spread of adjacent plants will provide the highest level of protection, although these options may be too slow to meet other objectives. A second option is to focus seed or plant collection in a small area adjacent to the site.

The spatial arrangement of planted material often is an important project consideration. Most revegetation projects include both transplants and seeding. Transplants provide instant cover and are visible evidence that restoration is under way. They also provide shelter for establishment of seedlings and an early seed source for further spread of plants. In small areas, plants should be distributed fairly evenly across the entire site. In the case of social trials or large sites, additional plants at access points might help dissuade recreational use and trampling. It is best to organize greenhouse plants or direct transplants in clumps. First, plants in groups appear to have higher survival rates and rates of spread, possibly because each plant has a protected aspect, and collectively they modify environmental conditions in their immediate area. Second, clumps of plants appear more natural to visitors and may assist in changing use patterns. Third, on large sites planting may take several years, and it is easier to add plantings between clumps than between individuals. If clumps of plants are distributed throughout and area, they can then provide seed sources all over.

Species selection is an important consideration. Subalpine and alpine plant communities are very diverse, even at the smallest scale. Reference communities provide models of species composition, relative dominance, and companion plant guidelines. For instance, in a wet sedge community black sedge and alpine aster are often are codominant, but cinquefoil, showy sedge, red mountain heather, and luetkea are present in smaller and more variable amounts. Greenhouse propagation may include all these species except heather, so planting guidelines are limited to the remaining species. Because black sedge and aster are codominant, planting guidelines for clumps might be seven to ten plants and three or four species. Black sedge or aster could make up two thirds of the total plants, with the remaining plants from the other species. Examples of this are four sedge, two aster, and two cinquefoil; two sedge, five aster, and two cinquefoil; or five aster, two cinquefoil, and one luetkea. There should also be guidelines for which species to plant singly or in clumps always greater than two. These guidelines may result from field observations, including size of plants available and expected growth rates (e.g., you can plant species that spread more slowly more densely than those with more rapid growth).

Timing of Planting

In the Pacific Northwest, most moisture in high-elevation areas comes in the fall or winter and usually in the form of snow. Snowmelt provides most of the water for summer plant growth, so seeding and planting must be timed so the plants are in place to take advantage of snowmelt. The only exception is in sites that can be watered in summer. Planting in the fall also gives transplants time for root growth and allows seeds to undergo natural stratification.

Erosion Mats and Watering

Erosion mats such as excelsior (aspen shavings), straw, or coconut fiber (coir) can be an important component of restoration. First, they modify soil temperatures and moisture levels. Second, they may help keep people off the restored site. Juelson (2001) evaluated the effects of watering and use of clear plastic on sites seeded with luetkea and black sedge, in the Enchantment Lakes of Washington. All plots were covered with an excelsior blanket. She found that double watering increased seedling emergence of the graminoids and that the plastic increased emergence of the luetkea. Similarly, Lester (1989) found that clear plastic increased seed germination of *Carex* species in North Cascades National Park. Using plastic necessitates checking the site frequently during the growing season and may require irrigation.

Restoration Costs

Typically the cost of restoration shapes project goals and objectives and may determine whether a project is feasible at all. Caution is advised with regard to modifying goals. Those who develop cost estimates for restoration must understand the objectives and know which objectives are flexible and which ones are fixed. Recontouring an abandoned roadway may be essential for a successful project, whereas reducing planting density may affect only the length of time needed for restoration or recovery of the aesthetics of the site. To assess the impacts of reduced work in relation to cost savings, a cost estimate of the restoration necessary to meet the goals should be developed and then optional objectives negotiated or modified to fit available funding. For example, at Mt. Rainier National Park, there is an established standard of planting eight plants per square foot in subalpine meadows. The natural density in undisturbed meadows is approximately eighty plants per square foot on average. Cost estimates for subalpine meadows are based on a minimum planting density of eight plants per square foot. If this is beyond available funding, new planting density objectives are developed. Based on project goals, it may be determined that some areas will receive seeding only, some areas will have lower density, and some areas will receive the target density.

The first step in preparing a cost estimate is to distinguish fixed costs from flexible ones (Table 11.2). Fixed costs are those that are absolutely necessary to success and cannot be deferred. Negotiable costs may be critical to long-term success but can be put off for future funding or implemented at lower levels. Fixed and nonfixed costs may vary by project based on the goals, but in general site work and stabilization are fixed costs. Planting density, seeding density, and mulching are typical examples of nonfixed costs. However, if a key goal is to have an immediate aesthetic benefit, then planting density may best be treated as a fixed cost.

Restoration cost estimating can be done in several ways: itemization of each component, total cost per plant, or total cost per area (generally based on a total cost per plant). Itemization of each component is recommended for projects that involve contracting, sites where there is no restoration history on which to base costs, or sites that need complex site work before plant installation. Estimating costs on a per-plant or area basis is best used for areas where there is a history of restoration and costs can be easily developed, projects are routine, and little or no contracting is involved. When developing restoration costs it is important to consider all necessary costs, some of which can be hidden. Therefore, itemized cost estimating is recommended initially.

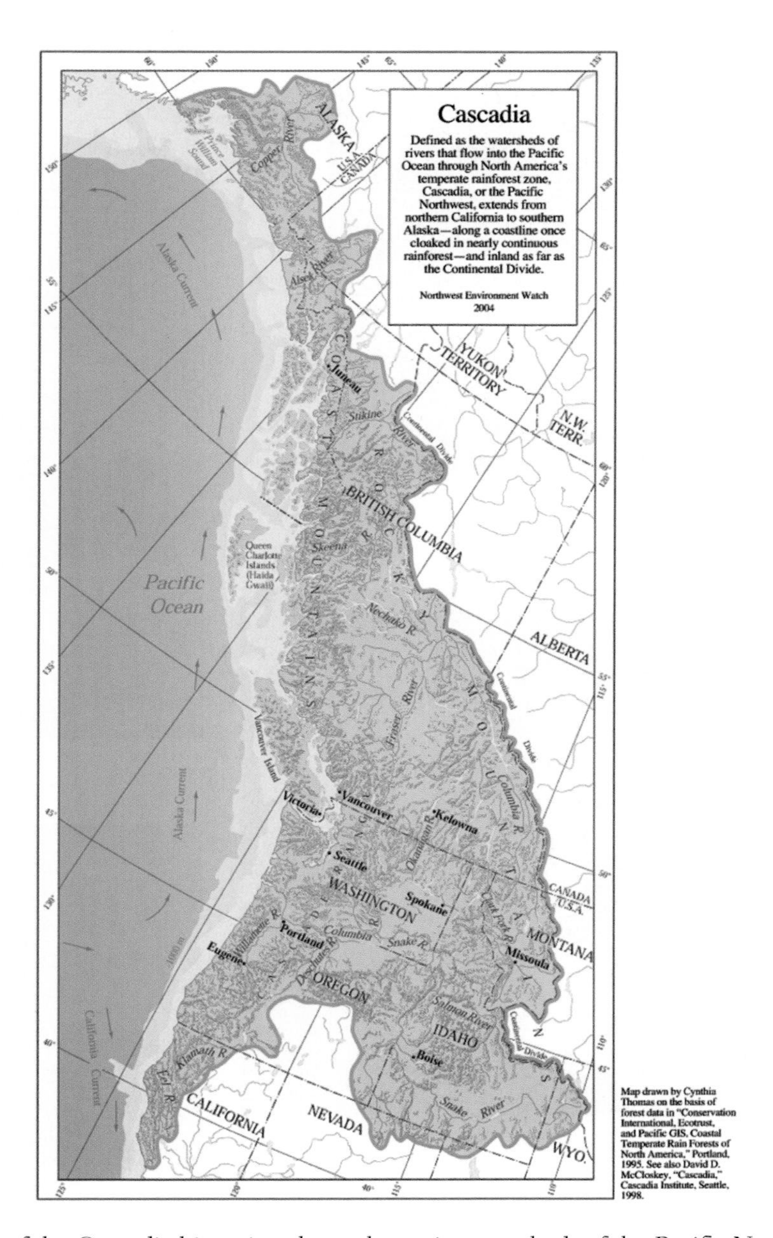

PLATE 1. This map of the Cascadia bioregion shows the main watersheds of the Pacific Northwest that flow through the temperate rainforest zone, from the Copper River in the north to the Eel River in the south, reaching east up the Snake River to Yellowstone National Park. (*Map drawn by Cynthia Thomas, using information from Conservation International, Ecotrust, and Pacific GIS. Used with permission from Northwest Environment Watch*)

(a)

(c)

(b)

(d)

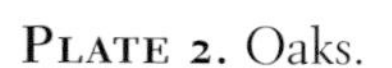

Plate 2. Oaks.
(a) Spring bloom in the Garry oak savanna understory of British Columbia. *(Photo by Tim Ennis, The Nature Conservancy of Canada)*
(b) A woodland skipper (*Ochlodes sylvanoides*) pollinates *Aster curtus*. *(Photo by Tim Ennis, Nature Conservancy of Canada)*
(c) Common camas (*Camassia quamash*) bulbs, commonly found in prairie and oak habitats, was a significant Native American trade item, second only to salmon. *(Photo by Ed Alverson)*
(d) Indian hyacinth (*Camassia quamash*), desert parsley (*Lomatium mohavense*), and oval-leaf shooting star (*Dodecatheon hendersonii*) are among the forbs that make up the crayon-colored prairie palette in spring. *(Photo by Tim Ennis)*

(a)

(b)

(c)

(d)

Plate 3. Prairie.
(a) Perennial wildflowers are visually dominant in spring in this Willamette Valley prairie. *(Photo by Bruce Taylor, courtesy of Defenders of Wildlife)*
(b) Great camas (*Camassia leichtlinii*).
(c) Rosy checkmallow (*Sidalcea virgata*).
(d) Bog deer-vetch (*Lotus pinnatus*) is one of two lotus varieties commonly found in prairie habitats.

(e)

(f)

PLATE 3. *(continued)*
(e) Willamette daisy (*Erigeron decumbens* var. *decumbens*) is listed as endangered throughout its range.
(f) Tough-leaf iris (*Iris tenax*).

(a)

Plate 4. Freshwater wetland communities.
(All photos by Fred Small of Pacific Habitat)
(a) Red alder—dominated forested wetland in central Washington Cascades, Mt. Rainier National Park.
(b) Low-elevation bog dominated by California pitcher plant (*Darlingtonia californica*) and Labrador tea (*Ledum glandulosum*).

(b)

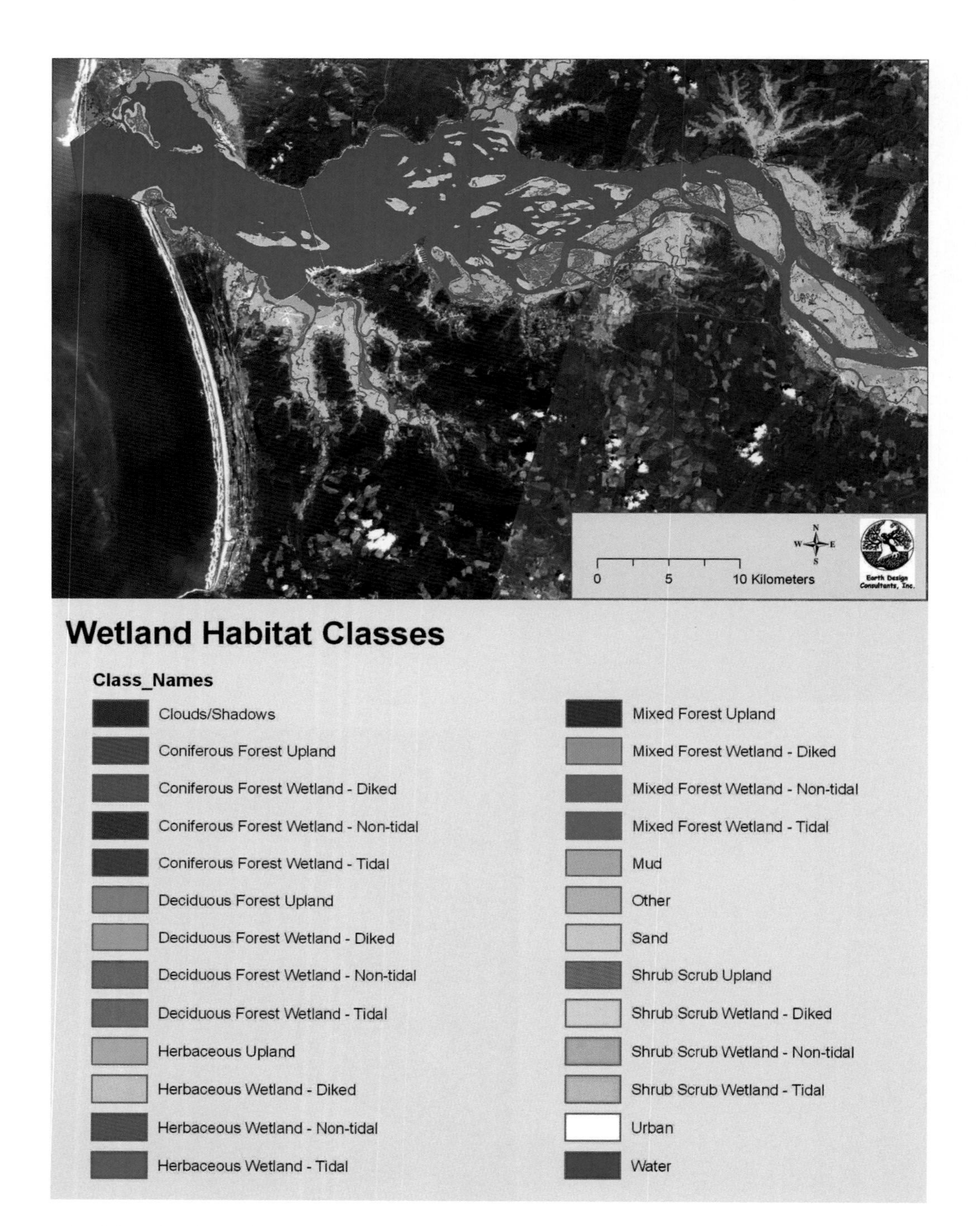

Plate 5. Tidal wetlands. Habitat cover classes in the Columbia River Estuary. This landscape perspective is very useful for prioritizing and planning tidal wetland restoration and conservation efforts across a wide area. (*From Earth Design Consultants*)

(a) (c) (b) (d)

Plate 6. Shrub steppe ecosystems. A sample of native species being restored to the shrub steppe.
(*All photos by Steve Link*)
(*a*) Big sagebrush (*Artemisia tridentata*).
(*b*) Bluebunch wheatgrass (*Pseudoroegneria spicata*).
(*c*) Purple sage (*Salvia dorii*).
(*d*) Pale evening primrose (*Oenothera pallida*).

(e)

(f)

(g)

(h)

PLATE 6. *(continued)*
(e) Indian rice grass (*Achnatherum hymenoides*).
(f) Bitterbrush (*Purshia tridentata*).
(g) Globe mallow (*Sphaeralcea munroana*).
(h) Sandberg's bluegrass (*Poa secunda*).

(a)

(b)

(c)

Plate 7. Ecosystem restoration at Mt. Rainier National Park. (*All photos by National Park Service*)
(a) A Paradise social trail site in 1986 before restoration.
(b) The same social trail site in 1987 after slope stabilization and initial planting.
(c) The same site in 2004, nearly indistinguishable from surrounding areas.

Plate 8. Urban restoration examples.
(*a*) Native wildflower meadow in the Boise, Idaho foothills. (*Photo by Tim Breuer, Ada County Parks*)
(*b*) Garry oak on rock outcrop on Vancouver Island, in the Victoria, British Columbia urban area. (*Photo by Carolyn Macdonald, City of Saanich, British Columbia*)

PLATE 8. *(continued)*
(c) Camas meadow near Portland, Oregon, restored through broom removal. (*Photo by Carolyn Macdonald, City of Saanich, British Columbia*)

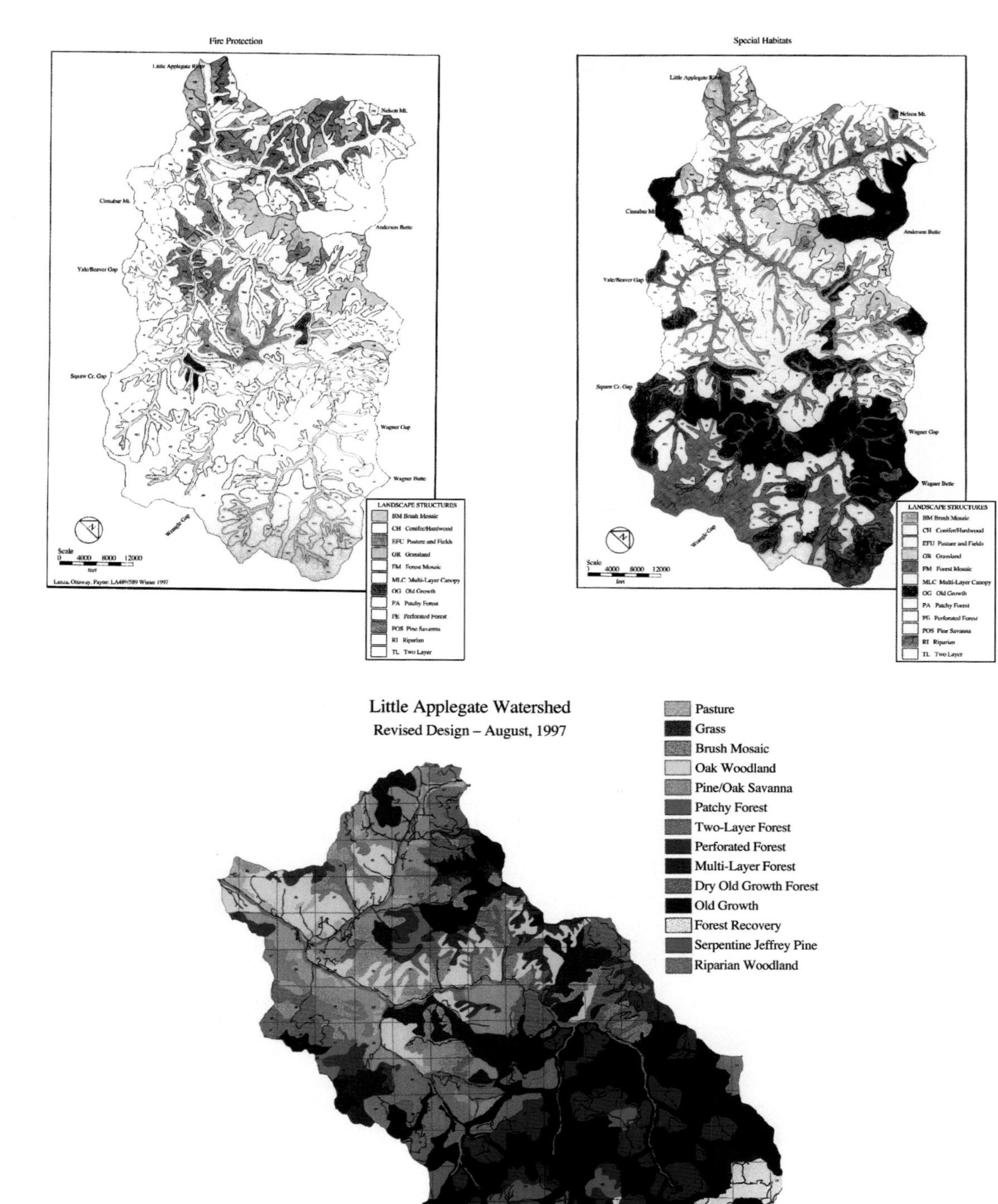

Plate 9. Watershed restoration.
(*a*) Little Applegate Watershed fire restoration emphasis areas. (*Courtesy of the University of Oregon Landscape Architecture Program*)
(*b*) Little Applegate Watershed old-growth restoration and other special habitat emphasis areas. (*Courtesy of the University of Oregon Landscape Architecture Program*)
(*c*) Final watershed design shows distribution of large-scale landscape structure types to guide restoration and continued resource extraction. (*Courtesy of the U.S. Forest Service*)

(a)

(b)

(c)

Plate 10. Wildlife restoration.
(a and b) White-breasted nuthatch (*photo by Bill Hubick*) and western bluebird (*photo by Nigel Milbourne*) both depend on mature oak woodlands and can use nest boxes as interim habitat.
(c) A gray wolf family unit in the Yellowstone area. Reintroduction of wolves has begun to reshape the ecosystem. (*U.S. Fish and Wildlife Service*)

(d)

(f)

(e)

(g)

PLATE 10. *(continued)*
d) Western pond turtle hatching at Fern Ridge in the Willamette Valley.
(e) Great blue heron, a species affected by the recovery of another, the bald eagle. (*Photo courtesy of Defenders of Wildlife*)
(f) Fender's blue butterfly, a species dependent on successful restoration and management of bunchgrass prairies. (*Photo by Alan D. St. John, Defenders of Wildlife*)
(g) Columbia Basin pygmy rabbit.

PLATE 11. Invasive species. Removing Scotch broom to restore oak woodland and savanna at Mt. Tzouhalem on Vancouver Island. (*All photos by Dave Polster*)
(a and b) Before invasive removal, taken in 1992 and 1994, respectively, all too typical of Garry oak ecosystems.
(c and d) After 5 years of broom removal efforts, taken in 1999. Note recovery of native forbs.

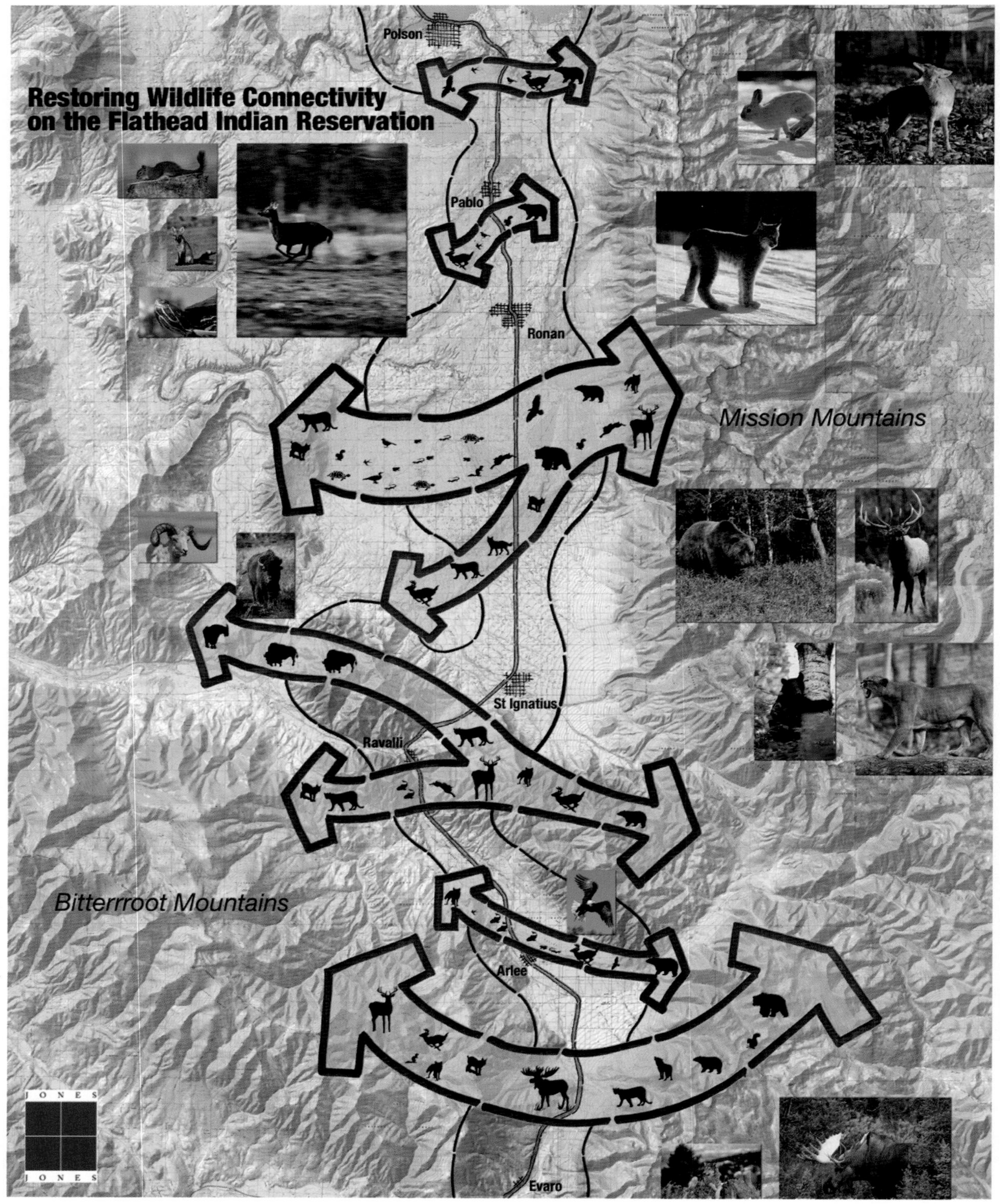

Plate 12. Restoring wildlife connectivity on the Flathead Indian Reservation across a major highway relies on local ecological knowledge. (*Concept graphic by Jones & Jones Architects and Landscape Architects*)

TABLE 11.2.

Typical costs for restoration activities (in 2004 dollars). These are not absolute costs but can be used to produce a rough estimate of project costs.

Item	In-House Costs[a,b]	Contracted Costs[c]
Restoration planning	$25–50/hour	$90–120/hour
Site work, contouring to grade		
With motorized equipment (e.g., front end loader)	NA	$22–35/hour
Hand labor	$10–22/hour	$15–25/hour
Reconstruction of soil bed	$10–22/hour	$15–25/hour
Soil (site-specific mix)	$50–75/cu. yard	$50–75/cu. yard
Nonnative plant control	$14–22/hour	$25–45/hour
Seed collection and cleaning	$10–15/hour	$30–45/hour
Hydroseeding and hydromulching	NA	$400–800/acre
Mulching and placement of erosion control matting	$10–22/hour	$1,000–2,000/acre
Erosion control matting	$116/roll, covers 720 s.f.	$116/roll, covers 720 s.f.
Plant propagation		
4-inch pot	$0.75–1.50/plant	$0.60–1.20/plant
1 gallon	$2–5/plant	$3–7/plant
5 gallon	$5–20/plant	$10–20/plant
5-foot tree	$20–50/plant	$30–120/plant
Plant delivery and installation		
4-inch pot	$0.75–1.50/plant[d]	$0.75–2.00/plant
1 gallon	$2–5/plant	$8–15/plant
5 gallon	$15–40/plant	$30–50/plant
5-foot tree	$25–50/plant	$75–150/plant
Postplanting plant care	$10–22/hour	$15–25/hour
Monitoring	$10–25/hour	$15–30/hour
Mobilization	NA	$500–3,000[e]
Per diem	NA	$50–100/day/person[f]

[a]In-house costs are based on federal government pay scales and do not include administrative overhead because these costs ordinarily are not paid directly from project funds. Costs assume existing program with necessary infrastructure (greenhouse, certifications, training).

[b]Except for plant propagation, costs do not include volunteer time, which can significantly reduce the overall cost of a project.

[c]Contracted costs are based on quoted rates from several contractors and include administrative overhead. Contractor costs vary by region and expertise.

[d]Includes helicopter transport where necessary. Contracted costs do not include helicopter transport.

[e]Mobilization costs vary greatly depending on equipment needs and distance from the project.

[f]Per diem varies greatly depending on location and housing availability.

A cost estimate based on a per-plant or per-area basis should be based on a history of similar projects with good cost tracking. To determine the costs per plant or area at Mt. Rainier National Park, the total cost for restoration related to subalpine meadows for one season was calculated. This was divided by the number of plants produced and planted to provide an estimate of approximately $3 per plant for subalpine meadow restoration. This estimate includes seed collection; plant propagation and growth; transport of plants, materials, and personnel to the site, including some helicopter transport of plants and materials; topsoil meeting park standards; some gravel for site stabilization; erosion control matting; watering; equipment replacement; and limited success monitoring. Volunteer time, which is substantial, is not included in this cost estimate because the volunteer base is reliable year to year. For estimating costs of routine subalpine restoration projects with similar needs, this per-plant figure is used. With larger projects that involve contracting this figure is used for an initial estimate to determine initial funding allocations, but an itemized cost estimate is developed for the final project.

Itemized restoration cost estimating is most beneficial for projects that include a lot of site-specific earthwork, plants needing long growing and holding times in the nursery, and contracted work. In contracting, mobilization costs vary greatly depending on the location of the contractor, location of soil and gravel materials, and location of plant materials. Additionally, contractors generally receive per diem costs. Table 11.2 lists typical restoration activities and estimates of costs in 2004 for work by a contractor and work conducted with existing staff.

Case Study: Restoration of a Small Impact: Paradise Social Trail

Paradise is a heavily used subalpine meadow in Mt. Rainier National Park. Each year approximately 2 million people visit Mt. Rainier, and most travel to Paradise to see the beautiful subalpine meadows that are easily accessed from the main park road. The Paradise Meadows extend from about the forest line at 5,400 feet (1,646 meters) up to the tree line at about 7,000 feet (2,134 meters) and encompass about 1,000 acres (416 hectares) of subalpine parkland. The area has been heavily used for more than 100 years, and by the mid-1980s, more than 900 social trails were documented in the Paradise area, in addition to the extensive, officially maintained trail system (Rochefort 1989). In 1986, Mt. Rainier National Park initiated a comprehensive program to document all impacts, upgrade the trail system, and restore unacceptable impacts to subalpine meadows. Because recreational use in the Paradise meadow was limited to day use along trails, most impacts were linear impacts or social trails. Social trails ranged from short trails with bare, compacted surfaces to long trails with erosion up to 3 feet (1 meter) deep. This case study presents the work conducted along one social trail in the Golden Gate area of Paradise.

Restoration Site and Goals

The Golden Gate trail climbs from about 6,000 feet (1,829 meters) to 7,000 feet (2,134 meters) through a heath–shrub, lush herbaceous meadow mosaic with about five switchbacks. This trail has been used since the early 1900s, and in this time sixteen shortcuts have developed in the lower part of the trail. We will present one of these switchbacks (GG12) as a typical example of restoration on a social trail. In 1986, the social trail was mapped and measured before initiation of restoration. At the time, the social trail was 116 feet (35.5 meters) long, 5.6 feet (1.7 meters) wide, and 1.2 feet (0.37 meter) deep. It extended between two sections of the switchbacked trail and was on a slope of 40° through a heath–shrub meadow. Dominant species in the area were mountain huckleberry (*Vaccinium deliciosum*) and red mountain heath, with lesser amounts of Cascades aster (*Aster ledophyllous*), alpine aster, western pasqueflower (*Anemone occidentalis*), sedges, and Cusick's speedwell (*Veronica cusickii*).

The goal for this area was to bring the trail up to the grade of adjacent areas, restore surface flow of water (it had become a channel for water draining from the trail), and restore the subalpine meadow. In 1986, the Mt. Rainier National Park greenhouse was not growing heather or huckle-

TABLE 11.3.

Comparison of plant cover and species diversity on a restored social trail and the adjacent, undamaged meadow.

	Percentage Plant Cover			Number of Species		
Site	1993	1998	2005	1993	1998	2005
Trail	16	27	42	10	16	14
Adjacent meadow	82	Unknown	72	16	Unknown	20

berry species in large quantities, so it was decided that revegetation would make use of seeds and greenhouse plugs of less common species in the surrounding communities, including broadleaf lupine, sedges, cinquefoil, aster, luetkea, and pasqueflower in order to restore native plant cover and impede erosion.

Restoration Methods

The first step in the process was to scarify the compact ground surface, then stabilize the site with subsurface erosion bars. Ninety wooden bars, 6 to 8 inches thick, were installed below the ground surface. After the bars were installed, 3 cubic yards of rock from an adjacent talus slope was placed on the scarified base, and the site was filled with a sterile soil mix (14 cubic yards). Finally, the site was planted, seeded, and covered with an excelsior mat to impede erosion and moderate soil surface temperatures and moistures. Planting and seeding continued over 3 years and involved 268 plants (ten species) and about 2 gallons of seeds (four species).

Results

Today the site is well on its way to recovery, and bare slopes are less apparent than in 1986. Geomorphically, the site is at the grade of the adjacent meadow, and water flow seems to be dispersed across the surface; no channels are visible on the site. Vegetatively, the site is dominated by minor components of the adjacent heath–shrub meadow, but there is some evidence of vegetative spread of heather and huckleberry along edges of the impact (Table 11.3). However, there is also evidence of recent use (footprints) that indicates that restoration success may decrease if the site is not protected more aggressively. During the first few years of restoration, the site was roped off to discourage cross-country travel; however, over the last few years protection efforts have been less diligent, and footprints were visible in bare areas during our monitoring in 2004. Although plant establishment has been successful, high mortality of plants on several nearby restoration sites provides evidence that restoration success relies as much on protection as on revegetation.

Case Study: Sunrise Campground, Mt. Rainier National Park

The Sunrise Campground was constructed in a subalpine meadow in Mt. Rainier National Park in the early 1930s. It was built in a valley adjacent to Yakima Park and Burroughs Mountain and provided beautiful views and easy hiking access to the subalpine parkland and alpine tundra. By the late 1960s, park managers realized that the beautiful setting of the campground was very sensitive to human use, and the popularity of the area was resulting in severe damage to natural resources (Figure 11.4). Curious campers, adventurous hikers, and vehicles wandered off the constructed roads and trails and caused numerous impacts and soil erosion in the valley adjacent to the campground. In 1973 the camp was closed to cars, and gradually camping was reduced to a small walk-in area. In the late 1970s some road surfaces were removed, but restoration funding was not available for large-scale rehabilitation. By 1980, plants were reestablishing only on cut-and-fill slopes along roads where soils were not compacted and seeds could germinate and survive. In 1996, Mt. Rainier National Park staff concluded that active restoration was the only way to restore the subalpine meadow ecosystem.

The campground is located on the east side of the park at about 6,400 feet (1,951 meters) elevation. A mosaic of tree clumps and herbaceous meadows characterizes vegetation of the area. Tree islands are dominated by subalpine fir, Engelmann spruce, and whitebark pine. The subalpine meadow is composed of green fescue, lupine, paintbrush (*Castilleja* spp.), and several asters. Vegetation in this area is sensitive to human use because of the short growing season and low resilience of plants to trampling. Typically the snow-free growing season extends for only 8–12 weeks in summer. Additionally, obscure Indian paintbrush (*Castilleja cryptantha*), an endemic plant, grows in wet meadows in the campground loops. This species is listed as sensitive on both the federal and

Figure 11.4. Sunrise Campground in 1931 illustrates intensity of recreation use beginning many decades ago. (*Photo by National Park Service*)

Washington State endangered species lists. It occurs only on twenty-five sites in Mt. Rainier National Park and two sites in the Wenatchee National Forest.

The former campground had four auto camp loops dispersed over about 50 acres. Restoration challenges included roadbeds, culverts, and sixty social trails. Although some natural revegetation had occurred over the preceding 25 years, it was limited to fill slopes along road edges. Based on the observed rate of plant succession on this site and other disturbed areas in the park's subalpine zone, park staff did not expect this area to revegetate naturally within the next 200 years. Even where plants had established naturally along the road corridor, the cut slope was still easily recognized by the altered topography and hydrologic patterns, resulting in different vegetation communities. In order to restore this site to subalpine parkland, it was necessary to restore the original land contours and hydrologic patterns, followed by revegetation of the disturbed areas. The plan included retention of a ten-site walk-in camp, with a new 1-mile access trail built in an existing roadbed.

In the 1980s, several restoration proposals were created, but it was not until 1996 that three partners came together to make the project feasible. In 1996, park personnel conducted surveys to quantify restoration needs and began meetings with the Washington National Guard 898th Engineer Battalion. Field surveys documented that 8,850 linear feet of roadway and approximately 10 acres of disturbed land would need to be treated. The Washington National Guard expressed an interest in using the road removal and recontouring as a training project as long as the park provided meals for the troops. Although personnel from Mt. Rainier National Park had previous experience in restoring small recreational impacts such as social trails and campsites (Rochefort and Gibbons 1992), complete restoration of large road surfaces had not been done. The park developed a funding proposal to the National Park Foundation. In January 1997, the foundation and the Canon Expedition into the Parks Program awarded a grant of $50,000. Work began immediately with strategic planning, development of interpretive materials, and initiation of restoration on the ground in July 1997.

Goals and Objectives

The broad goals of this project were to reestablish the natural subalpine ecosystem, protect cultural resources, interpret the project for the public, and retain a small walk-in campground to facilitate overnight recreation. More specific objectives included restoring the original contours or landforms, recreating natural hydrology, and using native plant communities. Cultural components included stabilizing and preserving a historic building and conducting surveys for archeological sites or artifacts. Interpretation included creating a brochure and videotape documenting the project. Interpretation was an essential component to help ensure future protection of the restored landscape.

Process

The process included planning, historic building stabilization, plant salvage, physical restoration, trail construction, revegetation, monitoring, and interpretation. Planning began in July 1996 concurrent with seed collection for future greenhouse propagation. In the fall and winter of 1996–1997, plant propagation was initiated and planning completed, with compliance documents completed by May 1997. Adjustments to planning details occurred during implementation and annual reviews. Plants were salvaged from the cut-and-fill slopes and stored in temporary onsite beds in July 1997, before road removal and recontouring. Once the roads were recontoured, the trail crew began building a 1-mile trail to the campground on a former roadbed. Once the physical restoration was complete, revegetation began.

Methods of Physical Restoration

Physical treatments were designed to correct the effects of road-related disturbance and included outsloping, decompaction and shaping, and road and stream crossing culvert excavation and removal. Supportive actions included scarifying, grading, or loading and transporting road surface materials to a specified fill site; loading and transporting portions of existing debris piles (i.e., soil and rock); and transporting and stockpiling old water line and other utilities for off-site disposal. All materials stored at fill site locations were buried under at least 2 feet of clean fill and appropriate cover material.

Outsloping moved sidecast fill from the downslope edge of roads up against cutbanks (Figure 11.5a). All areas where fill was placed were decompacted, or ripped to a 2-foot depth first. The finished surface was shaped to blend with the surrounding topography. Native topsoil was stockpiled and placed over the surface during final shaping.

Decompaction was also used on flat areas, such as road surfaces. Many of the road surfaces were originally covered with oiled gravel, 750 cubic yards of which were removed and used elsewhere as fill. Gravel and soil were tested to ensure that no contaminants were present. A sufficient number of passes were made to attain a maximum spacing of 2 feet between adjacent ripper shank paths. The decompacted surface was then reshaped, where necessary, to blend with the surrounding topography.

Road and stream crossing excavations involved the removal and disposal of fill, seven culverts, and other debris from five stream channels, followed by shaping of the completed excavation to blend with the surrounding land (Figure 11.5b). In most cases, the finished product closely mimicked the original (pre–road construction) stream channel and side bank configuration. All fill was stored in local areas (outsloped) or used as cover material at fill sites.

Work on all treatments was closely supervised to detect changes in the character of subsurface materials, with adjustments to specifications of the treatment to maximize physical treatment effectiveness. Identified native topsoils were selectively handled to ensure that they were placed on the surface during final shaping.

Revegetation

We developed our revegetation strategies by first examining the plant communities and landforms in the project area, identifying reference communities for each section of the disturbed area. Many of the cut-and-fill slopes were naturally recovering and were partially covered with native fescue and lupine communities or had early successional species such as Newberry's fleeceflower (*Polygonum*

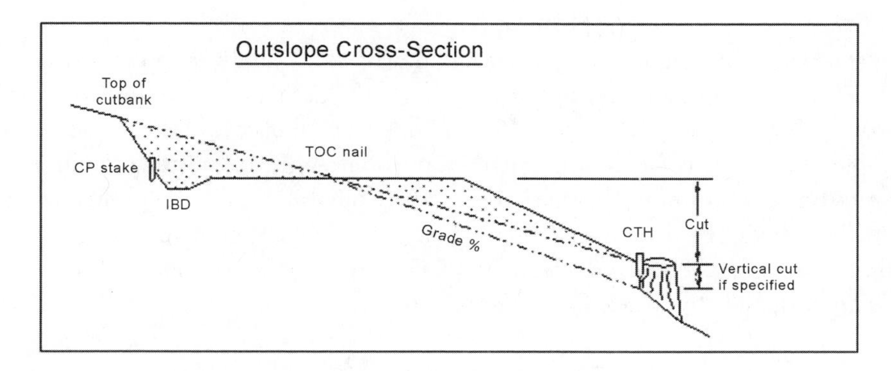

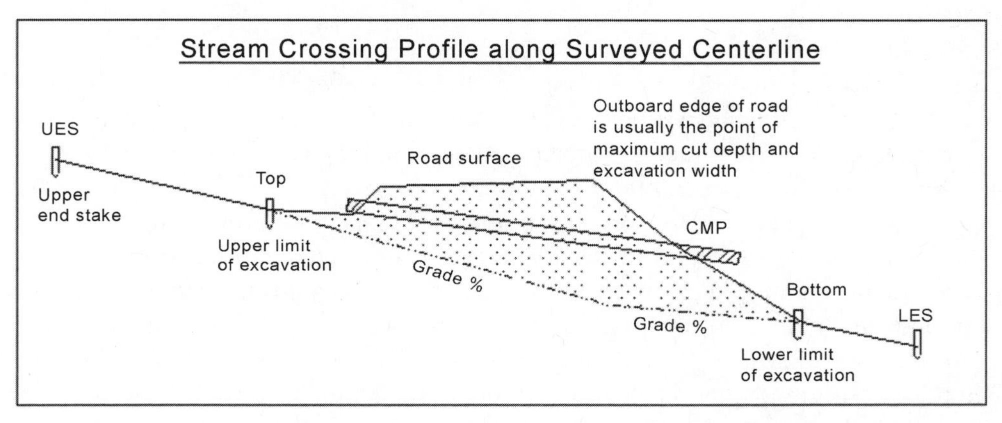

FIGURE 11.5. *(a)* Cross-section showing a typical outslope trail design. *(b)* Profile (along centerline) showing typical road and stream crossing treatment.

newberryi), *Aster ledophyllous*, or *Veronica cusickii*. A decision was made to supplementally seed these sites to assist natural reestablishment on these slopes. Much of the original road surface was on areas with a slight slope that historically would have been covered with one of three plant communities: dry grass (fescue)–lupine, heath–shrub, or low herbaceous (sedges and rushes). Despite decompaction and reshaping, we believed the soil was more compact than on the cut-and-fill slopes, and active planting and seeding were needed. We started closest to the new trail that led to the walk-in campground because this area had the highest risk of being trampled. Depending on the microtopography, these areas were planted with the dry grass or low herbaceous community. Heath–shrub communities were in areas farther from visitor use and were planted because of their very slow natural recruitment. Revegetation on these areas was delayed until after the areas closest to visitor access were completed.

After the contouring was complete, we collected soil samples in the three plant communities (dry grass, heath–shrub, and low herbaceous) to determine what was in the seedbank. We found only seven species at very low densities. *Juncus* species were found in all plant communities and at the highest volume. Other species were uncommon and included showy sedge, green fescue, fireweed

(*Epilobium* spp.), fanleaf cinquefoil, and two unknowns. This analysis confirmed that we would need to rely on plant propagation and direct seeding to establish cover within the first few years. Over the course of this project, propagation has included a total of seventeen species, although only seven to thirteen species have been propagated in any one year. The volume of plant materials collected depends on the community being restored, seed availability, and greenhouse staffing (volunteers or paid employees). Plants are seeded and planted in September, before the fall rains or snow. Showy and black sedge, green fescue, fleabane (*Erigeron peregrinus*), cinquefoil, wooly everlasting (*Antennaria lanata*), and alpine aster have been the most common species propagated (Table 11.4).

Monitoring Results

Surveys indicated that initial survival of planted material exceeded 90%, but there was uncertainty with regard to whether function was being restored. In order to quantitatively evaluate restoration of function and structure, we recruited a graduate student, Kim Frappier, from the University of Washington, who conducted her master's thesis on the site.

Frappier's study examined plant community and soil characteristics in the restoration area and compared them with nearby undisturbed subalpine meadows. Qualitative and quantitative monitoring is considered an integral component of the restoration program at Mt. Rainier National Park. However, only qualitative assessments of the Sunrise restoration had been conducted, focused on seeding success, seedling survival, and mulch condition. The objective of Frappier's study was to evaluate the Sunrise restoration ecologically through a quantitative approach by comparing soil chemical, physical, and biological properties between restoration sites of different ages; comparing vegetative cover and species composition in the restoration sites and adjacent undisturbed meadows; and evaluating the productivity of greenhouse-grown green fescue plugs in restoration sites.

Study Methods

Portions of road corridor were revegetated each year between 1997 and 2001, resulting in six separate sections. A chronosequence based on the time since planting was used to analyze change in plant and soil properties over time. A similar approach was used by Fyles et al. (1985) in their study of vegetation and soil development on coal mine spoil in the Canadian Rockies. In a chronosequence, sites of different ages are assumed to represent points in time in the development of an individual site. Because a longitudinal study was not possible, the chronosequence allows a space for time substitution.

In the summer of 2002, 100 sampling points in the restoration area and 30 sampling points in the reference area were established using a simple random sampling method. Total percentage cover and cover of individual species were collected using a variation of the cover class systems outlined by Elzinga et al. (1998). Aboveground biomass of green fescue was estimated using regression analysis. This reduced the number of plants harvested from the restoration sites.

TABLE 11.4.

Plant materials and labor used to restore Sunrise Campground.

Year	Number of Plants	Volume of Seeds (mixed species)	Paid Work Hours[a]	Volunteer Hours[a]
1997	17,112	119 cups (1,342 square feet)	2,216	2,269
1998	28,290 (5,383 square feet)	133 cups (1,967 square feet)	959	1,579
1999	27,077	11 cups (100 square feet)	1,439	931
2000	41,076	235 cups (3,587 square feet)	2,200	977
2001	21,582 (16,504 square feet)	114 cups (928 square feet)	2,097	952
2002	16,002 (4,825 square feet)	47 cups (786 square feet)	713	603
2003	12,219 (3,542 square feet)	100.5 cups (1,251 square feet)	1,520	2,911
2004	26,403 (5,705 square feet)	4 cups (44 square feet)	721	735

[a]Hours do not include greenhouse propagation, which ranged from 1,280 hours in 1997 to 2,480 hours in 2004.

Soil samples were collected in August and September 2002 from selected planted, unplanted, and reference sites. All samples were collected from the top 15 centimeters of the soil surface using a trowel and bulb planter. Samples were taken from the rooting zone of plant clumps. In cases where there was no vegetation, soil was collected to the same depth (15 centimeters) as samples beneath plants and included in analysis. Analysis conducted on these samples included soil moisture, carbon/nitrogen ratio, percentage carbon and nitrogen, microbial biomass and activity, pH, and available nitrogen.

Research Results

Analysis showed an overall increase in plant cover as the age of the planted site increased. There was an 18% increase in plant cover from the 2-year-old site to the 6-year-old site. Only a 5% difference was observed in cover between the 6-year-old restoration site and the reference site, indicating that plant cover target conditions had nearly been achieved. Ordination analysis confirmed the linear relationship between the age of the restoration site and plant cover. Despite the progress in plant cover, the 6-year-old site lacked the species diversity of the reference sites and continued to look like a planted landscape, as opposed to a natural one. Overall, a greater number of species was found on reference sites (twenty-six species) than on the five replanted sections, which had an average of twelve species. Green fescue cover and biomass in grams per square meter increased as the age of the restoration site increased. Green fescue cover was zero on the unplanted site, reflecting a lack of natural recruitment. The vegetation survey did not produce evidence of natural recruitment into revegetated sites except for two species: Newberry fleeceflower (*Polygonum newberryi*) and subalpine fir. These were the only two natural recruiters found on the unplanted site as well. This has important implications for restoration strategies, particularly when the restoration objective is to restore species composition and cover as quickly as possible.

Soil variables showed wide variation. Overall, carbon/nitrogen ratios, soil pH, and microbial biomass were similar across both reference and revegetated sites. The largest difference in carbon/nitrogen ratios was found between the unplanted restoration section (14 ± 0.87) and the reference sites (17 ± 1.62). Increases were observed for the total percentage of carbon in restoration sites (1.9% to 2.6%) and between unplanted (1.6%) and reference sites (3.9%). Soil moisture increased from the unplanted site (8.2%) to the 6-year-old site (13%). However, this pattern did not hold true on the reference sites, which equaled that found on the unplanted site. Available and total nitrogen were lowest in the unplanted site (net = 0.0013 mg/g, total = 1.13 mg/g), compared with the 6-year-old site (net = 0.0023 mg/g, total = 1.56 mg/g), and highest in the reference sites (net = .0093 mg/g, total = 2.25 mg/g).

The sparse cover on the unplanted restoration site and the low percentage of carbon and low levels of total and available nitrogen confirmed that revegetation efforts speed the recovery of disturbed subalpine plant communities, not simply in terms of vegetation cover but in terms of soil processes as well. Overall, results confirm that the restoration sites at Sunrise are on a trajectory toward recovery (Figure 11.6). The increase in vegetation cover and green fescue biomass, coupled with the increase in percentage of carbon, soil moisture, and total and available nitrogen on revegetated sites, reflects continued growth of greenhouse plants, litter accumulation, and the increase in soil nutrients.

Future Research and Restoration

To elucidate the viability of the plant community on the Sunrise restoration, a long-term demographic monitoring project similar to those conducted by Urbanska (1994) on restoration plots in the Swiss Alps would be useful. Demographic monitoring examines the age-state structure of plants, including flower and seed production. This coupled with further soil analysis would assist in analysis of the sustainability and trajectory of the restoration sites.

In the harsh subalpine ecosystem, survival and continued growth of greenhouse grown plants are an important first step to restoration success. We plan to continue revegetation of the site with greenhouse plants and direct seeding for at least the next decade. We also plan to continue to monitor the site to assess plant cover and soil ecological processes, refine current revegetation prescriptions, and evaluate the trajectory and success of the Sunrise restoration. See Plate 7 in the color insert for more subalpine restoration examples.

Figure 11.6. Road at Sunrise area *(top)* before, *(middle)* during, and *(bottom)* after restoration and conversion to a trail. (*National Park Service*)

Case Study: Whitebark Pine Restoration in Glacier National Park

Glacier National Park is situated in an area known as the Crown of the Continent. It is the meeting ground for four floristic provinces (Cordilleran, Boreal, Arctic-Alpine, and Great Plains) and is known for having all of its native predators intact. Unfortunately, three of these predators, the grizzly bear (*Ursus arctos horribilis*), gray wolf (*Canis lupus*), and Canada lynx (*Lynx canadensis*), are federally listed as threatened. A critical resource for the grizzly bear, the whitebark pine ecosystem, is seriously threatened in the park.

Historically, whitebark pine communities were significant components on 15–20% of forested lands in Glacier National Park, occurring on 75,000–100,000 acres. Whitebark pine is considered a keystone species because a wide variety of wildlife depend on the high-fat seeds for food, including not only the grizzly bear but also red squirrels (*Tamiasciurus hudsonicus*), Clark's nutcrackers (*Nucifraga columbiana*), and many others. The pines influence vegetation patterns across the landscape by pioneering dry, cold, exposed sites where no other tree species can establish. Their penchant for growing on windy ridges allows them to accumulate and retain snow, thus lengthening the snowmelt period. They also contribute high aesthetic value as their unique growth form characterizes the subalpine landscape. Whitebark pine stands also are important aspects of cultural and spiritual importance to neighboring tribal members.

Fire exclusion has hampered regeneration, and an exotic fungus, white pine blister rust (*Cronartium ribicola*), has decimated whitebark pine stands in Glacier National Park. Nearly half of all the trees have died. Of the remaining live trees, more than 75% are lethally infected and probably will die within a few decades (Figure 11.7). A quarter of the cone-bearing crowns on these trees have already died (Kendall and Keane 2001). White pine blister rust is a widespread problem in the Northwest and continues to spread. Stands surveyed in Washington, northern Idaho, northwest Montana, southern Alberta, and British Columbia were found to have 50–100% infection rates in live trees, with 40–100% of trees in each stand already dead (Kendall and Keane 2001).

A thorough inventory of whitebark pine in Glacier National Park was conducted in 1995–1997, resulting in a digitized map of stands and documented mortality rates and causes (Kendall et al. 1996). Based on these data, scientists and park managers agreed that whitebark pine will be functionally lost in the absence of active intervention. As a result, a whitebark pine and limber pine (*Pinus flexilis*, also affected by blister rust) restoration program was initiated in 1998. This case study focuses only on the whitebark pine.

Objectives

The primary goal was to maintain whitebark pine populations by planting seedlings that show genetic resistance to blister rust. Project components include the following:

- Forging cooperative agreements with neighboring agencies
- Conducting surveys to identify healthy trees in blister rust–decimated stands and collecting seeds from them
- Developing propagation protocols, using seed to grow seedlings

FIGURE 11.7. Whitebark pine dying among subalpine fir in Preston Park. (*National Park Service photo*)

- Identifying and selecting appropriate habitats to plant trees
- Monitoring to determine successes, failures, and needed improvements

Methods

In spring and early summer, trail crews and other field personnel were recruited to keep watch for and report locations of cone-producing trees. Primary collection sites identified included Preston Park, Oldman Lake, Bighorn Basin, and Siyeh Bend, all areas that had been hard hit by white pine blister rust. Pending genetic rust resistance studies being conducted by the U.S. Forest Service research station in Moscow, Idaho, the accepted approach is to assume that healthy cone-producing trees in these stands are likely to have some degree of natural rust resistance (Hoff et al. 2001).

In mid to late July, quarter-inch wire mesh cages were packed into the backcountry, trees climbed, and cones caged. Field technicians were trained in tree-climbing methods and used ropes and harnesses for safety. A contracted tree climber was also employed. In a few instances, cages broke or were blown off, but most stayed secure. Upon return in late September to collect cones, no seeds were found outside cages. They had all apparently been harvested by Clark's nutcrackers and red squirrels. In some areas animals had gotten inside the cages and raided the cones, highlighting the need for secure fastening.

Most seed was sent to the Forest Service's Coeur d'Alene Nursery to be raised as seedlings. A small percentage was kept to be raised in the park nursery along with cooperators in greenhouses at the Columbia Falls High School and the Blackfeet Community College. Because of its excellent facilities, the Coeur d'Alene Nursery was able to produce better stock than the park could on site. (Specific propagation protocols for whitebark pine can be obtained from the Coeur d'Alene Nursery.) If requests are submitted by the fall before sowing, the stock can be ready for planting within a year and a half after germination. In addition, resources were shared with the Waterton Lakes National Park across the border in Canada, where direct seeding field trials had been initiated recently.

Recent burn perimeters were overlain with a geographic information system layer of potential and existing whitebark pine habitat to aid selection of planting locations (Peterson 1999). Recent natural fires that burned through existing whitebark pine stands provide ideal planting conditions. Remaining black islands were selected for planting, and trees were placed near stumps or snags for shading. The trees were watered with a diluted B1 transplant solution and mulched with nearby light-colored duff. Trees targeted for future monitoring were planted and mapped in a series of circular plots, at various distances and compass bearings from a central point recorded using a global positioning system. Monitoring plots were installed to track about 15% of the planted trees.

In July 2001, 1,500 trees were planted on West Flattop Mountain in the 1998 Kootenai Complex burn area, and more than 1,400 were planted the next September with the help of Montana Conservation Corps volunteers. Because of the unpredictability of summer moisture, all subsequent plantings have been carried out in September. More than 2,000 additional whitebark were planted in the same area in 2002. In a couple of instances, groups of about 100 trees were flown to sites immediately after fires during resource monitoring. There has not been an opportunity to monitor survival at these sites because of their remote locations. In addition, more than 600 trees were shared cooperatively with the Blackfeet Indians, as students planted them on the reservation.

Results

Between 1997 and 2000, more than 22,500 whitebark pine seeds were collected. Germination results varied each year. The biggest cone crop was in 1998, with the percentage of fill ranging from 72% to 96%. In 1998, the seed averaged 67% germination (with individual tree lots as high as 81%). In contrast, in 1999 whitebark cones were very late in maturing, most of the seed molded in stratification, and there was almost no germination. It is likely that the seed never fully matured that year. It was encouraging to discover that 86% germination could be obtained from a lot of two-and-a-half-year-old seed. At least 75% of the germinants reached seedling stage.

Seedling Survival

Seedlings can be overwintered if they cannot be planted during the summer they are delivered, but overwintering is risky. High mortality was experienced in overwintered stock as a result of rodent predation and possible overheating of the conifers under protective foam and plastic used to cover stock. Use of rodenticides and discontinued use of white plastic cover has been successful in reducing overwinter mortality. Seedlings should not be held over for more than 1 year without repotting.

Seedlings planted in July 2001 were monitored for survival in September. There was a rainstorm shortly after the trees were planted, but the remainder of the summer was extremely dry. About one third succumbed to drought, a surprisingly good result considering the lack of precipitation in the preceding 2 months. The 2001 spring and fall plantings were surveyed in 2002 and showed a 45% survival rate after the first year. Trees were not surveyed in 2003–2004, but year 4 survival will be monitored in 2005.

Budget

From 1998 to 2003, more than 6,000 whitebark pine trees were planted. Although thousands of seeds can be collected during a good cone crop year, it was necessary to stagger the sowing of the seeds so as not to produce more material than could be handled for outplanting in one season. Approximately one third more seed was sown than the target for seedling production.

The overall costs were nearly $6 per tree planted. One third was in propagation, one sixth in planting, another sixth in transportation, and the rest in various categories. The cost of planning logistics, scouting, caging cones, collecting seed, processing cones, establishing monitoring plots, monitoring results, entering related data, and preparing summary reports added another $2 per tree. This does not include project supervision and oversight or preparation of funding proposals. Costs can vary, but these figures may provide a rough guide to others considering similar projects. Because most federal funding comes in 2- to 3-year blocks, the project was separated into phases, first for development of seed collection and propagation strategies and next for development of planting and overall restoration strategies.

Conclusions

Objectives were met through cooperative efforts between the Park Service, the U.S. Forest Service, the Blackfeet Tribe, and Waterton Lakes National Park. All worked together to achieve the common restoration goal for whitebark pine ecosystems. A number of healthy, cone-producing trees (sometimes called "plus" trees) have been located in heavily hit blister rust stands. Successful seed collection and propagation methods were pioneered on this project that can now be used elsewhere.

Fall planting, timed shortly before seasonal moisture arrives, is more successful than spring planting, when summer precipitation cannot be predicted. Fall moisture is more reliable. The young trees go into dormancy shortly after planting, then take advantage of spring moisture to give them a boost for the next season. Microsite planting is extremely important. The seedlings must have some shade, whether it is from a stump, a rock, or a snag, or they are likely to succumb to desiccation. Recently burned areas were ideal as planting locations. We continue to research additional criteria for selection of additional planting locations.

Monitoring is important for learning what strategies do or do not work well and determining the ultimate success of the program (Figure 11.8). Drought is a great inhibitor of survival, and the ideal time to complete restoration would be during a wetter cycle, which unfortunately is very difficult to predict when one is requesting funding for multiyear projects. Supplemental watering would be helpful but is not feasible in remote and widely scattered planting locations.

We have concluded that an active restoration program is critical for the long-term survival of whitebark pine in Glacier National Park. With modifications based on lessons learned, we hope that a long-term restoration program for whitebark pine will be established.

FIGURE 11.8. *Left*, Whitebark pine blister rust survey transect near Red Gap Pass trail, Glacier National Park. *Right*, Whitebark pine blister rust survey transect southeast of Numa Lookout, Glacier National Park. (*National Park Service photos*)

References

Agee, J. K. and L. Smith. 1984. Subalpine tree establishment after fire in the Olympic Mountains, Washington. *Ecology* 65(3): 810–819.

Bergland, E. O. 1984. *Olympic National Park Archeological Basemap Study: Summary Report to Superintendent Chandler*. Division of Cultural Resources, Pacific Northwest Region, National Park Service, Seattle.

Billings, W. D. 1973. Arctic and alpine vegetations: similarities, differences, and susceptibility to disturbances. *BioScience* 23(12): 697–704.

Billings, W. D. and L. C. Bliss. 1959. An alpine snowbank environment and its effect on vegetation, plant development, and productivity. *Ecology* 40: 388–397.

Billings, W. D. and H. A. Mooney. 1968. The ecology of arctic and alpine plants. *Biological Reviews* 43: 481–529.

Brink, V. C. 1959. A directional change in the subalpine forest–heath ecotone in Garibaldi Park, British Columbia. *Ecology* 40(1): 10–16.

Buol, S. L., F. D. Hole, and R. J. McCracken. 1980. *Soil Genesis and Classification*. Iowa State University Press, Ames.

Burtchard, G. C. 1998. *Environment, Prehistory & Archaeology of Mount Rainier National Park, Washington*. Unpublished report prepared for U.S. Department of Interior, National Park Service, Mt. Rainier National Park, and International Archaeological Research Institute.

Campbell, E. M. and J. A. Antos. 2000. Distribution and severity of white pine blister rust and mountain pine beetle on whitebark pine in British Columbia. *Canadian Journal of Forest Research* 30: 1051–1059.

Campbell, P. R. 1996. *Population Projections for States by Age, Sex, Race, and Hispanic Origin: 1995 to 2025*. PPL-47. Population Projections Branch, Population Division, U.S. Bureau of the Census, Washington, DC.

Canaday, B. B. and R. W. Fonda. 1974. The influence of subalpine snowbanks on vegetation pattern, production, and phenology. *Bulletin of the Torrey Botanical Club* 101: 340–350.

Catton, T. 1996. *Wonderland: An Administrative History of Mount Rainier National Park*. National Park Service, Seattle.

Choquette, W. T. 1981. The role of lithic raw material studies in Kootenay archaeology. *BC Studies* 48: 21–36.

Clague, J. J. and R. W. Mathewes. 1989. Early Holocene thermal maximum in western North America: new evidence from Castle Peak, British Columbia. *Geology* 17: 277–280.

Clague, J. J., R. W. Mathewes, W. M. Buhay, and T. R. Edwards. 1992. Early Holocene climate at Castle Peak, south Coast Mountains, British Columbia, Canada. *Palaeogeography, Palaeoclimatology, Palaeoecology* 95: 153–167.

Douglas, G. W. 1970. *A Vegetation Study in the Subalpine Zone of the Western North Cascades, Washington*. M.S. thesis, University of Washington, Seattle.

Douglas, G. W. and L. C. Bliss. 1977. Alpine and high subalpine plant communities of the north Cascades region, Washington and British Columbia. *Ecological Monographs* 47(2): 113–150.

Dunwiddie, P. W. 1977. Recent tree invasion of subalpine meadows by trees in the Wind River Mountains, Wyoming. *Arctic and Alpine Research* 9: 393–399.

Ecological Stratification Working Group. 1996. *A National Ecological Framework for Canada*. Agriculture and Agri-Food Canada, Research Branch, Centre for Land and Biological Resources Research and Environment Canada, State of Environment Directorate, Ottawa. Available at www.ec.gc.ca/soer-ree/English/Framework/framework.cfm.

Edwards, O. M. 1980. *The Alpine Vegetation of Mount Rainier National Park: Structure, Development, and Constraints*. Ph.D. dissertation, University of Washington, Seattle.

Elzinga, C. L., D. W. Salzer, and J. W. Willoughby. 1998. *Measuring & Monitoring Plant Populations*. BLM Technical Reference 1730-1.

Eng, M. 2003. *Large Scale Biogeoclimatic Mapping*. Available at www.for.gov.bc.ca/hre/becweb/subsite-map/largescale-01.htm.

Expert Committee on Soil Survey. 1987. *The Canadian System of Soil Classification*. 2nd ed. Agriculture Canada Publication 1646, Ottawa.

Franck, I. C. 2000. *An Archaeological Investigation of the Galene Lakes Area in the Skagit Range of the North Cascade Mountains, Skagit Valley Park, British Columbia*. Unpublished manuscript, Department of Archaeology, Simon Fraser University, Burnaby, BC.

Franck, I. C. 2003. *In the Shadow of Hozameen: An Archaeological Inventory of the Skyline Trail System in Manning and Skagit Valley Provincial Parks, Southwest B.C., Project Year Number One (2002)*. Prepared for Skagit Environmental Endowment Commission.

Franklin, J. F. and C. T. Dyrness. 1987. *Natural Vegetation of Oregon and Washington*. Oregon State University Press, Corvallis.

Fulkerson, C. 1988. *Predictive Locational Modeling of Aboriginal Sites in the Methow River Area, North-Central Washington*. Report submitted to the Okanogan National Forest, Purchase Order No. 00-05H7-7-2 10, Okanogan, WA.

Fyles, J. W., I. H. Fyles, and M. A. H. Bell. 1985. Vegetation and soil development on reclaimed mine land at high elevation in the Canadian Rockies. *Journal of Applied Ecology* 22(1): 239–248.

Galbraith, W. A. and E. W. Anderson. 1971. Grazing history of the Northwest. *Journal of Range Management* 24(1): 6–12.

Gavin, D. G. and L. B. Brubaker. 1999. A 6,000-year soil pollen record of subalpine meadow vegetation in the Olympic Mountains, Washington, USA. *Journal of Ecology* 87: 106–122.

Gehring, J. L. and Y. B. Linhart. 1992. Population structure and genetic differentiation in native and introduced populations of *Deschampsia caespitosa* (Poaceae) in the Colorado Alpine. *American Journal of Botany* 79(12): 1337–1343.

Goheen, E. M., D. J. Goheen, K. Marshall, R. S. Danchok, J. A. Petrick, and D. E. White. 2002. *The Status of Whitebark Pine along the Pacific Crest National Scenic Trail on the Umpqua National Forest*. Gen. Tech. Rep. PNW-GTR-530. U.S. Department of Agriculture, Forest Service, Pacific Northwest Research Station, Portland, OR.

Hamann, M. J. 1972. *Vegetation of Alpine and Subalpine Meadows of Mount Rainier National Park, Washington*. M.S. thesis, Washington State University, Seattle.

Henderson, J. A. 1974. *Composition, Distribution and Succession of Subalpine Meadows in Mount Rainier National Park*. Ph.D. dissertation, Oregon State University, Corvallis.

Hoff, R. J. 1992. *How to Recognize Blister Rust Infection on Whitebark Pine*. Research Note INT-406. USDA Forest Service, Intermountain Research Station, Ogden, UT.

Hoff, R. J., D. E. Ferguson, G. I. McDonald, and R. E. Keane. 2001. Strategies for managing whitebark pine in the presence of white pine blister rust. Pages 346–366 in D. F. Tomback, S. F. Arno, and R. E. Keane (eds.), *Whitebark Pine Communities: Ecology and Restoration*. Island Press, Washington, DC.

Ingersoll, C. A. 1991. *Plant Reproductive Ecology and Community Structure along a Subalpine Snowmelt Gradient*. Ph.D. dissertation, Oregon State University, Corvallis.

Juelson, J. L. 2001. *Restoring Subalpine Vegetation in the Enchantment Lakes Basin: Evaluating Restoration Treatments on the Seedling Emergence of* Juncus parryi, Carex nigricans, *and* Luetkea pectinata. M.S. thesis, Central Washington University, Ellensburg.

Kagan, J. and S. Caicco. 1992. *Manual of Actual Oregon Vegetation*. Available at www.gis.state.or.us/data/metadata/k250/GAPveg_250.pdf.

Keane, R. E. and S. F. Arno. 1993. Rapid decline of whitebark pine in western Montana: evidence from 20-year remeasurements. *Western Journal of Forestry* 8(2): 44–47.

Kendall, K. C. and R. C. Keane. 2001. Whitebark pine decline: infection, mortality, and population trends. Pages 221–242 in D. F. Tomback, S. F. Arno, and R. E. Keane (eds.), *Whitebark Pine Communities: Ecology and Restoration*. Island Press, Washington, DC.

Kendall, K. C., D. Schirokauer, E. Shanahan, R. Watt, D. Reinhart, R. Renkin, S. Cain, and G. Green. 1996. Whitebark pine health in northern Rockies national park ecosystems: a preliminary report. U.S. Forest Service Intermountain Research Station, Missoula, MT. *Nutcracker Notes* 7:16.

Lester, L. W. 1989. *Revegetation Efforts at North Cascades National Park Service Complex*. Presented at the Society of Ecological Restoration Management Annual Meeting, "Restoration: The New Management Challenge," January 16–20, Oakland, CA.

Linhart, Y. B. and J. L. Gehring. 2003. Genetic variability and its ecological implications in the clonal plant *Carex scopulorum* Holm. in Colorado tundra. *Arctic, Antarctic, and Alpine Research* 35(4): 421–433.

Linhart, Y. B. and C. A. Wise. 1997. *Genetic Variability in Populations of Revegetation Candidate Plants at Mt. Rainier National Park*. Unpublished report, EPO Biology Department, University of Colorado, Boulder.

Little, R. L., D. L. Peterson, and L. L. Conquest. 1994. Regeneration of subalpine fir (*Abies lasiocarpa*) following fire: effects of climate and other factors. *Canadian Journal of Forest Research* 24: 934–944.

McClure, R. H. Jr. 1989. Alpine obsidian procurement in the southern Washington Cascades. *Archaeology in Washington* 1: 59–69.

McNab, W. H. and P. E. Avers. 1994. *Ecological Subregions of the United States: Section Descriptions*. U.S. Department of Agriculture, U.S. Government Printing Office, Washington, DC.

Mierendorf, R. R. 1999. Precontact use of tundra zones of the northern Cascades range of Washington and British Columbia. *Archaeology in Washington* VII: 3–23.

Mierendorf, R. R. 2004. *Archeology of the Little Beaver Watershed, North Cascades National Park Service Complex, Whatcom County, Washington*. Report submitted to the Skagit Environmental Endowment Commission in fulfillment of Grant 02-01 by the National Park Service, North Cascades National Park Service Complex, Sedro-Woolley, Washington.

Moritsch, B. J. and P. S. Muir. 1993. Subalpine revegetation in Yosemite National Park, California: changes in vegetation after three years. *Natural Areas Journal* 13(3): 155–163.

Murray, M. P. and M. C. Rasmussen. 2003. Non-native blister rust disease on whitebark pine at Crater Lake National Park. *Northwest Science* 77(1): 87–91.

Nickels, A. M. 2002. *History under Fire: Understanding Human Fire Modification of the Landscapes of Mount Rainier National Park*. M.S. thesis, Central Washington University, Ellensburg.

Oregon Department of Forestry. 2003. *Land Ownership*. Available at www.gis.state.or.us/data/metadata/k24/public_ownership_draft.pdf.

Peterson, K. T. 1999. *Whitebark Pine (*Pinus albicaulis*) Decline and Restoration in Glacier National Park*. M.S. thesis. University of North Dakota, Grand Forks.

Pokotylo, D. L. and P. D. Froese. 1983. Archaeological evidence for prehistoric root gathering on the southern interior plateau of British Columbia: a case study from Upper Hat Creek Valley. *Canadian Journal of Archaeology* 7(2): 127–157.

Price, L. W. 1981. *Mountains and Man*. University of California Press, Berkeley.

Ramsay, M. J. 2004. *The Effects of Soil Amendment and Watering Regime on Germination and Establishment of Direct-Seeded Native Plant Species Used in Subalpine Restoration at Cascade Pass, North Cascades National Park*. M.S. thesis, University of Washington, Seattle.

Reeves, B. O. K. 2003. *Mistakis: The Archeology of Waterton–Glacier International Peace Park*. Vols. I and II. Submitted to the National Park Service, Intermountain Region, in fulfillment of Contract 290847, between Montana State University and the National Park Service, Bozeman.

Reimer, R. 2000. *Extreme Archaeology: The Results of Investigations at High Elevation Regions in the Northwest*. Unpublished MA thesis, Department of Archaeology, Simon Fraser University, Burnaby, BC.

Reimer, R. 2003. Alpine archaeology and oral traditions of the Squamish. In R. L. Carlson (ed.), *Archaeology of Coastal British Columbia: Essays in Honour of Professor Philip M. Hobler*. Publication No. 30, Archaeology Press, Simon Fraser University, BC.

Ripple, W. J. 1989. *Soil Textures for Selected Areas of Mount Rainier National Park*. Unpublished report, Oregon State University, Corvallis.

Rochefort, R. M. 1989. *Paradise Meadow Plan*. Unpublished report, Mount Rainier National Park, Ashford, WA.

Rochefort, R. M. 1990. *Mount Rainier National Park Revegetation Handbook*. Unpublished report, Mount Rainier National Park, Ashford, WA.

Rochefort, R. M. and J-Y. P. Courbois. In prep. *Status of Whitebark Pine in North Cascades and Mount Rainier National Parks, Washington*.

Rochefort, R. and S. Gibbons. 1992. Mending the meadow. *Restoration and Management Notes* 10: 120–126.

Rochefort, R. M., R. L. Little, A. Woodward, and D. L. Peterson. 1994. Changes in sub-alpine tree distribution in western North America: a review of climatic and other causal factors. *The Holocene* 4(1): 89–100.

Rochefort, R. M. and D. L. Peterson. 2001. Genetic and morphologic variation in *Phyllodoce empetriformis* and *P. glanduliflora* (Ericaceae) in Mount Rainier National Park, Washington. *Canadian Journal of Botany* 79: 178–191.

Shoal, R. Z. and C. A. Aubry. 2004. *The Status of Whitebark Pine on Four National Forests in Washington State*. Unpublished report, USDA Forest Service, Olympic National Forest, Olympia, WA.

Urbanska, K. M. 1994. Ecological restoration above timeberline: demographic monitoring of whole trial plots in the Swiss Alps. *Botanica Helvetica* 104: 141–156.

USGS. 2003. *National Land Cover Characterization*. Available at landcover.usgs.gov/natllandcover.asp.

Vale, T. R. 1981. Tree invasion of montane meadows in Oregon. *American Midland Naturalist* 105: 61–69.

Van Vechten, G. W. III. 1960. *The Ecology of the Timberline and Alpine Vegetation of the Three Sisters, Oregon*. Ph.D. dissertation, Oregon State University, Corvallis.

Vivian, B. C. 1989. *A Survey and Assessment of the Cultural Resources in Cathedral Provincial Park, B.C.* Report prepared for the British Columbia Ministry of Environment and Parks, Permit No. 1988-80, Southern Interior Region, Kamloops.

Washburn, A. L. 1956. Classification of patterned ground and review of suggested origins. *Bulletin of the Geologic Society of America* 67: 823–865.

Washington Department of Fish and Wildlife. 1999. *Washington GAP Data Products*. Available at wdfw.wa.gov/wlm/gap/landcov.htm.

Weidman, N. R. 1984. *Seed Banks of Three Adjacent Subalpine Meadows: Olympic National Park, Washington*. M.S. thesis, University of Washington, Seattle.

Whitlock, C. 1992. Vegetation and climatic history of the Pacific Northwest during the last 20,000 years: implications for understanding present-day biodiversity. *Northwest Environmental Journal* 8(5): 5–28.

Wischmeier, W. H. and D. D. Smith. 1978. *Predicting Rainfall Erosion Losses: A Guide to Conservation Planning*. USDA Handbook No. 537. U.S. Government Printing Office, Washington, DC.

Zabinski, C., T. Wojtowscz, and D. Cole. 2000. The effects of recreation disturbance on subalpine seed banks in the Rocky Mountains of Montana. *Canadian Journal of Botany* 78: 77–582.

Zeglen, S. 2002. Whitebark pine and white pine blister rust in British Columbia, Canada. *Canadian Journal of Forest Research* 32: 1265–1274.

Zwinger, A. H. and B. E. Willard. 1972. *Land above the Trees: A Guide to American Alpine Tundra*. Harper and Row, San Francisco.

PART III

Crossing Boundaries

The chapters of Part III explore categories of restoration that span the ecosystem boundaries encountered in Part II, present unique challenges, and impart important lessons. For example, urban restoration can include at least eight of the nine ecosystem types encountered earlier (all nine if one counts Whistler and other ski resorts as urban). Wildlife restoration can take place in any regional ecosystem or, in the case of the timber wolf, can actually shape and restore multiple ecosystems. Traditional ecological knowledge includes alternative approaches and cultural sensitivities that need to be better understood by most practitioners and by a wider conservation public. A single stream may course through several ecosystem types, and watersheds encompass all these and more. Invasive species are a bane to every regional ecosystem, yet techniques and specialized management expertise often are overlooked or not well understood. And much ecosystem restoration is nothing more than subtraction of that which does not belong.

Thus, the goal of Part III is to broaden our scope of knowledge so that practitioners, students, land managers, conservationists, and policymakers can connect more restoration dots. The information in these chapters is intended to help us recognize that what might seem like a stream restoration project really occurs in a watershed context that may include urban areas, that traditional knowledge may provide critical insights, that invasive species left unchecked may undo good work, and that the return of endangered or extirpated wildlife may be a critical component of a full restoration. Good restoration is comprehensive restoration, and every project can benefit from a better understanding of context.

Chapter 12

Urban Natural Areas

MARK GRISWOLD WILSON AND EMILY ROTH

Researchers in the Cliff Ecology Research Group at the University of Guelph in Ontario, Canada hypothesized that features of urban environments may reflect human preference for ancestral rock outcrop or cliff habitats. Therefore, they continue to provide opportunities for the same array of species with which humans have been associated for nearly a million years. They suggest that the human habitat is substantially one in which rock outcrop, cliff, and talus slope organisms have become cosmopolitan. They have displaced natural flora and fauna adapted to other habitats. Whereas in nature the surface area of cliff habitats typically occupies less than 1% over a given area, in cities the vertical surface area is several times larger (Larson and Lundholm 2002) (Figure 12.1).

Until recently, natural areas in cities have been thought to contain only unwanted cosmopolitan species and therefore to be unimportant as a focus for ecological restoration and conservation. This is in contrast to environments more highly valued for ecological research in remote "pristine" areas, undisturbed by humans. Trips to far-away places to experience ecosystems reflect and contribute to a deep seated belief that people and nature are separate (Stille 2002). In the past 25 years these foundations of ecological thought have been challenged by studies that demonstrate that even very remote "pristine" environments show clear signs of human disturbance or influence, including changed fire regimes, hunting of animals, and harvesting of plants.

Also, it turns out that urban areas may have much more conservation value than had previously been thought. Ecologists increasingly are finding cities to be interesting ecosystems with surprisingly high levels of biodiversity. Improved understanding and protection of these ecosystems may contribute greatly to wider conservation efforts (Stille 2002). A key advantage of restoration in urban areas is the higher social value that nearly any green, seminatural space has to local residents. This allows a greater concentration of resources, financial and labor, on a per-acre basis than is in the case in most remote settings.

But successful restoration or renaturalization of urban ecosystems requires careful thought and detailed planning in order to offset many challenges posed by vertical urban habitats. *Restoration* and *renaturalization* are used interchangeably throughout this chapter. In most urban areas, we are often talking about renaturalization: the creation of a natural area or habitat that was not historically at the site or does not contain all expected attributes. Complete restoration often is not possible because biotic and abiotic factors have been too permanently altered. Before one initiates an urban restoration project, it is advisable to identify cultural influences on urban ecology and unique technical challenges of renaturalization. "Traditionally, the

FIGURE 12.1. Peregrine falcon, an urban cliff dweller in Portland, Oregon. (*Photo by Bob Salinger of Portland Audubon*)

social scientists would go to the built part [of the cities] and ask what people were doing, how they made their decisions, and the ecologists would go over to the green spots and count the bugs. Now we have to ask how the people's decisions influence the green spots and how the green spots influence people's decisions" (Steward Pickett from Jensen 1998).

Renaturalization and Restoration of Nature in the City

Restoration of urban landscapes, from the micro to the meso and through the macro scale, offers unique opportunities and challenges (Dramstad et al. 1996). From residential-scale naturescaping to multiproperty watershed restoration, actions necessary to achieve a sustainable natural environment are multiscaled and culturally rooted. The new discipline of urban ecology is a blend of landscape ecology, urban planning and design, and long-term adaptive management. Urban green spaces exist in a matrix of a high-contrast, unnatural environment. If envisioned as part of a system, they can connect larger landscape elements through corridors or stepping stones and provide a diverse mosaic of habitat patches (see Chapter 14). Ideally, this system is designed and managed to reflect the character and species of a region and to play a constructive role in larger habitat conservation efforts and plans. Higher-functioning urban habitats can be linked with multiple wild and rural lands to form a tapestry of habitat that functions at least in part like the one that predated urban and agricultural transformation of the region.

The importance of nature in urban settings of the Pacific Northwest should not be overlooked. Every major city and most minor ones in this region are located at the confluence of major navigable waterways, including large and small rivers and estuaries. Historically, these probably were the areas of greatest regional biodiversity and productivity. Restricting conservation and wilderness to remote, high-elevation mountains will not be sufficient to conserve all the species that inhabit the Northwest. In addition, because most Northwest residents live in urban areas, it is critical that these play a leading role in integrating economic development and conservation and in serving as the main incubator for the environmental education of most of our citizens. Many still believe that nature begins "out there somewhere," rather than in our own backyards, and urban ecosystems are the place to change that perception.

However, there are clear limitations on the level of conservation that can be accomplished in urban areas. In 2003 researchers from the Heinz Center for Science, Economics, and the Environment estimated that about half of all natural lands in urban and suburban areas (nationally) are in patches smaller than 10 acres. A progressively smaller percentage of natural areas are found in larger patches, so that less than 5% of the total are found in patches of 1,000 acres or more. Very large patches (more than 10,000 acres) are found only in Western cities, and these patches account for 0.3% of all natural lands in urban and suburban areas. In addition to size, the quality of habitat and recreational value of urban natural areas are influenced by other factors, such as the shape of patches, the amount of edge and interior habitat, how isolated they are from other natural areas, and adjacent land uses (www.heinzctr.org 2003).

Patch size, structure, design, and location define habitat quality. Small patches of natural habitat generally are of lower quality for most native plants and animals (although this is not necessarily true for wetlands). They also provide less solitude, a circumscribed aesthetic experience, and fewer recreational opportunities for people. Small patches of habitat favor common, human-tolerant nonnative edge species such as fox squirrels, starlings, and English sparrows. Areas large enough to include interior habitat have a much greater potential for biodiversity for less common native species, including pileated woodpeckers, red-tailed hawks, warblers, and, if connected to source habitats, coyotes, bears, mountain lions, mink, otters, weasels, and many amphibians. In part it is the sheer size that matters, but larger size usually means more diversity and a greater range of niches.

Because urban projects are by definition embedded in a human matrix, they are the perfect laboratory to experiment with and improve community relations and support for ecological restoration. The scope, complexity, and success of urban restoration projects depend largely on community objectives, opportunities, and available partnerships. In recent years many urban projects have focused on reducing the amount and timing of stormwater runoff entering streams, lakes, and estuaries in order to aid aquatic restoration efforts. Small natural patches of vegetation, either restored or protected in place, can thus serve an important larger function of treating, detaining, or infiltrating water, in addition to whatever direct habitat value they provide. As opportunities for large-scale restoration projects are identified, partnerships and community outreach become increasingly important in linking multiple restored and conserved areas, thus establishing a large-scale, regionally significant network of natural areas.

Because of the fragmented nature of land ownership, high land values, and dense network of transportation systems, opportunities for very large-scale habitat patches are limited, and restored and conserved areas often are isolated or connected only by narrow corridors with high-contrast edges. With luck, these will link to larger patches outside the urban area to allow genetic mixing and repopulation, particularly of larger, more widely ranging wildlife species.

A typical range of urban restoration and conservation strategies includes the following:

- Protecting existing natural areas in and adjacent to urban areas to serve as core habitats
- Using restored and in some cases existing urban natural areas to help treat stormwater before it enters under ground or surface water systems
- Revegetating disturbed areas with native plants appropriate to altered conditions
- Forming partnerships at many levels to pool resources and facilitate larger restoration projects or beneficial effects
- Providing community education and stewardship opportunities to introduce and enlist many hands and minds in restoration and conservation

Urban restoration projects often exemplify the continuum of restoration goals as listed in the Society for Ecological Restoration (SER) *Guidelines for Developing and Managing Ecological Restoration Projects* (Clewell et al. 2000; see Chapter 2 of this volume). Full repair of a damaged ecosystem is considered the gold standard of restoration, in which a site is returned to its historic or predevelopment condition. Commonly, a few minor aspects of the preexisting ecosystem cannot be fully restored. These should be identified and accepted as exceptions. Restoration takes place at the same site where the initial damage occurred. Such restoration has been called in-kind (as the historic type of ecosystem is restored) and onsite (because restoration occurs at the same location where the historic ecosystem was damaged). Restoration projects in urban areas often are not necessarily onsite, and some are not in-kind. They tend to conform more to the following models (Clewell et al. 2000):

- *Creation of a new ecosystem of the same kind to replace one that was entirely removed.* This recognizes that a restored ecosystem must be entirely reconstructed on a site denuded of its historic ecological structure and composition. Creations are commonly done on surface-mined lands and brownfields (severely damaged and polluted urban and industrial lands). In this case, the goal may be to recreate a native forest ecosystem, but the path to full restoration is very long and uncertain, and initial efforts may be very indirect, such as soil building.
- *Creation of a different kind of regional ecosystem to replace the one that was removed from a landscape that became irreversibly altered.* This option is important for restoring natural areas in an urban context where, for example, original hydrologic conditions or functional soils cannot be restored (e.g., where a floodplain forest cannot be restored because it was created by entrenchment of a stream but where an upland forest could provide shade and other ecological functions).
- *Creation of a hybrid ecosystem on sites where an altered environment can no longer support any previously occurring type of regional ecosystem.* The replacement ecosystem may consist of novel combinations of indigenous and nonnative species that are assembled to suit novel site conditions, such as a mixed native wildflower and nonnative grass meadow at a closed solid waste disposal site (St. John's Landfill in Portland), which may provide important habitat for native wildlife but is far from a native prairie in composition and very far from the floodplain forest that occupied the area historically.

The full repair of urban natural areas may be possible only when large tracts of native vegetation have been set aside. For example, in Portland, Oregon the restoration of the 5,000-acre (2,000-hectare) Forest Park has been planned as a two-phase strategy. In the short term, the focus is on removing invasive nonnative plants, particularly English holly and ivy, from the maple, fir, hemlock, and cedar forest and adding understory native vegetation where needed. In the longer term, successful restoration will include protecting and repairing the upper watersheds (partly urbanized), repairing past soil disturbance, possibly selective removal of shade-tolerant maple trees to open the canopy enough for Douglas fir regeneration, and continuous management of nonnative vegetation, particularly on the park's outer edges (at the interface with urban development).

Most urban natural areas in the region are smaller sites with much higher levels of disturbance than Forest Park. Repairing highly disturbed sites and returning them to historic conditions is not possible in many cases. Severely altered hydrology and soils, poor water quality, and aggressive nonnative flora and fauna prevent full restoration. Partial renaturalization of these sites often is a more appropriate goal and should be considered experimental and adaptive. In some cases new ecosystem or hybrid creations are most appropriate. Examples include the following:

- Meadows of grasses and native wildflowers that provide habitats for small mammals and invertebrate assemblages and facilitate groundwater recharge
- Patches of managed savannas and woodlands in small parks linked with canopies of native or nonnative street trees to improve habitat for resident and migratory birds
- Native plant eco-roofs and created wetlands that pretreat or infiltrate stormwater, removing chemical pollutants before they enter urban waterways and thus facilitating conservation of aquatic ecosystems.

The challenge of planning, designing, and implementing restoration has engaged many Northwest urbanites who recognize the values of protecting and repairing nature close to home. Some communities have supported purchase of urban natural areas with bond measures and have lobbied successfully to protect remaining natural resources with more robust land use regulations, particularly stream setbacks. Limited funding has become available for small ecological restoration and management projects in several Northwest cities and towns, particularly where broad-based partnerships of federal and local agencies, nongovernment organizations, private citizens, and businesses have formed, such as the West Eugene Wetlands project. Unfortunately, to date there have been far fewer funding opportunities for the restoration of privately owned lands and publicly owned uplands. After projects are initiated, there is often public enthusiasm for site-based nature education and stewardship. Volunteers from "friends" groups assist with long-term maintenance and management of urban natural areas.

Urban Restoration Attributes

SER recently published *The SER International Primer on Ecological Restoration* (SER 2004), which includes a definition of restoration and nine expected attributes of restored ecosystems. These include characteristic plant assemblages, reference site descriptions, and the notion of self-sustaining ecosystems that will persist indefinitely with limited human involvement. The nine attributes incorporate both cultural practices and ecological processes that result in a functioning restored system. These are discussed in more detail in Chapter 2. They can be applied to urban restoration projects, although a few modifications are necessary to adjust for limited size, fragmentation, and the inability to completely mitigate or remove external stresses. The following modifications are suggested for attributes of urban restoration projects:

- The restored ecosystem consists of both indigenous and naturalized species that create an assemblage that is native to the greatest extent possible.
- A majority of the functional groups are present to maintain a stabilized ecosystem. Management plans must provide external inputs to maintain the ecosystem and supply colonizing species when needed because external sources are limited in the urban setting.
- The restored urban ecosystem functions normally but is closely monitored and modified when signs of dysfunction, stress, or loss of integrity are identified.
- The restored urban system is connected to larger functioning ecosystems through corridors or stepping stone habitats to provide biotic and abiotic flows and exchanges.
- The restored urban ecosystem adds to the urban fabric of the community, is aesthetically appropriate, and is viewed as a neighborhood asset.

Northwest Urban Restoration Case Studies

Cities throughout the Pacific Northwest have initiated ecological restoration programs and projects. From the West Eugene Wetlands in the south to Juneau Creek Watershed in the north, and from Victoria, British Columbia in the west to Boise and Missoula in the east, local communities, park departments, and regional, state, and federal agencies are restoring ecosystems. The challenges presented in any particular situation help define where on the restoration continuum a project should be located. The critical variables are the hydrology, biotic and abiotic conditions, neighborhood context, and level of continual disturbance expected. These shape the range of possibilities for a target habitat.

In the Columbia Slough case study that follows, the watershed has been so significantly altered by industrial and residential development that restoring the damaged ecosystem to its original condition is out of the question. However, creation of another type of regional ecosystem is a viable option, and that is what the City of Portland and its partners are pursuing.

Case Study: The Columbia Slough's Community Partnerships

The Columbia Slough's 32,000-acre watershed exemplifies the full range of impacts of urban development on natural ecosystems. It is home to more than 150,000 people, includes an international airport and port, and is a crucial regional economic center, yet it encompasses the nation's largest urban wetlands at the Smith & Bybee Lakes Wildlife Area and shelters multiple salmonid species and more than 150 bird species, including nesting bald eagles.

In short, it provides a case study for balancing economic development with natural resource conservation and habitat renaturalization. It is also an outstanding example of how cultivating multiple partnerships can lead to successful urban renaturalization.

Planning and Partnerships: The Columbia Slough Watershed Council

Restoration planning for the Columbia Slough watershed, a highly industrialized former floodplain of the Columbia River, began in 1993 when Portland was threatened with lawsuits over combined sewer overflows that dump polluted water into the Willamette and Columbia rivers during heavy rainstorms. Rapid development of the former floodplain, establishment of environmental zoning, designation of an "industrial sanctuary," continued threats to the area's unique natural resources, and unique recreation possibilities prompted a group of citizen advocates and agency representatives to organize a watershed council. Efforts were initially focused on sharing information and influencing and improving policies, practices, and projects in the watershed. A brief survey of major land use and infrastructure projects noted that more than 100 projects were planned in the watershed in 1994 alone. Watershed council members represent widely divergent constituencies, so formal council action requires 100% consensus. Coming to an agreement on the definition of *restoration* was a pivotal organizational milestone. Developing a glossary was also necessary before members could even agree on a mission and vision statement, a testament to the importance of language and communication. The work of the council now focuses on finding common ground, forming partnerships, and securing funds for restoration and enhancement projects while also advocating for the environmental and economic health of the watershed. The 2003 Action Plan is the blueprint, magnet, and information resource that attracts partners and funds for a wide variety of watershed projects. The council holds events in pubs and company lunchrooms and paddling events on the slough to attract the public, capture local knowledge, and nurture support (Figure 12.2). These partnerships enable the City of Portland to undertake watershed projects it would be unable to do on its own.

The Language of Partnerships

The council defines *restoration* as bringing a site or natural system in the Columbia Slough Watershed as nearly as possible back to its original condition, within the limits imposed by ownership, current use, resource availability and the degree of existing degradation. The council's mission is to protect, enhance, restore, and revitalize the Slough and its watershed.

FIGURE 12.2. Paddlers on the Upper Slough, Corps of Rediscovery outing, September 2005. (*Photo by Susan Barthel, Columbia Slough Program Coordinator, City of Portland*)

Revegetation: Businesses Lead the Way, 40 Miles and Counting

Restoration efforts moved into high gear when Portland's Watershed Revegetation program was established in 1996. Funding support from a federal Environmental Protection Agency grant subsidized costs for landowners and jumpstarted a vibrant city program. Endorsement by the Columbia Corridor Association, an influential local business advocacy group, and the Watershed Council legitimized the program for landowners. An aggressive grassroots landowner contact program resulted in high participation rates. Good-faith revegetation agreements were deemed sufficient for program participation. Because formal conservation easements were not required, Portland staff could optimize project funds and staff time. By 2004 this program had restored more than 650 acres and revegetated more than 40 miles of the slough.

Environmental Education: Learning in Many Styles, Places, and Languages

Environmental education is a vital element of Slough watershed activities. Environmental Protection Agency funds enabled the hiring of an educator for the watershed's forty-three schools and the purchase and conversion of a waterside junkyard into an environmental learning center that attracts more than 5,000 visitors yearly. Water quality testing, restoration plantings, and wildlife monitoring all occur a short distance from the city bus stop local students frequent. The same funds allowed the city and watershed council to develop hundreds of events, innovative programs, and tours for the watershed's adult residents, workers, and visitors. These include the widely copied Slough 101 class, Soup on the Slough (a lunchtime learning program), Wild in the City events and tours, and Eyes on the Slough, a citizen monitoring program. These events provide adults with free, watershed-specific, hands-on learning opportunities.

An active program of canoe and kayak events attracted more than 1,500 participants in 2004. Extensive media coverage, free boat rentals, and a family focus have made the Columbia Slough Regatta Oregon's largest one-day paddling event. For many participants it's their first nonmotorized boating experience and an introduction to the waterway and largely inaccessible riparian areas. *Exploring the Columbia Slough,* a guide to the slough's accessible natural resources and recreation opportunities, and a *Columbia Slough Paddlers' Guide* have attracted additional slough fans.

Explorando el Columbia Slough is a family-oriented environmental festival. Spanish-speaking kids, parents, and grandparents test water quality, identify plants and macroinvertebrates, and paddle the waterway. Hispanic restoration crews that have provided much of the contracted revegetation work in the slough watershed have also been honored at this event. This has strengthened the community's awareness of local watershed restoration efforts. (For more information about the Columbia Slough Watershed Council, contact Susan Barthel at susanb@bes.ci.portland.or.us.)

Case Study: Protecting the Backyard of Boise

One of the greatest challenges of ecologically connecting large urban greenspaces is the large number of property owners. Over the long term, successful projects include a strong coalition of both public and private landowners that share a common vision and pool resources. In Boise a coalition of federal, state, and local governments and private landowners formed the Boise Front Coalition to protect and restore the fragile foothill ecosystem and provide recreational opportunities on the Ridge to Rivers Trail. This group has undertaken large-scale habitat restoration and trail building to support their vision for the foothills.

The Boise Foothills rise from 2,610 feet at the valley floor to more than 6,575 feet at their top and form the northern backdrop to one of the fastest-growing urban areas in the nation. Through a partnership of public agencies and private landowners, the foothills' fragile environment is being protected and restored for wildlife, endangered plants, recreation, and other natural resources. An integrated system of public open space (37,835 acres) and more than 80 miles of trails are managed through numerous agreements with public agencies and private landowners. The trails provide an array of appropriate recreational uses for citizens and visitors to the Boise area, from hiking to all-terrain vehicle use, managed as part of consortium of agencies called the Ridge to Rivers Partnership. Protection of the watershed and creation of trails through the foothills are based on a community vision that first began in the 1940s.

The foothills are a high desert environment that receives only 12 inches of rain a year on average. Despite the proximity to Boise, largest city in the state, they provide winter range for Idaho's largest mule deer herd. Neotropical migrant birds use local riparian areas when flying between their summer and winter homes. Slickspot peppergrass (*Lepidium papillifoerum*) is one of the many rare plants found in foothills microenvironments. Key challenges include the susceptibility of high desert soils to erosion and the challenge of conservation and restoration in such an arid environment.

Early on, the citizens of Boise and Ada County realized the importance of protecting open space to maintain quality of life as the area experienced rapid growth. Once development began to creep up the foothills, off-road vehicle use of the area expanded, and an increasing number of people explored the foothills, creating a network of user trails. Off-road vehicle use in particular created visual scars on the landscape, and the fragile soils began to erode (Figure 12.3). In 1959 a large fire followed by a torrential rainstorm caused a wall of mud and debris to flow into Boise. This raised community awareness for watershed protection, which over the years led to the development of the protection and management partnership that exists today.

FIGURE 12.3. Mountain biking on the Ridge to River Trail near Boise, Idaho. (*Photo by Tim Breuer, Ada County Parks*)

In 1989 the Bureau of Land Management joined forces with off-road vehicle users to organize the Boise Front Coalition in order to better protect the watershed, restore land cover, and establish a legitimate trail system for recreation. This partnership grew to include numerous federal, state, and local agencies that pooled their resources to hire a coordinator, create a management plan, and set priorities for recreation and restoration. One of the first tasks was to work with private landowners to complete agreements on trail use through their lands. The partnership continued working together to sign a Memorandum of Understanding on the Boise Foothill Policy Plan in March 1997 and then to assist in the development of the Boise Foothill Open Space Management Plan for Public Lands in 2000.

The management plan identifies seven critical actions for protecting the watershed:

- Establish a coordinating agency for foothills open space stewardship.
- Educate citizens.
- Preserve public open spaces.
- Provide a range of sustainable public recreation opportunities.
- Conserve wildlife and beneficial vegetation.
- Provide for maintenance and conservation of public open space.
- Work cooperatively with private property owners.

The collaborative action of the partners has been successful because of their shared vision and the willingness of each agency to pool limited resources. Over the past 20 years, activities in the foothills have changed from unmanaged, harmful ones to appropriate use and recognition of the importance of preserving the watershed. Trails on steep slopes were removed or improved and the land restored, other trails were rerouted from erosion-prone areas, an education center was developed, big game wintering habitat was protected, and recreation visitors were taught to respect private lands. In 1996 a human-caused fire burned 15,300 acres. Rehabilitation started immediately, with the cooperating agencies spending $3.3 million to reduce erosion and restore the watershed. Recognizing the importance of protecting the foothills from future development, the residents of Boise passed a bond levy in 2001, which raised $10 million to purchase lands from willing landowners.

Education of landowners and the community on opportunities for ecological restoration is a key to project initiation and success. Naturescaping and similar backyard restoration projects are harnessed to educate the community about achieving sustainable land use that includes rehabilitating soil, maintaining and cleaning water onsite, and planting native flora that attracts wildlife. Numerous small backyard projects are gradually creating stepping stone habitats to larger neighborhood, community, and watershed restoration projects throughout urban areas of the Northwest. (For more information about the Ridge to Rivers trail system and the restoration work being completed, see www.ridgetorivers.org.)

Case Study: The Saanich Approach to Environmental Protection and Stewardship

The Municipality of Saanich, in the greater Victoria region of Vancouver Island, is committed to protecting important natural areas, species, and ecosystems at risk through a variety of approaches, from regulatory policies such as bylaws and development permits to nonregulatory education and stewardship programs. These different approaches often are the successful results of various types of partnerships with other agencies and in the community.

Saanich enforces a variety of bylaws and regulations governing watercourses, riparian areas, and marine shoreline protection. Development permits include guidelines to protect at-risk species and to manage stormwater runoff. Natural state and tree covenants are widely used during redevelopment and are legal tools that become registered on the land title. An example of a unique regulation in this region is the Tree Preservation Bylaw. This bylaw prohibits the removal of all Garry oak, arbutus, Pacific dogwood, and other significant native trees more than 5 meters (16.3 feet) tall and 10 centimeters (3.9 inches) in diameter without a permit. The bylaw includes penalties for tree removal or damaging activities and also protects trees within riparian areas.

The educational approach to environmental protection in Saanich is achieved through a variety of programs such as a community newsletter on the natural environment in Saanich, a Native Plant Salvage Program, the Naturescape BC program, and the Garry Oak Restoration Project (GORP).

GORP was formed in 1999 in response to the endangered status of the local Garry oak ecosystems. Garry oak ecosystems are one of the most endangered in Canada, with less than 5% of the original community remaining intact. The full extent of Garry oak ecosystems ranges along the West Coast of North America, from northern California to southwestern British Columbia (see Chapter 4). The region where Saanich is located once attracted European explorers and settlers such as Captain George Vancouver, who described the diverse oak landscape as enchantingly beautiful. Today, these ecosystems support more than 100 at-risk plant and animal species threatened by continued urban development, invasive species, and vegetation structure changes caused by fire suppression.

GORP is focused on both ecological restoration and education through active restoration of ten ecosystem remnants on municipal sites and support for a number of partnership sites in the community. The Municipality of Saanich has teamed up with the Georgia Basin Ecosystem Initiative (Environment Canada), the University of Victoria, and the Garry Oak Meadow Preservation Society to support a community-based volunteer stewardship program. The success of GORP is seen in the growth of volunteer involvement, similar community projects, and increased awareness on the part of the community, municipal staff, and officials, and is celebrated each spring blooming season. (For more information on Garry oak restoration efforts in Saanich, see the Saanich Web site at www.saanich.ca or the GORP Web site at www.gorpsaanich.com.)

Case Study: The High Point Redevelopment Project

Ecological restoration in urban areas often is indirect. For example, protection and restoration of urban creeks and streams includes retrofitting of upland neighborhood stormwater systems. Pollutants and high flows from stormwater are now recognized as major factors that reduce water quality and habitat of urban streams. The National Pollutant Discharge Elimination System designates stormwater as a source pollutant that is covered by municipal permits. Larger municipalities are

Figure 12.4. Design simulations showing grass and planted bioswales at High Point intended to manage stormwater naturally. (*Courtesy of Seattle Public Utilities*)

required to treat stormwater before it enters natural waterways. Many cities have developed stormwater treatment manuals detailing best management practices for developers and public entities. To ensure permit compliance, cities are providing incentives for private property owners and developers to reduce the amount of impervious surface and develop innovative solutions for treating stormwater before it enters aquatic environments. Activities include using bioswales instead of closed pipes, increasing tree cover and using native plant communities, reducing pollutant sources through eco-roofs and other adaptations, and educating the community and developers about how water and pollutants originate and travel throughout their watershed. The importance of treating urban runoff is highlighted by educational and partnership efforts public agencies are putting forth to reduce the nonpoint pollutants at the source. This case study demonstrates the innovative solutions a partnership in Seattle created to reduce pollutants in the Longfellow Creek Watershed.

Seattle Public Utilities collaborated with the Seattle Housing Authority to integrate a natural drainage system into the High Point project, 129 acres of mixed-income housing in west Seattle. This is one of the largest Seattle developments in recent history. It will include thirty-four blocks of new streets complete with utilities, street trees, sidewalks, and a total of 1,600 housing units. This extent of development is unique given the built-out conditions of Seattle, where typical new development averages only 3,000 square feet and creation of completely new right-of-ways is uncommon.

The High Point project contains a subbasin estimated to cover about 10% of the entire Longfellow Creek Watershed. Longfellow Creek is home to the largest salmon returns among all of Seattle's creeks, so this project provided an unprecedented opportunity to improve water quality and stream flows on a large scale. The natural system stormwater design integrates 22,000 lineal feet of swales throughout the development within the planting strip of street right-of-ways (Figure 12.4).

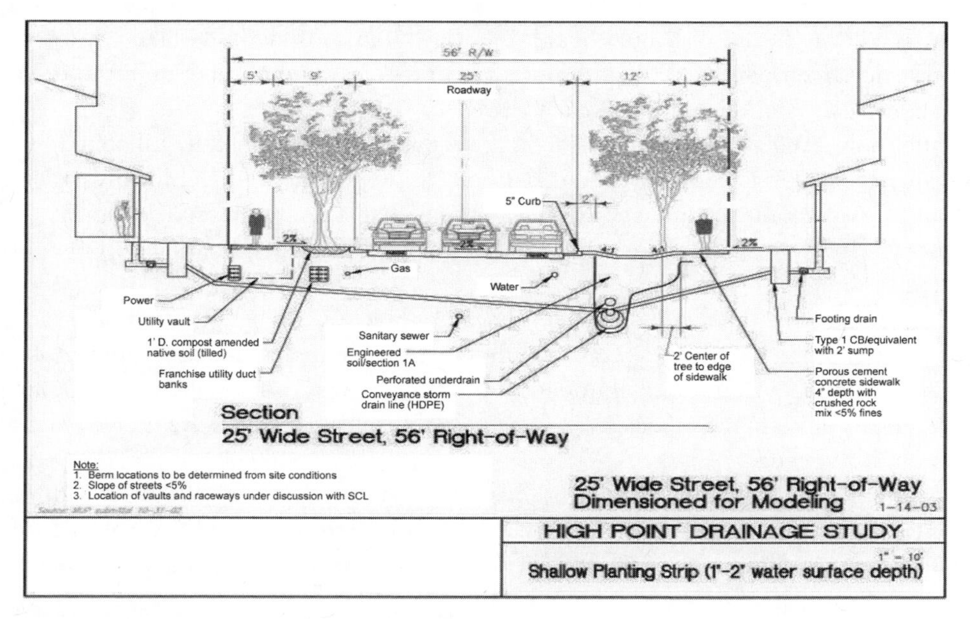

FIGURE 12.5. Street cross-section including the High Point swale. (*Courtesy of Seattle Public Utilities*)

These swales include subsurface engineered soil to provide storage and infiltration opportunities. The natural drainage system will provide water quality treatment for a storm expected to occur in a 6-month time frame and also attenuate some higher-intensity storms to a preurban pasture condition. Distributing the swale network through the street block system creates many more opportunities to cleanse, cool, and infiltrate stormwater than would be the case with a traditional pipe system (Figure 12.5).

The design team created a block-scale hydrologic model to predict how the system will perform under different storm conditions. The University of Washington will monitor performance of the stormwater system at both the block and subbasin scales. Although the High Point project will not restore the neighborhood to a preexisting native ecosystem, it will modify the urban landscape to act more like one in order to restore downstream creek habitats.

The High Point project pushes Seattle Public Utilities' Natural Drainage System Program to a new level as it is integrated into a large and dense area that has an estimated 65% impervious surface cover. It differs from other projects in that the redevelopment's street layout initially called for a traditional curb, gutter, and sidewalk approach. The challenge has been met, and this project will serve as an example that can be adapted to other large-scale redevelopment projects. (For more information on Seattle Public Utilities and natural drainage systems, see www.seattle.gov/util/About_SPU/Drainage_&_Sewer_System/Natural_Drainage_Systems/index.asp.)

Case Study: Birds as Indicators of Habitat Quality Along Urban Streams

Measuring habitat quality for both planning purposes and restoration success is an integral part of every urban restoration project. In larger, more pristine systems, there are usually biotic and abiotic factors that can be measured to indicate the improvements in ecosystem health. The challenges of urban restoration, particularly continual disturbance and very high edge-to-interior ratios, impose limits on the possible number of monitoring indicators. Resident and migratory native birds often are chosen as a key indicator of habitat health in urban areas. A study undertaken by a biologist at Metro Parks and Greenspaces in Portland shows how the variety and number of bird species can be used to measure ecosystem health in urban areas.

No matter how it may look to humans, habitat is good only if it is actually inhabited by the animals that are expected to use it. However, it is usually less expensive and technically easier to measure chemical or physical parameters to assess habitat than to conduct comprehensive fish and wildlife surveys. Indicator species are a compromise that reduces the cost and effort of conducting biological assessments.

Birds are valuable habitat quality indicators because they respond to conditions at a wide variety of spatial scales and may be easily detected without invasive methods such as trapping. Changes in certain bird species occurrence and abundance indicate changes in habitat cover, structural vegetation, and effects of nonnative species. Such habitat characteristics often are used to measure terrestrial restoration success.

In the Portland metropolitan region, bird habitat generally improves with forest width because wider forests contain more native plants and better structural complexity in the shrub and tree layers than narrower forests (Hennings and Edge 2003). Spring 1999 bird survey data were reassessed in fifty-four riparian park and greenspace study sites in the greater Portland, Oregon metropolitan region (urban, suburban, and rural parks) to develop sets of indicator species to aid habitat monitoring along small urban or suburban forested streams. Using statistical ordination techniques, the following indicator groups were positively related to the named habitat characteristic:

- *Good structural complexity*: Pacific-slope flycatcher (*Empidonax difficilis*), brown creeper (*Certhia americana*), winter wren (*Troglodytes troglodytes*)
- *Native shrub habitat* (low and high shrub): black-headed grosbeak (*Pheucticus melanocephalus*), song sparrow (*Melospiza melodia*), spotted towhee (*Pipilo maculatus*), Swainson's thrush (*Catharus ustulatus*), Pacific-slope flycatcher, winter wren
- *Native low-shrub habitat* (1.5 meters or less): Pacific-slope flycatcher, brown creeper, winter wren
- *"Bad" urban habitat* (nonnative species dominate in herbaceous and shrub layers, simplified woody vegetation structure): European starling (*Sturnus vulgaris*), house finch (*Carpodacus mexicanus*), house sparrow (*Passer domesticus*), western scrub-jay (*Aphelocoma californica*), barn swallow (*Hirundo rustica*)
- *Less disturbed conditions* (fewer roads, trails, houses, less imperviousness): Neotropical migratory songbirds, habitat interior specialists (see "good structural complexity" and "native habitat")

Increases in the occurrence and abundance of each species indicate a trend toward the habitat characteristic. With these simple biological assessment tools, inexperienced volunteers may be trained in just a few bird songs or call notes to provide valuable data with which to monitor the progress of an urban forest restoration project. If there is access to a substantial wildlife data set, there is an opportunity to develop a set of indicator species specific to that site and related to restoration expectations.

Monitored species must occur with sufficient frequency to provide a reliable indicator of restoration success; therefore, rare species are not appropriate indicators. Indicators often are specific to season, geographic area, and habitat type. Consideration must be given to the implications of spatial scale in restoration and monitoring; what happens adjacent to the site influences habitat quality. (For further information on this study, contact Lori Hennings at hennings@metro.dst.or.us.)

Measuring Success through Project Goals and Evaluation

The science and practice of urban natural area restoration are evolving, and land managers are still learning how to sustain the health of these landscapes over time. The potential for successful restoration of urban natural areas is limited by the extent of past damage and our lack of understanding about how urban landscapes systems function (Sauer 1998).

> A science-based habitat planning program not only provides a foundation for making the best decisions possible and the flexibility to modify them but also fosters confidence and consensus from a public that has to pay for and then

live with the decisions made during this process. A scientific framework also provides consistency in the planning and management process through time and staff changes (Mazzotti and Morgenstern 2002).

Land managers have generally used sound, science-based planning strategies on many of the projects described in this chapter to analyze and rectify the many demands, both conservation and use-related, on urban natural areas. Many plan projects and manage landscapes using an *adaptive management* strategy, which is the practice of collecting monitoring data, analyzing it, and making appropriate changes based on the information to determine long-term project success. The purposeful and systematic observation of the landscape over time is a vital component of natural area management because what happens at one stage of restoration dictates what needs to happen next (Clewell et al. 2000). In practice, adaptive management is a cyclic planning cycle composed of several steps and numerous feedback loops that are continuously informed by the collected monitoring data. A conceptual model of the adaptive management cycle is as follows:

- *Survey and inventory*: Identify existing ecological resources and human needs.
- *Identify the desired future condition*: Articulate a restoration target and conduct a public involvement process.
- *Research and assess*: Identify threats, articulate goals, and prescribe strategies.
- *Design, implement, and maintain* the project or the landscape.
- *Monitor and manage*: Assess results, review and revise goals and strategies.

The critical step in the adaptive management process in urban areas is the identification of a desired future condition. Components of this process include the following:

- Description of potential floral and faunal communities and actions to achieve them
- Determination of the level of human use recommended for the site
- Public discussion of key activities and management issues

Public participation and cooperation are essential for the successful renaturalization and management of urban natural areas. Although the public should be encouraged to use natural areas, most sites will not be able to withstand unregulated public use. Adaptive management monitoring can be used to determine the habitat value and the balance point between resource protection and public use.

Generally the adaptive management cycle begins with a survey and inventory, but in urban areas two or more of the steps can be conducted simultaneously or a step repeated in greater detail as planning proceeds and community awareness grows. Adaptive management planning is a long-term endeavor that provides meaningful data for dynamic decision making. This often results in dramatic changes to the urban natural area landscape. Designing an adaptive management program may seem daunting, but many students and university faculty are very interested in natural area research, management, and monitoring.

Insights for Urban Restoration Design and Management

The successful protection, restoration, or renaturalization of ecological integrity in urban natural areas depends on effective natural resource planning, adaptive management, and an understanding of the relationship between the local community and the landscape.

Preserving and using natural areas can be characterized as the dual horns of the dilemma of natural area management. We establish reserve systems because we want to preserve natural resources that have been identified as valuable and important to society. Natural area acquisition often is justified in terms of benefits and uses to humans; after all, humans are paying the cost of protection. However, adding the variable of human use to the already complex equation of managing an urban natural area system exacerbates the difficulties inherent in managing fragmented, isolated, and often disturbed habitat patches. It is the human presence in natural areas that provides the greatest challenge to resource managers (Mazzotti and Morgenstern 2002).

Protecting urban natural areas not only contributes to the conservation of biological diversity but also provides valuable opportunities for human enjoyment by improving water quality, nature-based environmental education, and urban natural area research. Many studies have shown that experiencing nature in the city has a strong appeal for urban residents. Views and direct experience of natural landscapes have been shown to provide spiritual sustenance and stress relief from urban life (Kaplan et al. 1998). Other research suggests that creativity improves within hours and days after a nature experience, thus linking nature conservation and restoration with economic development (USDA Forest Service 1995). A large majority of Pacific Northwest residents spend most of their time in urban areas and have increasingly less time to travel any distance to commune with nature. And many natural area volunteers also describe the positive psychological benefits of working to restore the ecological health of natural areas. Unfortunately, restoration activities are not automatically received with enthusiasm, in part because of misunderstandings and poor communications between the land managers and local residents, who may not understand why armies of workers have descended on their local wild area to kill plants and reduce their privacy (Gobster and Hull 2000).

One goal often expressed by urban natural area managers is to encourage people to develop a more direct attachment to nature. The thinking is that if site restoration and management goals are formulated to encourage this attachment, people will help protect existing areas and support efforts to acquire new ones (Gobster and Hull 2000). The following suggestions for co-managing natural areas for nature and people are adapted from the experiences of the Chicago Wilderness Coalition:

- *Step 1: Identify sensitive areas and features.* In order to protect sensitive natural resources and understand the community's strong attachments to the natural area site, the first step is to inventory and analyze the special features that are important ecologically and culturally.
- *Step 2: Develop a balanced management plan.* Consider establishing separate management zones within natural areas so that the site can be managed to address the opportunities and constraints identified in Step 1.
- *Step 3: Develop site-specific design and management strategies.* Establish management strategies that encourage community attachment and balance appropriate use with the protection of fragile environments. Some techniques include improving safety along trails, providing physical or visual access to unique features, and carefully designing the natural area landscape to address community landscape preferences.
- *Step 4: Allow and plan for incremental change.* The results of urban natural management research suggest that communities do not appreciate drastic change. Therefore, restoration projects should be planned to gradually introduce change to the landscape over time (Gobster and Hull 2000).

Managing small patches of natural area lands embedded in a larger matrix of urban development is very challenging. Difficulties arise from the complicated mix of ecological and cultural issues involved in management and the failure of managers to fully engage with the community in resolving those issues. Urban natural area managers are trained in ecology, not sociology. Choosing an appropriate level of "naturalness" often is very contentious because people experience natural areas in many different ways, based on how they use or experience them (Gobster and Hull 2000). Resolving issues such as the impacts of dogs or mountain bikes on wildlife habitat is very difficult. Managers must understand that different users have different expectations and should be prepared to work with users and the community to formulate thoughtful and explicit visions of naturalness that are both ecologically practical and culturally appropriate.

This vision should be the basis for an ecological restoration approach for any given site. Perhaps before decisions are made, potential activities could be evaluated in light of Aldo Leopold's well-worn statement, "A thing is right when it tends to preserve the integrity, stability and beauty of the biotic community, which includes humans" (particularly in urban areas). Thoughtful urban restoration provides an opportunity to reestablish our human relationships and heal our connection to the native landscape, flora, and fauna. See Plate 8 for more urban restoration examples.

References

Clewell, A., J. Rieger, and J. Munro. 2000. *A Society for Ecological Restoration Publication: Guidelines for Developing and Managing Ecological Restoration Project*.

Dramstad, W., J. Olsen, and R. Forman. 1996. *Landscape Ecology Principles in Landscape Architecture and Land-Use Planning*. Island Press, Washington, DC.

Gobster, P. and B. Hull. 2000. *Restoring Nature: Perspectives from the Social Sciences and Humanities*. Island Press, Washington, DC.

Heinz Center for Science, Economics, and the Environment. 2003. *State of the Nation's Ecosystems*. The Heinz Center for Science, Economics, and the Environment, Washington, DC. Available at www.heinzctr.org/ecosystems/report.html.

Hennings, L. A. and W. D. Edge. 2003. Riparian bird community structure in Portland, Oregon: habitat, urbanization, and spatial scale patterns. *The Condor* 105: 288–302.

Kaplan, R., S. Kaplan, and R. L. Ryan. 1998. *With People in Mind, Design and Management of Everyday Nature*. Island Press, Washington, DC.

Jensen, M. 1998. Ecologists go to town: investigations in Baltimore and Phoenix forge a new ecology of cities. *Science News On-Line*, available at www.sciencenews.org/pages/sn_arc98/4_4_98/bob2.htm.

Larson, D. W. and J. T. Lundholm. 2002. The puzzling implications of the urban cliff hypothesis for restoration ecology. *Society for Ecological Restoration News* 15(1).

Leopold, A. 1949. The land ethic. In *A Sand County Almanac*. Oxford University Press, Oxford, UK.

Mazzotti, F. and C. Morgenstern. 2002. *A Scientific Framework for Managing Urban Natural Areas*. University of Florida, Institute of Food and Agricultural Science, Gainesville.

Sauer, L. 1998. *The Once and Future Forest: A Guide to Forest Restoration Strategies*. Island Press, Washington, DC.

Society for Ecological Restoration International Policy and Science Working Group. 2004. *The SER International Primer on Ecological Restoration*. Society for Ecological Restoration, Tucson, AZ. Available at www.ser.org.

Stille, A. 2002. Wild cities: it's a jungle out there. *The New York Times*, November 23.

USDA Forest Service. 1995. *Landscape Aesthetics: A Handbook for Scenery Management*. Agricultural Handbook No. 701, Washington, DC.

Resources on the Web

Australian Research Centre for Urban Ecology, a Division of the Royal Botanic Gardens, Melbourne, Australia: arcue.botany.unimelb.edu.au/.

Center for Urban Ecology, National Park Service, U.S. Department of the Interior: www.nps.gov/cue/cueintro.html.

Center for Urban Restoration Ecology, a collaboration between Rutgers University and the Brooklyn Botanic Garden, NJ: www.i-cure.org/.

Chicago Wilderness, Chicago, IL: www.chiwild.org/index.cfm.

Columbia Slough Watershed Council, Portland, OR: www.columbiaslough.org/.

Ecological Cities Project, the University of Massachusetts at Amherst: www.umass.edu/ecologicalcities/.

Environmental Protection Agency, National Pollutant Discharge Elimination System information: cfpub.epa.gov/npdes/home.

Friends and Advocates of Urban Natural Areas (FAUNA), Portland, OR: www.urbanfauna.org/index.htm.

Institute of Urban Ecology, Douglas College, New Westminster, BC: www.douglas.bc.ca/iue/about.htm.

Long-Term Ecological Research: The Baltimore Ecosystem Study: www.beslter.org/.

Long-Term Ecological Research: The Central Arizona & Phoenix Ecosystem Study, Tucson: caplter.asu.edu/.

Management and Restoration of Natural Area Landscapes, U.S. Forest Service North Central Research Station,

Chicago, IL: www.ncrs.fs.fed.us/4902/focus/restoration/natural_areas/.

Metro Greenways, Minnesota Department of Natural Resources: www.dnr.state.mn.us/greenways/index.html.

Metro Regional Government (Portland), Regional Planning & Parks, Trails and Greenspaces: www.metro-region.org/.

Natural Area Park Vegetation Inventory, Portland Parks and Recreation: www.portlandparks.org/NaturalAreas/veg_survey/default.asp.

Restoration and the City, Manaaki Whenua Landcare Research, Lincoln, New Zealand: www.landcareresearch.co.nz/research/social/restorationandthecity.asp.

Seattle Public Utilities Natural Drainage Projects: www.seattle.gov/util/About_SPU/Drainage_&_Sewer_System/Natural_Drainage_Systems/index.asp.

Seattle Urban Nature Project: seattleurbannature.org/msngls.html.

Urban Ecology, Berkeley, CA: www.urbanecology.org/history.htm.

Urban Ecology Research Lab, University of Washington, Seattle: www.urbaneco.washington.edu/.

Urban Ecosystem Research Consortium, Portland State University & U.S. Fish & Wildlife Service, Portland, OR: www.esr.pdx.edu/uerc/.

Chapter 13

Stream Systems

Jack E. Williams and Gordon H. Reeves

Restored, high-quality streams provide innumerable benefits to society. In the Pacific Northwest, high-quality stream habitat often is associated with an abundance of salmonid fishes such as chinook salmon (*Oncorhynchus tshawytscha*), coho salmon (*O. kisutch*), and steelhead (*O. mykiss*). Many other native fish species, such as lampreys (*Lampetra* spp.), sturgeons (*Acipenser* spp.), sculpins (*Cottus* spp.), and suckers (*Catostomus* spp.), may be of less economic importance but are of high value to stream ecosystems. Northwest streams and rivers also are valuable for numerous beneficial uses besides fisheries, such as providing high-quality water to municipal and industrial users, ameliorating damage from high flood flows, recharging groundwater aquifers, and routing sediment. Benefits arising from healthy rivers may go unnoticed by the average person but nonetheless are important to society. Of course, healthy streams also are more attractive than degraded systems and often serve as focal points for water-based recreation.

Many streams in the Pacific Northwest, and across the entire nation, are so degraded that they no longer support these beneficial uses. Nearly one in three of our native fish species are so rare as to deserve some sort of protective classification, and hundreds of stocks of salmon and steelhead, the signature fish of the Pacific Northwest, are threatened with extinction or already have been extirpated (Nehlsen et al. 1991, Stein et al. 2000). For example, 80% of fish habitat in Oregon's upper Grande Ronde River failed to meet forest plan standards for water temperature, stream sediment, and riparian conditions. In Idaho's Clearwater National Forest, 70% of streams failed to meet similar standards for temperature, sediment, and riparian vegetation (U.S. Forest Service and U.S. Bureau of Land Management 1994). Many urban streams are channelized and contaminated by runoff containing road surface oils, fertilizers, and other chemicals. Clearly, there is much work to do.

If we seek to restore a stream, it is important to know what a healthy stream looks like and how it functions. Over the years, human activities in many watersheds have simplified stream channels, separated streams from their floodplains, fragmented streams, altered flows, and introduced toxic contaminants (Stanford and Ward 1992, Frissell 1997). Restoration seeks to reverse such changes. It may be impractical to restore streams to historical condition, even if we know what that condition was, but it is important to reverse the primary causal factors in the degradation process. Sometimes, the cause of stream degradation may be far upstream or upslope from where the symptoms of the problem appear. Regardless, stream restorationists often seek to reintroduce channel complexity, reconnect fragmented systems, restore natural flow regimes, revegetate riparian areas, and eliminate sources of chemical pollution.

The restoration potential of a degraded stream depends on its historical condition, cumulative impacts, soils, topography, and other factors. Nonetheless, many traits are common among healthy streams. Characteristics of a healthy stream include diverse habitat types, sufficient structural diversity provided by large woody debris and boulders, well-vegetated riparian areas, and deep pools (Table 13.1). However, a successfully restored stream is more than a mere compilation of these characteristics. It must function properly as well. Key concepts for stream function are outlined in the next section.

TABLE 13.1.

Characteristics of healthy stream systems in the Pacific Northwest.

Characteristic	Description
Habitat diversity	Roughly equal numbers of pools, riffles, and runs should be present. Complex braided channels are preferred over simple, straight streams.
Large wood	Downed trees and other large woody debris creates pools, stores sediments, and provides organic matter.
Water quality	Cool, pollutant-free water is critical to spawning, juvenile rearing, and adult resting habitat for many fish species. Generally, <16°C is needed for spawning and <18°C for rearing.
Flow regime	The hydrograph is similar in intensity and flow amounts to historical conditions. High flows may be needed at certain times to dig pools and move sediment.
Riparian vegetation	Adequate riparian vegetation is needed to shade streams, protect banks from severe erosion, and provide nutrients.
Deep pools	Sufficient deep pools are necessary as thermal refuges and holding habitat for many fish species.
Width-to-depth ratio	Ratio should be less than 10. Generally, deeper and narrower streams provide better habitat.
Bank stability	Banks should be 80–90% stable. Some erosion is needed, but too much is detrimental.
Fine sediments	Stream substrates should not exceed 20% fine materials (clay, silt, and sand) in riffles. Most streams suffer from high sediment loads.

Key Concepts for Stream Restoration

Rivers and streams represent a small percentage of the landscape, but their values to ecological functions and biological diversity far outweigh their size. Streams carry water and sediment downstream, eventually into estuaries and the ocean. They also provide habitat for many aquatic and semiaquatic organisms such as fish, amphibians, aquatic insects, mollusks, and plants. But this simple picture belies a surprising degree of dynamic movement and complexity. Water flows into streams from upstream channels, surface runoff, and groundwater aquifers. Streams that appear stable often may be eroding their banks and moving laterally over time. During floods, streams reconnect with their floodplains and may form new channels or reconnect with formerly abandoned ones. Erosion and sediment deposition are natural features of the stream channel. The species in streams may be equally mobile and subject to broad migratory patterns.

Successful stream restoration projects not only incorporate this complexity but also need to embrace it. Understanding the interactions between stream channels, riparian areas, and broader watersheds is an important component of stream restoration. That said, successful stream restoration projects may focus on a small stream reach or entire watershed, but even the small projects must be embedded in the broader context in which the work is occurring. Knowing some key hydrologic concepts will help get a project off on the right foot.

The size and flow of a stream are directly related to the size and shape of its watershed, or drainage area. Within this watershed, soil types, topography, and land use all play critical roles in further defining flow regimes. Most restoration projects are concerned with a particular stream or stream segment. However, because streams are defined by the conditions in their watersheds, the proper focus of restoration efforts often is much broader than one would initially imagine. As eminent hydrologist H. B. N. Hynes (1975) admonished, we must not divorce the stream from its valley in our thought process because to do so is to lose touch with reality.

Because understanding the proper size, or scale, of restoration work is so important in streams, it is necessary to understand how streams are organized, connected, and classified within watersheds. There are three separate classification schemes that look at streams from different perspectives. The first approach describes a hierarchical arrangement of streams in relation to broader valleys and watersheds. In this classification system, stream organization would proceed from a channel unit (such as a pool or riffle) to a stream reach, to valley segment, to watershed, to geomorphic province (Frissell et al. 1986). Each hierarchical level has independent drivers and sources of variability. A second approach is to classify streams according to function, as described by a hierarchy of channels from headwaters to larger, downstream sections (Figure 13.1).

A third classification system for streams is the channel type method of Rosgen (1996). Rosgen developed a system of naming streams by letters, Aa+ to G, in which slope, width-to-depth ratio, entrenchment, sinuosity (amount of curvature in the channel), and channel material are key defining characteristics (Figure 13.2). Type A streams are steep, with low sinuosity, whereas type E streams are meandering streams in flat valley bottoms. Rosgen lists four specific objectives against which the utility of any classification system for restoration efforts should be evaluated. Classification systems should inform restoration efforts in the following manner:

- Predict a stream's behavior from its appearance.
- Develop specific hydraulic and sediment relationships for a given channel type.
- Facilitate extrapolation of site-specific data to stream reaches having similar channel types.
- Provide a consistent frame of reference for communicating stream morphology and condition.

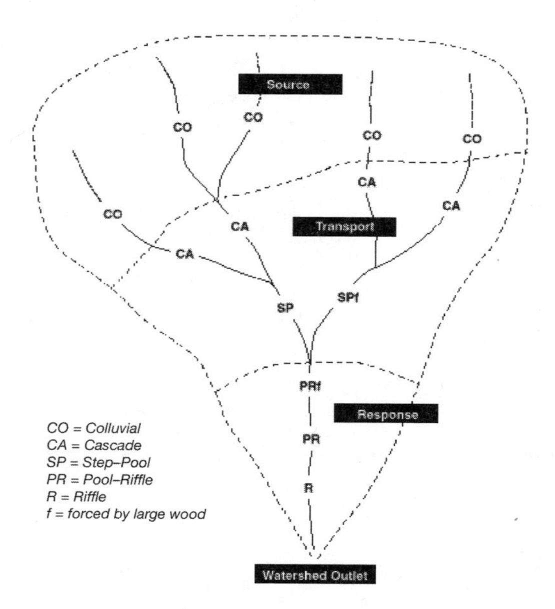

FIGURE 13.1. Watershed classification diagram based on stream gradient and stream function. Steeper headwater streams often serve as source habitats for wood and sediments that move downstream through middle-elevation transport reaches and ultimately arrive in valley bottom response streams. (*From Montgomery and Buffington 1993*)

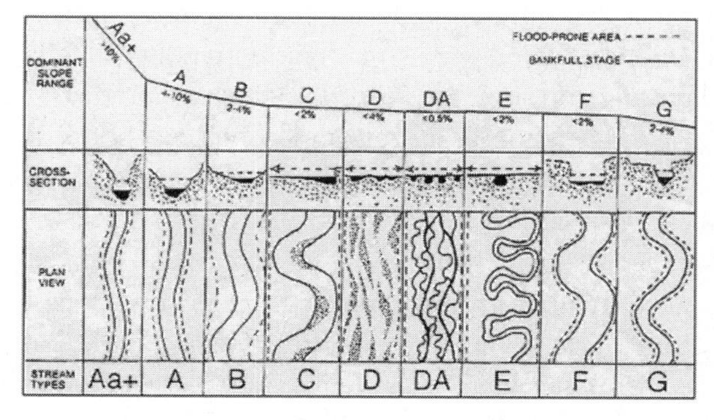

FIGURE 13.2. Diagram of Rosgen's stream classification system based on slope, cross-section channel configuration, and meandering. (*Figure provided by Wildland Hydrology and Dave Rosgen*)

Understanding the connections between streams and their surrounding lands also is critical for successful restoration. Water provides many of these connections. Obviously, streams are connected in a longitudinal sense, that is, from upstream to downstream. Streams also connect to land laterally, when floodwaters spill onto floodplains, and vertically, when surface flows connect to subsurface and groundwater systems. In many watersheds, human activities such as dam construction and stream channelization have severed these connections. Yet productive streams need woody debris from adjacent riparian areas and nutrients from upstream reaches. Restoring connectivity within watersheds often is needed to maintain healthy and productive river systems (Naiman et al. 1992, Williams et al. 1997). For example, connectivity could be restored between a river and its floodplain by removal of riprap and other hard channel structures. Flood energy is more likely to dissipate as high flows broaden out and silt is trapped to fertilize floodplains. One important caveat in reconnecting headwaters with downstream reaches involves nonnative species. If undesirable fish species or fishborne diseases occur in downstream reaches but not upstream, then efforts to reconnect stream systems could inadvertently facilitate spread of nonnatives or pathogens.

River channels seldom are straight. Normally, channels are curved, with the outside bend slowly eroding and the inside point bar building outward. In this way, the channel width remains roughly the same. The level of the floodplain typically is equal to the top of the point bar, except where channels have become significantly degraded and downcut over time.

Natural channels have sequences of riffles, runs, and pools that provide diverse habitats important to many aquatic species (Figure 13.3). Some species favor the shallow, rocky substrates often

FIGURE 13.3. Restoring habitat diversity in streams often depends on an adequate riparian zone. Riparian areas serve as a source of large woody material, which contributes to habitat complexity. In this photo of the Scott River near Ft. Jones, California, note that the riparian area is wide enough to allow natural steam meanders, which facilitates development of pool–riffle complexes. (*Photo by Brian Barr, World Wildlife Fund*)

associated with riffles, whereas others need the greater depths and often cooler conditions found in the pools along outside stream meanders. A run is a transitional habitat type found between riffles and pools and often is shallow but with smooth surface flow. The numbers, size, and shape of these habitat features may be critical to certain fish species or life history stages. One fascinating study documented a 33–73% loss of deep pool habitats and subsequent impacts to salmon and steelhead in eastern Washington and Oregon river systems over a 40-year period as large silt and sediment loads moved from upslope into downstream reaches (McIntosh et al. 1994). Such historical comparisons identify restoration needs that otherwise might not be obvious.

Floods and drought can be powerful drivers in stream shape and condition as erosion increases and sediment is deposited. Interestingly, climate change and urbanization alter the timing and intensity of such events. With increased urbanization in one well-studied watershed, the mean annual flood increased from 2,300 cubic feet per second to 3,490 cubic feet per second, and the 10-year flood increased from 7,300 cubic feet per second to 16,000 cubic feet per second (Leopold 1994). With more hardened surfaces in our urban and suburban areas, flood peaks occur sooner and at greater intensities, with significant stress to channel conditions.

Increasing frequency and intensity of flood events is not the only change wrought by urbanization. Stormwater runoff from city streets also carries a toxic brew of pesticides, herbicides, hydrocarbons, and other chemical contaminants. Riparian zones in urban areas often are highly constrained, further reducing the ability of vegetation to filter pollutants.

In the Pacific Northwest, many streams are dynamic and may change in condition every 1–10 years in response to floods and debris flows. Many stream systems are surprisingly well adapted to intense but short-duration disturbances. For example, in the central Oregon Coast Range infrequent stand-replacing wildfires may combine with intense winter rainstorms to produce landslides and debris flows. There may be short-term detrimental impacts such as reduction in food resources and higher sedimentation, but longer-term effects may be positive, with introduction of large wood and rebuilding of pools (Reeves et al. 1995). However, changes in the frequency, magnitude, or duration of such disturbances and the materials they leave behind can alter species composition, habitat features, and resilience of a stream system. Sometimes human activities result in longer-lasting disturbances that may force a system to a different endpoint. For example, roads built across streams may alter hydrology and sediment regimes for decades and may combine with natural disturbances for more severe impacts. The important point from a restoration perspective is to understand the broader context that is driving watershed condition and stream responses.

Choosing the Correct Restoration Approach

A key to successfully restoring streams is to properly understand the objectives of the effort and to use the correct management strategy, particularly with regard to expectations for the watershed in which the stream of interest resides. Recovery of fish populations generally is a primary motivator for undertaking restoration efforts. Expectations for a watershed with regard to recovery and the degree to which it has been modified influence potential courses of action and guide the management direction from restoration to rehabilitation, enhancement, or mitigation (National Research Council 1992). Systems that are currently in good condition and have intact native fish communities are candidates for *protection* rather than restoration. Habitat restoration should never be viewed as a substitute for or confused with habitat protection (Narver 1973, Reeves et al. 1991). There is no guarantee that restoration efforts will be successful, even with the most careful analysis, design, and execution. Preventing initial degradation is more economically and ecologically sound than repairing it, and some damage simply is not reversible. Protecting intact watersheds and streams should be part of any broader management effort.

For slightly altered systems, *restoration* can proceed either as natural restoration or as active restoration (National Research Council 1992). Natural restoration works best in systems that have the suite of key ecological processes intact but are slightly degraded biologically, physically, or chemically. The key is to remove or modify the causes of the degradation. Active restoration involves eliminating activities causing degradation and implementing additional efforts such as wood additions in channels or replanting of riparian zones. The latter efforts are viewed as catalysts that jumpstart the recovery process and are intended to be used only during the initial phases of recovery.

Rehabilitation occurs in systems where natural processes and conditions cannot be reestablished and restoration is not possible (National Research Council 1992). In highly developed urban streams, rehabilitation is the appropriate objective. These systems have diminished capability for developing and maintaining productive habitats and need more human effort and money to provide fish habitat and other beneficial uses. An example of rehabilitation would be continuous additions of wood into streams over time into riparian areas that no longer contain large trees. Rehabilitated watersheds can provide refuges for fish, but they are vulnerable to change and cannot be expected to supply long-term habitat (National Research Council 1992). Addressing the causes of degradation as part of the rehabilitation effort can improve the response of the system.

More degraded streams need extensive efforts to become productive. *Enhancement* is the intentional effort to increase a particular feature of the stream, often developing conditions that are outside the range normally found there (National Research Council 1992). Examples of enhancement include the use of gabions (wire baskets filled with rocks) to collect gravels and create pools and the introduction of wood into meadow streams. Successful enhancement efforts entail an understanding of the operation and behavior of stream systems. The risk of conducting enhancement without this understanding is that the system may shift from one degraded state to another.

Because they generally introduce materials to the system that may not be natural, enhancement efforts require long-term commitments of time, energy, and funding to maintain productivity and generally do not provide a good model for success.

Mitigation may be the primary approach in systems where there is habitat loss with little or no potential to recreate natural conditions. It attempts to minimize or offset the effects of loss of a habitat by creating new habitat, generally in restricted areas, within the system. This may seem straightforward, but it is not (National Research Council 1992). Generally, degradation or loss of habitat results from a suite of factors. Offsetting the loss of a particular habitat type may not mitigate for the loss of ecological processes and connections, and this may limit the potential success of any mitigation effort. Furthermore, evaluations of "new" habitat creation efforts have demonstrated limited success in our efforts to create substitute habitat that possesses the full suite of functions and benefits found in nature (Frissell 1997). Like enhancement, mitigation may entail a long-term commitment to maintain desired conditions.

Assessing Problems and Opportunities

Understanding the ecological context and potential of the watershed is another key to successful restoration. The types of habitats that may be found in a particular watershed and the types and productivity of fish populations are a function of its geomorphic features (Montgomery and Buffington 1997, Burnett et al. 2003). For example, the most productive habitats for coho salmon are low-gradient (less than 2%) watersheds with wide valleys and intermediate-sized streams (Burnett et al. 2003) (Figure 13.4a). The most productive habitats for steelhead, a sea-going rainbow trout, are in steep-gradient (3–5%) intermediate-sized systems with narrow valleys (Burnett et al. 2003) (Figure 13.4b). Prime areas for coho salmon tend to have more pools with slower velocities and large accumulations of gravel. Steelhead habitats have smaller pools with higher velocities. These fish have body forms and behaviors suited for each habitat type (Bisson et al. 1988). Coho salmon are surface oriented, with bodies that are able to maneuver quickly. In contrast, steelhead are more oriented toward the bottom, with large pectoral fins to help them hold position in the current and narrow, laterally compressed bodies.

Working in areas that have the inherent capacity to provide habitat for the fish of interest increases the likelihood that restoration efforts will be successful and maximizes the potential response of the fish. Recognizing the inherent potential of an area to provide habitat for a given species helps identify the types of habitats that can be created, thus increasing the likelihood that a restoration effort will be successful and cost effective (Burnett et al. 2003). It is important to recognize that not every watershed or portion of it can provide favorable conditions for every fish. Every stream system has limits that must be recognized.

Additional insight about identifying priority areas for restoration can be gained by looking at historical conditions or by comparisons to a similar type of stream in unaltered condition. Many systems have been altered so extensively that we have little appreciation for the extent of change or the range of potential restoration opportunities. A good source of information on historical conditions, particularly in the western United States, is the Secretary of War (the forerunner to the Department of Defense) records, which can be found in archives and libraries. These records contain the first surveys of areas and were made by the U.S. government before settlement. These reports offer verbal descriptions of physical and biological conditions and often are accompanied by detailed maps. Early land surveys, historical society photos and records, old aerial photos, and older agency files may contain valuable information on past conditions. It may not be feasible to recreate historic conditions in most restoration efforts, but this information can help develop potential approaches and options that best fit the system.

FIGURE 13.4. *Top*, Productive coho salmon habitat is found in low-gradient stream sections (*photo by Jack Williams, Trout Unlimited*). *Bottom*, Productive steelhead trout habitat is found in higher-gradient stream sections (*photo by Kelly Burnett, Pacific Northwest Research Station*).

Identifying the factors that limit the production of fish or the productivity of a given watershed is difficult (Reeves et al. 1991). Myriad factors influence the creation and maintenance of fish habitat, often in complex and little understood ways. Restoration efforts often are based on the best judgment of people most familiar with the watershed or stream of interest. These generally are educated guesses based on experience in and knowledge of the system. The likelihood that this approach will work depends on the people doing the assessment and how well they know and understand the system. Obviously, the better their knowledge and understanding, the more likely it is that they will be able to identify the factor or factors that must be addressed.

Decision support models offer another potential option for better identifying factors that should be addressed in restoration efforts. These models often are used to help determine courses of action when there are multiple factors to consider. They also have been used recently to determine options for managing fish harvests (Mackinson et al. 1999) and to assess environmental conditions (Meesters et al. 1998). Decision support models are being used to evaluate watershed condition as part of the monitoring program for federal lands in the Pacific Northwest of the United States and northern California that are managed under the Northwest Forest Plan (Reeves et al. 2004). These models are built on empirical evidence and, where that is not sufficient, on professional judgment. They also can be constructed to accommodate the variation in inherent conditions between watersheds from different areas. These models allow identification of the factors most responsible for the condition of a watershed or stream reach. These factors then can be prioritized within the restoration framework. Examples of these models can be found at www.reo.gov/monitoring/watershed/AREMP/2004/aremp.htm.

The models can be built collaboratively by interested groups and provide a framework for the group to articulate a common vision and understanding of how the stream or watershed of interest operates. The development of consensus actually may be the real strength of this approach. It fosters a dialogue that can lead to a better understanding and appreciation of different points of view and improved collaboration within a group.

Watershed analysis is another process for identifying features to be addressed in a restoration effort. Watershed analysis has taken many forms since its formal inception in 1993 (FEMAT 1993, Ziemer 1997). On federally managed lands, it is an interdisciplinary effort by physical and biological specialists that identifies, among other things, factors that influence the current condition of a watershed, portions of the watershed that may need activity to prevent or correct degradation, and priority sites where various management activities could occur. The *Federal Guide for Watershed Analysis* (U.S. Department of Agriculture and U.S. Department of the Interior 1995) provides a good description of the watershed analysis process. Like the development of decision support models, this is a collaborative undertaking that can foster communication between individuals and interest groups that, in turn, result in a better understanding of limiting factors and restoration opportunities.

Formal approaches for identifying potential factors that must be addressed in a restoration are limited. One example is a dichotomous key developed by Reeves et al. (1988) that helps identify habitat factors that limit the production of juvenile coho salmon. This approach estimates the amount of given habitats (e.g., spawning, early rearing) needed to maximize density-dependent survival from one life history stage to another. Restoration efforts then determine the type and amount of habitat that should be created. The shortcoming of this approach is that it does not foster an ecosystem perspective.

Long-term success of any restoration effort depends to a large extent on not just correcting habitat problems but addressing the causes of degradation. Factors or processes in one part of a watershed may be responsible for the condition in another. For example, pool numbers may decline because of increased sedimentation. Creating new pools may prove futile unless the sources of sediment are identified and corrected. In other cases, restoration may entail action in a part of the watershed away from where the symptoms of the problem are most obvious. For example, stream temperature may be high in a portion of stream that is important to fish. If the stream is large, improving riparian shade in the immediate vicinity may not solve the problem because the influence of riparian shade is minor on larger streams. The best solution in such a case would be to examine the area higher in the watershed or in tributaries that enter the segment of interest to locate the source of warm water and try to reduce water temperatures there so that cooler water is delivered to the reach of interest. Similarly, changing chemical practices on agricultural lands and addressing stormwater management issues are the only ways to resolve downstream water quality problems traceable to these sources of chemical contamination. Considering the watershed as a whole and the relationships between its pieces are important for a successful restoration effort.

Planning a Stream Restoration Project

The National Research Council (1992) provides a straightforward template to planning stream restoration projects and provides a more detailed discussion of what is summarized here. Stream restoration planning should begin with a careful analysis of the problem to be treated and an understanding of the broader watershed context in which the stream occurs. As described earlier, watershed analysis is a good tool for understanding appropriate restoration needs. Once the problem is understood clearly, the project's goal and objectives should be developed. Typical goals might include restoring water quality, channel condition, or flow regimes or might relate to restoring steelhead trout or other components of the biotic community. Objectives detail more specific components that must be addressed to reach the goals. For example, to restore water quality, objectives might include reducing stream temperature, reducing sediment input, and achieving other specific characteristics of water quality. Objectives for channel condition might include reestablishing deep pools or spawning beds. At some point, quantifiable performance indicators must be developed for each objective. Thus, the number of deep pools per stream reach and depth of pools should be determined as a part of the relevant objective. For reestablishment of spawning beds, a performance indicator might be the presence of five riffles with at least 30% substrates in gravel. These quantifiable measures are critical in effectiveness monitoring and in being able to evaluate the ultimate success of the project. A monitoring program should be an integral part of any restoration effort.

Stream Restoration Techniques

Riparian areas are critical to the health and integrity of stream systems and often are an appropriate place to start in stream restoration efforts (see Chapter 6). Riparian areas provide many benefits to streams. If sufficient riparian vegetation is present along streams, the riparian areas will provide shade to cool stream temperature, serve as a source of woody material into stream channels, filter large quantities of fine sediment from upland erosion, and provide an important source of nutrients in the form of leaves and smaller woody material. Guidelines for federal land management in the Pacific Northwest call for riparian buffer zones of 300 feet on each side of fish-bearing streams (FEMAT 1993). For private lands, establishing riparian buffer zones 100 feet wide on each side of streams is a good starting point for balancing ecological and social imperatives, but steepness of slope away from streams, unstable soils, and increased intensity of nearby human activity all are important factors that could necessitate wider riparian buffer zones.

Riparian protection zones provide space for stream meandering, buffer the impacts of upslope management activities, and facilitate vegetative growth that will shade streams and help filter out sediment. Buffers along small streams, even intermittent streams, are critical because these channels often greatly influence water quality and provide critical pathways for wood and sediment into larger, fish-bearing streams. Certain activities such as extensive tree harvest, removal of large trees, road building, and livestock grazing should be excluded from riparian buffers.

Another critical component of healthy stream systems is habitat diversity and patchiness. Streams with diverse habitat types, deep and shallow pools, pocket pools behind boulders, riffles, and runs provide a variety of physical habitats that in turn support diverse species and life history stages of fishes and macroinvertebrates. Many degraded streams lack this habitat diversity as silt has filled pools and channelization has eliminated riffles and structural diversity (Williams and Williams 2004). Habitat diversity often can be restored through reintroduction of large wood pieces and boulders and removal of hardened stream banks and berms, allowing streams to begin natural meanders.

Many stream restoration projects consist of placing logs, tree stumps, and boulders into stream channels in an effort to create pools and diversify habitats. The importance of large wood, boulders, and other instream structures relates to their ability to control routing of water and sediment, shape pools and riffles, and serve as a substrate for macroinvertebrates (Naiman et al. 1992). The amount of large wood needed in streams to meet proper functioning condition can be substantial. An analysis from reference streams in western Washington yielded the following large wood targets based on stream width classes to achieve a "good" rating (large wood is defined as greater than 10 centimeters in diameter and 2 meters in length): Streams 0–6 meters in bank full width need more than 38 pieces, streams 6–30 meters wide need more than 63 pieces, and streams 30–100 meters wide need more than 208 pieces (Fox et al. 2003). For forested streams east of the Cascade crest, the U.S. Forest Service and Bureau of Land Management established objectives for large wood of at least thirty-two pieces more than 10 meters long per kilometer (U.S. Forest Service and U.S. Bureau of Land Management 1994).

In some restoration projects, boulders, stumps, and logs have been placed into streams and anchored with cables for good long-term success (Schetterling and Pierce 1999). In other projects, high flows and unstable banks have undermined anchored wood pieces, resulting in the tearing away of cabled materials during peak flows and creation of new erosion problems (Frissell and Nawa 1992). The importance of understanding the broader watershed context of stream restoration work is evident. Cabled structures are more likely to fail in wider, Rosgen C channel types, which meander more and have a larger floodplain, than in B channel types that are more confined by valley walls (Schetterling and Pierce 1999). In the long run it may be more satisfactory to simply place boulders, stumps, and logs loosely into stream channels or along stream banks in riparian areas and let higher stream flows distribute them naturally. In general, we favor working with the forces of flow, erosion, and deposition inherent in particular stream systems. In this way, we can capitalize on the natural healing capacity of streams and allow the river to do most of the work.

In some cases, a compromise may be needed between the desire to let a stream meander naturally and the need to protect a particular site from erosion. Riprap and other hardened bank techniques are undesirable because they decrease habitat quality for fish and other species, and they may deflect erosion forces further downstream, creating additional concerns. An alternative that may be worth examining is the use of long-line cabled logs that are placed in slack water along eroding banks so that they remain floating during high flows (Nichols and Sprague 2003). The cabled logs will capture additional wood and debris during high flows and help stabilize erosion while improving habitat conditions.

Large amounts of fine sediment in stream substrates are a common concern for fishery biologists and land managers. Amounts of fine materials (clay, silt, and sand) exceeding 20% of stream substrates result in increased embeddedness and loss of interstitial space between gravels and cobble. The loss of this space diminishes the habitat for many aquatic insects, degrades spawning habitat, and reduces egg survival. Fine materials come from many sources, including row crop agriculture, livestock grazing, mining, timber harvest, road construction, and urban development. Many of the preferred methods to reduce the input of fine materials involve modifications to the upslope activities that cause the erosion in the first place and are summarized in Waters (1995). As discussed earlier, vegetation in riparian zones helps filter out silt and sediment before they enter streams. Once the causes of sediment production have been reduced, flushing or pulse flows may be needed to distribute fine sediments to channel margins or downstream and out of the system. In some situations, instream structures can be installed to increase current velocity and scour certain stream segments favorable to fish production (Waters 1995).

Bridges, culverts, and other stream crossings modify hydrologic processes by restricting flows and sediments and alter movement of fish and other stream biota. Stream crossings result in direct loss of instream and riparian habitat. Poorly designed or undersized culverts may impede fish by creating vertical barriers or velocity barriers. In addition, culverts and other water crossings pose a risk to downstream habitat if structural failure occurs during high flows. For all these reasons, stream crossings provide many restoration opportunities. Common restoration work may reengineer placement of the culverts, install larger culverts, or install baffles in the culverts to provide more suitable flows for fish passage. Ideally, circular culverts are replaced with bridges or bottomless archway culverts that retain natural stream substrates. The Washington Department of Fish and Wildlife provides a useful handbook for proper design of road culverts that is available online (Washington Department of Fish and Wildlife 2003). When designing fish passage through road crossings it is important to consider flow variation and its impact on movement of the full range of native fishes, juveniles and adults, and macroinvertebrates present in the system. A properly designed stream crossing does not impede any species or substrate.

Many thousands of dams and diversions exist in the Pacific Northwest and are the single leading cause of stream degradation and loss of fisheries (Collier et al. 1996). Although many dams provide important economic and social benefits, for others cost–benefit analyses indicate that removal should be considered. Like any constructed feature, dams have finite lifetimes and costs associated with their safety or maintenance that may present opportunities for removal. In some cases, pumps or other technologies that do not entail complete stream blockage can provide the service that a dam provides, such as irrigation water diversion. This would address concerns with fish passage but not flow problems. In other cases, the benefit accruing from the diversion may be so minimal that acquisition of the water right and conversion to an instream right is feasible. Numerous case studies of dam removal in the Pacific Northwest, including Jackson Street Dam and Marie Doran Dam in Oregon, four dams along Butte Creek in California, and the Rat Lake Dam in Washington, detail successful stream restoration efforts. At least 177 dams were removed in the 1990s in the United States (Friends of the Earth et al. 1999). Several excellent information sources exist on the various costs and benefits associated with maintaining or removing dams and the numerous environmental concerns related to either choice (River Alliance of Wisconsin and Trout Unlimited 2000, American Rivers and Trout Unlimited 2002).

CASE STUDY: RESTORING LARGE WOOD STRUCTURE IN TENMILE CREEK

Addition of large wood to stream channels is one of the most common components of habitat restoration efforts. Large wood is an important element of good habitat and provides structure for scouring pools, storing and sorting gravel, and retaining organic material that forms the energy base for organisms that are eaten by fish. Many streams currently lack large wood because of past activities that removed instream structure or because of the harvest of trees from riparian areas.

Tenmile Creek, on the central Oregon coast, was the site of a major restoration effort that began in 1996 and was undertaken by local landowners, the Oregon Department of Fisheries and Wildlife, Siuslaw National Forest, and the National Audubon Society. The stream had been affected by activities such as channel clearing during initial settlement and by more recent activities, primarily timber harvest. Surveys conducted at the start of the restoration program found that the amount of large wood was low, only 4–33% of that found in a reference stream in a nearby wilderness area. The consequence was that large, deep complex pools were lacking in Tenmile Creek, and the stream channel was isolated from the floodplain in many sections.

A total of 240 pieces of large wood were added to fifty-four locations in the channel. Two hundred of the pieces were whole trees, including branches (31-inch butt diameter and 110 feet long). The remainder were primarily root wads. Whole trees were taken from adjacent ridges, away from the stream. The other pieces were taken from landslide deposits high in the watershed. All wood was moved with a helicopter and was placed in clusters of three to eight pieces in lower-gradient portions of the network, near the upper or lower entrance to old side channels or in natural bends where wood naturally accumulates (Figure 13.5). Some pieces were wedged between live trees in the riparian zone to increase stability, but the majority were not cabled or attached so that they could move downstream and settle in areas where they could provide the most benefit. Allowing natural wood

FIGURE 13.5. Large wood restoration in a coastal Oregon stream. Log jams were placed where they could increase the stream's connections with its floodplain during moderate stream flow. (*Photo by Jack Sleeper, Siuslaw National Forest*)

movement is preferred, but in situations where such movement is undesirable, pieces can be cabled together or attached to streamside trees. Cost for the wood placement in Tenmile Creek was $140,000, about 3% of the total project costs.

Wood placement increased the number of pools deeper than 3 feet in the summer and the amount of side channel area available in winter, important rearing habitat for juvenile salmonids (Figure 13.6). The abundance of steelhead smolts and the freshwater survival of coho salmon and steelhead increased in Tenmile Creek after wood placement.

The other components of the restoration effort in Tenmile Creek contributed to the success of the wood placement. Roads were realigned and stabilized, which allowed the stream to move onto floodplains. Riparian zones that were in poor condition were planted with conifers or thinned where conifers were present but showed poor growth. The comprehensive nature of this effort was the key to its success.

FIGURE 13.6. Habitat conditions in Tenmile Creek *(a)* before and *(b)* after the addition of large wood. (*Photos by Jack Sleeper, Siuslaw National Forest*)

Rehabilitating Urban Streams

Urban streams present some of the most important but challenging restoration opportunities. Although urban lands make up less than 2% of the land base in the Pacific Northwest (Pease 1993), they have a disproportionately large influence on fisheries because their valley bottom locations historically were occupied by complex, braided-channel stream systems. Such areas were very important for rearing juvenile salmon, providing cold water refuges for adults, and facilitating migration. Urban stream systems present a formidable challenge to land managers because they have been so dramatically altered. Because of the extent of this alteration and infeasibility of returning these systems to historic conditions, it typically is more appropriate to speak of rehabilitating urban streams rather than achieving full restoration.

Urban streams typically are affected by stream channelization and loss of instream habitat, changes in hydrologic regimes, pollution, and loss of riparian habitat. Urban areas have higher densities of roads and more parking lots, rooftops, and other impermeable surfaces and often are characterized by extremely constructed floodplains. This means that rains run off more quickly, often carrying pollutants such as hydrocarbons, phosphorus, insecticides, and heavy metals. In some urbanized Puget Sound streams, for example, what used to be a 10-year flood event before urbanization now occurs every 1 to 4 years (Moscrip and Montgomery 1997). Although such changes have profound impacts on hydrologic regimes and macroinvertebrate and fish communities, there have been several recent case studies of successful urban stream rehabilitation (Fresh and Lucchetti 2000, Konrad 2003).

The first step in restoring urban streams is properly diagnosing problems and identifying their causal factors. Rehabilitation projects should be designed to address as many of these causal factors as possible. Often, problems include degraded instream habitat, loss of riparian habitat, reduced water quality, and altered flows. Although rehabilitation of urban streams may be a compromise between social constraint and ecological need, watershed-scale analysis, application of scientific principles, monitoring, and adaptive management will benefit those working in urban zones as much as it benefits those focusing on wild areas.

Perhaps because of the extent of modifications, incremental improvements to almost any urban streams are achievable. However, restoration of natural flows in urban streams is difficult because of the extent of development in the watershed and construction on the floodplain. Recreating wetlands and constructing sedimentation basins and stormwater retention systems are effective in slowing down urban runoff (Konrad 2003). Stream habitats in urban areas often are channelized and lack the structural diversity found in more natural environments. Pool habitats often are lacking. Efforts to slow flows in higher-gradient (more than 2%) streams can be achieved by reintroducing boulders and large wood that reduce flows and recreate pool conditions. Riprap and concrete stream banks should be removed when practical. One of the most effective rehabilitation techniques is to provide a protected riparian buffer and reestablish streamside vegetation. If riparian areas cannot be widened, removing riprap and concrete and replacing them with woven mats and other materials that will protect soils from erosion but still allow planting and revegetation of stream banks will aid rehabilitation of urban streams.

Commercially produced polyethylene grids (known by trade name Geoweb) are available in 8- by 4-foot sections with 8-inch honeycomb cells that can be anchored in place, filled with substrate, and planted (Hunt 1993). Revegetating with native plants should be emphasized. Nonnative plants, including Himalayan blackberries, purple loosestrife, and reed canary grass, should be controlled because they often are detrimental to native wildlife and overtake native plant species. Typical impacts to urban streams and appropriate rehabilitation measures are described in Table 13.2.

TABLE 13.2.

Common problems in urban streams and appropriate rehabilitation responses.

Modification	Resulting Impact	Rehabilitation Techniques
Channelization and filling of instream habitat	Loss of instream habitat; loss of instream structure and habitat diversity	Where practical, remove hardened stream bank material and widen channel.
Channelization and lack of stream meandering	Loss of gravel bars and spawning habitat	Where practical, remove hardened stream bank or introduce spawning gravels.
Dams, diversions, and instream blockages	Impeded movement of fish, invertebrates, and stream substrates; altered flow regimes	Replace dams and diversions with pumps or install fish ladders if blockage cannot be removed.
Increased pavement and other impermeable surfaces	Increased runoff; increased intensity and frequency of floods	Construct sediment basins to slow inflow; increase channel width, culvert size, and other stream crossings to accommodate higher flows.
Loss of riparian habitat and streamside vegetation	Increased erosion; increased water temperature; substrates >20% fine materials	Provide undisturbed riparian buffers; replant native vegetation along streams; introduce wood and boulders into streams.
Increased pollution from streets and industrial sources	Reduced water quality and impaired fish and macroinvertebrate communities	Install bioswales and sedimentation basins to reduce runoff; treat inflows; educate public about need to reduce polluted runoff.

Case Study: Removing a Small Irrigation Dam in Bear Creek

Thousands of small, older dams impede fish movement, degrade water quality, and alter temperature and sediment regimes in streams across the West. Some of these dams have been torn down as they became unsafe or obsolete. In recent years, the need to restore streams and rehabilitate fisheries has provided additional incentives for dam removal.

Breaching of the Jackson Street Dam, a small irrigation diversion dam on Bear Creek in southwestern Oregon, provides a case study of dam removal for the purpose of restoring natural stream conditions and anadromous fish passage while maintaining the initial purposes of the dam (Smith et al. 2000). The dam was constructed in downtown Medford in 1960 to divert water from Bear Creek into irrigation canals operated by the Rogue River Valley Irrigation District. The dam consisted of an 8-foot-high concrete structure that was raised another 3 feet with the addition of stoplogs during irrigation season. When full, the 11-foot-high structure created a 2-acre reservoir. Over time, sediment had filled the reservoir to the top of the 8-foot concrete dam level.

Despite its location in the middle of a growing urban zone, steelhead, coho salmon, chinook salmon, and Pacific lamprey all ascend Bear Creek from the Rogue River to spawn. A fish ladder existed on the dam, but it was constructed on the opposite side of the channel from the main flow, creating a near total blockage of adult fish moving upstream under certain flows (Smith et al. 2000). In addition to fish passage problems, the Jackson Creek Dam contributed to poor water quality in Bear Creek, where high water temperature is a chronic summer problem. The dam and silt-filled reservoir also were considered an eyesore in downtown Medford (Smith et al. 2000).

In the 1980s, there was broad community consensus that removing or breaching the dam was in the community's best interest, but there also was recognition that any such project should incorporate an alternative that would provide irrigation water to surrounding agricultural lands. In 1985, the Rogue Valley Council of Governments approached the irrigation district about conducting a feasibility study to remove the dam. The irrigation district agreed to a study provided that the district could still obtain irrigation water, would not have to pay for dam removal, and would not have to pay higher maintenance costs for its irrigation system. The feasibility study, completed in 1987, found that removing the dam would improve water quality and restore stream habitat but also would entail relocation of a buried fiber optic cable, which would add $350,000 to the cost of the project. Instead of a total removal, project cooperators opted to breach the dam and retain 3 feet of the foundation in order to protect the fiber optic cable. Breaching reduced the dam to a series of three 1-foot drops. A new, 3-foot-high irrigation diversion was proposed nearby, with an effective fish passage structure for adults and a rotary screen to prevent loss of juvenile fish. The new structure would be removed from October through April (nonirrigation season), preventing buildup of sediment. An underground pipeline also was needed from the new diversion to the existing irrigation delivery system. The feasibility study estimated a cost of $500,000 for breaching the old dam, creating the new seasonal diversion with appropriate fish passage and screens, and connecting the new diversion to the existing irrigation delivery system. Ultimately, the cost reached $1.2 million to address all environmental, irrigation, operational, and engineering requirements. The new 3-foot diversion, consisting of collapsible steel supports, wooden stoplogs, fish screens, and fish passage facilities, was completed in 1995. The Jackson Street Dam was breached in 1998 after the creek was diverted around the dam to provide a dry workplace.

Thirteen years elapsed from initial proposal of the feasibility study to completion of the project. Twenty major stakeholders were involved, including five local government agencies, the irrigation district, the local watershed council, seven state agencies, two federal agencies, the project contractor, and local partners. Smith et al. (2000) provided the following lessons learned from efforts to remove the Jackson Street Dam:

- Select candidate dams whose functions can be accomplished by alternative means.
- Identify objectives for the stream and determine how removal of the dam can contribute to their accomplishment.
- Identify a project champion who will lead the effort.
- Identify all potential stakeholders, often a longer list than first anticipated.

- Obtain informal agreement to the project individually from each stakeholder.
- Strive to understand all stakeholder perspectives.
- Implement a public education campaign about the project that is supported by all stakeholders.
- Seek support for the project from elected officials.
- Develop a detailed project budget, including feasibility study and engineering estimates.
- Ensure that all stakeholders understand that the project will take time, effort, and money and that unexpected complications are likely to arise.

Monitoring and Evaluating Stream Restoration

Monitoring and evaluation determine the effectiveness of restoration efforts, suggest appropriate modifications to future restoration actions, help ensure the best results for restoration dollars, and communicate results to interested parties. Despite their importance, monitoring and evaluation often are overlooked, in part because streams pose some unique problems and opportunities that can confound efforts to design an appropriate monitoring program.

Special considerations in stream monitoring include seasonal and annual variability in streams, caused by changes in flows and weather, and natural variability in fish populations, which often are the target of stream restoration efforts. Monitoring the response of salmon or steelhead or other migratory fishes may take a long time because of the 2- to 5-year return interval, depending on life history of the species in question. For instance, it may take more than 15 years of monitoring to gauge the response of three fish generations to restoration work. Another special consideration in streams is whether to focus monitoring at the level of the stream reach or watershed scale (Roni et al. 2003). Monitoring the effectiveness of physical habitat changes, such as reductions in stream temperature or increases in pool frequency, is easier than determining success of desired increases in fish populations or other biological changes. Clearly, the inherent complexity of many streams and aquatic communities necessitates careful planning for monitoring information to be useful.

Monitoring must be designed to detect changes and be able to distinguish the effects of restoration treatments from other sources of variability. This may entail comparison of treated areas with control areas (i.e., areas without restoration efforts) and the collection of multiple samples across broader watershed areas or over longer time periods. Kershner (1997) provides some good examples of monitoring programs in stream settings. Each example includes the following elements:

- A concise goal statement for the restoration project
- Detailed objectives of the restoration treatments
- Implementation monitoring to determine whether the restoration treatments were implemented as planned
- Identification of control stream reaches that are not influenced by restoration activity

- Repeated measurements of treatment and control areas to determine effectiveness of restoration
- Analysis of monitoring results

Collecting monitoring information without proper analysis and reporting on the analysis to interested parties amounts to wasted effort. The concept of adaptive management is central to working in complex and naturally variable ecosystems. Successful adaptive management depends on sound monitoring and proper use of monitoring results to inform future restoration and management. Most problems in stream restoration can be attributed to failures to understand the watershed context of the problem, failure to understand and treat the root causes of degradation, or failure to adequately monitor and adapt management accordingly (Williams et al. 1997).

Stream restoration is a long-term investment in understanding a watershed. Observing streams during periods of high and low flow can inform the astute restoration practitioner. Streams that appear small and tame during times of drought may overflow their banks during spring runoff. Accelerated erosion and sources of siltation may be visible only during high flows. What are stressors in periods of high flows may not be noticeable at other times. Viewing restoration actions over sufficiently long time periods that include drought and flood may be necessary before success or failure can be determined. Merely understanding the variability of stream systems may not be sufficient to ensure success. Successful restoration will embrace variability, complement natural recovery, and work with changing flow regimes. By patiently observing the complex stream ecosystem, we may learn what restoration practices work best and why.

Acknowledgments

We thank Nathaniel Gillespie and Cindy Deacon Williams for their reviews of an earlier version of this chapter. We also would like to thank Eric Dittmer, Johan Hogervorst, and Jack Sleeper for providing information used in developing the case studies.

References

American Rivers and Trout Unlimited. 2002. *Exploring Dam Removal: A Decision-Making Guide*. American Rivers, Washington, DC.

Bisson, P. A., K. Sullivan, and J. L. Nielson. 1988. Channel hydraulics, habitat use, and body form of juvenile coho salmon, steelhead trout, and cutthroat trout. *Transactions of the American Fisheries Society* 117: 262–273.

Burnett, K. M., G. Reeves, D. Miller, S. Clarke, K. Christiansen, and K. Vance-Borland. 2003. A first step toward broad-scale identification of freshwater protected areas for Pacific salmon and trout in Oregon, U.S.A. Pages 144–154 in J. P. Beumer, A. Grant, and D. C. Smith (eds.), *Aquatic Protected Areas: What Works Best and How Do We Know?* Proceedings of the World Congress on Aquatic Protected Areas, Cairns, Australia, August 2002. Australian Society for Fish Biology, North Beach, Western Australia.

Collier, M., R. H. Webb, and J. C. Schmidt. 1996. *Dams and Rivers: A Primer on the Downstream Effects of Dams*. Circular 1126. U.S. Geological Survey, Tucson, AZ.

FEMAT (Forest Ecosystem Management Assessment Team). 1993. *Forest Ecosystem Management: An Ecological, Economic, and Social Assessment*. U.S. Forest Service, Portland, OR.

Fox, M., S. Bolton, and L. Conquest. 2003. Reference conditions for instream wood in western Washington. Pages 361–393 in D. R. Montgomery, S. Bolton, D. B. Booth, and L. Wall (eds.), *Restoration of Puget Sound Rivers*. University of Washington Press, Seattle.

Fresh, K. L. and G. Lucchetti. 2000. Protecting and restoring the habitats of anadromous salmonids in the Lake Washington watershed, an urbanizing ecosystem. Pages 525–544 in E. E. Knudsen, C. R. Steward, D. D. MacDonald, J. E. Williams, and D. W. Reiser (eds.), *Sustainable Fisheries Management: Pacific Salmon*. Lewis Publishers, New York.

Friends of the Earth, American Rivers, and Trout Unlimited. 1999. *Dam Removal Success Stories: Restoring Rivers through Selective Removal of Dams That Don't Make Sense*. American Rivers, Washington, DC.

Frissell, C. A. 1997. Ecological principles. Pages 96–115 in J. E. Williams, C. A. Wood, and M. P. Dombeck (eds.), *Watershed Restoration: Principles and Practices*. American Fisheries Society, Bethesda, MD.

Frissell, C. A., W. J. Liss, C. E. Warren, and M. D. Hurley. 1986. A hierarchical framework for stream habitat classification: viewing streams in a watershed context. *Environmental Management* 10: 199–214.

Frissell, C. A. and R. K. Nawa. 1992. Incidence and causes of physical failure of artificial habitat structures in streams of western Oregon and Washington. *North American Journal of Fisheries Management* 12: 182–197.

Hunt, R. L. 1993. *Trout Stream Therapy*. University of Wisconsin Press, Madison.

Hynes, H. B. N. 1975. The stream and its valley. *Internationale Vereinigung für Theoretische und Angewandt Limnologie Verhandlungen* 19: 1–15.

Kershner, J. L. 1997. Monitoring and adaptive management. Pages 116–131 in J. E. Williams, C. A. Wood, and M. P. Dombeck (eds.), *Watershed Restoration: Principles and Practices*. American Fisheries Society, Bethesda, MD.

Konrad, C. P. 2003. Opportunities and constraints for urban stream rehabilitation. Pages 292–317 in D. R. Montgomery, S. Bolton, D. B. Booth, and L. Wall (eds.), *Restoration of Puget Sound Rivers*. University of Washington Press, Seattle.

Leopold, L. B. 1994. *A View of the River*. Harvard University Press, Cambridge, MA.

Mackinson, S., M. Vasconellas, and N. Newlands. 1999. A new approach to the analysis of stock–recruitment relationships: "model-free estimation" of fuzzy logic. *Canadian Journal of Fisheries and Aquatic Sciences* 56: 686–699.

McIntosh, B. A., J. R. Sedell, J. E. Smith, R. C. Wissmar, S. E. Clarke, G. H. Reeves, and L. A. Brown. 1994. Historical changes in fish habitat for selected river basins of eastern Oregon and Washington. *Northwest Science* 68 (special issue): 36–53.

Meesters, E. H., R. P. M. Bak, S. Westmacott, M. Ridgley, and S. Dollar. 1998. A fuzzy logic model to predict coral reef development under nutrient and sediment stress. *Conservation Biology* 12: 957–965.

Montgomery, D. R. and J. M. Buffington. 1993. *Channel Classification, Prediction of Channel Response, and Assessment of Channel Conditions*. Timber/Fish/Wildlife Report TFW-SH10-93-002. Washington State Department of Natural Resources, Seattle.

Montgomery, D. R. and J. M. Buffington. 1997. Channel-reach morphology in mountain drainage basins. *Geological Society of America Bulletin* 109: 596–611.

Moscrip, A. L. and D. R. Montgomery. 1997. Urbanization, flood frequency, and salmon abundance in Puget Lowland streams. *Journal of the American Water Resources Association* 33: 1289–1297.

Naiman, R. J., T. J. Beechie, L. E. Benda, D. R. Berg, P. A. Bisson, L. H. MacDonald, M. D. O'Connor, P. L. Olson, and E. A. Steel. 1992. Fundamental elements of ecologically healthy watersheds in the Pacific Northwest coastal ecoregion. Pages 127–188 in R. J. Naiman (ed.), *Watershed Management: Balancing Sustainability and Environmental Change*. Springer-Verlag, New York.

Narver, D. W. 1973. *Are Hatcheries and Spawning Channels Alternatives to Stream Protection?* Circular 93. Fisheries Research Board of Canada, Nanaimo, BC.

National Research Council. 1992. *Restoration of Aquatic Ecosystems*. National Academy Press, Washington, DC.

Nehlsen, W., J. E. Williams, and J. A. Lichatowich. 1991. Pacific salmon at the crossroads: stocks at risk from California, Oregon, Idaho and Washington. *Fisheries* 16(2): 4–21.

Nichols, R. A. and S. G. Sprague. 2003. Use of long-line cabled logs for stream bank rehabilitation. Pages 422–442 in D. R. Montgomery, S. Bolton, D. B. Booth, and L. Wall (eds.), *Restoration of Puget Sound Rivers*. University of Washington Press, Seattle.

Pease, J. R. 1993. Land use and ownership. Pages 31–39 in P. L. Jackson and A. J. Kimerline (eds.), *Atlas of the Pacific Northwest*. Oregon State University Press, Corvallis.

Reeves, G. H., L. E. Benda, K. M. Burnett, P. A. Bisson, and J. R. Sedell. 1995. A disturbance-based ecosystem approach to maintaining and restoring freshwater habitats of evolutionarily significant units of anadromous salmonids in the Pacific Northwest. Pages 334–349 in J. L. Nielsen (ed.), *Evolution and the Aquatic Ecosystem: Defining Unique Units in Population Conservation*. American Fisheries Society Symposium 17, Bethesda, MD.

Reeves, G. H., F. H. Everest, and T. Nickelson. 1988. *Identification of Physical Habitats Limiting the Production of Coho Salmon in Western Oregon and Washington*. General Technical Report PNW-GTR-245. USDA Forest Service, Pacific Northwest Research Station, Portland, OR.

Reeves, G. H., J. D. Hall, T. D. Roelofs, T. L. Hickman, and C. O. Baker. 1991. Rehabilitating and modifying stream habitats. Pages 519–557 in W. R. Meehan (ed.), *Influences of Forest and Rangeland Management on Salmonid Fishes and Their Habitats*. Special Publication 19. American Fisheries Society, Bethesda, MD.

Reeves, G. H., D. B. Hohler, D. P. Larsen, D. E. Busch, K. Kratz, K. Reynolds, K. F. Stein, T. Atzet, P. Hays, and M. Teehan. 2004. *Effectiveness Monitoring for the*

Aquatic and Riparian Component of the Northwest Forest Plan: Conceptual Framework and Options. General Technical Report PNW-GTR-577. USDA Forest Service, Pacific Northwest Research Station, Portland, OR.

River Alliance of Wisconsin and Trout Unlimited. 2000. *Dam Removal: A Citizen's Guide to Restoring Rivers*. River Alliance of Wisconsin, Madison.

Roni, P., M. Liermann, and A. Steel. 2003. Monitoring and evaluating fish response to instream restoration. Pages 318–339 in D. R. Montgomery, S. Bolton, D. B. Booth, and L. Wall (eds.), *Restoration of Puget Sound Rivers*. University of Washington Press, Seattle.

Rosgen, D. L. 1996. *Applied River Morphology*. Wildland Hydrology Books, Pagosa Springs, CO.

Schetterling, D. A. and R. W. Pierce. 1999. Success of instream habitat structures after a 50-year flood in Gold Creek, Montana. *Restoration Ecology* 7: 369–375.

Smith, L. W., E. Dittmer, M. Prevost, and D. R. Burt. 2000. Breaching of a small irrigation dam in Oregon: a case history. *North American Journal of Fisheries Management* 20: 205–219.

Stanford, J. A. and J. V. Ward. 1992. Management of aquatic resources in large catchments: recognizing interactions between ecosystem connectivity and environmental disturbance. Pages 91–124 in R. J. Naiman (ed.), *Watershed Management: Balancing Sustainability and Environmental Change*. Springer-Verlag, New York.

Stein, B. A., L. S. Kutner, and J. S. Adams, eds. 2000. *Precious Heritage: The Status of Biodiversity in the United States*. Oxford University Press, New York.

U.S. Department of Agriculture and U.S. Department of the Interior. 1995. *Ecosystem Analysis at the Watershed Scale: Federal Guide for Watershed Analysis*. Version 2.2. Regional Interagency Executive Committee, Portland, OR.

U.S. Forest Service and U.S. Bureau of Land Management. 1994. *Environmental Assessment for the Implementation of Interim Strategies for Managing Anadromous Fish-Producing Watersheds in Eastern Oregon and Washington, Idaho, and Portions of California*. U.S. Forest Service and U.S. Bureau of Land Management, Washington, DC.

Washington Department of Fish and Wildlife. 2003. *Design of Road Culverts for Fish Passage*. Periodically updated and available online at www.wa.gov.wdfw/hab/engineer/cm/toc.htm (accessed February 2, 2005).

Waters, T. F. 1995. Sediment in streams: sources, biological effects, and control. *American Fisheries Society Monograph* 7, Bethesda, MD.

Williams, J. E. and C. D. Williams. 2004. Oversimplified habitats and oversimplified solutions in our search for sustainable freshwater fisheries. Pages 67–89 in E. E. Knudsen, D. D. MacDonald, and Y. K. Muirhead (eds.), *Sustainable Management of North American Fisheries*. American Fisheries Society, Bethesda, MD.

Williams, J. E., C. A. Wood, and M. P. Dombeck. 1997. Understanding watershed-scale restoration. Pages 1–13 in J. E. Williams, C. A. Wood, and M. P. Dombeck (eds.), *Watershed Restoration: Principles and Practices*. American Fisheries Society, Bethesda, MD.

Ziemer, R. R. 1997. Temporal and spatial scales. Pages 80–95 in J. E. Williams, C. A. Wood, and M. P. Dombeck (eds.), *Watershed Restoration: Principles and Practices*. American Fisheries Society, Bethesda, MD.

Resources on the Web

Many agencies and organizations maintain valuable Web sites on stream restoration techniques. The following sites maintain a long list of useful references that are linked for quick access.

Links to various publications and reference materials on stream restoration provided by California Department of Water Resources: www.watershedrestoration.water.ca.gov/urbanstreams/reference/bibliography.cfm.

Links to various stream restoration manuals and case studies from throughout the Pacific Northwest: wildfish.montana.edu/resources/manuals.asp.

Trout Unlimited maintains a Web site about small dam removal, including case studies and a citizens' guide: www.tu.org/small_dams.

Recommended Books

Montgomery, D. B., S. Bolton, D. B. Booth, and L. Wall, eds. 2003. *Restoration of Puget Sound Rivers*. University of Washington Press, Seattle. Eighteen chapters on salmon recovery, fluvial processes, floodplain forests, instream wood, urban issues, and monitoring. Available at www.washington.edu/uwpress or by calling (800) 441-4115.

National Research Council. 1992. *Restoration of Aquatic Ecosystems: Science, Technology, and Public Policy*. National Academy Press, Washington, DC. Top science on restoration of lakes, rivers, streams, and wetlands.

Riley, A. L. 1998. *Restoring Streams in Cities: A Guide to Planners, Policymakers, and Citizens*. Island Press, Washington, DC.

Waters, T. F. 1995. *Sediment in Streams: Sources, Biological Effects and Control.* Monograph 7, American Fisheries Society, Bethesda, MD. Practical guidance and techniques for controlling inputs of sediment from agriculture, forestry, mining, and urban development. Available at www.fisheries.org or by calling (678) 366-1141.

Williams, J. E., C. A. Wood, and M. P. Dombeck, eds. 1997. *Watershed Restoration: Principles and Practices.* American Fisheries Society, Bethesda, MD. Ecological principles and numerous case studies of what works and what does not for watershed-scale restoration, with an emphasis on stream systems. Available at www.fisheries.org or by calling (678) 366-1141.

Chapter 14

Landscape and Watershed Scale

DEAN APOSTOL, WARREN WARTTIG, BOB CAREY, AND BEN PERKOWSKI

Landscape-scale and watershed restorations have emerged as key aspects of conservation in the Pacific Northwest. Landscape-scale restoration includes planning, design, and strategies that encompass areas large enough to include several ecological types or a single type extending over thousands of hectares. Most ecological restoration takes place at individual patch or small landscape scales, a few to perhaps a few hundred hectares. At this scale, restoration projects can be planned and executed as discrete projects, usually within the bounds of one land ownership. Although multiple ecosystem types may be involved, these usually are closely related and overlapping. When the scale ranges from several hundred to tens of thousands of hectares, restoration can involve multiple landowners and encompass very different ecosystem types. The number of issues, complexity, and timeframe all must be expanded to account for larger scales.

Watershed restoration also is done at large scales but uses a common drainage area as the template, spanning a few hundred to tens of thousands of hectares. The health of aquatic ecosystems usually is the main objective, with upland areas viewed primarily in the context of how they relate to streams or other water bodies. The links between landscape-scale and watershed restoration are as follows:

- Large areas, hundreds to tens of thousands of hectares
- Highly complex issues, multiple ecosystems, and many variables
- The need to engage multiple partners over long timeframes.

Restoration is not practiced so much at landscape scales, except in large wilderness areas, where prescribed natural fire is the main tool (see Chapter 9). In most cases restoration is planned or coordinated, then used to set a context and priorities for site-scale restoration projects. Landscape ecology, conservation biology, and watershed assessment are key underpinnings for restoration practiced at large scales.

Important Concepts

Landscape ecology is a subset of the older field of ecology and is concerned with understanding the interactions of organisms and their environment across kilometer-wide mosaics. It is an interdisciplinary field that includes contributions from biogeography, anthropology, and landscape architecture. Landscape ecology has been applied across a wide range of land uses, including agriculture, urban areas, wilderness, and forests (Dramstad et al. 1996). It is a value-neutral field, with practitioners generally more interested in studying landscapes dispassionately or analytically rather than as conservation advocacy. Key concepts from landscape ecology that are used in restoration include patches, corridors, matrixes, disturbances, natural variability, structure, function, and resilience.

Conservation biology is a multidisciplinary science that formed to address the worldwide crisis of declining biological diversity. It is part of the larger field of applied ecology, which includes forestry and game management and also draws on many of the same analytical sources and methods as landscape ecology. But conservation biology is not a neutral science; it is explicitly mission oriented. It is driven by a core value that biodiversity is a good thing and that society has a responsibility to conserve it (Noss and Cooperrider 1994). Conservation biologists advocate for biodiversity protection and restoration but also use scientific methods and empirical research to craft technical solutions such as identifying core conservation areas, delineating appropriate buffers, locating corridors, and designing comprehensive reserve systems.

Watershed analysis includes efforts to diagnose and prescribe a course of action aimed primarily at promoting or recovering healthy aquatic ecosystems. Some approaches, such as the *Oregon Watershed Assessment Manual* (Watershed Professionals Network 1999), are stream focused, whereas others, such as the *Ecosystem Analysis at the Watershed Scale: Federal Guide for Watershed Analysis*, used by the Forest Service and the Bureau of Land Management, give more attention to upland habitats (U.S. Department of Agriculture and U.S. Department of the Interior 1995). Like conservation biology, watershed analysis can be described as mission oriented, at least in the sense that the value of restoring healthy aquatic systems is assumed. The practice has evolved to focus more on long-term processes and functions and less on short-term habitat structure. The challenge has been to move from mere analysis to creating conservation and restoration strategies and actions, particularly because watershed analysis often is embedded in a broad stakeholder social framework.

The importance of these emerging fields to ecological restoration lies in the way they help synthesize complex information with multiple variables over long timeframes. Each provides a conceptual framework for generating, evaluating, and selecting a restoration strategy for large areas.

Landscapes often are described and analyzed as vegetation mosaics that consist of patches, corridors, and a matrix. Various processes, including small- and large-scale disturbances, interact with this mosaic, shape it, and in turn are shaped by it. Over time the mosaic fluctuates within a variable range. For example, moist conifer forests ranged from 25% to 75% being in an old-growth state at any given time (at very large landscape scales), with the precise location of old growth shifting about in response to wind, insects, and fire.

Landscapes structures are physical, tangible ecosystem elements, including forest stands, wetlands, rock outcrops, and prairies. Ecological functions interact with these structures and can be classified in a variety of ways. Five main functions generally recognized are capture, production, cycling, storage, and output. Structures and functions are codependent and act to change each other. As Forman and Godron (1986) state, "Past functioning has produced today's structure; today's structure produces today's functioning; today's functioning will produce future structure."

Landscape patches are of particular importance to ecological restorationists because they represent the field of action. They include areas of vegetation or barrens (i.e., tidal mud flats) that are homogeneous in terms of composition, three-dimensional structure, and age (Forman 1995). Common Northwest patch types include conifer, mixed, and hardwood forest, prairies, oak woodlands, emergent wetlands, lakes, steppe, farm fields, and urban districts with varying vegetation. The scale of analysis determines the degree of refinement in defining patches. In all cases, very rare or unique elements, such as aspen or juniper groves, are noted. Patches vary a great deal in terms of size and shape. Size is important because it relates to the habitat needs of wildlife. The greater the patch size, the greater the proportion of inte-

FIGURE 14.1. A landscape mosaic with an urban matrix surrounding forest and farm patches near Oregon City, Oregon. (*Courtesy of Portland Metro Newell Creek Watershed Project*)

rior habitat available, assuming that the edge zone width is constant.

A landscape matrix normally is defined as the most dominant patch in the landscape being analyzed or planned (Forman 1995). Old-growth conifer forest was the historic matrix in the coastal and Cascade ranges before Euro-American settlement. In interior valleys, bunchgrass prairie and oak woodlands were the matrix. In urban areas, the matrix is buildings, streets, parking lots, and mostly nonnative vegetation, and native plant communities are in the form of remnant patches. Defining a matrix depends on scale of resolution. At one scale, mature forest could be the matrix, whereas at a larger scale mature forest could be a patch within a grassland or urban matrix (Figure 14.1).

Corridors are a particularly important feature of landscape ecology and conservation biology. These are narrow, linear features that may connect similar patches through a dissimilar matrix or aggregation of patches. Streams are corridors that connect aquatic habitats, often across very large areas. A protected riparian zone may function as a corridor that connects patches of mature upland forest in an otherwise heavily logged, farmed, or urbanized landscape. The functional importance of corridors is their ability to allow flows of materials or organisms across significant parts of a landscape. The effectiveness of a corridor depends on its width, the frequency and severity of interruptions, and the presence of nodes, or wide points.

The value of habitat corridors as a conservation tool has become an article of faith among many restoration planners, conservation biologists, and environmental activists, especially those working in urban and agricultural areas. The concept of wildlife corridors has been expanded to regional and even continental scales through the Wildlands Project (see www.wildlandsproject.org/roomtoroam/endangered). However, there is very little empirical

FIGURE 14.2. A stream and riparian corridor in the Oregon Cascades. (*Photo courtesy of the U.S. Forest Service*)

information to help design or retain a functional corridor for specific species. Some ecologists are concerned that narrow habitat corridors can be death traps or population sinks for species that have nowhere to go to escape from predators. Even so, corridors are a powerful conceptual landscape planning tool and an increasingly important component in most landscape-scale restoration projects (Figure 14.2).

Stepping stones are a companion to corridors. These are clustered patches (small islands) that act like corridors by allowing wildlife to move between habitats that would otherwise be unreachable. They may be particularly important for birds in urban and agricultural landscapes but can also be important in managed forests (Dramstad et al. 1996).

Nested scales and ecological context also are very important for ecological restoration practitioners and planners. No ecosystem is completely isolated, so full understanding entails investigation into its place and role with respect to both larger and smaller scales. For example, is the riparian area being restored part of a critical regional corridor, or is it isolated? If the former, the design should consider the needs of wide-ranging species. If the latter, the focus could be on local needs.

As is discussed in Chapter 2, the field of ecological restoration has embraced the importance of processes in natural landscapes. An important aspect of processes is the natural resistance and resilience of ecosystems. *Resistance* is the ability of an ecosystem to absorb and contain disturbance in ways that prevent large-scale alteration. For example, an open ponderosa pine stand holds fires at ground level and thus resists damaging crown fires, so this is a way of resisting abrupt change.

Resilience is the ability of an ecosystem to return to a reference state or composition and structure after a significant disturbance (Kohm and Franklin 1997). For example, a chaparral plant community in the Siskiyou Mountains is conditioned to burn to a crisp every 30 years or so but then recovers its former composition and structure within a few years. On the other hand, an old-growth conifer forest takes about 250 years to recover its former state. The retention of key refugia, such as very cool, moist pockets protected by topography, can be the key to initiating recovery of an ecosystem.

Fragmentation is an important concept, particularly in the Northwest. This is the breaking up of a large habitat into smaller bits, to the point where the remnants lose their ecological integrity. It was the recognition of fragmentation of old-growth forest habitat, and the impacts this was having on the northern spotted owl, that initiated monumental changes in the way public forestlands are managed regionally. Dispersed clearcuts within an unbroken matrix of mature or old-growth forest create openings and edges that change ecosystem functions. As more clearcuts are added, the remnant forest becomes all functional edge, meaning that it no longer has interior habitat conditions. Models have shown that logging 50% of an area with dispersed clearcuts completely eliminates interior habitat (Perry 1994).

Edge effects are found at the interface between contrasting landscape elements, such as between a clearcut and closed canopy forest, or where an urban neighborhood butts up against a riparian woodland. Edges have environmental conditions (temperature, light, atmosphere, humidity, wind) that are different from those of adjacent elements. They can be abrupt or gradual. Very often, the plant and animal composition of an edge is a mixture of species belonging to adjacent components. Many animals thrive in edges, but others need deep interiors, particularly for breeding.

An *interior habitat* is far enough away from an adjacent patch that it is not much influenced by it. Some mammals and birds nest deep in forest interiors in order to be safe from competitors, predators, or brood parasites (such as the brown-headed cowbird). Generally, edge effects penetrate several hundred feet into a forest edge, depending on shape, aspect, slope, and vegetation density (Perry 1994).

Connectivity is a measure of how easy or difficult it is for organisms and energy to move through a landscape without encountering insurmountable barriers. Landscapes are connected both within a matrix and via corridors. The degree of connectivity varies from species to species. Some may need completely connected corridors, whereas others, such as birds, may fly from one patch to another, using them as stepping stones, as long as the patches are not too far apart.

Corridors and stepping stones have become popular concepts among conservation planners, but another idea that has emerged more recently is the *landscape continuum model*. This idea was developed in Australia, where the boundaries between separate patch types are not very sharp and are not readily distinguished from a background matrix. Variegated, low-contrast landscapes should be distinguished from fragmented, high-contrast ones. This idea has become a very important feature of ecological forestry, which seeks to blur boundaries and establish a more seamless landscape (Lindenmayer and Franklin 2003).

A related idea is that of *ecological linkages*. These are places where one landscape or watershed connects to adjacent areas or those farther away. They are usually easily identifiable features, such as the mouths of streams or ridgeline gaps.

Landscape flows are phenomena that move across and interact with landscapes. They respond to various elements (water bodies, forests, roads), which may promote or inhibit them. Flows include both organic and inorganic phenomena, including mammals, seeds, birds, water, air, fire, and soil. As each flow passes through or interacts with an element, one or more ecological functions (capture, production, cycling, storage, or output) occur. For example, an elk (the landscape flow) might use a wetland for food and a mature forest stand for thermal cover and calving (production and storage). Water might flow to a pond, where some of it is used by aquatic plants, and some evaporates back into the atmosphere (production and cycling). Understanding the relationships between key landscape flows and elements is an important tool in developing large-scale restoration plans designed around functions.

Landforms often are overlooked in analyses of landscape ecology. In mountainous and hilly areas (much of the Northwest), landforms often are the most important driver of the landscape pattern. Ridgelines and valleys exert strong influence on vegetation, water flows, species movement, and human land use. Valley floor riparian areas receive energy, nutrients, and material that gradually move downhill. They also may have areas of slow drainage where bogs or wet prairies form. Ridgelines are natural movement corridors for many species, including raptors and large mammals. Landforms create niche conditions that allow some specialized patch types to occupy a landscape. For example, south-facing slopes with shallow soils in transition zones may provide space for prairies or oak woodlands in an otherwise continuous conifer forest.

Natural landscapes are increasingly acknowledged as dynamic and disturbance driven rather than climax ecosystems. Ecologists now incorporate the concept of *historic variability* to understand how the disturbance regimes have shaped landscape mosaics over long periods of time. Understanding historical variability takes good detective work, including old photographs, early survey records, pioneer journals or ancient literature, oral tradition, fire scar dating, and pollen analysis. Change induced by disturbance may be crucially important to maintaining ecosystem functioning and resilience. Some change is gradual and imperceptible over a human life span, but disturbances such as large wildfires usually are noticeable and can be unsettling and very difficult to integrate into ecological restoration planning.

Natural disturbances include fire, flood, wind, avalanches, landslides, insects, and pathogens. Cultural disturbances are very important to consider and include logging, farming, introduction of invasive species, and flooding caused by unwise development or mismanagement of stormwater. Some disturbances affect very large areas, whereas others tend to be more localized. A windstorm may do its work in a few hours, whereas root rot can take many years. Disturbances may follow predictable cycles of occurrence, with the most extensive and severe usually having a low frequency. Disturbance cycles can take place within larger climatic cycles, such as warming and drying, or increased wetness and cooling, and the cycles themselves are subject to change over time.

Biological legacies are traces and features that remain in a landscape after a disturbance. They include organisms, organic structures (such as downed logs and standing dead trees), soil, and patterns that persist and link the old landscape structure with the emerging one (Lindenmayer and Franklin 2003). Legacies are a key concept in ecological forestry and have been much studied in the wake of the 1980 eruption of Mt. St. Helens in Washington State (Figure 14.3) and the 1988 wildfires of Yellowstone Park. Researchers in both areas have found that biological legacies were very important in colonizing surrounding areas from deep within the disturbed zones (Perry 1994).

Core reserves are strictly protected habitat zones, usually managed with biodiversity conservation as the primary goal. Natural disturbances are allowed to occur or may be mimicked through management, such as prescribed fire. Core reserves can be fairly small or quite large, depending on landscape context. Generally, they are established at confluences of highest biodiversity, such as *biological hotspots* and *anchor habitats*.

Refugia occupy positions protected from large-scale disturbance and have an important role as repositories of biological legacies. These may include areas on the lee side of hills or mountains sheltered from prevailing winds or areas where soils are unusually deep and rich. The Klamath–Siskiyou Mountains are believed to have served as critical refugia in past periods of climate change, which in part explains the very high biodiversity of this region.

Keystone species are those that play prominent roles in ecosystems, affecting the habitat of other

Figure 14.3. Mt. St. Helens blast zone illustrates the potential for infrequent but very large-scale natural disturbance. (*Photo by Dean Apostol*)

species by their presence or absence. For example, large carnivores affect the populations of their prey, which in turn alter grazing pressures. Recent research in Yellowstone Park since reintroduction of the timber wolf demonstrates that some riparian woodlands are recovering because elk spend less time grazing in risky places. Beavers, another keystone species that creates wetlands, are returning to riparian areas to dine on the willows that have grown back in these woodlands (Ripple and Beschta 2003; see Chapter 15). At the opposite scale, the presence of mycorrhizal fungi facilitates more rapid tree growth by extending the effective reach of root systems. Keystone species are those that affect the links in food webs so strongly, and their role is so specialized, that their removal may result in a radical reorganization of the food web (Mills et al. 1993).

Wildlife guilds are an important tool used in large-scale restoration planning. These are groups of species that have similar habitat and dispersal characteristics. If well selected, they provide a wider perspective on ecosystem function than does relying on a single species (Mellen et al. 1997).

Together these concepts, most developed over the last 20 years, help ecological restoration planners and designers to operate at landscape and watershed scales. Not all projects merit the time and expense needed for detailed analysis over thousands of hectares, but some understanding of landscape context and the role of a particular restoration site is more than merely interesting. The following section reviews four case studies, all of which have tapped the concepts and approaches pioneered by colleagues in landscape ecology, conservation biology, and watershed assessment.

Case Study: Building a Culture of Restoration in the Mattole River Watershed

At the far southern tip of the Pacific Northwest region is the beautiful and very rural Mattole River watershed (Figure 14.4). This 300-square-mile, 192,000-acre area is in many ways our best regional laboratory of watershed restoration. It includes a number of the ecosystem types covered in this book, and the local community has built a restoration culture neighbor by neighbor and agency by agency over a 30-year period. The Mattole story has been told in much greater detail and with more eloquence elsewhere, but no book on regional restoration is complete without at least a summary (House 1999, Freedlund 2005).

The Mattole landscape is characterized by very steep, rugged mountains that rise up from sea level to more than 1,200 meters in the King Range. It is one of the most tectonically active areas in North America, lying at the margin of three great plates, the Pacific, Gorda, and North American. Much of this area is uplifting quite rapidly, particularly the King Mountain Range. Torrential rates of rainfall, among the highest in the United States, combine with complex and highly erodible, soft terrain to result in an extremely high natural rate of erosion and sedimentation (Larson 2005).

Vegetation historically was a complex mosaic of open prairie, oak woodlands, and conifer forests, much of which was old growth. Complex topography, coastal fog penetration, seasonal summer droughts, and cultural land practices all conspired to shape a fine-grained pattern of vegetation. The Mattole and Sinkyone people were active land managers for hundreds or thousands of years. They burned the prairies and woodlands and harvested wood for many purposes. But by 1864 they had been completely driven out of the area. The dominant vegetation historically was mixed conifer and hardwood forest, at about 57% of the total area. Pure conifer and hardwood forests occupy less than 20% and 10%, respectively. Grasslands cover most of the remainder (Isom 2005).

Euro-American settlers initiated changes beginning in the 1850s. Early on they introduced domestic animals to the grasslands and girded trees at the margins to expand them. They planted orchards and gardens where the soil was best. Around the turn of the century widespread harvest of tanoak bark was common and lasted for two decades. Sheep ranching was the mainstay of the economy until intensive harvest of conifer trees began after World War II, spurred in large part by a tax placed on standing timber by the State of California. Between 1945 and 1961 more than 72,000 acres, or 38% of the entire watershed, was logged, primarily using poorly constructed cat roads and skid trails on steep, unstable terrain.

The consequences to the river system have been well documented. Two huge rainstorms during the period of heaviest logging (1955 and 1964) triggered numerous slides and released large amounts of sediment (Figure 14.5). The result was filling of formerly deep pools and widening of stream channels, particularly in lower-gradient stream reaches (Figure 14.6). This in turn led to lethal summer stream temperatures and the collapse of salmon populations. In the late 1960s and throughout the 1970s, a number of back-to-the-land hippies moved to northern California and southern Oregon, with some landing in the Mattole watershed. A number of large upland ranches had been subdivided down to 40- to 160-acre segments and were available for quite reasonable prices (Freedlund 2005). Long-term residents had noted the degradation of stream habitat and salmon numbers, but newcomers were at first unaware. A 1973 state fishery report identified the prime cause of habitat degradation as excessive sediment, tied to logging and poor road construction.

Two of the back-to-the-landers, Freeman House and David Simpson, began organizing the community around the idea of restoring the native king salmon, which House has called a totem species

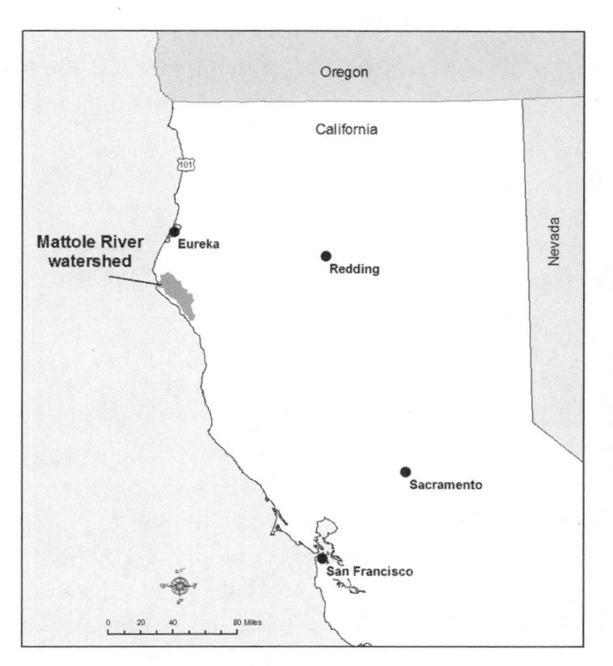

FIGURE 14.4. Mattole River Watershed location map. (*Courtesy of Mattole River Council*)

FIGURE 14.5. Landslides and erosion are common occurrences because of the steep terrain of the land, local weather, and environmental degradation that has taken place. (*Photo courtesy of Mattole River Council*)

FIGURE 14.6. Sedimentation in the Mattole watershed has created large gravel bars along the mainstem, which contribute to the dynamic channel morphology of the Mattole River. (*Courtesy of Mattole River Council*)

(House 1999). The local population was one of only six on the California coast that had not been influenced by hatchery-bred fish, giving it genetic significance. Outreach was neighbor to neighbor, and various projects were initiated. One of the first was a hatchbox program, in which a small number of wild fish were trapped and induced to lay eggs in artificial boxes with clean gravel, thus bypassing the problem of sediment suffocating eggs laid on muddy stream bottoms (Figure 14.7). Restoration workers used weirs to force migrating salmon into enclosed areas, where they were trapped and netted. Young salmon were carefully reared and kept in cool water pools until they reached optimal size, then released to make their way to the ocean only after fall rains had allowed the river to breach the sandbar at the estuary (House 1999).

Mill Creek, a 1,200-acre sub-watershed, was chosen as the site of a coho restoration effort. It had some of the best remaining water quality in the area because it still had a mostly intact old-growth forest cover. A badly designed culvert had blocked coho runs in 1964. Efforts to restore coho focused attention on removing migratory blockages. One technique at the time involved removing large wood from streams, which biologists mistakenly believed caused dense jams that inhibited migration. A 1985 research study showed that large logs are critical to salmon habitat and that jams only rarely and temporarily blocked fish migration. But over a 10-year period much large wood had already been removed and a lot of damage done.

Upriver residents tended to focus on land restoration projects such as tree planting and bank stabilization, whereas downriver ones concentrated on population increases through hatchbox and release programs. Another problem that surfaced was the condition of the estuary. In 1987 a summer die-off of more than 100,000 juvenile chinook salmon was attributed to lethal water temperatures that had resulted from the buildup of silt and loss of deep water pools. Since that time, a number of estuary mitigation projects have been tried, including willow planting, use of floating log structures, and log–boulder structures along the shoreline.

FIGURE 14.7. Mattole salmon restorationist Gary "Fish" Peterson places fertilized chinook salmon eggs in streamside hatchbox at Squaw Creek, 1992. (*Photo by Maureen Roche*)

Instream projects to improve habitat were initiated in 1991, beginning in Honeydew Creek. As has been the case in many other coastal watersheds, initial projects were poorly engineered and did not survive high seasonal flows in lower stream reaches. In upper reaches, where flows are not so great, structures have survived (Freedlund 2005).

An admirable trait in the Mattole restoration community has been the willingness to recognize what works and admit what does not and to apply the lessons learned. For example, a major stream reengineering project was attempted in the early 1990s on Mattole Canyon Creek, which had been severely affected by logging and road construction. Its lower gradient reach near the mouth was sediment choked, and the energy during high flows appeared unable to transport the sediment out. A series of artificial S-curves were carved, many acres planted, and spawning gravel placed in an effort to predict what the river wanted to do and give it a boost. But Mattole Canyon Creek chose to meander in a completely different direction, leaving the newly constructed channel high and dry and useless, at a cost of several hundred thousand dollars. In the meantime, one local resident had initiated a gradual, patient willow and cottonwood planting effort to stabilize the barren banks. The creek has begun to move sediment out and now has a new channel lined with thousands of trees planted by one family.

Snorkel surveys, another recent innovation in fishery biology, were initiated in the 1990s as a nonlethal and informative method for tracking fish presence and studying how habitat is used throughout the year. A key discovery has been that summer rearing habitat probably is the single most limiting factor in the Mattole, largely because of high temperatures and few cold water refugia.

Upland Conservation and Restoration

Upland landscapes typically have been something of an afterthought in many watershed-scale efforts, which have been stream and riparian focused. The natural tendency is to attack the most visible signs of a problem directly and quickly. Additionally, fish biologists spend a lot more time in streams than they do on steep uplands, explaining why instream habitat projects appeared first, followed by riparian. In the Mattole, because denuded upland slopes were an evident problem early on, reforestation and slope stabilization projects were initiated roughly simultaneously with work closer to streams in the 1970s. As a result, upland forests have recovered quite quickly, and attention has shifted to management of young, crowded, fire-prone stands and grassland habitats. Early reforestation efforts used species not native to the watershed, particularly bishop and knobcone pine (Freedlund 2005). By the early 1980s a shift was made to using Douglas fir, raised from local seed sources. By the mid-1990s, the reforestation program had run out of upland acres to plant and turned attention back to riparian areas. Replanting projects often are integrated with gully repair, slope stabilization, and other efforts to keep soil on the slopes.

Conservation of remaining old-growth forests remains an important part of watershed conservation and recovery. Research indicates that logging, land conversion to homesteads, and wildfire resulted in a 92% loss of the original old-growth cover. Remaining public old growth, primarily on Bureau of Land Management land, has been designated for protection, but some remaining privately owned old growth, primarily by Pacific Lumber Company, continues to be logged despite multiple efforts on the part of local conservationists.

New forestry initiatives have been aimed at thinning thousands of acres of overcrowded second-growth forests that dominate much of the watershed. As is the case with salmon and stream restoration, this is a popular initiative supported by nearly every resident of the area. In large part this is because fire-prone young forests threaten everyone's homes and the livelihood of many.

One concern first raised by older watershed residents was shrinking grassland acreage. A lack of prescribed fire, along with a shift from sheep to cattle, had allowed fir seedlings to encroach on meadows. A mapping project revealed that grassland acreage had shrunk by one third over a 50-year period, threatening both local biodiversity and the ranching economy (Freedlund 2005).

Sociology of Restoring the Mattole

Although this chapter and volume deal mainly with the ecological aspects of restoration, the experience of the Mattole community provides valuable lessons on sociology and politics. This is a very rural community and offers a microcosm of regional tensions between property rights, increasing environmental awareness, and the importance of building sustainable relationships across deep political divides. A key lesson has been that the effort to build a watershed community identity is easily thwarted by people gathering into groups of like-minded neighbors and forming "team identity" around specific issues that inhibit wider cooperation. At its core, this problem probably represents competing views of the human–nature relationship. The two major competing teams in the Mattole are represented by the old-timers, or families that often go back to pioneer days, and the back-to-the-landers who began arriving in the late 1960s, seeking a rural lifestyle and an ecological Eden. But there are many subdivisions as well: college educated versus land educated, individualists versus community builders, private landowners against government agency staff, and so forth (Freedlund 2005). Even restoration advocates divide into camps over the best approaches, methods, or priorities. It takes a lot of work, attention, and leadership to overcome a community's tendency to divide into smaller groups and begin working against one another.

The initial concern over the king salmon united the community. Nearly everyone supported the hatchbox program and pitched in to help in various ways that best suited them. But as attention turned to wider restoration goals the community fragmented because longer-term residents felt threatened by the aims of newcomers. In particular, efforts to better regulate timber harvest were viewed as an economic threat to the larger ranch landowners and timber companies. It all came to a dangerous head during the "Redwood Summer" of 1990, when the Mattole became base camp for hundreds of protesters who blocked logging roads, occupied trees, and generally threw the community into turmoil (Figure 14.8). A year later the Mattole Watershed Alliance was created, which brought ranchers, timber companies, agencies, and the back-to-the-land restorationists together in a forum where all interests were represented (see www.mattole.org/index.shtml).

Over time, many wounds that opened up over the regulatory battle have begun to heal, aided partly by a growing number of landowner incentive programs, including funding for fuel reduction, road repair, and culvert replacement. The Mattole restoration community has initiated many creative community-building approaches, the best known of which is the development and staging of a play called *Queen Salmon*, which dramatizes the early years of efforts of people with very different

FIGURE 14.8. Community protest to protect Goshawk Grove in Baker Creek, 1989. (*Photo by Ali Freedlund*)

world views and experiences in coming together to help save a community resource. The story is nonjudgmental and allows each side to tell its perspective in its own words.

The social center of restoration efforts has been the Mattole Restoration Council, which formed in 1983 and has since spawned and helped coordinate a number of other organizations and initiatives, including education and outreach programs, the Sanctuary Forest, various land trusts, and local subgroups such as the Mill Creek Watershed Conservancy. Altogether the trusts have managed to acquire 6,000 acres of land, including unlogged old growth and six miles of high-quality riverside habitat.

As in much Northwest restoration, Mattole workers and planners have discovered that ecological restoration is a very long-term endeavor and that finding ways to get the land condition back in synch with natural processes is more important than direct benefits of structure improvements. The key natural process in the Mattole is the sediment budget. The stream system can digest and process only so much at a time, and unless the watershed can be stabilized back to a near-natural rate of erosion, all the other work in streams, replacing culverts, and so forth cannot do much for salmon. Therefore, addressing upslope areas and poorly constructed older roads has become a key focus of work. A watershed-wide program called Elements of Recovery was initiated in 1986 to train local residents to survey and recognize erosion-related problems in twelve sub-watersheds. Despite the controversy over initial mapping inaccuracies, this proved to be a very important effort in raising awareness of road-related sediment problems across the watershed. The Bureau of Land Management pitched in by undertaking the largest road restoration project in the watershed, removal of a

3.5-mile stretch of midslope road on Honeydew Creek in 1994. The success of this project spawned a wider Road to Recovery program that used heavy equipment to recontour unneeded problem roads in other areas. On roads that are needed for various purposes, the goal is to reduce chronic sediment problems, and another program called Good Roads, Clear Creeks was initiated for this purpose, with a goal of addressing all problem roads by 2009.

After nearly 30 years of effort, Mattole residents agree that the landscape and streams look a lot better than they did, but efforts to return salmon remain stuck about at the level they were when the restoration work began. An estimated average return run of 30,000 king salmon had been reduced to a few hundred by the late 1970s and has rebounded to only around a thousand today. Hundreds of thousands of dollars, countless hours of restorative work on all aspects of the land, and the hatchbox program have clearly begun to heal the watershed and probably prevented extinction. But restoration will not be complete, according to the Mattole Watershed Council, until it is no longer necessary. Most expect this to take decades.

Case Study: Landscape-Scale Restoration Design in the Little Applegate Watershed

Forest landscape design is a process that conceptually organizes the pattern and structure of large areas over long periods of time. It was developed by the U.S. Forest Service in 1992 as a way to apply key principles of landscape ecology and conservation biology to forest management (Diaz and Apostol 1992). Since the 1970s, the Forest Service had been managing land at two basic scales: that of an entire forest (hundreds of thousands to a few million hectares) and individual projects (a few to a few hundred hectares). At the larger scale, forests were zoned according to a highest and best use principle. Areas could be designated as scenic viewsheds, wilderness, recreation areas, general forest, and so forth. Intensive timber harvest was confined to the general forest zone, with lesser amounts of harvest allowed in other zones but excluded from congressionally designated wilderness areas.

The individual project scale included clearcuts of up to 60 acres, campgrounds, roads, trails, ski runs, and myriad other activities. This operational scale is where the structure of the forest was created, but the legal basis rested in the adopted forest plan. Nowhere was there any landscape-scale analysis or plan that showed how the forest was to be shaped over time. Because of controversies over clearcutting dating from the late 1960s, the Forest Service generally limited individual harvests to 40–60 acres, called staggered setting. Clearcuts, usually rectangular in shape to simplify management, were spaced out over a given area, with uncut or regenerated forest left as buffers. The goal was to have an even range of forest ages from freshly cut up through age 70–90 years, depending on estimated growth rates. Foresters called this a "fully regulated" forest, meaning that every patch was to be a human artifact, or plantation, maximizing timber productivity.

The landscape pattern and structure that resulted from this system was not designed; it simply happened by default as the individual projects added up. As noted earlier in this chapter, a key ecological impact of this approach was fragmentation (Figure 14.9). By the second or third harvest, an

FIGURE 14.9. Forest fragmentation in Cascade Mountains of Oregon's Mt. Hood National Forest. (*Photo by Pat Greene*)

area would be transformed entirely into edge habitat, even where half the forest was still old growth. This placed stress on species that needed interior forest conditions. It also had mounting impacts on aquatic ecosystems, in large part because of extensive road networks but also because of unforeseen changes in the melting rates of snowpack.

The idea of deliberately designing the forest pattern and structure originated with the British Forestry Commission in the 1960s. The "forests" of Great Britain by that time were mostly plantations of nonnative conifers, established as strategic reserves in case of war. By the 1960s, the British public had grown restive over the aesthetic consequences of geometrically shaped plantations seemingly attached by Velcro to the open hills and mountains. Landscape architects developed a design approach that tailored the shapes of these plantations to natural contours, streams, and cultural features of the land.

In the early 1990s, the British approach was adapted at the Mt. Hood National Forest and was integrated with landscape ecology. A 3,000-hectare pilot project called Leoland was chosen as a landscape design unit. Vegetation was mapped and analyzed as a series of patches, matrixes, and corridors, using the terms and techniques described earlier. The primary idea was to create a designed landscape structure related to key flows, including wildlife, water, wind, and fire. This mosaic was time indeterminate but provided a general direction for landscape managers to create supportive projects (timber harvest, replanting, prescribed fire) that would help shape the desired condition.

The Little Applegate Watershed Forest Design used the Leoland approach. It was initiated in the fall of 1996 by the Rogue River National Forest. A watershed analysis provided a detailed assessment of landscape and environmental conditions but stopped well short of providing a map of the future.

Forest managers wanted a better picture of how the landscape should look and function across the entire watershed. They wanted this picture to include federal and private lands and hoped it could be used to establish priorities regarding silviculture, restoration, and fire management. They also wanted to use the process to build better relations with the local community and other stakeholders.

Environmental Context

The Little Applegate watershed is located south and a bit west of Medford, Oregon. It includes about 29,000 hectares, of which two thirds are in federal ownership, the remaining one third private. Private lands include large timber industry holdings in the middle to upper watershed and ranches, small farms, and forested homesteads in the lower end. The pattern of land ownership follows a township, range, and section line grid that has been superimposed onto irregular mountain topography.

The landscape is quite diverse, ranging from a low elevation of 1,466 feet at the Little Applegate mouth up to nearly 7,500 feet at the Siskiyou Mountain crest. The rugged topography includes high mountain glacial features, steep slopes, dense stream networks, and gentle valley floors. Vegetation includes farm fields, pastures, bunchgrass prairies, oak woodlands, chaparral, pine–oak woodlands, mixed conifer forests, and high mountain subalpine parklands.

This diversity of landform and vegetation leads to a great diversity of wildlife habitat, including chinook and coho salmon, steelhead trout, spotted, flammulated, and great grey owl, black-backed and white-headed woodpecker, northern goshawk, red tree vole, western pond turtle, Siskiyou Mountain salamander, and several important bat species. A number of these are listed or are candidates for listing under the federal Endangered Species Act.

An ad hoc citizen and agency task force developed a list of issues and objectives to guide the design. These were drawn from three main sources: the watershed analysis, the document *Words into Action* by the Rogue Institute for Ecology and Economy (Priester n.d.), and the direct knowledge of those on the task force. A key assumption was that landscape objectives had to be consistent with existing policy direction for both federal and private lands. Even though some members of the task force did not support all existing policies, they agreed to try and work within their framework rather than spend time debating policies that this group could not easily change.

The watershed analysis had good information on technical issues related to fire risk, water quality, wildlife habitat, and vegetation. Recommendations were incorporated into design issues and objectives. *Words into Action* provided a social value framework for the entire Applegate watershed and documented how the local community felt about the natural resources of the area. The ad hoc team included members very knowledgeable about the ecology and social dynamics of the area. Draft issues and objectives included ecological restoration but also continued economic production.

Landscape Pattern and Structure

All agreed that a key issue was the large-scale changes in landscape pattern and structure that have occurred over the past 150 years as a result of displacement of Native Americans and subsequent settlement by Euro-Americans. Some key changes included the following:

- Less acreage and fragmentation of mid- to high-elevation old-growth conifer forest
- Loss of large, old open-grown pine and oak stands
- Loss of bunchgrass prairies to brush encroachment
- Damage to streams and riparian woodlands from road building, agriculture, and logging
- Degradation of subalpine meadows as a result of historic sheep grazing
- Overall crowded and hazardous forest conditions caused by an abundance of small-diameter Douglas fir and grand fir trees

The task force identified eighteen patch types that formed the basis for the design. These represented broad descriptions of major plant communities. There was a lot of variety within these broad categories, but this was left for the project level of planning. The patch types identified included farm fields, prairies, pine–oak savannas, old-growth conifer, riparian woodlands, and several types of forests associated with timber production. Sixteen key flows were listed, and an analysis was done to determine how they related to the patch types. Flows included salmon, water, spotted owl, northern goshawk, amphibians, fire, and weeds. A basic design goal was to develop an arrangement of patches that facilitated desirable flows (salmon) while discouraging problematic ones (fire, weeds). For example, old-growth forest patches were strategically located to block weed flows while enhancing salmon habitat.

Four overriding restoration issues needed to be resolved in the design. The first was to identify a logical network of areas where old-growth conifer forest could be retained and restored. The design organizes old growth into a connected east–west band that links prominent gaps in mountain ridgelines. This old-growth band was concentrated on Forest Service–administered lands but incorporated upland riparian and unstable slopes on private holdings. The result of this approach is a fairly organic-shaped old-growth pattern that follows landforms rather than strictly adhering to property boundaries.

The second was the threat of stand-replacing wildfire. Decades of fire suppression and logging had created dangerously high small fuel loads. The threat to homes and habitats is serious, and the area experienced a 4,000-acre stand-replacing fire in 2000. The restoration design strategically located open patch types, such as restored pine and oak woodlands that could interrupt fire and even benefit from it. These were concentrated along south-facing ridges in the central part of the watershed, where they had been more common historically. If this project was successful, crown fires would be prevented from sweeping across the entire watershed, no matter what their direction or origin. These open woodlands would also improve habitat for many species and return a good portion of the landscape to its historic, pre–European settlement condition. In addition to savanna restoration, a more proactive prescribed fire approach in brush fields was also called for. These brushy areas have chaparral-type vegetation that becomes quite flammable as it ages (see Chapter 4).

The third restoration issue was aquatic habitat. The listing of several fish as threatened or endangered made it critical to design a landscape that offered a way to improve water quality and quantity in streams. Ideas included improving the ability of upland areas to retain winter snowpack, reducing fine sediment erosion, and increasing riparian shading. Wider riparian buffers in the lower watershed would replace some pasture and fields that were no longer economically important. In addition, redundant roads in the middle to upper watershed, particularly those on erodible granitic soils, would be decommissioned to reduce sedimentation.

A fourth issue that greatly influenced the design was the desire to integrate restoration with continued economic use of the land. Traditional land-based economic opportunities included small-scale farming, ranching, timber harvest, larger-scale industrial forestry, collection of mushrooms, berries, and greens, and protection of aesthetic qualities that attracted new residents with capital and ideas. The Little Applegate Watershed is not a wilderness, nor is it a park. It is and has been a working landscape for thousands of years. The challenge to the designers was to harness economic uses to help shape a better-functioning landscape where salmon could spawn, owls could hunt and raise their young, and fire would once again be more friend than foe. Designers integrated economic uses with restoration objectives by retaining some open, agricultural land in the lower watershed and by retooling forest practices in the middle watershed to reduce erosion, thin overcrowded woodlands, and restitch the forest matrix. Time will tell whether this strategy proves to be successful. (See Plate 9 in the color insert for maps.)

Case Study: Landscape-Scale Restoration in the Skagit River Basin

Geographic Context and Landscape History

The Skagit River Basin in northwest Washington State is the largest drainage in Puget Sound, supplying more than 30% of its total freshwater input (North Cascades Institute 2002). Altogether, the Skagit drains more than 8,000 square kilometers of the North Cascade Mountains of Washington and British Columbia. Elevations range from sea level to 3,275 meters at Mt. Baker. The area has a marine climate with mild winters and drier summers. Average annual rainfall ranges from about 90 centimeters in the lower floodplain to 460 centimeters at higher elevations (Drost and Lombard 1978). Lowland subbasins are located in the western valley and are more developed with urban and agricultural land use than are the forested mountain basins.

Approximately three quarters of the basin is publicly owned. Of that, 44% is within federally protected national parks or recreation or wilderness areas. An additional 24% is in the Mt. Baker–Snoqualmie National Forest. About 19% is in private and State of Washington ownership, with the remaining 13% in British Columbia, most of which is part of the provincial forest system. More than 157 miles of the Skagit, including its Cascade, Sauk, and Suiattle tributaries, are included in the National Wild and Scenic River System. Protected federal lands are concentrated at the highest elevations, and timber harvest predominates in the low and middle elevations of the middle part of the watershed. Agriculture is the major land use in the delta, which also contains the major population centers.

Numerous species, from the secretive old-growth–dependent marbled murrelet to the wide-ranging grizzly bear, inhabit forested uplands. The Skagit hosts one of the largest wintering bald eagle concentrations in the lower United States. The delta and adjoining bays are important stopovers along the Pacific flyway, supporting tens of thousands of shorebirds, waterfowl, and raptors. The Skagit is one of the most critical salmon-producing rivers in the lower forty-eight states and is vital to the recovery of federally threatened Puget Sound chinook salmon. It produces eight species of salmon and 30% of all anadromous fish that enter Puget Sound (North Cascades Institute 2002). It is the only river in the lower forty-eight that produces substantial numbers of all six species of Pacific salmon, including the most abundant wild chinook salmon populations in Puget Sound (Smith 2003). It also produces some of the most abundant chum and pink salmon populations in the United States and has the largest population of federally threatened bull trout in Puget Sound and possibly in the entire state (Beamer et al. 2000b, City of Seattle 2001).

Several Native American tribes also call the Skagit watershed home. Land development by Euro-Americans began around 1860. Commercial fishing and salmon canneries in Anacortes were among the first export industries to develop in Skagit County. Around this time, many of the delta's floodplain and tidal wetlands were diked, drained, and cleared to enable farming. Agriculture became an important industry, with oats, barley, hay, and other crops grown on the rich floodplain soils (Bourasaw 2002). Logging also grew in importance by the late 1800s. Railroad and road construction expanded rapidly in the early 1900s, leading the way for increased mining and logging in upland and mountainous areas (Smith 2003). Five major hydroelectric dams have been constructed, three on the upper Skagit River and two on the Baker River. Still, the area remains largely rural, with a total population of fewer than 100,000. Its industries include agriculture, logging, quarrying, oil refining, manufacturing, and tourism.

Despite nearly 150 years of landscape change, the river system remains in surprisingly good ecological health, largely because of the preponderance of highly protected federal lands in the upper basin. In the middle basin, timber management, road building, hydroelectric dams, and residential development have led to moderate levels of habitat loss, interrupting natural processes. Timber management and road building have altered hydrologic and sediment regimes. Dams have blocked or drowned habitats, created lakes where rivers once flowed, and trapped sediment. Residential and agricultural development have caused direct losses to stream and riparian habitats.

The most intensive development has occurred in the Skagit delta, where most of the human population settled and the land was converted to farm and urban uses. An estimated 72% of the historic lowland wetlands have been lost. Tidal habitats have been hardest hit, reduced by about 84% from their presettlement extent. Estuary forested transition and emergent marsh habitat have been reduced by 66% and 68%, respectively (Beamer et al. 2002). Approximately 75% of slough habitat has also been lost (Beechie et al. 2001). Dikes have been a major cause of estuary habitat loss. From the mouth of the river to the town of Sedro-Woolley, an estimated 62% of the mainstem channel edge has been diked, bank hardened, or both (Beamer et al. 2000a). Within the range of tidal influence, nearly all of the channel length is diked.

Reduction in tidal habitat has had significant impact on salmon and other species. Estuaries are considered critical nursery habitats for many species of fish and shellfish (Beck et al. 2003). Historically, the estuary contained an extensive network of blind and open channels, very important areas for salmon rearing. Tidal marshes also produce vast quantities of biomass annually that support a complex food web, including macroinvertebrates, shellfish, fish, and birds.

Because of the size and importance of the Skagit Basin, there has been a good deal of interest in habitat protection and restoration over the last few decades. However, conservation and restoration typically were being done in the absence of a coordinated, strategic framework.

Planning, Priorities, and the Skagit Watershed Council

In 1997 thirteen organizations came together to form the Skagit Watershed Council to improve coordination and effectiveness of salmon habitat protection and restoration. The council has since grown to include forty member organizations, including private industrial and agricultural interests, state and federal agencies, local governments, tribes, environmental groups, and citizen-based groups. The council was officially designated as the lead entity for salmon recovery under State of Washington legislation. All state and federal salmon recovery funding is routed through the State Salmon Recovery Funding Board. As lead entity, the council screens and prioritizes all Skagit basin habitat protection and restoration proposals first submitted to the funding board. It has competed very effectively for this funding through creation of a landscape-scale, natural process–based approach. Three major documents guide habitat restoration and protection efforts.

First, a *Habitat Protection and Restoration Strategy* (Skagit Watershed Council 1998) lays out a scientific framework for analyzing landscape processes and provides a set of procedures for screening and prioritizing restoration and protection actions. It identifies the natural landscape processes active in a given watershed, the effects land use has on these processes, and the relationships between land use and habitat conditions. It focuses not on the symptoms of watershed degradation but on the causes.

This approach is based on previous research that has tied natural processes to creation and maintenance of habitat conditions to which native aquatic and riparian species are adapted (e.g., Peterson et al. 1992, Doppelt et al. 1993, Reeves et al. 1995, Ward and Stanford 1995, Beechie et al. 1996, Kauffman et al. 1997). A conceptual model was developed to illustrate how watershed controls and natural landscape processes form various habitat conditions. Watershed controls such as geology, climate, and scale are independent of land management and act over large areas and long timeframes to shape the range of possible habitat conditions within a watershed (Naiman et al. 1992). Landscape processes, including sediment supply, hydrology, and riparian functions, create and maintain habitats over the shorter term and are influenced by land uses. Natural rates of landscape processes are defined as those that existed before widespread timber harvest, agriculture, and urban development. Threshold levels of disturbance for each process were established to help identify the process in need of protection or restoration and to evaluate whether restoration is likely to succeed in areas where multiple processes are disturbed.

Second, the *Application of the Skagit Watershed Council's Strategy. River Basin Analysis of the Skagit and Samish River Basins* (Beamer et al. 2000a) identified site-specific levels of impairment of landscape processes throughout the basin and called for specific actions to restore and protect salmon. For this document, a combination of field-based inventories and geographic information system data identified the degree of disturbance to habitat-forming landscape processes such as peak flows, sediment supplies, riparian functions, blockages to salmon migration, and water quality. For each process, the council classified the disturbance level for each river reach or tributary basin as functioning, moderately impaired, or impaired based on the thresholds identified in the *Habitat Protection and Restoration Strategy*. For instance, about 46% of the areas in upper mountain subbasins of the Skagit have impaired sediment supply, and 23% probably are impaired with respect to peak flow hydrology. This information helped to identify more 400 potential projects and provided an overview of the spatial pattern of disturbance.

Third, the *Strategic Approach* (Skagit Watershed Council 2004) combines the scientific information developed in the *Application*, with recent data on particular species to target certain project types and specific geographic areas for restoration and protection. The *Strategic Approach* relies on three principles to guide the council's objectives for project identification and prioritization. The first states that the best available information should be used to target the most biologically important areas for salmon restoration and protection. The council prioritizes the habitats most important to listed and depressed salmon stocks, such as chinook.

The second principle states that within the identified target areas, the council should protect the highest-quality habitats first and then restore key habitats. Protecting highly functioning habitats is done through various tools, including fee simple acquisition and conservation easements. This approach is essential for anchoring highly productive spawning and rearing areas for long-term recovery. It is also more cost-effective than first attempting to restore highly degraded habitats. However, given the degree of salmon habitat degradation locally as well as regionally, protecting key habitats and refugia alone would not be sufficient to ensure long-term survival or recovery of salmon. So the council also encourages reestablishment of key habitats in target areas using a variety of restoration tools, including sediment reduction, riparian planting, fencing, installation of log jams, reconnection of isolated habitats, and water quality enhancement. The goal is to expand existing refugia and establish new ones where these have largely been lost. The selected projects depend on the suite of natural process impairments. For example, instream work such as the installation of log jams is not done until underlying sediment regime issues are addressed.

The third guiding principle supports focusing on the most cost-effective projects first in order to ensure the best and most efficient use of limited funding. There are cost effectiveness formulas for certain project types, such as fish passage, riparian planting, and acquisition, that are applied to assist in prioritization of the final project proposals before each funding round.

In addition to these three guiding principles, the council also considers the following elements as critical for the successful implementation of its *Strategic Approach*:

- Using comprehensive and current information, such as updates from field surveys on fish productivity
- Maintaining a fully functioning organizational infrastructure, including communication strategy, data management, monitoring, and reporting methods
- Maintaining sufficient support capacity within the council and its member organizations
- Identifying and working with willing landowners in priority areas
- Building a knowledgeable, supportive community

The most recent edition of the *Strategic Approach* identifies and delineates four target areas: near shore, estuary, floodplain, and watershed processes in the Skagit and Samish River basins (Figure

14.10). It requires that all projects submitted for funding be located in one of these four target areas and be consistent with the stated objectives. These target areas and the objectives were selected as essential for the top priority target species, Skagit chinook.

Because the lack of estuarine habitat is considered a major limiting factor for chinook, the estuary is identified as the top priority among all first-tier target areas. The council also supports actions that will restore and protect landscape processes in the Skagit floodplain, selected tributaries and upland areas, and pocket estuary habitats in the nearshore environment.

Results

Since the council began reviewing and prioritizing project proposals in 2000, fifty-one projects have received funding totaling more than $15.9 million (Table 14.1). These projects include a wide range of types, sponsors, objectives, and geographic locations, but each was reviewed to ensure consistency with the strategic focus on landscape process restoration and protection and with the specific objectives and target areas identified in the council's *Strategic Approach*.

The Skagit Watershed Council has encouraged protection of existing high-quality habitat and cost-effective projects. Projects that have ranked high include the Skagit Land Trust's Middle Skagit and The Nature Conservancy's Upper Skagit acquisition projects. Both used quantitative analysis of financial costs and ecological benefits along large reaches of the Skagit River floodplain to generate a prioritized list of potential projects. They targeted only the most cost-effective projects rather than relying on a more opportunistic willing landowner approach. The board awarded more than $1.9

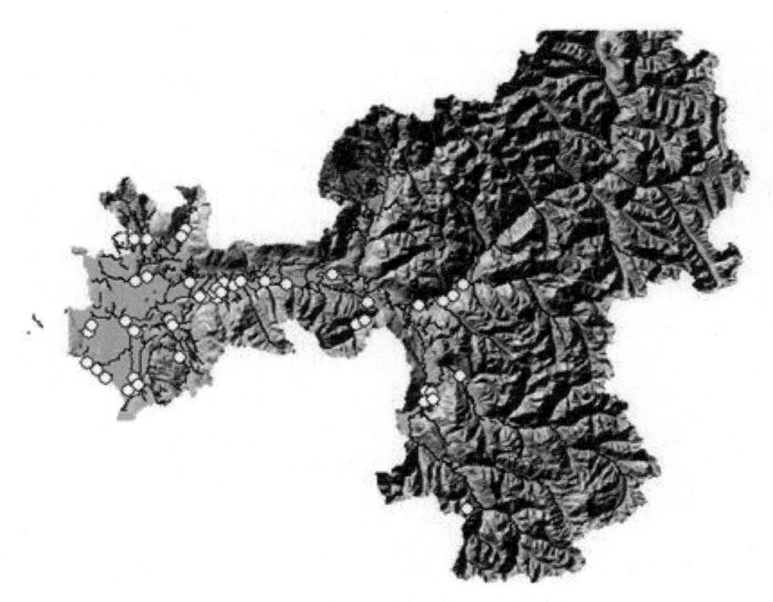

FIGURE 14.10. Map of Skagit River Basin with fifty-one restoration project locations since 1999. Note the heavy concentration in the lower watershed. (*Map by Ben Perkowski*)

TABLE 14.1.

Summary of Skagit River watershed restoration and protection projects funded by the state Salmon Recovery Funding Board (SRFB) since 2000.

	SRFB 1	SRFB 2	SRFB 3	SRFB 4	SRFB 5	Total of 5 Rounds
Number of projects	5	20	14	6	6	51
Total SRFB funding	$1,632,628	$2,719,710	$3,967,321	$2,072,420	$1,591,299	$11,983,378
Total match	$637,072	$723,963	$848,756	$953,050	$849,346	$4,012,187
Total cost	$2,269,700	$3,443,673	$4,816,077	$3,025,470	$2,440,645	$15,995,565
Project types	3 protections 1 riparian restoration	2 protections 5 restoration and protection combinations 4 studies 2 riparian restorations 2 invasive species 1 sediment reduction 4 habitat reconnections	3 sediment reductions 3 protections 4 studies 2 isolated habitat reconnections 1 riparian restoration 1 instream restoration	1 protection 1 restoration and protection combination 1 design 1 off-channel restoration 2 habitat reconnec-tions	3 estuarine restorations 2 studies 1 riparian restoration	9 protections 11 studies 4 estuarine restorations 5 riparian restorations 6 combinations 6 habitat reconnections 2 invasive species 4 sediment reductions 1 instream restoration 2 habitat reconnections 1 off-channel restoration
Average number of projects funded per lead entity by SRFB	4.0	6.4	4.9	3.1	4.0	NA
Average SRFB funding	$633,333	$1,382,609	$1,415,385	Amount to come	Amount to come	NA

million, including both acquisition and assessment funds. Project proponents were able to secure additional match funds.

The council's strategic framework also strongly encourages restoring impaired landscape processes in upland areas rather than waiting to deal with damage to downstream areas. An example is projects that address elevated sediment levels. Four sediment reduction projects focused on upgrading or decommissioning roads that had been identified as having high potential to cause mass wasting events and transport sediment to important aquatic habitats. Road treatments generally included removing or upgrading stream crossings, adding and upgrading drainage structures, and stabilizing or removing fills. Significant progress has been made in alleviating elevated sediment in the mainstem Skagit and major tributaries. In Finney Creek basin alone, the Forest Service has treated more than 69 miles of road.

Reconnecting isolated habitat has been a major part of the strategic focus, in part because of a high likelihood of success. Most have focused on improving passage to tributary habitats. A comprehensive inventory of blockages has been extremely valuable for project identification. Council members have developed methods for prioritizing potential habitat gains using a combination of geographic information system data and field-based data. A number of partners have executed fish passage projects. The Skagit Fisheries Enhancement Group has initiated a variety of restoration projects, working directly with landowners and volunteers. Since 1998, they have improved fish passage to more than 25 miles of habitat, and they monitor twenty-eight fish passage restoration sites. Volunteers and staff visit more than 12 miles of streams to document returning salmon, redds, and carcasses. Last year more than 7,400 live salmon were observed spawning, including chinook, coho, chum, and pink. This compares to 1999, when 7 miles were surveyed for spawning, and fewer than 700 salmon were observed. Although the Enhancement Group does not take credit for all of this increase, the gains indicate that passage projects have benefited salmon by increasing the amount of total habitat available and by targeting areas that benefit fish.

Reconnection of floodplain habitat in the delta and estuary also is a high priority. These projects can be complex and expensive, often involving removal or reconstruction of dikes and other flood control and drainage infrastructure. Nonetheless, several projects are in the acquisition, design, or feasibility phase, including one aimed at restoring more than 175 acres of estuarine habitat. The Deepwater Slough project at the Skagit Wildlife Area restored more than 200 acres of estuarine habitat in the South Fork. It represents one of the largest such projects in the Pacific Northwest. Initial monitoring shows that juvenile chinook are using the restored areas in comparable abundance to natural reference sites (Hood et al. 2003). Long-term monitoring will be critical for planning future projects.

Summary

The Skagit River is one of the largest on the Northwest coast and is still home to some of the largest and healthiest populations of salmon in the region. Because of its size and importance, many nonprofit organizations, government agencies, tribes, and private individuals and businesses are engaged in various aspects of conservation and restoration, all with their own missions, tactics, and interests. Despite this multitude of actors and interests, the Skagit Watershed Council has provided a forum for a strategic, collaborative approach to landscape-scale restoration. Through the establishment of a natural process–based scientific framework and the rigorous screening of projects proposed for salmon recovery funding, the council has made overall basin-wide restoration efforts more effective. Projects that have been screened by the council compete effectively for funding because of the strong scientific scrutiny. More salmon recovery funding has been passed through the Skagit Watershed Council than through any other comparable entity in Washington. The result of this collaborative approach is that hundreds of acres of high-quality habitat have been protected, dozens of miles of forest roads have been treated, many miles of stream have been made accessible to fish, hundreds of acres of riparian forest have been replanted, and a few hundred acres of old farm fields have been restored to tidal marsh.

There are also many habitat projects that have been implemented without state funding and independent of the council. Although the council has no direct control over these projects, their effectiveness probably has been influenced by the collaborative work of multiple council partners.

For example, standards and expectations for the types, location, and quality of restoration projects have been raised, and few entities continue to pursue purely opportunistic projects. Duplication of effort between groups has been reduced by improved communication and the existence of a restoration strategy that exceeds the planning and analytical capacities of individual organizations. There is less negative competition between groups as they become more aware of what each is doing. And a growing number of collaborative partnerships go beyond the reach of salmon recovery. In 2004, a number of Skagit groups received the President's Coastal America Partnership Award in recognition of the depth and effectiveness of the partnerships pursuing basin-wide conservation of this important watershed.

Case Study: Kennedy Flats Watershed Restoration

Kennedy Flats is located in Clayoquot Sound on the west coast of Vancouver Island, British Columbia, Canada (Figure 14.11). Altogether there are ten stream systems and nine subbasins between Ucluelet and Tofino. This area has been the focus of a watershed-based ecosystem restoration plan since 1994 (Warttig et al. 2000). The ten streams flow through several different private timber company management areas, or tenures. Lost Shoe Creek flows through areas managed by Interfor and Isaak, then through Pacific Rim National Park Reserve before reaching the Pacific Ocean.

Restoration efforts were made a priority by local communities after severe declines in salmon were documented in the 1980s. In response, management standards for logging around streams were established in 1988, and restrictions on commercial and recreational fisheries were put in place a few years later. By 1995, restrictive logging guidelines came into force for all publicly owned forestland in British Columbia.

In Kennedy Flats additional rules governing logging practices were developed by the Clayoquot Sound Scientific Panel, which was the first ecosystem-based management strategy brought into Canada (Mabee et al. 2004). Among other things, the panel used a watershed-based planning approach, which had a strong influence on the format of the Kennedy Flats Restoration Plan.

The entire Kennedy Flats watershed is 12,937 hectares and includes nine subbasins: Kootowis, Hospital, Sandhill, Staghorn, Trestle South, Trestle, Indian/Harold, Salmon, and Lost Shoe (Table 14.2). These streams are unique on the west coast of British Columbia by virtue of their very low gradients, which average less than 0.4%. This low gradient allows nearly the entire length of the stream system to support populations of chum, coho, steelhead, sea run and resident cutthroat, and rainbow trout (Clough et al. 1996). In a 1995 survey, chinook were not present in the streams, but since then they have returned, consistent with traditional ecological knowledge of the area.

The lower Lost Shoe Creek component of Pacific Rim National Park had experienced heavy logging before park establishment in 1970. Logging dates back to the early 1900s, with acceleration of harvest in the 1960s and 1970s. As a result, the riparian and floodplain areas of many of the creek systems now consist almost entirely of either high-density brush or alder-dominated forests with poor

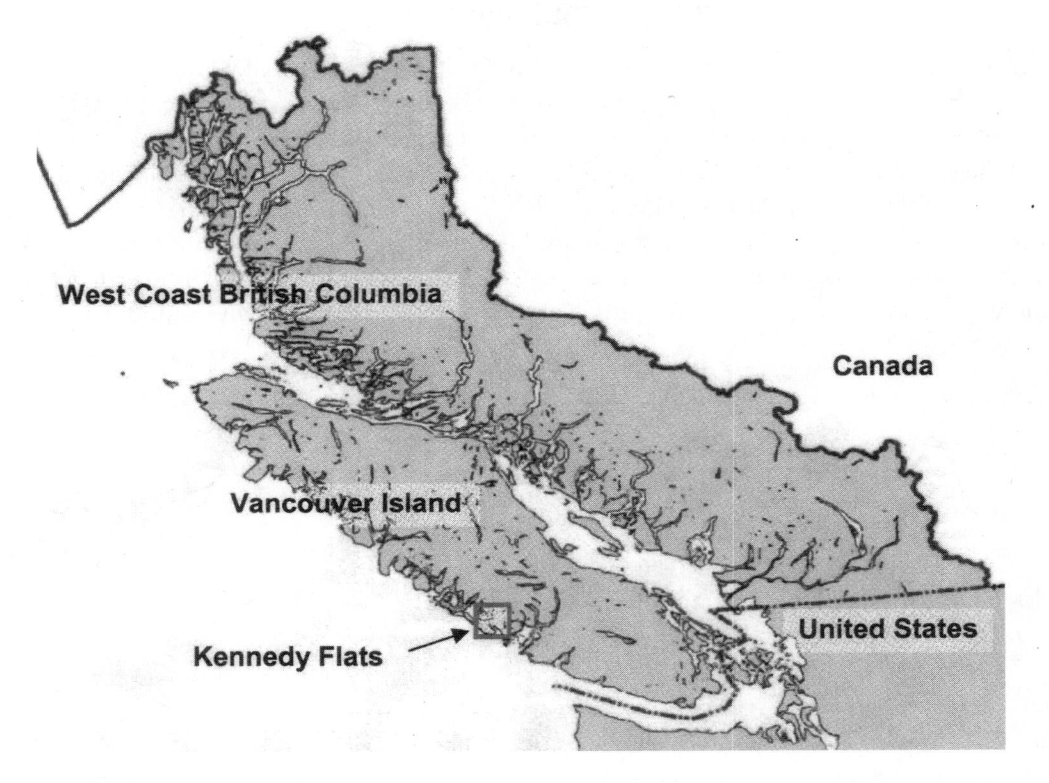

Figure 14.11. Kennedy Flats location, on the west coast of Vancouver Island. (*Courtesy of International Forest Products Limited*)

TABLE 14.2.

Kennedy Flats watershed summary.

Stream	Watershed Area (ha)	Mainstem and Major Tributary Length (km)	Mainstem Gradient (%)	Stream Order
Hospital	653	10.36	0.3	3rd
Kootowis	3,489	77.59	0.4	4th
Staghorn	2,533	47.92	0.3	4th
Trestle	852	14.70	0.3	2nd
Trestle South	634	13.10	0.3	2nd
Indian/Harold	455	12.95	1.5	3rd
Lost Shoe	2,053	32.90	0.4	3rd
Salmon	575	6.90	0.5	3rd
Sandhill	1,693	24.90	0.4	3rd
Total	12,937	241.32		

conifer understory (Poulin and Simmons 2001). Large wood is lacking in area streams, largely because of an ill-advised stream cleanout program in the 1980s. Given the number large conifers, future recruitment of large woody debris is limited.

Bank erosion is apparent in several areas. The dense brush and alder that line the banks have poor rooting strength and thus do not do a good job of stabilizing banks. Instream placement of large wood was deemed necessary to mitigate bank erosion, reestablish the stream thalweg (deepest channel), increase stream complexity, and provide cover.

Historic logging practices often left logs in the stream, often spanning the channel to facilitate easier yarding. The logs left behind were of economically low value, which at the time included western red cedar. Some of these were later salvaged and milled or split on site for shakes and shingles. The residual waste from the processing released large amounts of small woody debris, which, combined with remaining cross-stream logs, created large debris jams. These jams caused streams to spread out across the valley floor, lowered water quality, and inhibited fish passage. Clearly, Kennedy Flats streams were a mess.

Two community groups played key roles in the initial restoration efforts at Kennedy Flats. First, the Tofino and Thornton Creek hatcheries are run primarily by volunteers. They focused their efforts on augmenting wild populations and limited hatchery origin fish to less than 10% of the wild fish numbers.

The second organization was the Central Westcoast Forest Society. The society is a registered charity based in the local town of Ucluelet. Because of multiple tenures, watershed-based restoration would have been difficult if not for the society's ability to create and nurture partnerships. These include four First Nations (Hesquiat, Ahousaht, Tla-o-qui-aht, and Ucluelet), the Ucluelet and Tofino town councils, the Alberni–Clayoquot Regional Forest District, the International Woodworkers of America, Interfor, and the Pacific Rim National Park Reserve.

The Forest Society focused on ecosystem restoration projects in the Ucluelet and Tofino areas for the first nine seasons. Labor was provided by local First Nations members and displaced forestry and commercial fishery workers. The society's funding comes from a variety of sources, including government, foundations, corporations, and individual donations.

Restoration funding often comes with conditions. For example, Forest Renewal and the Forest Investment Account funds must be used on provincial Crown lands. Therefore, affected stream reaches in the Pacific Rim National Park Reserve (lower Lost Shoe Creek and Sandhill Creek) generally lacked funding for restoration. Also, annual funding cycles have not allowed completion of multiyear projects. For example, Fisheries Renewal BC provided funding to complete phase one of instream restoration in 1999, but that funding source was no longer available for phase two.

The society has tapped some unrestricted funds through the Vancouver Foundation, the Home Depot Foundation, Eco-Action 2000, and other sources. These grants allowed completion of restoration activities on Lost Shoe Creek within the national park.

The restoration plan developed for Kennedy Flats was a top-down approach (Slaney and Zaldokas 1997). Between 1994 and 1996 issues of slope stability and sediment sources associated with roads and landslides were mitigated through best management practices, deactivation of unneeded roads, hydroseeding, planting, and slope bioengineering. The highest-priority items were identified through risk assessments. For example, each road segment was assessed using a formula of Hazard × Consequence = Environmental Risk. Overall risk was calculated as Environmental Risk × Feasibility, with each component having a maximum value of 9. Hazard is the likelihood of failure, consequence is the measurement of impact if the road fails, and feasibility is the success a given amount of investment would have in reducing the risk. With this scoring system, the higher the score, the greater the return on restoration investment.

Once upslope road restoration components were completed, a similar risk assessment was used to prioritize stream restoration. Stream restoration followed a four-step process:

1. Removal of embedded small woody debris and reorientation of large woody debris to increase channel complexity.
2. Maintenance and annual monitoring for 3 years, then every 5 years thereafter. Large wood was added in deficient areas (based on Slaney and Zaldokas 1997).
3. Addition of spawning gravel to deficient areas.
4. Restoration of the riparian areas (including thinning and planting to establish a better balance of conifer and deciduous trees).

Adding large wood to many sites posed a problem because often there was no road access (Figure 14.12). In some cases material and ballast had to be brought to sites by helicopter. Given the cost and maximum lift capacity, all log structure and ballast requirements had to be designed around the available helicopter. All logs and ballast are estimated for weight to ensure that every piece is less than 1,100 kilograms.

Because the helicopter costs more than $2,000 per hour, the restoration team needs excellent organization and plans with little room for error. Therefore, each piece of material is drilled, cabled, and numbered for exact placement. At Lost Shoe Creek sites were all predetermined. Bank protection and scour hole structures were given the highest priority.

The Department of Fisheries and Oceans reviewed plans, and for areas within the park, Parks Canada personnel completed required environmental assessments. The Lost Shoe Creek system (and most of the other Kennedy Flats systems) contrasts with what is the norm on most coastal systems in British Columbia. About 15% is classified as upslope, with the remainder basically flat and low gradient. There is no orographic lift factor, nor any rain-on-snow events. Lost Shoe is also characterized by a maze of forest swamp systems with significant water storage capacity. This limits peak discharge and is a primary reason for the highly tannic water after high-rain events.

Part of the mandate of the Pacific Rim National Park Reserve is to maintain and restore ecological integrity to formerly disturbed ecosystems. The challenge of the Lower Lost Shoe was beyond the technical and financial resources of Parks Canada. By partnering with the society, they obtained both funding and technical expertise. This project is a prime example of the cooperation and innovation needed to restore damaged aquatic ecosystems, particularly where habitats cross jurisdictional boundaries.

FIGURE 14.12. *(a)* Before and *(b)* after log jam removal and large wood placement. *(Photos by Warren Warttig)*

Since the initiation of restoration, there has been a significant increase in salmon escapements in areas where the restoration work has taken place. There has also been significant improvement in water quality, including temperature, dissolved oxygen, and nutrient levels. Although much has been accomplished, approximately $3.5 million (Canadian) is still needed to complete restoration work on Kennedy Flats.

Summary

Since we began researching and writing this chapter, there has been an impressive acknowledgment of the importance of paying attention to key landscape processes, as opposed to focusing too much on snapshots of habitat structure. This has led many watershed efforts to place more emphasis on upslope challenges, particularly sediment delivery, before investing in downstream habitat improvements. We know a great deal more about how landscapes and watersheds work and ecosystems interact at large scales and how they vary over time.

Of course, the negative side is that we have just begun to learn how degraded our ecosystems are and how much restorative work will be needed to recover them. In the Mattole example, we can also see the importance of restoring damaged social relationships as a prerequisite to restoring watersheds. Many restoration ecologists are not adequately trained or oriented to this critical aspect of our work. Our advice to the reader who is contemplating or engaged in a large-scale restoration effort is to pay attention to sociology, political science, and environmental psychology as well as to the technical aspects of restoration.

References

Beamer, E., T. Beechie, B. Perkowski, and J. Klochak. 2000a. *Application of the Skagit Watershed Council's Strategy. River Basin Analysis of the Skagit and Samish River Basins: Tools for Salmon Habitat Restoration and Protection*. Skagit Watershed Council, Mt. Vernon, WA.

Beamer, E., R. Henderson, and K. Larsen. 2002. *Evidence of an Estuarine Habitat Constraint on the Production of Wild Skagit Chinook*. Presentation at Western Division AFS Meeting in Spokane, April 29–May 1, 2002. Skagit System Cooperative, LaConner, WA.

Beamer, E. M., R. E. McClure, and B. A. Hayman. June 2000b. *Fiscal Year 1999 Skagit River Chinook Restoration Research*. Skagit System Cooperative, La Conner, WA.

Beck, M. W., K. L. Heck, K. W. Able, D. L. Childers, D. B. Eggleston, B. M. Gillanders, B. S. Halpern, C. G. Hays, K. Hoshino, T. J. Minello, R. J. Orth, P. F. Sheridan, and M. P. Weinstein. 2003. The role of nearshore ecosystems as fish and shellfish nurseries. *Issues in Ecology* 11: 1–12.

Beechie T., E. Beamer, B. Collins, and L. Benda. 1996. Restoration of habitat-forming processes in Pacific Northwest watersheds: a locally adaptable approach to salmonid habitat restoration. Pages 48–67 in D. L. Peterson and C. V. Klimas (eds.), *The Role of Restoration in Ecosystem Management*. Society for Ecological Restoration, Madison, WI.

Beechie, T. J., B. D. Collins, and G. R. Pess. 2001. Holocene and recent geomorphic processes, land use, and salmonid habitat in two north Puget Sound river basins. Pages 37–54 in J. M. Dorava, D. R. Montgomery, B. B. Palcsak, and F. A. Fitzpatrick (eds.), *Geomorphic Processes and Riverine Habitat*. Water Science and Application Volume 4. American Geophysical Union, Washington, DC.

Bourasaw, N. 2002. *Skagit River Journal of History & Folklore*, Sedro-Woolley, WA. Available at www.geocities.com/skagitjournal/Towns-West/MVHistory1940.html.

City of Seattle. 2001. Seattle's watersheds outside the municipal boundaries. In *Seattle's Urban Blueprint for Habitat Protection and Restoration*. City of Seattle, Seattle Public Utilities, Seattle.

Clough, D. C., K. Eakins, J. Hillaby, and K. Rude. 1996. *Kootowis, Staghorn, Lost Shoe Watershed Restoration Program*. Prepared for International Forest Products Ltd., Vancouver, BC.

Diaz, N. and D. Apostol. 1992. *Forest Landscape Analysis and Design: A Process for Developing and Implementing Land Management Objectives for Landscape Patterns*. USDA Forest Service, Pacific Northwest Region, Portland, OR.

Doppelt, B., M. Scurlock, C. Frissell, and J. Karr. 1993. *Entering the Watershed*. Island Press, Covelo, CA.

Dramstad, W. E., J. D. Olsen, and R. T. T. Forman. 1996. *Landscape Ecology Principles in Landscape Architecture and Land Use Planning*. Island Press, Washington, DC.

Drost, B. W. and R. E. Lombard. 1978. *Water in the Skagit River Basin, Washington*. Water Supply Bulletin 47. Washington State Department of Ecology and U.S. Geological Survey, Olympia.

Forman, R. T. T. 1995. *Land Mosaics: The Ecology of Landscapes and Regions*. Cambridge University Press, Cambridge, England.

Forman, R. T. T. and M. Godron. 1986. *Landscape Ecology*. Wiley, New York.

Freedlund, A. March 2005. *Mattole Watershed Plan Chapter One: The Evolution of Conservation and Restoration in the Mattole River Watershed*, Unpublished report by the Mattole River Watershed Council.

Hood, W. G., E. Beamer, and R. Henderson. 2003. Preliminary results from monitoring of the Deepwater Slough Restoration. *Skagit River Tidings*. Skagit Watershed Council, Mt. Vernon, WA.

House, F. 1999. *Totem Salmon: Life Lessons from Another Species*. Beacon Press, Boston, MA.

Isom, J. April 2005. *Mattole Watershed Plan Chapter Six: Restoring Wild and Working Forests*. Unpublished report by the Mattole River Watershed Council.

Kauffman, J. B., R. L. Beschta, N. Otting, and D. Lytjen. 1997. An ecological perspective of riparian and stream restoration in the western United States. *Fisheries* 22(5): 12–24.

Kohm, K. A. and J. F. Franklin. 1997. *Creating a Forestry for the 21st Century*. Island Press, Washington, DC.

Larson, C. 2005. *Mattole Watershed Plan Chapter Five: Sediment*. Unpublished report by the Mattole River Watershed Council.

Lindenmayer, D. B. and J. Franklin. 2003. *Conserving Forest Biodiversity*. Island Press, Washington, DC.

Mabee, W. E., E. Fraser, and O. Slaymaker. 2004. Evolving ecosystem management in the context of British Columbia resource planning. *BC Journal of Ecosystems and Management* 4: 2–5.

Mellen, K., M. Huff, and R. Hagestedt. 1995. *HABSCAPES: Reference Manual and User's Guide*. Unpublished manuscript, U.S. Forest Service.

Mills, L. S., M. E. Soule, and D. F. Doak. 1993. The keystone-species concept in ecology and conservation. *BioScience* 43(4): 219.

Naiman, R. J., D. G. Lonzarich, T. J. Beechie, and S. C. Ralph. 1992. General principles of classification and the assessment of conservation potential in rivers. Pages 92–124 in P. J. Boon, P. Calow, and G. E. Petts (eds.), *River Conservation and Management*. Wiley, New York.

North Cascades Institute. 2002. *Skagit Watershed Education Project*. North Cascades Institute, Sedro-Woolley, WA.

Noss, R. F. and A. . Cooperrider. 1994. *Saving Nature's Legacy*. Island Press, Washington, DC.

Perry, D. 1994. *Forest Ecosystems*. The Johns Hopkins University Press, Baltimore, MD.

Peterson, N. P., A. Hendry, and T. Quinn. 1992. *Assessment of Cumulative Effects on Salmonid Habitat: Some Suggested Parameters and Target Conditions*. TAW-F3-92-001. Washington Department of Natural Resources, Olympia.

Poulin, V. A. and B. Simmons. 2001. *Lost Shoe Creek Riparian Assessments and Recommendations for Riparian Restoration*. Prepared for International Forest Products Ltd., Vancouver, BC.

Priester, K. n.d. *Words into Action: A Community Assessment of the Applegate Valley*. Unpublished report. Executive summary available online at www.naturalborders.com/applegate.htm.

Reeves, G. H., L. E. Benda, K. M. Burnett, P. A. Bisson, and J. R. Sedell. 1995. A disturbance-based ecosystem approach to maintaining and restoring freshwater habitats of evolutionary significant units of anadromous salmonids in the Pacific Northwest. *American Fisheries Society Symposium* 17: 334–339.

Ripple, W. J. and R. L. Beschta. 2003. Wolf reintroduction, predation risk, and cottonwood recovery in Yellowstone National Park. *Forest Ecology and Management* 184: 299–213.

Skagit Watershed Council. 1998. *Habitat Protection and Restoration Strategy*. Skagit Watershed Council, Mt. Vernon, WA.

Skagit Watershed Council. 2004. *Year 2004 Strategic Approach*. Skagit Watershed Council, Mt. Vernon, WA.

Slaney, P. A. and D. Zaldokas. 1997. *Fish Habitat Rehabilitation Procedures*. Watershed Restoration Technical Circular No. 9. Ministry of Environment, Lands and Parks, UBC, Vancouver, BC.

Smith, C. 2003. *Salmon and Steelhead Habitat Limiting Factors Water Resource Inventory Areas 3 and 4, The Skagit and Samish Basins*. Washington State Conservation Commission, Lacey.

U.S. Department of Agriculture and U.S. Department of the Interior. 1995. *Ecosystem Analysis at the Watershed*

Scale: Federal Guide for Watershed Analysis. Version 2.2. Regional Interagency Executive Committee, Portland, OR.

Ward, J. V. and J. A. Stanford. 1995. Ecological connectivity in alluvial river ecosystems and its disruption by flow regulation. *Regulated Rivers: Research and Management* 11: 105–119.

Warttig, W. R., D. Clough, and M. Leslie. 2000. *Kennedy Flats Restoration Plan*. Prepared for International Forest Products Ltd., Central Westcoast Forest Society, Pacific Rim National Park, Ministry of Forests, Ministry of Environment, Lands and Parks, and FRBC, Victoria, BC.

Watershed Professionals Network. June 1999. *Oregon Watershed Assessment Manual*. Prepared for the Governor's Watershed Enhancement Board, Salem, OR.

Chapter 15

Restoring Wildlife Populations

Bruce H. Campbell, Bob Altman, Edward E. Bangs, Doug W. Smith, Blair Csuti, David W. Hays, Frank Slavens, Kate Slavens, Cheryl Schultz, and Robert W. Butler

This chapter has two goals: to provide a background on general concepts, considerations, and limitations of fish and wildlife restoration and to provide a sampling of species or animal guilds that are being restored or are candidates for restoration in the Pacific Northwest.

Wildlife Restoration Background

Wildlife restoration in the United States began in the late nineteenth century when private groups such as the Boone and Crockett Club and Izaak Walton League of America began translocating animals in response to declining game populations. Further developments came in the early 1930s with the establishment of modern wildlife management and passage of the Federal Aid in Wildlife Restoration Act (Kallman et al. 1987). Restoration efforts focused on game species until the late 1960s and early 1970s, when the environmental movement and passage of the Endangered Species Act broadened public policy to include other animals. Public interest in restoration further increased in the 1990s, as has interest among wildlife professionals. This has led to new concepts of biodiversity, ecosystem-based management, new technical journals such as *Restoration Ecology* and *The Journal of Restoration Ecology*, and the growth of conservation biology as a new subdiscipline.

Simberloff et al. (1999) define restoration as the exact reproduction, with human intervention, of a community or ecosystem that once was present. Following this definition, wildlife restoration is the exact reproduction, with human intervention, of a wildlife population or community that was once present. This is typically accomplished by three methods: population restoration, population augmentation, and planned recolonization. A fourth method, reintroduction, is used when exotic game species are involved.

Population restoration is the active reestablishment of a self-sustaining, free-ranging population. It is accomplished by translocation of genetically similar animals from neighboring wild or ex situ populations. Generally, this is done only after the limiting factors such as habitat loss or degradation, interspecific competition, and overexploitation are rectified. Reestablishment of bighorn sheep (*Ovis canadensis*) in Oregon is an example of population restoration (Oregon Department of Fish and Wildlife 2003).

Population augmentation is the active addition of genetically similar animals to a natural population that is below or in danger of falling below a minimum viable size. It is generally accomplished through translocation after limiting factors have been addressed. Augmentation is also used to boost selected age classes, sex ratios, or genetic variability in a population. Animals for augmentation may come from either wild or ex situ populations. Examples of the latter are the augmentation of western pond turtle (*Emys marmorata*) populations in western Washington and augmentation of Oregon silverspot butterfly (*Speyeria zerene hippolyta*) populations in Oregon (Andersen et al. 2002). The western pond turtle program is profiled later in this chapter.

Recolonization is the natural reestablishment of a species in an area through immigration. Like restoration, this generally occurs after reestablishment of appropriate habitat, enhancement of existing habitat, or enactment of species protection. Reestablishment of osprey (*Pandion haliaetus*) in the Pacific Northwest after the ban on DDT is an example of recolonization (Marshall et al. 2003).

Reintroduction is the reestablishment of a nonnative species in an area where it had previously been introduced. An example is reintroduction of wild turkeys (*Meleagris gallopavo*) in parts of the Pacific Northwest after previously established populations died out. Reintroduction will not be further addressed in this chapter.

Goals of Restoration

As previously defined, wildlife restoration ideally results in the exact reproduction of a wildlife population that was once present. However, this is almost never a fully achievable goal (Simberloff et al. 1999). What previous condition is to be reproduced: the one that existed before human colonization of North America, the one that existed before arrival of European settlers, or the one that existed a century ago? To confound the issue, what would a population look like today had it continued to exist in a self-sustaining, free-ranging state? Environmental and selective pressures are dynamic, and even in pristine systems undisturbed by humans, the process of succession causes continual change. Finally, exact replication is impossible because the unique genotypes that existed at the reference time probably no longer exist.

Given the temporal limitations on wildlife restoration, a more realistic goal is to approximate conditions at a point on a trajectory of change rather than to precisely replicate conditions that were once present. Specifying what the trajectory would have been is an enormous challenge but, in principle, should be the goal.

Restoring Biodiversity

In recent years, there has been a shift in focus from restoration of individual species to restoration of biodiversity. Biodiversity is defined as the variety and variability among living organisms and the ecological complexes in which they occur (Office of Technological Assessment 1987). Because biodiversity is such an all-inclusive concept, monitoring the progress of a biodiversity restoration project is difficult, expensive, and in some cases not possible. Consequently, a subset of one or more species often is used as a surrogate for broader biodiversity. These are called keystone, indicator, umbrella, or flagship species (Scott et al. 1999, Kunkel 2003).

An *indicator species* typically is used as a surrogate for ecosystem health or areas of high species richness (Landres et al. 1988). An example is the black-backed woodpecker (*Picoides tridactylus*), which has been recommended as an indicator species for mature and old-growth lodgepole pine forests in central Oregon (Goggans et al. 1988). The black-backed woodpecker depends on trees with heartrot for nesting, diseased trees or decayed snags for roosts, and adequate decaying substrate to provide a prey base of wood-boring insects, all characteristics of mature and old-growth lodgepole pine forests.

Where the goal is to protect a habitat or community, an *umbrella species* sometimes is used to delineate the size of an area or type of habitat over which protection should occur (Caro and O'Doherty 1999). An example is the mountain lion (*Puma concolor*) in southern California (DeNormandie 2000). Umbrella species may also be used to delineate connectivity or thresholds between patch distances in a fragmented environment. The assumption is that sufficient landscape connection for the umbrella species will also be sufficient for other species. Roberge and Angelstam (2004) provide a good review of umbrella species.

Flagship species are charismatic and often are used to raise awareness, build public support, and attract funding (Caro and O'Doherty 1999). Flagship species may also be umbrella species, such as the American bald eagle (*Haliaeetus leucocephalus*) and gray wolf (*Canis lupus*).

Size Matters

Regardless of whether a single species or biodiversity is the restoration target, spatial scale is important. Restoration areas must be of proper size and quality to meet the needs of target species, including ecological processes and biological phenomena, at all life history stages, in all seasons, and in the face of possible environmental disturbances (Scott et al. 1999). Proper size may range from several hundred square kilometers for the gray wolf (Mech and Boitani 2003) to a several dozen acres for Fender's blue butterfly (*Icaricia icariodes fenderi*) if the habitat is of high quality and the risk of disturbance is low (Schultz et al. 2003).

Although not always possible, it is desirable to have sufficient habitat to allow development of subpopulations or widely dispersed populations. This increases the likelihood that there will be a source of pioneering individuals to recolonize areas after local extinctions or environmental disasters such as severe storms or wild fire. The risk of conserving a small, localized population is demonstrated by the disastrous event that severely jeopardized the recovery of the Puerto Rican parrot (*Amazona vittata*). The Luquillo forest in Puerto Rico supports the only remaining population of parrots. Hurricane Hugo severely damaged this forest in 1989, nearly wiping out the species (Simberloff 1994).

The importance of large areas does not mean that small-scale projects are unimportant. In fact, many large-scale restoration projects actually are a mosaic of small-scale patches of suitable habitat interspersed with other land uses. As long as adequate connectivity between the patches of suitable habitat is maintained, several fragmented habitats can function as a larger one.

Wildlife corridors are linear habitats that differ from the more extensive matrix in which they are embedded (Forman and Gordon 1986). A central idea in landscape ecology is that corridors can provide connectivity in fragmented landscapes. Corridors facilitate daily or seasonal movements of animals, ensuring that subsections of a population have access to all the resources they need and maintaining the potential for all individuals in the population to interbreed. Movements can be daily, such as between nesting, roosting, and brooding sites, or could be seasonal migrations by large mammals and migratory birds. Corridors also allow dispersal of animals from their place of birth to rearing or adult home ranges. At the landscape level, there may be a network of different-sized corridors rather than a single one.

Key characteristics of corridors are width, length, vegetative cover, habitat quality, location, human influence, noise, light, edge effects, degree of connectivity, and the presence of barriers (Duke et al. 2001). Corridors are more likely to be used by a species if they include its key habitat components (Rosenberg et al. 1997). Availability and quality of corridors are dynamic and may vary seasonally and from year to year.

Predation and Invasive Species

Predation can be a major issue in wildlife restoration. Excessive predation on translocated or recolonizing animals can reduce or eliminate age classes and skew sex ratios, jeopardizing the success of restoration. For example, mountain lion (*Puma concolor*) predation has hampered reintroduction of desert bighorn sheep (*Ovis canadensis mexicana*) in parts of its historic range in the West (Krausman et al. 1999). Conversely, predation or the threat of predation on domestic animals and wild game can adversely affect support for the restoration of large predator species such as the gray wolf or grizzly bear (*Ursus arctos*).

Predation can also reduce interspecies and intraspecies competition between prey species (Terborgh et al. 1999) and increase biodiversity. In a healthy system a feedback process maintains a stable cycle in which numbers of predators and prey come to equilibrium or oscillate within limits. However, once this equilibrium is lost by removal of predators or extirpation of a prey species, the feedback system breaks down, and both the health of remaining species and biodiversity decline. The removal of large predators from much of the West has broken the predator–prey equilibrium, allowing ungulate populations to expand. In many locations, increased grazing by native ungulates has resulted in declining plant diversity, plant community structure, and biodiversity. In Yellowstone National Park, for instance, overgrazing and browsing by elk (*Cervus elaphus*) has caused a more than 50% decline in aspen and willow patches since the 1930s. Formerly tall willow patches have been converted to short hedges, and plant richness has declined (Singer et al. 2003). Declines in willow and aspen have contributed to loss of beavers and subsequent loss of wetland ecosystems. These changes undoubtedly have affected natural wildlife communities in the park.

Invasive wildlife species can also influence restoration by competing for space and resources and sometimes preying on native species. Invasive species are those that have evolved elsewhere and have been introduced by humans or expanded their range naturally. They could include native species that aggressively compete for resources if not kept in check. Nonnative species have contributed to an estimated 68% of all fish extinctions over the past 100 years and to the decline of 70% of fish species listed as threatened or endangered (Li 1995). Well-known Pacific Northwest introduced exotics include the European starling (*Sturnus vulgaris*), house sparrow (*Passer domesticus*), nutria (*Myocastor coypus*), opossum (*Didelphis virginiana*), bullfrog (*Rana catesbeiana*), and large-mouth bass (*Micropterus salmoides*). Native species that have pioneered into the Northwest or have become invasive include the cattle egret (*Bubulcus ibis*) and brown-headed cowbird (*Molothrus ater*), respectively (Marshall et al. 2003). The cattle egret is a pioneer from other continents that has expanded its range to North America and the northwest in recent years. The range of the brown-headed cowbird, originally associated with large grazing ungulates on interior North American prairies, has expanded in response to livestock grazing in western North America (Marshall et al. 2003).

In most of the Northwest invasive species are so ubiquitous that their complete removal from an area is not feasible. However, maintenance control at some reduced level may be necessary before a target species can be restored. Where invasives constitute an isolated population or have a low dispersal and population growth rate, they may be completely eliminated before they become more widespread.

There are three general methods of control or removal of invasive species: mechanical, chemical, and biological (Simberloff et al. 1999). Mechanical control of invasive wildlife usually is accomplished with traps, nets, or shooting. The first two, in conjunction with humane euthanasia, often are more publicly acceptable than the third. Accepted trapping techniques can be found in *Research and Management Techniques for Wildlife and Habitat* (Bookhout 1994) and *Prevention and Control of Wildlife Damage* (Timm 1994). Euthanasia methods are available from the American Veterinary Medical Association (AVMA Panel on Euthanasia 2000). Removal of bullfrogs from parts of the western pond turtle's historic range in western Washington before the release of captive hatched young turtles is an example of mechanical control of an exotic predator.

Chemical control usually is not acceptable biologically or socially. Poisons and gases are not host specific and are lethal to any animal coming into contact with them. However, the U.S. Department of Agriculture Wildlife Services has dispensing mechanisms that limit nontarget wildlife exposure to lethal chemicals. Timm (1994) identifies chemicals that are effective in the control of exotic or pest wildlife species.

Although it is a popular method of controlling invasive plants, biological control is not a common method of controlling exotic animals. The Australian government is investigating biological controls for the cane toad (*Bufo marinus*; see www.nrme.qld.gov.au/factsheets/pdf/pest/PA21.pdf). However, this toad demonstrates the risks associated with biological control. The cane toad originally was introduced into Australia from South America as a biological control for the grey-backed cane beetle (see www.jcu.edu.au/school/phtm/PHTM/staff/rsbufo.htm). The toad did not eat the beetles but did eat anything else it could catch, including native amphibians, reptiles, birds, and small mammals. The cane toad was also introduced into Hawaii as a biological control for sugar cane pests, with the same outcome as in Australia (see www.columbia.edu/itc/cerc/danoff-burg/invasion_bio/inv_spp_summ/Bufo_marinus).

Source Populations

The following discussion draws heavily from Morrison (2002), Schaal and Leverich (2004), and Honeycutt (2000). The latter provides a good discussion of genetic applications for the conservation and management of biodiversity. There are two sources of animals for restoration: in situ, or from the wild, and ex situ, or from captivity. Donor population source and size play important roles in the success of restoration efforts. Small, isolated donor populations often have undergone genetic alteration and may have a genetic structure different from the species over much of its evolutionary history. The processes of isolation and declining population size lead to genetic drift and inbreeding depression. As a result, alleles are lost and genotypic distribution moves toward homogenization. Loss of genetic variability leads to reduced population fitness, sometimes expressed in poor reproductive rates, low progeny fitness, and malformations. Attempts at captive breeding and reintroduction of the pygmy rabbit, discussed later in this chapter, illustrate the challenges of genetic loss.

Because the factors that affect a source population's genetics will be imported into a new population, donors with a high level of genetic variability and low level of relatedness between individuals should be used where possible. Unfortunately, in some instances the only remaining animals in the wild are in a small, isolated population. In these cases genetic variation can be enhanced by introducing members of genetically similar animals from other populations, such as the introduction of western mountain lions into Florida to enhance genetic variability of the Florida panther (*Puma concolor coryi*) (U.S. Fish and Wildlife Service 1995).

Ex situ populations have additional challenges. Because captive breeding is not the ideal method for achieving recovery of a rare species or population (because of the likelihood that genetic variation and evolution potential will be lost), it typically is not used until the species is at serious threat of extinction. Consequently, donor stock usually comes from a small, isolated population that suffers from severe inbreeding. In addition, once the ex situ population is established and begins to reproduce, each generation is a genetic subset of previous generations. Genetic variability becomes more limited, and more rare alleles are lost through genetic drift.

Besides the problem of genetic drift, there is also the problem of selection. The captive environment may place selective forces on the population that remove alleles maladapted for captivity but important for survival in the wild. These same forces may also select for genetic adaptation to the captive environment rather than the wild.

Even with the genetic challenges associated with captive breeding, ex situ populations have played an important role in reestablishing wildlife populations. Captive-reared black-footed ferrets (*Mustella nigripes*) were used to restore wild populations in Wyoming, South Dakota, and Montana (Miller et al. 1993, 1997). Birds produced ex situ were used to restore the California condor (*Gymnogyps californianus*) in California and Arizona in 1992 and 1996, respectively (U.S. Fish and Wildlife Service 1996, Arizona Game and Fish 2003). Captive-reared birds were released in the Aleutian Islands as part of the Aleutian Canada goose (*Branta canadensis leucopareia*) recovery effort (Pacific Flyway Council 1999).

Several actions can be taken to improve the genetic health of donor populations and ex situ populations. The captive population should be large enough to minimize genetic change (see Morrison 2002 for discussion and review). Source animals should come from wild populations. If there is a choice of wild populations for translocation or founder stock for ex situ populations, the source should be the one most closely related genetically to the native stock and show similar ecological characteristics (morphology, physiology, behavior, habitat preference). If captive bread stock is to be used, it must be from a population that has been soundly managed both demographically and genetically according to the principles of contemporary conservation biology (IUCN 1998).

Animal Health

Historically, wildlife translocations were undertaken with little regard for disease, mainly because of a lack of awareness of potential health risks. Fortunately, since that time much work has been devoted to the analysis of health risks associated with translocation of animals (Leighton 2002). Wildlife ecologists now realize that translocation of a wild animal never represents movement of a single species but rather translocation of an entire community of organisms (Davidson and Nettles 1992). Unless guidelines for health analysis of source populations are in place, there is substantial risk that hazardous disease agents may be released into new environments along with the species of interest. Failure to establish and follow a health monitoring program can negate the tremendous personal, financial, and political efforts involved in a wildlife restoration project. There are a number of cautionary examples, including bighorn sheep and scabies in Oregon (Thorne et al. 1992) and desert tortoise (*Gopherus agassizii*) and mycoplasma in the Mojave Desert (Jacobson et al. 1991).

The potential for exposure of translocated animals to pathogens already present in the release area is also a concern. This can be especially devastating if the translocated animals are from a population with no previous exposure. The reintroduction of the whooping crane (*Grus americana*) in Idaho has been hampered by avian tuberculosis (Dein et al. 1995).

Handling of translocated animals influences the survival and successful reproduction of founder animals. Trauma, capture myopathy (weakening of the muscles), and infectious diseases are associated with the stress of capture, quarantine, transport, and release. A translocation project should be designed to minimize the potential for such trauma and stress (Gaydos and Corn 2001). Several good reference sources are available, including *Guidelines on the Care and Use of Wildlife* (Canadian Council on Animal Care 2003) and *Research and Management Techniques for Wildlife and Habitat* (Bookhout 1994).

There are two basic techniques for releasing animals: soft and hard release (Morrison 2002). In soft release animals are held in captivity, in cages or fenced enclosures, at the eventual release site for an extended time. This allows them to become accustomed to the site and allows biologists to monitor animal condition. The biggest disadvantage of soft releases is the maintenance of the captive animals and risk of their becoming habituated to humans. Soft releases often are used in captive breeding programs or where animal health is a concern.

Hard release involves immediate transfer of animals into the wild without prior conditioning. This is done to eliminate the stress that accompanies captivity. Hard releases generally are more economical because holding facilities and feeding aren't necessary. There appears to be no clear association of translocation success with soft or hard release methods (Griffith et al. 1989, Thompson et al. 2001).

Health problems can be minimized through a project risk assessment. These assessments include rigorous application of common sense and professional experience to determine whether health-related risks are associated with a particular restoration project and, if so, how they can be minimized. Assessment guidelines are available from Leighton (2002) and the IUCN (1998).

Human Dimensions of Wildlife Restoration

Practicing ecologists have known for many years that wildlife management and restoration include people management. Without support from the public and key stakeholders, restoration projects have little chance of succeeding. Failure to recognize or consider the concerns of people, particularly those affected disproportionately, often results in litigation or other actions that jeopardize the effort. As a result of contention, poor information, conflicting values, differing views on costs and benefits, and basic mistrust, it took more than 22 years to reintroduce the gray wolf into Yellowstone National Park. The first wolf recovery plan for the Rocky Mountains, completed in 1980, was a technical design developed without public involvement. It wasn't until 1985, when the human dimensions of wolf management were finally considered and addressed, that progress in restoring wolves to the Yellowstone ecosystem was made (Enck and Bath 2001).

Wildlife restoration can have major impacts on ecosystems and human communities. These impacts may be manifested in many ways, both positive and negative. Removal of invasive species may be opposed by those concerned about animal welfare or trapping. Restoration of large predators, such as wolves or grizzly bears, generates understandable fears of safety or economic loss.

Enck and Bath (2001) provide an excellent discussion on the human dimensions of wildlife restoration. To avoid common pitfalls associated with decisions made without public input, they suggest a nine-step process to determine the social feasibility of restoring a wildlife species or populations. Briefly, these steps are as follows:

1. An agent (individual, local community, non-government organization, or agency) raises the issue of restoring a species.
2. A wildlife agency provides policy guidance so the agent can conduct research and answer questions about minimum levels of social feasibility and possible mitigation plans.
3. The agent conducts research to identify potential areas for restoration that account for social issues.
4. The agent initiates discussions about restoration with identified communities. Communities deliberate the proposed restoration to determine whether it is in their best interest.
5. More information is collected for or by the communities to identify or further explore problems and opportunities associated with the proposal.
6. The agent and community leaders develop a communication and education program to help community members develop more informed opinions about the proposal.
7. Communities further discuss whether restoration is in their best interest, in light of new information.
8. Communities decide whether they will accept restoration of the species and under what conditions.

9. The agency determines whether the project is socially feasible, mitigation plans are sufficient, and the proposed restoration (within social limits) is still biologically feasible. If so, restoration is undertaken.

This discussion sets a common framework for wildlife restoration projects. The following section summarizes six case studies that describe wildlife restoration programs in the Pacific Northwest.

Case Study: Restoring Populations of Cavity-Nesting Oak-Associated Bird Species

The loss and degradation of oak habitats in the Pacific Northwest has had a profound effect on associated bird species. Three cavity-nesting species have suffered local or regional extirpations, including Lewis's woodpecker as a breeding species from western Oregon, western Washington, and southwestern British Columbia; white-breasted nuthatch, recently extirpated from the Puget lowlands of western Washington; and western bluebird, extirpated from southwestern British Columbia and northwestern Washington since the 1970s.

Awareness of bird population declines and overall loss and degradation of oak habitats has helped spark recent interest in oak conservation. Numerous habitat management and restoration projects have been undertaken (see Chapter 4), but examples of restoration of bird populations are scant. This is especially true where an emphasis of the restoration activity is to provide large, open-growth oaks in a savanna or woodland setting for cavity-nesting birds. Because oak growth is slow and natural cavity development is realized only in older oaks, the cavity-nesting bird response to restoration can take decades.

However, some characteristics of the extirpated species provide opportunities to consider reestablishment of populations in the near future. A variety of factors must be considered in the process of restoring populations of these species, including the amount or configuration of habitat, density of trees, and competition for cavities. A principal factor for all three species is that their specialized nest site (i.e., cavities) limits their presence or abundance. Restoration techniques needed to address cavity availability include habitat management or augmentation, coupled with recruitment of individuals into restored areas. Differences in species ecology and other factors necessitate different degrees and combinations of these activities (Table 15.1).

One principal component of habitat management for all three species is promotion of the presence of cavities through stochastic events, natural tree aging, or the susceptibility of oak trees to cavity excavation by primary excavators (i.e., woodpeckers). Cavity development can be enhanced through "release" of suitable oaks from competition. However, this is generally a long-term process that relies on tree aging after release.

While waiting for natural cavities to develop, oak area managers can augment habitat by providing nest boxes, especially for secondary cavity-nesters such as western bluebird and white-breasted nuthatch. The establishment of western bluebird nest box trails has brought this species back from near extirpation in parts of the Pacific Northwest, including Fort Lewis, Washington, and the Willamette valley southwest of Portland. The use of nest boxes to enhance or establish bird populations generally is considered a short-term tool, although in oak habitats it may need to be long term unless very mature oaks are already present.

TABLE 15.1.

Summary of habitat and recruitment needs to reestablish populations of three oak-associated cavity-nesting birds where they have been extirpated.

Species	Association with Oak in Extirpated Area	Habitat Management Needed	Nest Box Habitat Augmentation	Recruitment of Individuals into Extirpated Area
Western bluebird	Highly associated where oak occurs but will use other open savanna conditions without oak, including clearcuts	Low: openness with few scattered trees	Yes: readily accepts, almost to the exclusion of looking for natural cavities	Passive in southwest Washington, working off local populations and migratory movements; active translocations in British Columbia and northwest Washington where there are no populations or migratory pathways near extirpated area
White-breasted nuthatch	Obligate	Moderate: needs few big, old trees in savanna or woodland setting	Yes: will use nest boxes but also seeks natural cavities even if boxes available	Active translocations: resident species with no populations near extirpated areas
Lewis's woodpecker	Highly associated but will use some other types (e.g., cottonwood, open burned forest with snags)	High: needs few big, old trees but only in savanna setting	Weak excavator, but the need to excavate a cavity may be accommodated with nest box full of shavings	Passive in southwest Oregon because of wintering birds; active translocations in Willamette Valley, WA and BC because of limited or no migratory pathway and little or no wintering

One issue that must be considered when using nest boxes is the manner of recruitment. Successful efforts for western bluebird restoration have used the passive recruitment philosophy, "If you build it, they will come," which has worked where there is an existing population nearby or where a migratory population naturally passes through and is looking for nesting opportunities.

If there is no nearby source or migratory population, an active approach involving translocation of individuals may be appropriate. There are numerous examples of reestablishment of extirpated bird populations through reintroduction of individuals captured elsewhere and translocated to a project area. This approach is likely to be necessary in southwestern British Columbia and northwestern Washington for western bluebird and in western Washington for the white-breasted nuthatch. White-breasted nuthatch reintroduction into western Washington is likely to be needed because it has a nonmigratory behavior that precludes natural recruitment of individuals from out of the area.

Lewis's woodpecker presents a unique restoration challenge for several reasons. First, the spatial extent of extirpation was very great. Second, its ability to adapt to nest boxes is not well established. Third, competition with European starlings for available nest sites is a further impediment to natural reestablishment. In western Oregon, reintroduction through passive means may be effective where there is consistent occurrence of nonbreeding birds, primarily in the Rogue Valley, which has a regular wintering population. In the Willamette Valley, and especially in the Puget Trough of Washington, translocation of individuals probably will be needed. However, the migratory nature of Lewis's woodpecker adds another layer of complexity to reintroduction.

Restoration of populations of these three oak-associated cavity-nesting species in the near future probably will entail appropriate habitat management to help development of cavities but also habitat augmentation through the use of nest boxes. Both passive and active strategies may be needed, depending on the ecology of the species and location of the restoration site.

Case Study: Gray Wolves

In recent history the gray wolf (*Canis lupus*) had the greatest natural distribution of any land mammal on earth, with the exception of humans (Mech and Boitani 2003, Paquet and Carbyn 2003). Like humans, wolves benefit by being ecologically adaptable, highly mobile top predators, having a tight-knit family social structure that cooperatively gathers and shares resources and the ability to defend territory, cooperatively raise young, and aggressively colonize new areas (Mech and Boitani 2003). Wolves are adept social learners and can effectively prey on many ungulates, including livestock (Figure 15.1). As a consequence, agricultural societies have persecuted wolves with great passion and determination. Throughout the northern hemisphere but particularly in North America, organized human persecution extirpated wolves from vast areas of their historic range. Wolf populations in the western United States and much of southwestern Canada were extirpated or drastically reduced by 1950. Human social values changed, and wolf populations began to be restored as game animals in Canada in the 1960s and in the northwestern United States under the Endangered Species Act of 1973 (Bangs et al. 1998). Recent restoration of wolves to historic habitat in the western United States allows us to speculate on the ecological role of wolves and how their restoration might affect ecosystem structure (Smith et al. 2003, Robbins 2004).

Wolves function ecologically as an extended family unit or pack (Mech and Boitani 2003, Paquet and Carbyn 2003). A wolf is born in spring, then raised and taught by parents until it reaches adult size. Wolf packs fiercely defend their territories from other packs. A single wolf has no special powers compared with other predators, but in packs they effectively hunt large ungulates and reduce competition from other species. Pack boundaries naturally abut one another, and it is in such saturated environments that wolves exert their most powerful ecological effects.

The primary effects of wolf predation are simple. Wolves kill young, old, sick, weak, stupid, and unlucky ungulates throughout the year and leave pieces of them scattered throughout the landscape. Wolves must be selective hunters because their prey evolved to hurt, avoid, or outrun them.

FIGURE 15.1. Wolves chase down an elk and begin to change the entire Yellowstone area ecosystem. (*U.S. Fish and Wildlife Service*)

Wolves typically test many more prey animals than they are able to kill. In Yellowstone only a small fraction of encounters between wolves and prey result in chases, and only one in five chases are successful. Wolves defend their territory and harass or kill their competitors, including other wolves, mountain lions, bears, and coyotes, as well as scavengers such as ravens, eagles, and mesocarnivores that may try to steal their prey. Wolves reduce their range in spring and summer, when they are raising their young.

What do other animals do to avoid becoming wolf scat? The long-term evolutionary effect of this relationship is simple: "What but the wolf's tooth whittled so fine/the fleet limbs of the antelope" (Jeffers 1947). Antelope are fast, elk form herds, moose and bison are big and tough, and mountain goats and bighorn sheep can walk on cliffs. The slower, less alert, weak, and flat-footed ungulates were killed off by wolves or similar pointed-toothed species thousands of years ago. Mountain lions and black bears have claws, in part, to climb trees to avoid wolf packs. Grizzly bears benefit from being big and tough enough to routinely steal wolf kills, especially when important foods such as crops of white-bark pine nuts fail. Caribou calve in the far north tundra and elk at high elevations to put distance between them and wolves. White-tailed deer are so named because of their wolf-alert system. Coyotes are able to avoid wolves in thick bush, in deserts, or near towns, but without some refuge, there are generally fewer coyotes where there are wolves. If one has a taste for decaying protein, wolves provide a temporally and spatially distributed source of food that permits a diverse scavenger guild to exist (Berger 1999, Wilmers et al. 2003). In Yellowstone twelve different species of vertebrate scavengers have been observed at a single wolf-kill.

Ungulate hunting by modern humans does not have the same effects as wolf predation. Hunting does not remove ungulates based on their relative fitness (Soule et al. 2003, Berger et al. 2001), affect prey behavior, distribution, or survival throughout the year (Hebblewhite et al. 2002), or provide a year-round food base for diverse scavengers (Smith et al. 2003, Robbins 2004, Wilmers et al. 2003).

Wolves cause prey to act and be wild (Hebblewhite and Pletscher 2002, Berger 1999). The ecological effect of this natural behavior has been called "the ecology of fear" (Ripple and Beschta 2004). Predation reduces prey density, but more importantly, wary prey move more often or avoid certain areas, allowing some preferred food plants to escape herbivory (Beschta 2003, Kay et al. 2000, Ripple et al. 2001). Wary prey change their habitat use pattern, and the competitive edge between ungulate species shifts. Recovery of plant community structure provides more habitat diversity for migrant birds, beavers, or other species (Berger et al. 2001, Nietvelt 2001). Less fit prey are selectively removed, thus lowering average herd age and increasing the proportion of prime age breeding females while reducing the proportion of less fit, diseased, or parasitized individuals (Smith et al. 2003, Miller et al. 2001). Prey remains provide a year-round supply of protein for a multitude of scavengers and smaller predators (Smith et al. 2003, Robbins 2004, Wilmers et al. 2003).

These changing components and relationships interact and cascade down the food chain. In just one example, fewer coyotes may mean higher antelope, ground squirrel, and fox survival. Fewer elk and less browsing on willows combined with more foxes and ground squirrels might mean less and more fragmented open areas, thus more predation on ground-nesting birds. More riparian vegetation may mean more beaver, wetlands, ponds, and streams with more shade and structure, leading to more fish or amphibian rearing sites and certain migratory birds (Beschta 2003, Nietvelt 2001, Ripple et al. 2001). When wolves were first reintroduced in 1995 there was only one beaver colony in the northern tier of Yellowstone National Park. By 2003 there were seven colonies, largely because of a recovery of a key beaver food: willow. Although it is too early to be certain that recovery of willows is linked to wolf recovery, most researchers believe wolves outweighed the effect of climate, geomorphology, chemical plant defenses, and a 1988 fire-initiated decline in moose. The increased willows and newly formed beaver ponds probably will affect even more species of animals and plants, and ultimately wolves will touch almost everything in Yellowstone. There are dozens of more theoretical interacting ecological scenarios, some canceling out or contradicting others, but wolves have almost certainly fostered greater biological diversity.

As one tries to predict the outer rings of the ecological ripples caused by dropping wolves back into the pond, clear relationships become more difficult or impossible to detect or scientifically defend. A simple observation is that Yellowstone without wolves was scenic, but with them it is a dynamic, living landscape, worthy of the world's first National Park. The full effects of wolves on natural ecological processes and relationships in Yellowstone National Park and elsewhere will reveal themselves not in the next few years but over the next century.

There is a great deal of speculation about whether the astounding wolf-initiated ecological changes being revealed in Yellowstone National Park will manifest themselves elsewhere (Smith et al. 2003, Robbins 2004). Wolf restoration occurred because of changing societal values about how wildlife contributes to the quality of human life. These values are continuing to evolve toward increased public desire for the restoration of wildlife and natural ecological processes and will almost certainly result in an expansion of wolf range in the northwestern United States.

Currently more than 835 wolves live in Montana, Idaho, and Wyoming. Lone wolves dispersing from these populations have been documented in Washington, Oregon, Utah, and Colorado, and it is only a matter of time before a wolf pack becomes established in these states. Wolves exert their most powerful ecological effects in places such as Yellowstone National Park, where human influences are minimized and pack boundaries are contiguous. The park now has all the species it did when Columbus first stepped ashore. It is also saturated with wolf packs. Natural processes have the greatest opportunity to dominate ecological relationships in large systems such as Yellowstone. But even there human impacts such as fire suppression, roads, recreation, mortality of wolves venturing outside the park, and control of migrating ungulates to reduce private property damage moderate nonhuman processes, including the ecological ramifications of natural predation.

Outside the park, in the Greater Yellowstone Ecosystem (72,800 square kilometers), the impact of humans is much greater. Wolves are very susceptible to human-caused mortality, and humans cause about 85% of all known wolf deaths (Bangs et al. 1998). Conflicts between wolves and people disrupt natural processes to various degrees, and in many areas wolves are not tolerated by people. In most parts of southwestern Canada and the northwestern United States, including most public lands, wolf presence is sporadic at best because of conflicts with livestock producers and, at times, with hunters over surplus wild ungulates. Thus the potential for large-scale ecological effects caused by wolves outside huge contiguous blocks of public land is greatly diminished. For wolf packs (which use an average of 800 square kilometers each), there are few areas in the Pacific Northwest west of Idaho that are large or unfragmented enough to support them. Wolves may expand across the Northwest, but packs will be widely scattered, often temporary, and so affected by human-caused mortality that widespread ecological effects, as witnessed in Yellowstone, are unlikely to occur at a landscape level. The most dramatic effect of wolves on humans may be their uniquely powerful symbolism and inspiration, which occurs even when wolves are not physically present.

Case Study: Restoration of the Columbia Basin Pygmy Rabbit

The pygmy rabbit (*Brachylagus idahoensis*) is endemic to the temperate desert ecoregion of western North America (Bailey 1998). It is found in eight western states: California, Nevada, Utah, Wyoming, Montana, Idaho, Oregon, and Washington. It is the smallest North American rabbit and among only a few that dig their own burrows. Pygmy rabbits are closely associated with deep soils and dense stands of big sagebrush, which provide food and cover (Wilson and Ruff 1999). An isolated population of pygmy rabbits is located in the Columbia Basin of eastern Washington (Dobler and Dixon 1990). Genetic analyses showed significant differences between pygmy rabbits in the Columbia Basin and those in neighboring states (U.S. Fish and Wildlife Service 2003). Genetic and archeological evidence suggests that Columbia Basin pygmy rabbits have been isolated for at least

10,000 years, possibly much longer (Lyman 1991). Comparison with museum specimens indicates that these isolated subpopulations have experienced inbreeding and loss of genetic diversity in the past 50 years.

Few historical records of pygmy rabbits from the Columbia Basin are available. Six small subpopulations were discovered between 1987 and 1996, but five of these have been lost (U.S. Fish and Wildlife Service 2003). No Columbia Basin rabbits are known to survive in the wild as of this writing. Habitat loss and fragmentation over the past 100 years are the likely cause of population decline. Thousands of hectares of Columbia Basin shrub–steppe habitat were converted to irrigated agriculture, particularly since the 1940s, with the development of the Columbia Basin Project (Vander Haegen et al. 2001). Causes of the more recent population decline are unknown, although two populations were lost after wildfires. Other potential factors are disease, inbreeding, continued habitat degradation, and predation. Pygmy rabbits in other Western states are also experiencing declines. The Columbia Basin pygmy rabbit has been listed as endangered in Washington State since 1993 and was listed as federally endangered in 2003 (U.S. Fish and Wildlife Service 2003).

In 2000, the Oregon Zoo began working with the Washington Department of Fish and Wildlife to breed pygmy rabbits in captivity in order to assist recovery. Initially, pygmy rabbits captured in Idaho were used to develop husbandry techniques. The captive breeding program was initiated by Washington State University and the Oregon Zoo in 2002, with eighteen individuals from sixteen original founders divided between the two institutions. Idaho pygmy rabbits continue to be reared for research purposes, to compare reproductive performance and test reintroduction techniques. Additional captive breeding facilities housing pygmy rabbits from Idaho have been constructed at Northwest Trek Wildlife Park in Eatonville, Washington.

Eight-foot tanks filled with soil and planted with sagebrush created mini-habitats to foster natural behavior. Video cameras were installed in the enclosures to observe behavior, nesting, and reproductive activity. Pygmy rabbits are spring seasonal breeders, and females may have up to three litters per year. Females dig natal burrows near the end of a 25-day gestation period. Young emerge at 2 weeks of age and immediately begin feeding on grasses and sagebrush.

A Columbia Basin pygmy rabbit recovery team has been formed and is drafting a federal recovery plan. Recovery actions under consideration include introduction of genetic material from pygmy rabbits in other states through captive breeding. Experimental intercross breeding of Washington and Idaho rabbits has been conducted to prepare for its possible use in recovery. Research is also being conducted on genetics, diet, behavior, rearing techniques, and immune system function.

The first 2 years showed significant differences in average breeding success between Washington and Idaho rabbits (Elias 2004). Only about 50% of Columbia Basin rabbits successfully bred, compared with 100% of those from Idaho. Because of their low pregnancy rates, Columbia Basin pygmy rabbits produced half as many offspring as Idaho pygmy rabbits. Washington rabbits had poor immune systems compared with their Idaho cousins. Poor reproductive performance could be attributed to low genetic diversity in the small captive population. A number of captive pygmy rabbits have died from various diseases, especially mycobacteriosis and coccidiosis (an intestinal infection).

Washington State University is developing techniques to reintroduce rabbits to appropriate habitat. With support from the Idaho Department of Fish and Game, wildlife biologists from the Stoller

Corporation conducted an experimental release of forty-two captive-reared Idaho pygmy rabbits in late summer of 2002–2004 at the Idaho National Engineering and Environmental Laboratory. The Idaho pygmy rabbits were closely monitored after their release, and valuable information was generated for future planning. Several techniques were investigated, including large prerelease pens, temporary containment fencing, supplemental feeding, and artificial burrows (Westre 2004).

Case Study: Restoration of Western Pond Turtles

The western pond turtle (*Emys marmorata*) declined to fewer than 200 individuals in Washington State by 1990 because of a combination of wetland destruction, human development, and introduced predators such as bullfrogs (*Rana catesbeiana*). In 1993 the pond turtle was listed as an endangered species by the Washington Department of Fish and Wildlife, and a recovery plan was crafted.

The recovery program began with a twofold approach: address the ecological causes of decline while increasing the numbers of turtles through protection of wild nests and selective captive breeding. This program has included surveys, acquisition of critical habitat, captive breeding, head starting, release of wild and captive-bred hatchlings, mark and recovery follow-up, bullfrog control, habitat enhancement, and public education. Babies are head-started at the Woodland Park Zoological Society and the Oregon Zoo, with captive breeding at Woodland Park. The program has grown to include the Washington Department of Fish and Wildlife, the Nature Conservancy, the U.S. Fish and Wildlife Service, Bonneville Power Administration, the U.S. Forest Service, Point Defiance Zoo, the University of Puget Sound, and numerous volunteers.

Bullfrog removal includes adults, tadpoles, and egg masses by any means possible. Both adult frogs and tadpoles are captured in turtle traps, adults are gigged or caught by fishing lure, and, most importantly, egg masses are systematically removed from three ponds and one small lake. Egg mass removal began with 186 egg masses removed in 1998. In 2003, only one egg mass was found. Each body of water is surveyed for egg masses every other day, and the eggs are scooped out of the water and thrown on shore to dry out. Other habitat enhancements have included mowing nesting areas in late May to provide a shorter grass habitat that improves turtle nesting success, providing additional basking areas, and enhancing water flow to the ponds.

Turtles are trapped each year from April 1 through May 15. A combination of baited hoop traps and basking traps allows capturing in most parts of the aquatic environment. All captured turtles are weighed and measured. Transmitters are attached to the carapace of adult females. Telemetry to locate these females during nesting is conducted from mid-May to mid-July. From 1990 to 2003, sixty-eight females were monitored and 183 nest sites located. These are protected from predators with wire exclosures, allowing natural incubation of the eggs. Hatchlings are collected in the fall (after 90–130 days incubation) and are brought to the zoos for head-starting.

As of this writing 714 hatchlings have been raised in the head-start programs. Nearly 700 have been released back to the wild, enhancing two populations discovered in 1985. A new population has been established in the Columbia Gorge from babies collected from wild nests, and another has been established in western Washington from captive-bred turtles.

Released hatchlings get pit tags and are notched with numbers for future identification. Trapping in subsequent years has shown that they have adapted to the wild habitat successfully. Growth rates after release are consistent with growth rates of wild hatched turtles, and juvenile survival is estimated at 90%. Many of the head-started turtles released in 1991 and 1992 have achieved adult status, and at least three of these females have successfully nested.

In 2001, 53% of previously released head starts were recaptured at one site. Four head-started females nested in the wild in 2003, including one that double-clutched. Natural recruitment has begun again in three of the four ponds where bullfrog removal has been most intense. In 2003 twelve wild hatchlings, one yearling, and three 2-year-old wild-born turtles were captured. These are the first natural recruitments that have been captured or sighted since work began in 1990.

The state's total estimated population of western pond turtles increased to more than 800 by 2003. Program managers are optimistic that eventual recovery of western pond turtle in Washington State will be successful.

Case Study: The Teeter-Totter Effect of Eagle Recovery on Great Blue Herons

The recovery of bald eagles (*Haliaeetus leucocephalus*) in North America is reason to celebrate. Eagle numbers have tripled in some parts of the continent after decades of very low numbers that probably resulted from a combination of persecution by humans, habitat degradation, and eggshell thinning from pesticides (Elliott and Harris 2001). Enforcement of laws to protect eagles, elimination of many contaminants from the food chain, and habitat protection have clearly improved the survival and reproductive success of eagles in much of North America. Eagles are now showing strong recovery in the Pacific Northwest, other Western states, Lake Superior, Nova Scotia, Chesapeake Bay, and Florida. In the Pacific Northwest, the number of eagles began an upward climb in the late 1980s and nearly tripled by the end of the twentieth century (Elliott and Harris 2001).

One effect of the recovery of eagles has been increased disturbance and predation on the great blue heron (*Ardea herodias fannini*) (Vermeer et al. 1989, Butler 1997a). Concern over disturbance of nesting colonies of the coastal subspecies of heron endemic to the Pacific Northwest resulted in it being designated a species of special concern by the Committee on the Status of Endangered Wildlife in Canada (Butler 1997b). Predation and disturbance of herons by eagles lowers breeding success by causing herons to abandon their nests (Vennesland and Butler 2004). By the start of the twenty-first century, eagles had put many herons on the run at colonies that had a long history of successful nesting. The ongoing fragmentation of forested lands near the seacoast probably exacerbates the heron's plight because they breed in large, obvious colonies near abundant food sources. Some eagles take no notice of herons, even where they nest within a few meters of one another, whereas others attack and consume adult and young herons. Why some eagles choose to hunt herons while others do not is unknown.

The choice of foraging habitat for breeding herons is influenced by the presence of an abundant nearby supply of small fish (Butler 1992). Extensive eelgrass (*Zostera marina*) meadows are especially attractive. Herons also forage along shallow beaches, from rock jetties and docks, and in freshwater marshes, fish farms, and garden ponds. Herons need quiet forested areas to build their nests. Where human disturbance is prevalent, herons may need several alternative locations to hide their nests. Boundary Bay in southern British Columbia has an eelgrass meadow several square kilometers in area. However, fewer than 100 herons use the bay, presumably because of a shortage of nearby nesting sites. Herons there have had poor success nesting in low crabapple trees and a forest plantation. Restoration of herons at Boundary Bay will entail allowing some young forest to mature.

The Heron Working Group is a consortium of government biologists, academic scientists, conservation groups, and individual citizens that established a Heron Network in the Pacific Northwest. They advocate that landscapes near good heron foraging habitats be managed to provide nesting opportunities for herons. This includes ensuring that there is adequate forested land for nesting herons near foraging areas, a simple idea that builds on research of the foraging and nesting needs of the herons. A key question is, "How much forest is needed, and where should it be?" The number of herons that use an eelgrass meadow is related to its size (Butler 1997a). Nearly all herons nest within 10 kilometers of their foraging grounds. Ensuring that at least 50 hectares of contiguous mature forest is within 10 kilometers of major foraging areas should provide for the minimum needs of herons. Of course, more forest would provide them with more options, but the 50 hectare–10 kilometer rule is considered the minimum.

Case Study: Restoration of Butterflies in Northwest Prairies

Butterflies in Pacific Northwest prairies are in decline because of habitat loss, cessation of succession, and invasion by exotic species. These threats have resulted in a greater than 99% loss of prairies and oak savannas in Willamette Valley in Oregon, south Puget Sound prairies in Washington, and coastal prairies in Oregon and Washington (Noss et al. 1995). The few remaining prairies habitats harbor at least seven at-risk butterfly species: two that have been protected under the Endangered Species Act (Fender's blue and Oregon silverspot), two candidates (Taylor's checkerspot and Mardon skipper), two species of concern (island marble and valley checkerspot), and one Washington State species of concern (Puget blue). Recovery of all of these species will entail restoration at both the landscape and site levels (Table 15.2). At the landscape level, conservation entails restoration of metapopulations. At the site level, it includes management actions to enhance remaining prairie remnants, by reducing invading exotics and encroaching woody species, and active restoration of areas that no longer maintain butterfly populations or critical larval and adult plant resources. Fender's blue butterfly is the species for which the greatest amount of restoration work has been initiated (Schultz 2001, Schultz et al. 2003).

Restoration of prairie butterfly populations should be viewed through a metapopulation lens, which means a network of interacting semi-independent subpopulations (Hanski and Gilpin 1997).

TABLE 15.2.

At-risk butterfly species in Pacific Northwest prairies.

Species	Status[a]	Current Distribution	Habitat Association	Host Plant[b]	Major Restoration Needs
Fender's blue (*Icaricia icarioides fenderi*)	Endangered	Willamette Valley, Oregon	Upland prairies	Kincaid's lupine (*Lupinus sulphureus* var *kincaidii*), spur lupine (*Lupinus arbustus*)	Additional habitat to restore network, reduction or elimination of exotic species, fire or other disturbance to reverse succession
Oregon silverspot (*Speyeria zerene hippolyta*)	Threatened	Oregon and northern California coast	Coastal prairies	Early blue violet (*Viola adunca*)	Additional habitat to restore network, reduction or elimination of exotic species, fire or other disturbance to reverse succession
Taylor's checkerspot (*Euphydryas editha taylori*)	Candidate	Willamette Valley, Oregon and south Puget Sound, Washington	Upland and glacial outwash prairies	Paintbrush (native, *Castilleja hispeda*) and plantain (exotic, *Plantago lanceolata*)	Additional habitat to restore network, reduction or elimination of exotic species, fire or other disturbance to reverse succession
Mardon skipper (*Polites mardon*)	Candidate	Willamette Valley and Siskiyou Mountains, Oregon; south Puget Sound, and Mt. Adams, Washington; Del Norte County, California	Upland and glacial outwash prairies and mountain meadows	Roemer's fescue, Idaho and red fescue (*Festuca romeri*, *F. idahoensis*, *F. rubra*)	Additional habitat to restore network, reduction or elimination of exotic species, fire or other disturbance to reverse succession
Island marble (*Euchloe ausonides insulanus*)	Species of concern	San Juan Island, Washington	Island prairies	Field mustard (exotic, *Brassica campestris*); native host plant unknown	Unknown; probably same as others
Valley checkerspot (*Speyeria zerene bremneri*)	Species of concern	South Puget Sound prairies	Glacial outwash prairies	Early blue violet (*Viola adunca*)	Additional habitat to restore network, reduction or elimination of exotic species, fire or other disturbance to reverse succession
Puget blue (*Icaricia icarioides blackmorei*)	Washington State species of concern	South Puget Sound prairies	Glacial outwash prairies	Sickle-keeled lupine (*Lupinus albicaulis*)	Additional habitat to restore network, reduction or elimination of exotic species, fire or other disturbance to reverse succession

[a]Status reflects federal status unless noted otherwise.

[b]Host plants are native unless noted otherwise.

Habitat for a metapopulation includes both occupied and unoccupied patches. Metapopulation dynamics are a shifting balance of extinction and colonization that is maintained across a landscape. Some subpopulations may go extinct at any given time, but other populations exist to expand and recolonize suitable habitats. Two key theories are useful in thinking about restoration of butterfly populations. First, habitat patches should be close enough together to support dispersal between the occupied patches (to reduce the likelihood of patch extinction) and new colonization of unoccupied patches (to reestablish populations at patches that have gone extinct). Thus, if habitat fragmentation has caused substantial isolation of remaining populations, a landscape view that includes creating new stepping stones between remaining patches may be important to restore a functioning population across a landscape (Schultz 1998). Second, empty habitat patches may be acceptable but are currently unoccupied. This habitat may not need much active restoration but may be unoccupied because of natural extinction. However, if it is close enough to an occupied patch it may soon be recolonized.

Experience with the Fender's blue butterfly demonstrates the use of metapopulation ideas for restoration. Current efforts in the West Eugene Wetlands are focusing on creating a functioning network of butterfly habitat. A substantial amount of recently purchased public land is managed by agencies interested in restoring butterfly habitat. Careful dispersal studies have concluded that existing habitats are too far apart to support successful dispersal between the remnant patches (Schultz 1998). Therefore, efforts are under way to determine the extent and location of new prairie patches that will support connectivity throughout the network. Surveys of Fender's blue populations have been conducted annually since 1993, and in that time both patch extinctions and recolonizations have been observed (Fitzpatrick 2004). Recolonizations generally have occurred within close proximity of existing patches.

At the site level, restoration can be broken down into two categories: enhancement of existing habitats and active restoration of substantially degraded habitat. Enhancement of existing habitat includes removal of exotic species and reintroduction of natural disturbance processes. To enhance existing habitat, almost all management alternatives (e.g., fire, mowing, herbicides, grazing, and manual weed removal) have some positive effects on target invasive species but often are accompanied by unintended side effects. For example, elimination of one invasive plant often leads to expansion by another. Thus, the biggest challenge in habitat enhancement is finding the technique that allows removal of undesirable species while having limited impact on native species. Severely degraded habitats that no longer maintain butterfly populations or native plant communities can be treated more aggressively by removal of all undesirable species before revegetation begins. For butterfly habitat, the goal is to establish populations of important larval host plants and adult nectar resources as well as a broader plant community that supports them (Schultz 2001). If an appropriate plant community can be established, it is hypothesized that nearby butterfly populations will disperse in and establish a population. As of this writing, there are no large-scale habitat restorations far enough along to validate this hypothesis for Fender's blue butterfly. However, several habitat restorations are under way or planned at which butterfly arrival is expected in the next decade, assuming nearby populations are sustained over that period.

An additional issue is the role of captive breeding. The Oregon and Woodland Park zoos have teamed up to breed Oregon silverspot butterflies (Shepherdson et al. 2001). For this species, four isolated populations remain, two hanging on by a slender thread (U.S. Fish and Wildlife Service 2001). Captive breeding efforts probably have prevented extinction of one of these populations, but the long-term success is unclear because the silverspot's coastal habitat is so degraded. Captive rearing has the potential to substantially augment declining populations and to create a stock with which to establish new populations at restored sites. However, it must be done in tandem with site-level efforts to restore degraded habitat and landscape-level planning efforts that include consideration of both demography and genetics of the captive-reared species.

See Plate 10 in the color insert for photographs of some of the species discussed in this chapter.

Summary

Much restoration across the Northwest is done with at least a partial goal of improving habitat for native wildlife. "Build it, and they will come" is the basic idea. But in some cases they may not come. Remaining populations are too dispersed (butterflies, oak woodland birds), they are outcompeted by introduced species (western pond turtle), or their numbers have shrunk below their ability to breed successfully (pygmy rabbits). In other cases, such as that of the gray wolf, restoring a keystone species may trigger restoration of the ecosystem. Although much ecological restoration is rightly focused on plant communities, policymakers, land managers, and practitioners also need to consider working directly with wildlife populations as part of their strategy. Wildlife biologists have developed a number of tools over the years for augmenting populations, and although these have usually focused on game species, they are increasingly being used for restoration.

References

American Veterinary Medical Association Panel on Euthanasia. 2000. 2000 report of the Panel on Euthanasia. *Journal AVMA* 218(5): 669–696. Available at www.avma.org/resources/euthanasia.pdf.

Andersen, M. J., B. Csuti, and D. Shepherdson. 2002. *The Oregon Silverspot Butterfly Project*. Oregon Zoo, Portland. Available at www.zooregon.org/Conservation Research/silverspotcon.htm.

Arizona Game and Fish. 2003. *California Condor Recovery*. Arizona Game and Fish, Phoenix, AZ. Available at www.gf.state.az.us/w_c/california_condor.shtml.

Bailey, R. G. 1998. *Ecoregions Map of North America: Explanatory Note*. Miscellaneous Publication 1548. USDA Forest Service, Washington, DC.

Bangs, E. E., S. H. Fritts, J. A. Fontaine, D. W. Smith, K. M. Murphy, C. M. Mack, and C. C. Niemeyer. 1998. Status of gray wolf restoration in Montana, Idaho, and Wyoming. *Wildlife Society Bulletin* 26: 785–798.

Berger, J. 1999. Anthropogenic extinction of top carnivores and interspecific animal behavior: implications of the rapid decoupling of a web involving wolves, bears, moose and ravens. *Proceedings of the Royal Society of London* B 2261–2267.

Berger, J., P. B. Stacey, L. Bellis, and M. P. Johnson. 2001. A mammalian predator–prey imbalance: grizzly and wolf extinction affect avian neotropical migrants. *Ecological Applications* 11: 947–960.

Berger, J., J. E. Swenson, and I. Peterson. 2001. Recolonizing carnivores and naive prey: conservation lessons from Pleistocene extinctions. *Science* 291: 1036–1039.

Beschta, R. L. 2003. Cottonwoods, elk, and wolves in the Lamar Valley of Yellowstone National Park. *Ecological Applications* 13: 1295–1309.

Bookhout, T. A., ed. 1994. *Research and Management Techniques for Wildlife and Habitat*. The Wildlife Society, Bethesda, MD.

Butler, R. W. 1992. Great blue heron. In A. Poole, P. Stettneheim, and F. Gill (eds.), *The Birds of North America*, No. 25. Academy of Natural Sciences, Philadelphia and American Ornithologists' Union, Washington, DC.

Butler, R. W. 1997a. *The Great Blue Heron*. University of British Columbia Press, Vancouver.

Butler, R. W. 1997b. *Status of the Great Blue Heron*. Committee on the Status of Endangered Wildlife in Canada, Ottawa.

Canadian Council on Animal Care. 2003. *Guidelines on the Care and Use of Wildlife*. Canadian Council on Animal Care, Ottawa.

Caro, T. M. and G. O'Doherty. 1999. On the use of surrogate species in conservation biology. *Conservation Biology* 13: 805–814.

Davidson, W. R. and V. F. Nettles. 1992. Relocation of wildlife: identifying and evaluating disease risks. *Transactions of the North American Wildlife and Natural Resource Conference* 57: 466–473.

Dein, J., K. Converse, and C. Wolf. 1995. Captive propagation, introduction, and translocation programs for wildlife vertebrates. In E. T. LaRoe, G. S. Farris, C. E. Puckett, P. D. Doran, and M. J. Mac (eds.), *Our Living Resources: A Report to the Nation on the Distribution, Abundance, and Health of U.S. Plants, Animals, and Ecosystems*. USDI National Biological Service, Washington, DC. Available at biology.usgs.gov/s+noframe/u219.htm.

DeNormandie, J. 2000. *The Umbrella Species Concept and Bioregional Conservation Planning: A Comparative Study*. Master's thesis, Utah State University, Logan.

Dobler, F. C. and K. R. Dixon. 1990. The pygmy rabbit *Brachylagus idahoensis*. Pages 111–115 in J. A. Chapman and J. E. C. Flux (eds.), *Rabbits, Hares and Pikas: Status Survey and Conservation Action Plan*. IUCN/SSC Lagomorph Specialist Group, Gland, Switzerland.

Duke, D. L., M. Hebblewhite, P. C. Paquet, C. Callaghan, and M. Percy. 2001. Restoring a large-carnivore corridor in Banff National Park. Pages 261–275 in D. S. Maehr, R. F. Noss, and J. L. Larkin (eds.), *Large Mammal Restoration*. Island Press, Washington, DC.

Elias, B. A. 2004. *Behavior, Reproduction, and Survival in Captive Columbia Basin and Idaho Pygmy Rabbits (*Brachylagus idahoensis*)*. M.S. thesis, Washington State University, Pullman.

Elliott, J. E. and M. L. Harris. 2001. An ecotoxicological assessment of chlorinated hydrocarbon effects on bald eagle populations. *Reviews in Toxicology* 4: 1–60.

Enck, J. W. and A. Bath. 2001. Restoration of wildlife species. Pages 307–328 in D. J. Decker, T. L. Brown, and W. F. Siemer (eds.), *Human Dimensions of Wildlife Management in North America*. The Wildlife Society, Bethesda, MD.

Fitzpatrick, G. S. 2004. *Status of Fender's Blue Butterfly (*Icaricia icarioides fenderi*) in Lane County, Oregon: Population Estimates and Site Evaluations (Part 1) and Effects of Mowing on the Fender's Blue Butterfly: Implications for Conservation and Management (Part 2)*. Report to the Oregon Natural Heritage Program and the U.S. Fish and Wildlife Service, Eugene, OR.

Forman, R. T. T. and M. Gordon. 1986. *Landscape Ecology*. Wiley, New York.

Gaydos, J. K. and J. L. Corn. 2001. Health aspects of large mammal restoration. Pages 149–162 in D. S. Maehr, R. F. Noss, and J. L. Larking (eds.), *Large Mammal Restoration*. Island Press, Washington, DC.

Goggans, R., R. D. Dixon, and L. C. S. Seminara. 1988. *Habitat Use by Three-Toed and Black-Backed Woodpeckers, Deschutes National Forest, Oregon*. Nongame Project Number 87-3-02. Oregon Department of Fish and Wildlife, Bend.

Griffith, B., J. M. Scott, J. W. Carpenter, and C. Reed. 1989. Translocation as a species conservation tool: status and strategy. *Science* 245: 477–480.

Hanski, I. and M. Gilpin. 1997. *Metapopulation Biology*. Academic Press, San Diego, CA.

Hebblewhite, M. and D. H. Pletscher. 2002. Effects of elk groups size on predation by wolves. *Canadian Journal of Zoology* 80: 800–809.

Hebblewhite, M., D. H. Pletscher, and P. C. Paquet. 2002. Elk population dynamics in areas with and without predation by recolonizing wolves in Banff National Park, Alberta. *Canadian Journal of Zoology* 80: 789–799.

Honeycutt, R. 2000. Genetic applications for large mammals. Pages 233–259 in S. Demarais and P. R. Krausman (eds.), *Ecology and Management of Large Mammals in North America*. Prentice Hall, Upper Saddle River, NJ.

IUCN. 1998. *IUCN Guidelines for Re-introductions*. Prepared by the IUCN/SSC Re-introduction Specialist Group, IUCN, Gland, Switzerland and Cambridge, England.

Jacobson, E. R., J. M. Gaskin, M. B. Brown, R. K. Harris, C. H. Gardiner, L. La Pointe, H. P. Adams, and C. Reggiardo. 1991. Chronic upper respiratory tract disease of free-living desert tortoises. *Journal of Wildlife Diseases* 27: 296–316.

Jeffers, R. 1947. The bloody sire. Page 76 in *Robbinson Jeffers Selected Poems*. Vintage Books, New York.

Kallman, H., C. P. Agee, and W. R. Goforth, eds. 1987. *Restoring American Wildlife 1937–1987*. USDI, Fish and Wildlife Service, Washington, DC.

Kay, C. E., B. Patton, and C. A. White. 2000. Historic wildlife observations in the Canadian Rockies: implications for ecological integrity. *Canadian Field Naturalist* 114: 561–583.

Krausman, P. R., A. V. Sandoval, and R. C. Etchberger. 1999. Natural history of desert bighorn sheep. Pages 139–191 in R. Valdez and P. R. Krausman (eds.), *Mountain Sheep of North America*. University of Arizona Press, Tucson.

Kunkel, K. E. 2003. Ecology, conservation, and restoration of large carnivores in western North America. Pages 250–295 in C. J. Zabel and R. G. Anthony (eds.),

Mammal Community Dynamics: Management and Conservation in the Coniferous Forests of Western North America. Cambridge University Press, Cambridge, England.

Landres, P. B., J. Verner, and J. W. Thomas. 1988. Ecological use of vertebrate indicator species: a critique. *Conservation Biology* 2: 316–327.

Leighton, F. A. 2002. Health risk assessment of the translocation of wild animals. *Revue Scientifique et Technique* 21: 187–195.

Li, H. 1995. Non-native species. Pages 427–428 in E. T. LaRoe, G. S. Farris, C. E. Puckett, P. D. Doran, and M. J. Mac (eds.), *Our Living Resources: A Report to the Nation on the Distribution, Abundance, and Health of U.S. Plants, Animals, and Ecosystems*. USDI, National Biological Service, Washington, DC.

Lyman, R. L. 1991. Late quaternary biogeography of the pygmy rabbit (*Brachylagus idahoensis*) in eastern Washington. *Journal of Mammalogy* 72(1): 110–117.

Marshall, D. B., M. G. Hunter, and A. L. Contreras. 2003. *Birds of Oregon*. Oregon State University Press, Corvallis.

Mech, L. D. and L. Boitani. 2003. Wolf social ecology. Pages 1–34 in L. D. Mech and L. Boitani (eds.), *Wolves: Behavior, Ecology and Conservation*. University of Chicago Press, Chicago.

Miller, B., D. Biggins, L. Hanebury, and A. Vargas. 1993. Reintroduction of the black-footed ferret. Pages 455–463 in P. J. S. Olney, G. M. Mace, and A. T. C. Feister (eds.), *Creative Conservation: Interactive Management of Wild and Captive Animals*. Chapman and Hall, London.

Miller, B., D. Biggins, A. Vargas, M. Hutchins, L. Hanebury, J. Godbey, S. Anderson, J. Oldemeier, and C. Wemmer. 1997. The captive environment and reintroduction: the black-footed ferret as a case study. Pages 97–112 in D. J. Sheperdson, J. D. Mellen, and M. Hutchins (eds.), *Environmental Enrichment for Captive Animals*. Smithsonian Institution Press, Washington, DC.

Miller, B., B. Dugelby, D. Foreman, C. Martinez del Rio, R. Noss, M. Phillips, R. Reading, M. Soule, J. Terborgh, and L. Wilcox. 2001. The importance of large carnivores to healthy ecosystems. *Endangered Species Update* 18: 202–210.

Morrison, M. L. 2002. *Wildlife Restoration. Society for Ecological Restoration*. Island Press, Washington, DC.

Nietvelt, C. G. 2001. *Herbivory Interactions between Beaver (*Castor canadensis*) and Elk (*Cervus elphus*) on Willow (*Salix *spp.) in Banff National Park, Alberta*. M.S. thesis, University of Alberta, Edmonton.

Noss, R. F., E. T. LaRoe III, and J. M. Scott. 1995. *Endangered Ecosystems of the United States: A Preliminary Assessment of Loss and Degradation*. National Biological Service, Washington, DC.

Office of Technical Assessment. 1987. *Technologies to Maintain Biological Diversity*. O.T.A.-F-330. U.S. Government Printing Office, Washington, DC.

Oregon Department of Fish and Wildlife. 2003. *Oregon's Bighorn Sheep and Rocky Mountain Goat Management Plan*. Oregon Department of Fish and Wildlife, Salem.

Pacific Flyway Council. 1999. *Pacific Flyway Management Plan for the Aleutian Canada Goose*. Pacific Flyway Study Committee, Sub-committee on Aleutian Canada Geese (c/o U.S. Fish and Wildlife Service), Portland, OR.

Paquet, P. C. and L. N. Carbyn. 2003. Gray wolf. Pages 482–510 in G. Fledhamer, B. C. Thompson, and J. A. Chapman (eds.), *Wild Mammals of North America*. Johns Hopkins University Press, Baltimore, MD.

Ripple, W. J. and R. L. Beschta. 2004. Wolves and then ecology of fear: can predation risk structure ecosystems? *BioScience* 54(8): 755–766.

Ripple, W. J., E. L. Larsen, R. A. Renkin, and D. W. Smith. 2001. Trophic cascades among wolves, elk, and aspen on Yellowstone National Park's northern range. *Biological Conservation* 102: 227–234.

Robbins, J. 2004. Lessons from the wolf. *Scientific American* 290(6): 76–81.

Roberge, J. and P. Angelstam. 2004. Usefulness of the umbrella species concept as a conservation tool. *Conservation Biology* 18: 76–85.

Rosenberg, D., B. R. Noon, and E. C. Meslow. 1997. Biological corridors: form, function and efficiency. *BioScience* 47: 677–686.

Schaal, B. and W. J. Leverich. 2004. Population genetic issues in ex situ plant conservation. Pages 267–285 in E. O. Guerrant, K. Havens, and M. Maunder (eds.), *Ex Situ Plant Conservation*. Island Press, Washington, DC.

Schultz, C. B. 1998. Dispersal and its implications for reserve design in a rare Oregon butterfly. *Conservation Biology* 12: 284–292.

Schultz, C. B. 2001. Restoring resources for an endangered butterfly. *Journal of Applied Ecology* 38: 1007–1019.

Schultz, C. B., P. C. Hammond, and M. V. Wilson. 2003. Biology of the Fender's blue butterfly (*Icaricia icarioides fenderi*), an endangered species of western Oregon native prairies. *Natural Areas Journal* 23: 61–71.

Scott, J. M., E. A. Norse, H. Arita, A. Dobson, J. A. Estes, M. Foster, B. Gilbert, D. B. Jensen, R. L. Knight, D. Mattson, and M. E. Soule. 1999. The issue of scale in selecting and designing biological reserves. Pages 19–37 in M. E. Soule and J. Terborgh (eds.), *Continental Conservation*. Island Press, Washington, DC.

Shepherdson, D., B. Csuti, M. Anderson, and J. Steel. 2001. Oregon silverspot butterfly (*Speyeria zerene hippolyta*) Cascade Head population supplementation.

Pages 217–221 in L. Cadigan and K. Girton (eds.), *AZA Annual Conference Proceedings*. St. Louis, American Zoo & Aquarium Association (AZA), Silver Springs, MD.

Simberloff, D. J. 1994. Habitat fragmentation and population extinction. *Ibis* 137: 105–111.

Simberloff, D. J., D. Doak, M. Groom, S. Trombulak, A. Dobson, S. Gatewood, M. E. Soule, M. Gilin, C. Martines del Rio, and L. Mills. 1999. Regional and continental restoration. Pages 65–98 in M. E. Soule and J. Terborgh (eds.), *Continental Conservation*. Island Press, Washington, DC.

Singer, F. J., W. Guiming, and N. T. Hobbs. 2003. The role of ungulates and large predators on plant communities and ecosystem processes in Western national parks. Pages 444–486 in C. J. Zabel and R. G. Anthony (eds.), *Mammal Community Dynamics: Management and Conservation in the Coniferous Forests of Western North America*. Cambridge University Press, Cambridge, England.

Smith, D. W., R. O. Peterson, and D. B. Houston. 2003. Yellowstone area wolves. *BioScience* 53: 330–340.

Soule, M. E., J. A. Estes, J. Berger, and C. Martinez del Rios. 2003. Ecological effectiveness: conservation goals for interactive species. *Conservation Biology* 17: 1238–1250.

Terborgh, J., J. A. Estes, P. Paquet, K. Ralls, D. Boyd-Herger, B. J. Miller, and R. F. Noss. 1999. The role of top carnivores in regulating terrestrial ecosystems. Pages 39–64 in M. E. Soule and J. Terborgh (eds.), *Continental Conservation*. Island Press, Washington, DC.

Thompson, J. R., V. C. Bleich, S. G. Torres, and G. P. Mulcahy. 2001. Translocation techniques for mountain sheep: does the method matter? *Southwestern Naturalist* 46: 87–93.

Thorne, E. T., M. M. Miller, D. A. Jessup, and D. L. Hunter. 1992. Disease as a consideration in translocating wild animals. *Proceedings of the American Association of Zoo Veterinarians* 18–25.

Timm, R. M., ed. 1994. *Prevention and Control of Wildlife Damage*. Nebraska Cooperative Extension Service, University of Nebraska, Lincoln.

U.S. Fish and Wildlife Service. 1995. *Second Revision Florida Panther (*Felis concolor coryi*) Recovery Plan*. U.S. Fish and Wildlife Service, Atlanta, GA.

U.S. Fish and Wildlife Service. 1996. *California Condor Recovery Plan, Third Revision*. U.S. Fish and Wildlife Service, Portland, OR.

U.S. Fish and Wildlife Service. 2001. *Draft Recovery Plan for the Oregon Silverspot Butterfly*. U.S. Fish and Wildlife Service, Portland, OR.

U.S. Fish and Wildlife Service. 2003. Endangered and threatened wildlife and plants: final rule to list the Columbia Basin Distinct Population Segment of the pygmy rabbit (*Brachylagus idahoensis*) as endangered. *Federal Register* 68: 10388–10409.

Vander Haegen, W. M., S. McCorquodale, C. R. Peterson, G. A. Green, and E. Yensen. 2001. Wildlife of eastside shrubland and grassland habitats. Pages 292–316 in D. H. Johnson and T. A. O'Niel, managing directors, *Wildlife–Habitat Relationships in Oregon and Washington*. Oregon State University Press, Corvallis.

Vennesland, R. G. and R. W. Butler. 2004. Factors influencing great blue heron nesting productivity on the Pacific coast of Canada from 1998 to 1999. *Waterbirds* 27: 289–296.

Vermeer, K., K. H. Morgan, R. W. Butler, and G. E. J. Smith. 1989. Population, nesting habitat, and food of bald eagles in the Gulf Islands. Pages 123–131 in K. Vermeer and R. W. Butler (eds.), *The Ecology and Status of Marine and Shoreline Birds in the Strait of Georgia, British Columbia*. Canadian Wildlife Service Special Publication, Ottawa.

Westre, R. D. 2004. *Behavior and Survival of Reintroduced Idaho Pygmy Rabbits*. M.S. thesis, Washington State University, Pullman.

Wilmers, C. C., D. R. Stahler, R. L. Crabtree, D. W. Smith, and W. M. Getz. 2003. Resource dispersion and consumer dominance: scavenging at wolf- and hunter-killed carcasses in the Greater Yellowstone, USA. *Ecology Letters* 6: 996–1003.

Wilson, D. E. and S. Ruff. 1999. *The Smithsonian book of North American mammals*. Smithsonian Institution Press, Washington, DC.

Chapter 16

Managing Northwest Invasive Vegetation

DAVID F. POLSTER, JONATHAN SOLL, AND JUDITH MYERS

This chapter provides an overview of the role of invasive species management in ecological restoration in the Pacific Northwest. Information on invasive species problems associated with each of the nine major ecosystems discussed earlier is provided, along with some promising control techniques that have been applied in the context of ecological restoration projects. We have also included a section on the need to establish an effective monitoring and management program after initial restoration.

Invasive species, including plants, animals, and microorganisms, pose a significant threat to native ecosystems (Vitousek 1990). Murray and Pinkham (2002) suggest that along with habitat conversion caused by urban development and agriculture, invasive species pose the largest threat to native ecosystems. In the context of ecological restoration, invasive species are those that have the ability to become established and displace native species or change the ecological processes that shape an ecosystem. Invasive species may be introduced from other regions, such as knapweeds (*Centaurea* spp.) or cheatgrass (*Bromus tectorum*), or they may be native species such as Douglas fir (*Pseudotsuga menziesii*) or snowberry (*Symphoricarpos albus*) that spread into ecosystems where they had been absent as a result of changes in ecological processes, particularly fire suppression.

Most of the species that are invasive in the Northwest are not invasive in their native ranges because of a variety of factors, including natural predators and diseases, which keep populations in check. Some species that are invasive here are controlled in their native ranges by competition from other plants (Callaway et al. 1999). In the absence of natural controlling factors in the new location, the populations of invasives grow unchecked.

Invasive species have arrived from a variety of locations, with the majority from Europe, largely because until recently our major trade has been with Europe. They have come by a wide variety of means, and many were intentionally introduced (Myers and Bazely 2003). For example, species such as dandelions (*Taraxacum officinale*) were introduced from Europe accidentally, in the ballast of ships. Scotch broom (*Cytisus scoparius*) is native to Europe and was first brought to the Northwest as an ornamental, having become well established by the early part of the twentieth century. Approximately one third of Pacific Northwest invasive species have been intentionally introduced as ornamentals. Some that are now considered invasive were previously used in agriculture, such as crested wheatgrass (*Agropyron cristatum*). Others came in through land reclamation, including orchardgrass (*Dactylis glomerata*). Contaminated seed is another source of many of our most obnoxious invasive species, including the knapweeds. Although to an extent we can limit the establishment of invasive species by better managing pathways of introduction, further movement of weeds is inevitable, given increasing international trade.

Regardless of their means of introduction, invasive species can cause significant and fundamental changes to natural ecosystems. Scotch broom often forms a shrub layer in oak savanna ecosystems that previously lacked shrubs, thus reducing native species diversity and cover. Similarly, knapweeds can kill understory herbaceous species in sagebrush steppe ecosystems, creating bare soils prone to erosion (Bais et al. 2002). Saltcedar (*Tamarix ramosissima*) replaces native cottonwood trees and willows in riparian areas, altering the hydrology and structure of plant communities (Myers and Bazely 2003). Many invasive grasses have established in Garry oak (*Quercus garryana*) woodlands and have significantly modified their fabric, often creating a dense litter layer that prevents regeneration of native species (Garry Oak Ecosystems Recovery Team 2003).

Invasive species may also modify processes to the detriment of the ecosystem. Changes in nutrient status, including increased soil nitrogen, accompany growth of Scotch broom in Garry oak woodlands (Garry Oak Ecosystems Recovery Team 2003). This can favor invasive grass, which further compromises natural ecological processes and suppresses native species growth. Introduced grasses can alter regeneration of forests after logging by establishing a grass–fire feedback system that prevents the reestablishment of trees (D'Antonio and Vitousek 1992). Cheatgrass modifies natural fire regimes by expanding the fire season and extending fires that otherwise would be extinguished naturally. Repeated burning associated with cheatgrass-dominated ecosystems in former sagebrush steppe results in a loss of native grasses and shrubs until eventually only cheatgrass is present (Zouhar 2003a). Seeds of some invasive species, including Scotch broom, are stimulated to germinate after fire, so reintroduction of fire must be coupled with additional control treatments. Velvet grass and blackberry may also increase in cover after fire. Thus the reestablishment of natural fire in the presence of some invasive species may actually make the problem worse (MacDougall 2002).

Native invaders pose a special problem for ecological restoration. The ecological consequences of changes in natural disturbance regimes can be devastating. Many Western forest ecosystems, particularly ponderosa pine and other interior forests, have been severely modified by fire suppression or overgrazing. Fire suppression has led to forests becoming very dense in drier areas of the Pacific Northwest, resulting in catastrophic wildfires that consume everything in their path. Overgrazing reduces fine fuels and prevents low-intensity surface fires that would otherwise inhibit forest stem and brush density. Eventually small conifers not only create an explosive fire situation that burns everything but also reduce the diversity and cover of native understory species. Restoration efforts in ponderosa pine forests have sought to ameliorate the impacts of a century or more of land mismanagement (Friederici 2003).

Some Northwest ecosystems are subjected to both alien invasive species and native species that are encroaching as a result of human interventions. For example, Garry oak ecosystems are largely overrun with alien invasive species (MacDougall and Turkington 2004). The elimination of traditional burning by First Nations people has resulted in the shift of many deep-soiled sites to Douglas fir forests (Boyd 1999). Oak restoration often initially includes removing invasive species, followed by recreating or mimicking traditional land use practices such as burning and possibly digging camas (*Camassia* spp.). Ecologists and land managers are only beginning to come to grips with the extent of ecological damage that has been done, and unraveling the problems we have created will take dedication, persistence, and creativity.

Invasive Species of Northwest Ecosystems

Prairie and Grasslands

Most Northwest prairie and grassland ecosystems have been modified extensively by agricultural activities. Grazing and cultivation open sites up for the establishment of invasive species. Many of the weeds commonly associated with agriculture also plague remnant prairies. Perennial species such as toadflax (*Linaria* spp.) and Canada thistle (*Cirsium arvense*) spread from agricultural areas into native grasslands and can be very difficult to eradicate. Annual species such as Russian thistle (*Salsola kali* L.) and summer cypress (*Kochia scoparia*) are common in disturbed areas and can make restoration difficult. Their quick growth in the spring can monopolize available moisture and nutrients at the expense of newly planted native species struggling to get established.

Prairie sites typically have numerous small roads and tracks, including the tracks of all-terrain vehicles. Vehicles and domestic animals provide transport and disturb soil, allowing invasive plants to easily establish in the open, sunny conditions. Restoration projects in grassland and prairie areas must contend with an abundance of invasive species.

Old-Growth Conifer Forests

Invasive plants have a difficult time establishing in the moist, undisturbed old-growth conifer forests west of the Cascade and Olympic mountains. The dense cover of native vegetation often includes an almost continuous carpet of bryophytes. However, in drier forests, such as those in the rain shadow of the Olympic Mountains, invasive species such as holly (*Ilex aquifolium*), ivy (*Hedera helix* and *H. hibernica* [*Hedera helix* includes *H. hibernica*]), and spurge laurel (*Daphne laureola*) have little difficulty establishing a toe-hold even in undisturbed forests. In disturbed areas diverse invasive species can establish and create restoration challenges. Species such as Scotch broom and Himalayan blackberry (*Rubus discolor*) and herbaceous species such as hairy cats-ear (*Hypochaeris radicata*) can be found along many backroads in mature conifer forests. These can readily invade restoration sites and, absent a well-developed strategy for addressing these species, can derail or at least delay restoration. False brome (*Brachypodium sylvaticum*) is becoming a problem in the southern parts of the Pacific Northwest and may be able to invade undisturbed old-growth forests. As invasive species occupy early seral stages in natural succession, restoration strategies that aim at reestablishing the natural successional trajectory can be effective. Shade-tolerant invasives such as English ivy, holly, and spurge laurel often compete effectively against late seral native species and therefore are more difficult to control. For these species, initial removal must be coupled with continued monitoring and prompt treatment of reestablishing plants. As the current prevalence of these species in surrounding areas is reduced, reinfection rates should decrease as well.

Ponderosa Pine and Interior Forests

Two types of invasive species threaten ponderosa pine and interior forest ecosystems: native species that increase in cover and abundance in the absence of low-intensity ground fires and alien species that capitalize on anthropogenic disturbances associated with these ecosystems, including logging, grazing, and urban development. Invasive species that establish in ponderosa pine and interior forests include those intentionally planted and some that establish accidentally (Friederici 2003). Crested wheatgrass (*Agropyron cristatum*), desert wheatgrass (*Agropyron desertorum*), smooth brome (*Bromus inermis*), orchardgrass, and a wide variety of other nonnative grasses and legumes have been planted intentionally in pine woodlands in order to improve forage quality for livestock. In some cases, these introduced species have become so firmly established that they are excluding natives and are now a major impediment to ecological restoration. Alien invasive species that have established in the ponderosa pine and interior forests unintentionally can be even more devastating. These include the knapweeds, which have spread onto thousands of hectares of open forest habitat, seriously degrading native pine woodlands.

Oak Woodlands and Savannas

Invasive species dominate many oak woodland and savanna ecosystems. Murray and Pinkham (2002) polled a range of oak ecosystem experts and determined the top ten invasive species in oaks (Table 16.1).

The inclusion of four grasses in this list reflects the concern experts in the field have for the degradation of the fabric of these ecosystems. The annual grasses cheatgrass and medusahead (*Taeniatherum caput-medusae*) are more common problems in southern parts of oak woodlands. Bulbous bluegrass (*Poa bulbosa*) is a short-lived perennial that may become a problem in the future. Yellow star thistle is becoming one of the most common problem species in the southern oak woodlands. Management of this diverse invasive species assemblage takes ingenuity and persistence. The conversion of oak woodlands to urban development, impacts associated with fire suppression, heavy grazing, and agricultural activities have all increased invasive species problems.

TABLE 16.1.

Top ten invasive species in Garry oak woodlands in order of greatest to least concern.

Orchardgrass	*Dactylis glomerata*
Scotch broom	*Cytisus scoparius*
Gorse	*Ulex europaeus*
English ivy	*Hedera helix*
Velvet grass	*Holcus lanatus*
Spurge laurel	*Daphne laureola*
Common hawthorn	*Crataegus monogyna*
Sweet vernalgrass	*Anthoxanthum odoratum*
Himalayan blackberry	*Rubus discolor*
Hedgehog dogtail grass	*Cynosurus echinatus*

Freshwater Marshes

Management activities such as changes in natural hydrologic cycles, whether intentional or not, have resulted in opportunities for invasive species to establish in freshwater marsh ecosystems in the Pacific Northwest. Species such as reed canarygrass (*Phalaris arundinacea*) that have been widely used to enhance hay production on wetland sites have become a significant problem, smothering native species and choking drainage channels. Although a reed canarygrass is native to the western United States, the European ecotype of reed canarygrass typically found is much more aggressive and is the cause of the problem. Purple loosestrife (*Lythrum salicaria*) is starting to become a major problem in western North America, having already been a significant problem in the east for many years. Purple loosestrife can seriously degrade natural wetland habitats by creating a monoculture of dense, stiff stems that inhibits access by some wildlife. Whitt et al. (1999) found that although avian abundance was greater in purple loosestrife–dominated habitats, diversity was reduced.

Riparian Woodlands

Degradation of riparian woodlands by invasive species can have wider ecosystem consequences because riparian areas often serve as important biological corridors. A wide variety of invasive species can be found in riparian woodlands. Shade-tolerant species such as holly and English ivy are ready invaders in shady areas. In more open areas along stream banks, grassy species such as reed canarygrass and Himalayan blackberry can become problems, modifying successional trajectories with long-term consequences on riparian processes and wildlife habitat. Japanese (and giant) knotweed (*Polygonum cuspidatum* and *P. sachalinense*) and giant cow-parsnip or giant hogweed (*Heracleum mantegazzianum*) are starting to become problems in riparian woodlands throughout the coastal

Pacific Northwest. The giant hogweed threatens native riparian ecosystems, and the sap contains a highly toxic group of substances that sensitize skin to ultraviolet radiation, causing painful burns that resist healing. As with all ecosystems, disturbance opens the ecosystem up to invasion by alien species, and riparian zones are naturally prone to disturbance from flooding.

Tidal Wetlands

Tidal wetlands are harsh environments that support only a few specialized plant species. Nevertheless, these areas are vitally important. The principal invasive plants of tidal wetlands are spartina grasses (*Spartina patens*, *Spartina alterniflora*, and *Spartina angelica*) and common reed (*Phragmites australis* and *Phragmites communis*). Spartina grasses alter tidal wetlands by trapping sediments, thus turning biologically rich mud flats into visually attractive but ecologically depauperate wet meadows. This change significantly reduces foraging areas for shorebirds that feed on invertebrates found in open mud flats. The impact of spartina grasses on areas such as Willapa Bay has been substantial, with restoration costs very high. Common reed has similar but less dramatic effects on tidal wetlands.

Mountain Ecosystems

Invasive plant species are not a significant problem in undisturbed mountain ecosystems. However, many alpine and subalpine areas have been disturbed by grazing; road, track, or trail development; and grading of ski slopes. These actions have allowed exotic species such as toadflax and knapweeds to establish and persist. In addition, vehicle, horse, and hiker use introduces invasive species inadvertently through movement of seed-infested mud. Marsh plume thistle (*Cirsium palustre*), a European native, is beginning to become a major problem in moist montane ecosystems and can even invade undisturbed sites, displacing native species.

The past use of many alien species for erosion control, fire rehabilitation, and forage enhancement has led to the preponderance of these species on disturbed sites. Species such as bluegrasses (*Poa* spp.), timothy (*Phleum pratense*), orchardgrass, and alsike clover (*Trifolium hybridum*) are persistent. In some cases, seed contaminated with thistles, hawkweeds, and other invasive species complicates matters.

Sagebrush Steppe

Cheatgrass is the most troublesome and widespread invasive species associated with sagebrush steppe ecosystems, although medusahead is beginning to replace it in parts of Oregon. Knapweeds, leafy spurge (*Euphorbia esula*), and skeleton weed (*Chondrilla juncea*) also become problems in some locations. Cheatgrass can significantly affect natural successional trajectories by shortening fire return intervals. Big sagebrush, the keystone shrub in this ecosystem, can be completely eliminated under the frequent fire regimes associated with cheatgrass infestations. Healthy sagebrush ecosystems are resistant to invasive species, but heavy grazing reduces native bunchgrasses and forbs, disturbs the microbiotic crust, and opens areas up to cheatgrass, setting in motion a fire cycle that the sagebrush cannot survive (Table 16.2).

Management of Invasive Species

Containment or elimination of invasive species is a major component of many restoration projects, particularly in urban and settled areas, where invasive species removal often is a primary activity. In some projects, removal of invasives is the sum total of what is needed, an approach called restoration by subtraction. Where abundant invasive species seeds are in the soil seed bank, disturbances associ-

TABLE 16.2.

Major Northwest invasive plants and their occurrences in ecosystems.

	Prairies	Conifers	Interior	Oaks	Freshwater Wetlands	Tidal Wetlands	Riparian	Steppe	Mountains
Toadflax	•								
Canada thistle	•								
Russian thistle	•								
Summer cypress	•								
Scotch broom	•	•		•			•		
English holly		•							
English ivy		•		•					
Spurge laurel		•							
Himalayan blackberry		•		•			•		
Hairy cats-ear		•							
False brome		•							
Crested wheat grass			•						
Smooth brome			•						
Orchardgrass	•		•	•					•
Knapweeds	•		•	•					
Gorse				•					
Velvet grass				•					
Spurge laurel				•					
English hawthorne				•					
Sweet vernal grass				•					
Hedgehog dogtail				•					
Cheatgrass			•	•				•	
Medusahead				•				•	
Bulbous bluegrass				•					
Yellow star thistle				•					
Reed canarygrass					•		•		
Purple loosetrife					•				
Japanese knotweed							•		
Giant hogweed							•		
Spartina patens						•			
Spartina alternafolia						•			
Spartina angelica						•			
Common reed						•			
Marsh plume thistle									•
Timothy grass									•
Clover									•

ated with restoration activities, particularly brush clearing and reintroduction of fire, can result in a healthy crop of weeds and little else. Similarly, where the adjacent weed flora is composed of species with windborne seed such as Canada thistle, an unvegetated restoration site can provide an ideal seed bed.

A number of methods are used to eliminate or reduce the presence of invasive plants. The application and success of these methods depend on many factors, including the overall condition and ecology of the ecosystem being restored, available budget, experience, presence of surface water, and labor force. Managers of every project must assess these factors before choosing one or more methods. The discussion in this section and Table 16.3 provide a brief summary of some of the main methods being used.

TABLE 16.3.

Best control strategies for major Northwest invasive species.

	Competition	Shade	Biological	Mowing	Tilling	Thermal	Flooding	Chemical	Physical	Grazing
Toadflax				•						•
Canada thistle					•					
Russian thistle				•						
Summer cypress				•						
Scotch broom	•	•		•		•	•			
English holly									•	
English ivy								•	•	•
Spurge laurel									•	•
Himalayan blackberry	•	•		•		•	•	•	•	•
Hairy cats-ear		•								
Crested wheat grass	•									
Smooth brome						•				
Orchardgrass			•			•				
Knapweeds										•
Gorse	•	•		•		•	•		•	•
Velvet grass						•				
English hawthorne									•	
Sweet vernal grass						•				
Hedgehog dogtail						•				
Cheatgrass	•							•		•
Medusahead	•							•		•
Yellow star thistle	•							•		•
Reed canarygrass				•						
Purple loosetrife									•	
Japanese knotweed									•	
Giant hogweed									•	
Spartina alternafolia			•					•		
Spartina angelica								•		
Timothy grass						•				

Assessing Ecological Conditions

Ecological restoration projects that kill invasive species but fail to address the ecological conditions that encouraged invasion in the first place typically have limited and short-term success. Project planners need to determine what has caused the presence of invasive species as part of their overall site assessment (see Chapter 2). If it was a one-time site disturbance followed by a rapid invasion of, say, Scotch broom or blackberry, then removal followed by replanting may be sufficient. But if disturbance is from off-road vehicles that opened the soil up and transported weed seeds into an area, removing the weeds but failing to prevent future vehicle incursions clearly will not solve the problem.

Site Preparation

Where there is a threat of invasive species establishing on a site that is being restored, careful site preparation is a wise investment. For instance, restoration of riparian areas often occurs in tandem with reshaping of channelized stream banks. Timing removal of invasive plants or seed heads (such as thistle) might reduce the risk of weeds quickly occupying the site once construction is completed.

Once planted native vegetation is well established, weed seeds are less likely to find suitable microsites for germination and growth.

Most invasive species are adapted to establishing on bare ground. Therefore, if restoration treatments can minimize the extent and duration of bare ground, invasives from afar may not get much of a chance to find places to become established. Where planted native materials are small and widely spaced, consideration should be given to using thick mulch to deny a seed bed to weeds. Mulch can be organic, such as chips, semiorganic, such as cardboard or newspaper, or inorganic, such as plastic or gravel.

Solarization and sheet mulching are two methods used to prepare sites where all existing vegetation is considered undesirable. There are many different approaches to these methods, but in all cases the basic principle is to completely cover existing vegetation and deprive it of light. In a few months, or in some cases up to a year (depending on the vigor of the roots of the plants), the unwanted vegetation will die. If there is evidence of abundant weed seed in the topsoil, it is advantageous to cover the bare soil with clear plastic, which desiccates the seeds and prevents germination. Depending on the time of year, this can take a few weeks to several months. Some restorationists (and organic gardeners) wet the soil first, to encourage germination of weed seeds, allowing young, vulnerable sprouts to be baked or steamed to death. Others do a shallow tilling to bring up even more viable seed. It is not uncommon to have to repeat the solarization process several times to truly exhaust the seed bank. Maintaining some soil moisture is necessary for solarization to be effective.

Sod stripping and removal is another site preparation method that has been successfully used, particularly on grassland and reed canarygrass restoration sites. The invasive vegetation and several inches of topsoil are cut, rolled up, and hauled away (preferably very far away), leaving bare soil immediately ready for reseeding or planting.

Herbicides can be used to selectively kill invasive species or to completely kill vegetation over a given area to prepare a site for restoration. Herbicide use is discussed in greater detail later in this chapter.

Invasive Species Control Methods

We have identified thirteen basic methods of invasive species control and provide a brief description in this section.

Competition and Shade

Increasing competition and creating overhead shade can reduce the presence of invasive species in native ecosystems. Restoration practitioners can increase competition or shade by taking steps to promote successional advancement, which strengthens the natural ability of an ecosystem to develop and maintain healthy native plant populations over time. The basic idea is that all ecosystems have successional stages and that early stages tend to be more vulnerable to invasion by weedy species programmed to occupy bare ground, when competition and shade are at their lowest. As ecosystems mature and competition for space, light, water, and nutrients increases, many weedy species lose vigor and give way. There are simply fewer resources available to invasive plants, which often are not effective when the competition is intensive. This may be one reason that invasives often form dense thickets, which keep competition at bay. Experience has shown that some weeds of prairies and grasslands, such as Russian thistle and summer cypress, can be reduced or eliminated simply through establishment of a healthy cover of native perennial bunchgrasses and forbs.

Invasive species often are shade intolerant. Management can establish or encourage later-successional species that will eventually overtop and shade out the offending invasive. This method may include temporary control or cutting back of invasives to allow room for shade-creating species to establish. For instance, dense thickets of Himalayan blackberry can be controlled in forest ecosystems by interplanting of red alder (*Alnus rubra*) or big-leaf maple (*Acer macrophyllum*). These fast-growing mid-successional plants can overtop and shade out blackberry and advance the ecosystem along a successional path toward conifer forest. Deciduous trees are more effective than slower-growing conifers, which have a difficult time breaking clear of the blackberry. Later, conifers can be planted under a thinned deciduous canopy.

Biological Control

These methods are targeted to remove offending plants without causing other impacts on natural ecosystems and can be very effective in controlling invasive species. Although a broad definition of biological control includes grazing, fungi, and microbes, grazing is examined elsewhere in this section. Microbiological control methods are based on the idea that invasive plants have natural enemies and diseases in their native habitats that they have managed to escape. Importing one or more of these enemies can be effective at controlling the invasive. Clearly, bringing more nonnative species in carries its own set of risks and necessitates major investments in research before such introduction can be deemed safe. Such research is expensive and is rarely targeted exclusively at ecological restoration. However, land grant universities do investigate biological control methods for species that are agricultural or economic threats, and at times these same plants pose natural ecosystem threats. A classic successful case of biocontrol is the tansy ragwort (*Senecio jacobaea*). There has also been a great deal of promising research into finding biocontrol agents for purple loosestrife and leafy spurge. Efforts to control Canada thistle, spotted and diffuse knapweeds, and yellow star thistle have not been successful, although research continues (Tu et al. 2001).

Repeated Mowing and Tilling

Some invasive species that have extensive rhizome systems, such as quackgrass (*Elymus repens*) and Japanese knotweed, can be controlled or eliminated through repeated mowing or tilling. Mowing knocks plants back to ground level but does not disturb root systems. It can kill rhizomatous invasives by depleting energy reserves, as shown in Figure 16.1. The goal is to deprive the roots of energy resupplied in photosynthesis. When invasive plants are growing in ecologically sensitive areas, such as reed canarygrass in a wetland, repeated cutting at the right times of year can reduce vigor or completely kill the offending plants while retaining or encouraging desired native vegetation, such as willows and dogwoods. Mechanical weed whackers are the tool of choice, but armies of volunteers with hand tools can accomplish the same result. The key is to understand the phenology and physiology of the invasive plant, allowing cutting efforts to be timed during the weakest point in its life cycle.

Multiple tilling has a similar effect to repeated mowing and can be used where terrain and soil conditions are favorable. It may take four or more repeated tillings to exhaust the weed seed bank in the topsoil.

Restoring Historic Disturbance Regimes

Many invasive plants need some level of disturbance to persist. However, elimination or radical change in historic disturbances has also opened the way for both native and nonnative ecosystem invaders. In oak and pine woodlands and savannas, native species such as Douglas fir and snowberry invade in the absence of frequent, low-intensity fires. In some cases it may be possible to simply

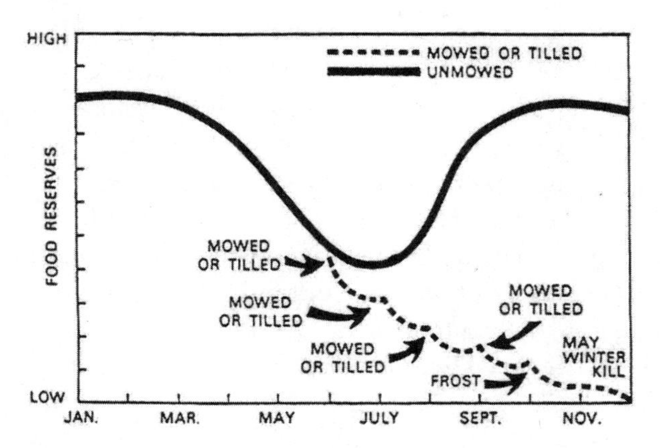

FIGURE 16.1. Graph of plant energy reserve depletion by mowing or tilling. (*Redrawn from Polster and Landry 1993*)

restore the historic frequency of fire, setting in motion restoration. This depends on the extent to which the basic structure of the original ecosystem is still intact. In most cases, some initial site preparation, including cutting the invasive species, raking needles or duff away from the bases of old trees, and other actions are needed before fire can be restored safely.

An opposite example involves cheatgrass in sagebrush steppe ecosystems. In this case, the invasive cheatgrass may have initiated more frequent fire than was present in the historic ecosystem. Fire kills sagebrush and perpetuates the cheatgrass, so direct removal of cheatgrass or prevention of fire is needed to restore the historic infrequent fire regime.

Ecological degradation and subsequent weed invasion have also been associated with altered flood frequencies and magnitudes. Saltcedar is found more often in Southwestern and Great Basin riparian areas of flood-controlled rivers and streams than on rivers and streams that experience naturally high fluctuations in flow (Zouhar 2003b). Reed canarygrass has spread widely in the upper reservoir of the High Arrow Dam on the Columbia River near Revelstoke, British Columbia because of substantial changes to local hydrology. In some cases, it may be possible to mimic historic flows as one means of controlling invasive species. For example, increasing spring water release from Columbia and Snake River dams to aid fish migration may also help restore willow bars downstream by flooding out broom or blackberry patches that have occupied former floodplains. In Portland, water control structures were installed at Oaks Bottoms and Smith & Bybee Wetlands Natural Area, in part to better mimic historic flows and make conditions less suitable for reed canarygrass and other invasives (R. Rodgers, personal communication, 1996).

Herbicides

Natural-based or synthetic herbicides have been conscripted into the ecological restoration war for some time. Some, such as 2,4-D and sethoxydim, are specific to certain classes of plants (dicots and grasses, respectively), whereas others, such as glyphosate, kill or damage all plant types. Some, such as picloram, have residual effects in the soil, whereas others, such as glyphosate and triclopyr, are only foliar active. Herbicides can be applied selectively to target plants or broadcast to cover an entire area.

Herbicides are particularly useful for controlling deep-rooted and rhizomatous species, both of which tend to resprout after burning, mowing, cutting, or hand-pulling. When used as part of an integrated pest management program and when the chemical, the application method, and the timing are carefully selected, herbicides can provide excellent control of invasive species, usually at a small fraction of the cost of manual removal. As with any method, there may be side effects. Care must be taken to avoid or minimize nontarget effects, especially to desirable vegetation but also to other classes of organisms, including the herbicide applicator. Furthermore, some people may object to herbicide use on philosophical grounds, and such social factors should be factored into restoration practice decision making.

Each herbicide has particular strengths and weaknesses. Advice on the control of particular species using herbicides and related literature reviews are available from many sources, including The Nature Conservancy's Wildland Weeds Web site (tncweeds.ucdavis.edu). Before using any herbicide, always read and follow the directions on the manufacturer's label and material safety data sheet. When in doubt, contact your local Department of Agriculture or extension agent for advice before taking action.

Physical Methods

Manual control of invasive species is favored by many restoration practitioners, particularly where volunteer labor is available (Figure 16.2). Unwanted plants are cut back, dug up, or yanked out of the ground. A variety of approaches and tools have been developed to help in manual weed control. Some, such as spades, hoes, and hoedads, were developed for agricultural or forestry purposes and have been adopted for use in restoration. Others, such as the Weed Wrench, flamers, and heaters, were developed specifically for restoration or organic gardening. Spades are effective at digging up small plants with shallow and compact root systems. Hoes are designed primarily to remove small herbaceous plants, usually annuals. The choice of tool may depend on the scale and nature of the problem.

The Weed Wrench is a patented metal tool originally designed by Tom Ness to pull broom in the Marin Headlands of California. It has a set of jaws that clamp strongly onto the target plant's stem as one pushes down on a long lever, allowing the plant to simply be lifted out of the ground, roots and all. Four models are available (heavy, medium, light, and mini). They are somewhat heavy but also very sturdy, and they are repairable if damaged. The largest model can clamp around and uproot plants up to 2.5 inches in diameter. More details on Weed Wrenches can be found at www.weedwrench.com. Be aware that Weed Wrenches can result in soil disturbance that allows weeds to sprout.

FIGURE 16.2. Removing ivy manually from Newell Creek Watershed in Oregon City, Oregon. (*Photo by Marcia Sinclair*)

Hand power tools include chainsaws and weed eaters of many types. Of course, there are safety concerns involved in the use of sharp, fast-moving metal blades or heavy strings. A skilled chainsaw operator can effectively clear heavy brush, particularly on steep slopes or where soil compaction is a concern. Weed eaters generally are used in lighter brush or to knock down herbaceous vegetation.

Use of wheeled or tracked machines is the most intensive form of physical control. Backhoes, bulldozers, slash busters, and other implements have been drafted into the service of ecological restoration. Generally machine methods are much more economical to use in the short run than hand methods but can introduce unwanted side effects that raise costs in the longer run. Soil compaction, disturbance to existing native plants, excessive slash and debris, and opening up of bare areas to future weed invasion all can result from use of large machines in restoration.

Grazing

There have been many attempts to use domestic grazing animals, primarily goats and sheep, to assist in restoration (Tu et al. 2001). Although grazing does not eliminate invasive species, combined with other methods it can reduce their occurrence and help with site preparation for restoration. In moist sites west of the Cascade Mountains goats have been used to graze down and trample blackberry, ivy, and Scotch broom with mixed success. A fundamental challenge with goats is that they are very clever and curious and will eat almost anything, including plants that the restorationist does not want eaten, as well as those that may be poisonous to the goats (such as rhododendrons). Goats prefer browsing woody species to eating grass and forbs, which can make them useful in oak and pine woodland or savanna restorations, where reduction of native and nonnative brush is a key goal. On the other hand, sheep are easier to control and prefer to eat grass and forbs.

Both goats and sheep will eat weedy forbs toxic to other wildlife, including leafy spurge, knapweed, and toadflax. Sheep can pass the seeds of spurge through their digestive systems and infect other areas, so care must be taken in deciding where they go after they leave a restoration site. Temporary fencing, salt licks, and other measures usually are needed to manage herds. Goats and sheep are vulnerable to predators, including coyotes.

Finding goats and sheep to bring to a restoration site can be a challenge. Only a few herders in the region lease their animals out. Cattle have also been used to prepare sites for restoration, primarily through deliberate heavy-impact grazing to knock back weedy species, then with follow-up seeding of natives. Ranchers in the Oregon Country Beef cooperative of central Oregon have altered grazing approaches to encourage native bunchgrass, streamside willows, and aspen groves, but not directly as weed control (www.pccnaturalmarkets.com/producers/orcountrybeef.html).

Grazing can be an economical method for invasive species control, but restoration planners who are considering this approach need to be ready to make a long-term, multiyear commitment to it. Many projects have tried grazing for a single season or even a few weeks, been disappointed with the results, and thrown in the towel, perhaps too soon.

Monitoring and Adaptive Management

The most effective way to control invasive species is to prevent their establishment in the first place. Most Pacific Northwest ecosystems are susceptible to invasion by a variety of alien species against which they have few defenses. Recently restored ecosystems may be particularly prone to invasion by alien species because they may lack the robust vegetation cover of mature ecosystems. For this reason, systems of ecosystem monitoring and maintenance should be considered as a part of all restoration projects.

Adaptive management has become a standard approach to dealing with the uncertainty associated with many ecological restoration projects. Invasive species introduce a significant amount of uncertainty to restoration projects. Therefore, adaptive management approaches are effective for addressing ongoing invasive species management in the context of ecological restoration projects (Murray and Jones 2002). Figure 16.3 presents an outline of adaptive management.

Monitoring is an essential part of all adaptive management programs and must be part of any effective invasive species management program. Three basic types of monitoring are implementation, effectiveness, and validation monitoring. Implementation monitoring is conducted to ensure that the work is carried out as specified. For instance, if a broom removal project is undertaken by a local stewardship group as part of an ecological restoration project, implementation monitoring is conducted to ensure that the work is undertaken in a systematic manner according to the plans that were developed before the work began. Effectiveness monitoring is conducted to ensure that the work is resulting in improved conditions relative to the original objectives (Gaboury and Wong 1999). Following our example of broom removal, effectiveness monitoring would be undertaken to determine whether removal of the broom was assisting in the restoration of the site. Validation monitoring, which is rarely undertaken in the context of most restoration projects, tests the validity of the original hypothesis that broom removal will expedite recovery of the degraded ecosystem. Most restoration projects involving invasive species include elements of both implementation and effectiveness monitoring. Rarely do we test the hypothesis that invasive species removal is an important part of ecological restoration, but for most restorationists, this question seems too self-evident to bother about.

Monitoring to ensure that invasive species do not reestablish or that new invasive species do not establish is essential for the long-term health of restored ecosystems. Understanding the conditions that lead to the establishment of invasive species, (disturbances, changes in disturbance regimes, and increased numbers of invasive species in adjacent areas) can help in determining monitoring frequency and intensity. For instance, highly dynamic ecosystems such as shifting sand dunes in sagebrush steppe ecosystems or the shorelines in riparian woodlands where water levels fluctuate widely should be monitored more frequently and with greater intensity than mature old-growth conifer forest ecosystems, where ecological conditions are stable over long periods of time.

Monitoring should be conducted more frequently early in the restoration process. In the early years, frequent monitoring allows identification and rectification of problems before they become unmanageable. Once the restored site has been established for some time, monitoring frequency can be reduced. The absolute frequency and intensity of the monitoring program needed for each restored ecosystem differ and should be developed early in the restoration planning process so that adequate resources are allocated to this part of the restoration project.

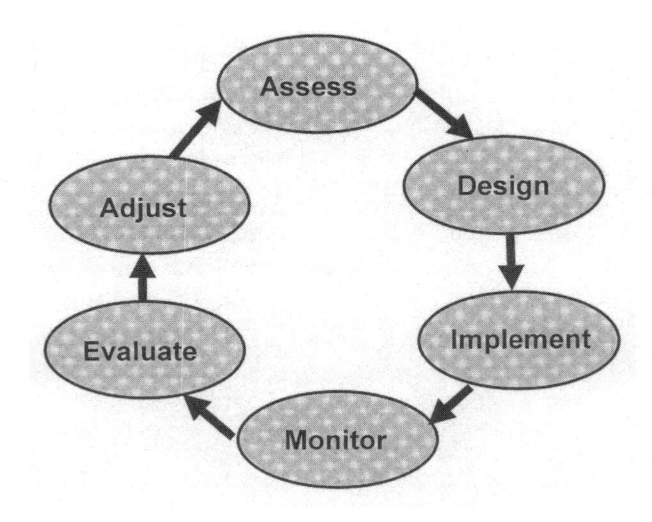

FIGURE 16.3. Adaptive management framework. *(Adapted from Murray and Jones 2002)*

Monitoring can reveal interesting and useful information about the state of the ecosystem being restored. Invasive species often are an indicator of ecosystem health. Stressed and degraded ecosystems are more susceptible to the establishment of invasive species. Managing the elements that are causing the stress will allow the ecosystem to recover more quickly and fully than allowing the stresses to continue. Monitoring can identify restoration problems in a timely manner.

Case Study: Repeated Cutting to Tame Reed Canarygrass

Reed canarygrass had taken over and was threatening to choke Koksilah Creek, a small salmon stream on the east coast of Vancouver Island. A local elementary school class was interested in restoring the creek and reestablishing a viable coho salmon (*Oncorhynchus kisutch*) population. Because of the engagement of children and the streamside location of the reed canarygrass, the use of herbicides was ruled out. A decision was made to try repeated cutting to reduce the vigor of the grass. Initially the whole class (and a few moms and dads) turned out to remove as much reed canarygrass biomass as possible. Pickup trucks and trailers were filled with the thatch that had developed over many years of unchecked reed canarygrass growth. Of course, this just released the grass to resprout with a vengeance, so a regular program of repeated cutting was needed to keep the grass under control. The decision was made to cut the grass whenever it grew to 20 centimeters or more. Cutting was done as close to the soil as possible using weed-eaters. That first year it seemed as if the canarygrass would retain the upper hand, but slowly native willows, red-osier dogwood, cattails, and sedges began to emerge along the creek. Then the cutting had to be done more carefully to avoid damage to the native species while inflicting as much stress as possible on the reed canarygrass. By fall, the canarygrass was hardly growing at all, and the native riparian species were doing well. A program of riparian species planting was planned, and again the whole class turned out. Additional riparian species were planted, and by the next spring, the stream was ready to accept the young salmon the children had raised in the classroom over the winter. Some additional grass cutting was conducted that year, but the growth of the woody species was taking over the canarygrass by providing shade and competition. This simple project has been shared with the wider community, and the students now bring their parents to show them the creek they helped to restore.

Case Study: Eliminating Scotch Broom

The Mount Tzuhalem Ecological Reserve, established in 1984, is one of a system of 151 ecological reserves established in British Columbia to preserve representative examples of the diverse ecosystems of the province. In some areas (including Mount Tzuhalem), reserves shelter rare and endangered species. In other cases, reserves protect important biological or geological phenomena while also protecting important genetic resources. Although destructive sampling or testing is not allowed, research and education are an important part of many ecological reserves.

The Mount Tzuhalem Ecological Reserve was established to protect an excellent example of Garry oak woodlands and spring wildflower meadows. This 18-hectare site of beautiful flower meadows was invaded by Scotch broom, and by the late 1980s the flower meadows were rapidly disappearing. Recognizing the urgency of the situation, the Cowichan Valley Naturalists' Society initiated an all-out assault on the broom. Weekly gatherings of four or five hearty souls, led by longtime local resident, mountaineer, and naturalist Syd Watts, started to tackle the broom problem (Figures 16.4 and 16.5). Because of the rocky terrain and the wish to avoid damage to whatever flowers remained under the broom, the clearing was all done by hand, although helicopters were enlisted to remove the mountains of cut broom from the reserve. Syd encouraged his team with stories of exploring the intricacies of the area as a young boy. He told of the struggles to win the hearts and minds of local politicians to create the reserve from local municipal land, having convinced them it would never make productive forest.

By the late 1990s, much of the broom had been cleared, and the flower meadows were starting to reappear. However, there were still some areas where the steep, rocky terrain prevented maintenance access, and the seeds in the soil seed bank started to reestablish the dominance of the broom once again. Just as it appeared that all of the efforts of the naturalists would be lost, funding was secured from the Canadian government's Habitat Stewardship Program, and a crew of strong young people took over the fight from the tired and rapidly aging naturalists. Syd continued to lead the new crew, often showing the willing participants where an obscure patch of broom was growing or pointing out some unique microsites with some special species. In recent years, flowering Scotch broom has not been found in the reserve, and the seedlings that sprout from the soil seed bank are easily pulled in the winter when soils are soft, and there is little damage to the fragile native vegetation. Persistence has paid off, and each spring the show of native wildflowers, including many rare and endangered species, gets better and better.

FIGURE 16.4. A hapless restoration worker swallowed by Scotch broom. (*Photo by Patrick Dunn*)

FIGURE 16.5. Removing Scotch broom from an oak woodland before it goes to seed. (*Photo by Eric Delvin*)

Case Study: Biological Control

Biological control is the introduction of natural enemies from the native habitat of the weed to the area where it has become invasive. The rationale is that by reestablishing some of the natural enemy complex, the competitive advantage of the invasive weed will be reduced. Biological control is a slow and expensive process, but when successful it can provide continuous control over large areas with little further input. Programs are initiated in Canada by Agriculture and Agri-food Canada and carried out with the help of provincial agencies. A biological control program starts with foreign exploration to find likely candidates for introduction. This is followed by extensive testing of the host plant specificity of the potential agents. A comprehensive report is circulated for evaluation in both Canada and the United States, and a careful approval process is followed before introductions are permitted. Species of agents that are considered to be safe must then be reared to eliminate parasitoids and diseases before introduction. Once introduced and established, these agents are redistributed. Evaluation of their impacts does not receive sufficient attention, so in many cases we do not learn as much as we might from these large-scale experiments. Because the introduction of each new species of agent could result in nontarget effects, it is important to release as few species as possible (Louda et al. 2003). Reviews of biological control indicate wide variation in the success rate of biological weed control, ranging from 20% to 80% in different parts of the world (McFadyen 1998, Myers and Bazely 2003).

Recently several biological control programs have begun to achieve success using beetles that are capable of killing Northwest invasives. Dalmatian toadflax is a rangeland weed in the interior of British Columbia. The stem-boring weevil *Mecinus janthinus* has been introduced as a control agent (De Clerck-Floate and Harris 2002). This beetle is having a major and rapid impact on plant survival and density. In other cases, various species of *Apthona* are successfully controlling leafy and cypress spurge in different microhabitats in western North America. *Apthona* species vary in their abilities to thrive in different environments (Bourchier et al. 2002a). The stem-boring weevil *Mogulones cruciger* is reducing the densities of houndstongue (*Cynoglossum officinale*) in rangelands (De Clerck-Floate and Schwarzländer 2002). Populations of purple loosestrife have been greatly reduced at many locations, primarily by the leaf-feeding beetle *Galerucella calmariensis* (Lindgren et al. 2002, Denoth and Myers 2005). Unfortunately, loosestrife often is replaced by reed canarygrass. Tansy ragwort has been controlled in pastures in the lower mainland of British Columbia and Oregon by the beetle *Longitarsus jacobaeae*, but this insect does not survive at higher elevations, where tansy persists at high densities (McEvoy and Coombs 1999).

After 30 years of effort and the introduction of twelve different species of biological control agents, diffuse knapweed (*Centaurea diffusa*) has declined in many areas of the southern Okanagan since the introduction of a weevil, *Larinus minutus* (Bourchier et al. 2002b). Feeding by adult *Larinus* in the spring kills knapweed plants, but it appears to be replaced by cheatgrass in many sites, and recent fires have disrupted biological control at some locations. Further information on these and other biological control programs is available at res2.agr.ca/lethbridge/weedbio/plant/bdifknap_over_e.htm.

Two Case Studies on Herbicide Use

Japanese and Giant Knotweed

The invasive species Japanese and giant knotweed (*Polygonum cuspidatum* and *P. sachalinense*) are deep-rooted rhizomatous perennial plants originally from Asia that are aggressive invaders and pernicious pests of riparian and floodplain habitats throughout much of the temperate world. Because their roots extend down as much as 3 meters, cutting or pulling even modest-sized patches (say 50–200 stems) is effective only if repeated many times annually over several years. Experiments conducted by The Nature Conservancy Oregon Field Office and the Metro Parks and Greenspaces Program (Portland, Oregon) have shown that foliar application of triclopyr and foliar application or stem injection of glyphosate-based herbicides can be highly effective, producing 70–100% reduction in stem number and 90–100% reductions in biomass with a single application. When treatment is done correctly, complete eradication of small patches (less than fifty stems) is almost always achieved with 2–3 years of treatment. Although very large patches (more than 1,000 stems) may need 4 or 5 years of treatment, most patches produce only a few stunted stems after the first year of herbicide treatments, and replanting may safely occur after the first or second year of treatment.

English Ivy

English ivy (*Hedera helix*) is nearly omnipresent in urban and near-urban open spaces in the Pacific Northwest, where it climbs and topples or strangles trees and buries smaller-stature forest floor vegetation. The resulting ivy monoculture is poor habitat for all categories of native species and can allow more sun-loving weeds entry into the system. Although ivy is generally shallow-rooted and can be hand-pulled, the effort is labor intensive (i.e., expensive). Especially in areas with little or no remnant native vegetation, herbicides offer a more economical alternative. Applied with appropriate penetrants and surfactants to get through ivy's waxy coating, both glyphosate- and triclopyr-based herbicides can produce 95% reductions of ivy cover in a single treatment. Because most Pacific Northwest native perennial species are dormant in winter, while ivy remains active, careful winter application of herbicide can be particularly effective at killing ivy while protecting native vegetation. See Plate 11 in the color insert for additional photographs.

Summary

Management and control of invasive plants are necessary elements in ecological restoration of most Pacific Northwest ecosystems. In some areas, removal of exotic plants is all that is needed to reestablish a healthy, functioning ecosystem. In other cases invasive control is done in conjunction with additional strategies, including planting and prescribed burning. Northwest ecologists and land managers are still developing effective control strategies for many species, and continued field experimentation, including documentation and sharing of results, is vital. Given the long list of exotics, the forgiving climate, and the increase in international trade, which results in new exotics reaching our shore, it is clear that control of invasive plants will be a permanent feature of ecological restoration work.

References

Bais, H. P., T. S. Walker, F. R. Stemitz, R. A. Hufbauer, and J. M. Vivanco. April 2002. Enantiomeric-dependent phytotoxic and antimicrobial activity of (±)-Catechin. A rhizosecreted racemic mixture from spotted knapweed. *Plant Physiology* 128: 1173–1179.

Bourchier, R. S., S. Erb, A. S. McClay, and A. Gassmann. 2002a. *Euphorbia esula* (L.), leafy spurge, and *Euphorbia cyparissias* (L.), cypress spurge (Euphorbiaceae). Pages 346–358 in P. G. Mason and J. T. Huber (eds.), *Biological Control Programmes in Canada, 1981–2000*. CABI, Wallingford, UK.

Bourchier, R. S., K. Mortensen, and M. Crowe. 2002b. *Centaurea diffusa* Lamarck, diffuse knapweed, and *Centaurea maculosa* Lamarck, spotted knapweed (Asteraceae). Pages 302–313 in P. G. Mason and J. T. Huber (eds.), *Biological Control Programmes in Canada, 1981–2000*. CABI, Wallingford, UK.

Boyd, R., ed. 1999. *Indians, Fire and the Land in the Pacific Northwest*. Oregon State University Press, Corvallis.

Callaway, R. M., T. H. DeLuca, and W. M. Belliveau. 1999. Biological-control herbivores may increase competitive ability of the noxious weed *Centaurea maculosa*. *Ecology* 80(4): 1196–2101.

D'Antonio, C. and P. M. Vitousek. 1992. Biological invasions by exotic grasses, the grass/fire cycle, and global change. *Annual Review of Ecology and Systematics* 23: 63–87.

De Clerck-Floate, R. A. and P. Harris. 2002. *Linaria dalmatica* (L.) Miller, *Dalmatian toadflax* (Scrophulariaceae). Pages 368–374 in P. G. Mason and J. T. Huber (eds.), *Biological Control Programmes in Canada, 1981–2000*. CABI, Wallingford, UK.

De Clerck-Floate, R. A. and M. Schwarzländer. 2002. *Cynoglossum officinale* (L.), Houndstongue (Boraginaceae). Pages 337–343 in P. G. Mason and J. T. Huber (eds.), *Biological Control Programmes in Canada, 1981–2000*. CABI, Wallingford, UK.

Denoth, M. and J. H. Myers. 2005. Variable success of biological control of *Lythrum salicaria* in British Columbia. *Biological Control* 32: 269–279.

Friederici, P., ed. 2003. *Ecological Restoration of Southwestern Ponderosa Pine Forests*. Arizona Board of Regents, Island Press, Washington, DC.

Gaboury, M. and R. Wong. 1999. *A Framework for Conducting Effectiveness Evaluations of Watershed Restoration Projects*. Watershed Restoration Technical Circular No. 12. Watershed Restoration Program, Ministry of Environment, Lands and Parks and Ministry of Forests, Victoria, BC.

Garry Oak Ecosystems Recovery Team. 2003. *Invasive Species in Garry Oak and Associated Ecosystems in British Columbia*. Garry Oak Ecosystems Recovery Team, Victoria, BC.

Lindgren, C. J., J. Corrigan, and R. A. De Clerck-Floate. 2002. *Lythrum salicaria* L., purple loosestrife (Lythraceae). Pages 383–390 in P. G. Mason and J. T. Huber (eds.), *Biological Control Programmes in Canada, 1981–2000*. CABI, Wallingford, UK.

Louda, S. M., A. E. Arnett, T. A. Rand, and F. L. Russell. 2003. Invasiveness of some biological control insects and adequacy of their ecological risk assessment and regulation. *Conservation Biology* 17: 73–82.

MacDougall, A. 2002. *Invasive Perennial Grasses in* Quercus garryana *Meadows of Southwestern British Columbia: Prospects for Restoration*. Gen. Tech. Rep. PSW-GTR-184. USDA Forest Service, Portland, OR.

MacDougall, A. S. and R. Turkington. 2004. Relative importance of suppression-based and tolerance-based competition in an invaded oak savanna. *Journal of Ecology* 92: 422–434.

McEvoy, P. and E. Coombs. 1999. Biological control of plant invaders: regional patterns, field experiments, and structured population models. *Ecological Applications* 9: 387–401.

McFadyen, R. E. 1998. Biological control of weeds. *Annual Review of Entomology* 43: 369–393.

Murray, C. and C. Pinkham. 2002. *Towards a Decision Support Tool to Address Invasive Species in Garry Oak & Associated Ecosystems in BC*. Prepared by ESSA Technologies Ltd. for the GOERT Invasive Species Steering Committee, Victoria, BC.

Murray, C. and R. K. Jones. 2002. *Decision Support Tool for Invasive Species in Garry Oak Ecosystems*. Prepared by ESSA Technologies Ltd. for the Garry Oak Ecosystems Recovery Team. Available at www.goert.ca/docs/goe_dst.pdf.

Myers, J. H. and D. R. Bazely. 2003. *Ecology and Control of Introduced Plants*. Cambridge University Press, Cambridge, England.

Polster, D. F. and L. M. Landry. 1993. *An Ecological Approach to Vegetation Management*. Unpublished report prepared for CP Rail, Office of the Chief Engineer, Calgary, Alberta.

Tu, M., C. Hurd, and J. M. Randall. 2001. *Weed Control Methods Handbook: Tools and Techniques for Use in Natural Areas*. Available at tncweeds.ucdavis.edu/handbook.html.

Vitousek, P. 1990. Biological invasions and ecosystem processes: towards an integration of population biology and ecosystem studies. *Oikos* 57: 7–13.

Whitt, M. B., H. H. Prince, and R. R. Cox Jr. 1999. Avian use of purple loosestrife dominated habitat relative to other vegetation types in a Lake Huron wetland complex. *Wilson Bulletin* 111: 105–114.

Zouhar, K. 2003a. *Bromus tectorum*. In *Fire Effects Information System*. U.S. Department of Agriculture, Forest Service, Rocky Mountain Research Station, Fire Sciences Laboratory (Producer). Available from www.fs.fed.us/database/feis/ (accessed January 12, 2004).

Zouhar, K. 2003b. *Tamarix* spp. In *Fire Effects Information System*. U.S. Department of Agriculture, Forest Service, Rocky Mountain Research Station, Fire Sciences Laboratory (Producer). Available from www.fs.fed.us/database/feis/ (accessed May 2, 2004).

Chapter 17

Traditional Ecological Knowledge and Restoration Practice

René Senos, Frank K. Lake, Nancy Turner, and Dennis Martinez

Ecological restoration is a process, a directed action aimed at repairing damage to ecocultural systems for which humans are responsible. Environmental degradation has impaired the functioning of both ecological and cultural systems and disrupted traditional practices that maintained these systems over several millennia. Indigenous and local peoples who depend on the integrity and productivity of their immediate environments more than the global, urbanized society are directly affected by ecosystem damage. Conversely, ecosystems have become further diminished in the absence of the cultural practices that once sustained them. Despite this clear connection between cultural and ecological integrity, however, the knowledge and interests of indigenous peoples typically are not considered in attempts to restore degraded ecosystems.

We propose that the incorporation of traditional ecological knowledge (TEK) and practices of indigenous people into contemporary restoration projects will greatly enhance the success of restoration efforts. Successful restoration in this view means not only the capacity of TEK-based restoration to enhance ecosystem functioning but also the ability to sustain indigenous or local peoples' economies and cultural practices. We explore the role of traditional ecological knowledge in restoration theory and practice and discuss key topics such as methods, reference systems, cultural values, and management practices. Differences and correlations between traditional and Western science practices and perspectives are considered as we advocate an integrated approach to ecological restoration.

TEK-based restoration projects encompass various processes and management strategies such as prescribed fire and enhancement of native species and span a wide range of systems including fisheries, riverine and estuarine environments, forest and savanna ecosystems, and wildlife and native plant species (Figure 17.1). We present a variety of case studies that demonstrate the application of TEK to restoration projects in the Pacific Northwest. These examples show how TEK has been successful not only in restoring ecosystems, habitats, or species but also in fostering the interrelationships of people and place. Finally, we suggest future directions and potential expansion of the role of traditional ecological knowledge in Pacific Northwest restoration.

A Restoration World View That Incorporates Traditional Ecological Knowledge

What principles, knowledge, and practices should modern restoration programs and projects use, whether they are implemented by government agencies, nongovernment organizations, or First Nations and American Indian tribes across the

FIGURE 17.1. Hoopa Tribal Forestry staff burning beargrass for basket material enhancement. (*Courtesy of Hoopa Tribal Forestry*)

Pacific Northwest? Should best available science be the default guiding knowledge? What is the role of indigenous knowledge and practices in restoring Pacific Northwest ecosystems?

TEK may be defined as "a cumulative body of knowledge, practice, and belief, evolving by adaptive processes and handed down through generations by cultural transmission, about the relationship of living beings (including humans) with one another and with the environment. . . . [TEK] is both cumulative and dynamic, building on experience and adapting to changes" (Berkes 1999:8; see also Johnson 1992).

Indigenous peoples around the world have used knowledge of their local environment to sustain themselves, maintain their cultural identity, and deliberately manage a wide range of resources and ecosystems for thousands of years. The Western scientific community has only recently recognized this accumulated knowledge as a valuable source of ecological information for understanding and managing various ecological processes, individual animal and plant species, and entire habitats and landscape systems. Restoration philosophy and practice have similarly evolved in the last two decades by incorporating cultural dimensions into restoration definitions, methods, and applications.

In the Pacific Northwest, indigenous ecologist Dennis Martinez, Eric Higgs, and other colleagues led the charge for creating a restoration world view that included aboriginal peoples' knowledge and practices in ecological restoration. In 1995, at the Society for Ecological Restoration's (SER) meetings in Seattle, Washington, Martinez organized a symposium on "Indigenous Peoples: Knowledge and Restoration," in which several practical and theoretical aspects of this issue were discussed. Although the concepts of TEK and traditional land and resource management (TLRM) had received much attention before this time (cf. Anonymous 1992, Berkes 1993, Blackburn and Anderson 1993, Freeman 1979, Freeman and Carbyn 1988, Inglis 1993, Johnson 1992, Williams and Baines 1993), the SER symposium was one of the first venues that explicitly recognized the desirability—and in some cases the necessity—of TEK input in restoration work. One important outcome of the SER Seattle meeting was the formation of the Indigenous Peoples' Restoration Network (for more information about the IPRN, see www.ser.org/iprn). Largely because of this essential groundwork and the thoughtful reconsideration of SER's definition of ecological restoration (Higgs 1997, 2003, McDonald 2003), cultural and historical dimensions are now explicitly ingrained in the general framework of restoration theory and practice.

TEK restoration practitioners have referred to the deliberate incorporation of cultural aspects into ecological restoration as ecocultural restoration. In his book *Nature by Design*, Higgs (2003) explores the cultural, ethical, and philosophical shifts that have promoted good ecological restoration. Reestablishing human relationships with the land being restored is critical for future environmental sustainability. Higgs argues that for restoration endeavors to be truly successful, they must engage people at the community level. Technologically or scientifically based restoration activities alone are not enough, and he refers to projects that bring community participants together as focal restoration. Higgs maintains that focal practices will, in the end, restore more than the environment; they will restore human relationships with that environment and thus will contribute to long-term human health and well-being.

By including traditional knowledge and practices, restoration can be a powerful vehicle for reprising or sustaining cultural vitality. As aboriginal land bases have diminished, a corresponding decrease in indigenous languages, cultural practices, and rituals has transpired. TEK-based restoration provides a mechanism for capturing and applying traditional knowledge that is often held by community elders and transmitted through oral tradition. "For First Nations/American Indians, incorporating traditional ecological knowledge can mean the maintenance and preservation of important aspects of their heritage and natural resources. Partnerships between indigenous people and restoration ecologists focusing on the reintroduction of traditional management practices and techniques could provide avenues for tribal elders and communities to pass on knowledge that is not readily available in published literature" (Ruppert 2003:4).

Barriers to Integrating TEK with Contemporary Restoration Efforts

TEK is a holistic integrative approach that incorporates the metaphysical with the biophysical. The spiritual aspects of TEK have been an awkward component of restoration that is not readily accessible to most nonindigenous people. For many indigenous people, the implementation of TEK in restoration treatments is as much a spiritual reconciliation and restitution as it is a biological and physical restoration of ecological processes, habitats, or species. The genesis of TEK is rooted in creation accounts and natural laws entrusted to human beings from the spirit beings or Creator on how to respectfully and sustainably use natural resources. This same philosophy and practice carries over into restoration ecology, where it is recognized that human beings were and continue to be a part of the natural world. In the view of many indigenous people, restoration involves a continued intimate relationship with "Nature as a hardware store, pharmacy, supermarket, and church" (F. Lake, personal communication, 2005).

Relying on best available science and adaptive management is the prevailing norm among restoration ecologists in the Pacific Northwest. However, most professionally trained and educated practitioners, as well as grassroots community restorationists, have little first-hand experience or understanding of TEK. Restoration managers' lack of familiarity and exposure to TEK in practice and academic curricula often leads to its unintentional exclusion from restoration projects (Kimmerer 2002). TEK is generally viewed as "folk knowledge" lacking the substantive quality of Western science. In many regions the reservoir of TEK resides in the collective practices and minds of indigenous people and is less documented in published or Web-based data facilities. Information relating to TEK is published more often in social science publications than in natural science journals, thus limiting exposure of science professionals. As the study and practice of TEK becomes more visible in science-based publications, restorationists may more readily perceive its application to their own projects.

Even among First Nations and American Indian tribes, where it would be assumed that TEK would readily be integrated into restoration efforts, land managers and resource specialists often overlook TEK as a valid source of information. Many restoration projects on tribal lands apply standard scientific reasoning and gauge project success based on biological measures alone, often overlooking sociocultural indicators. For example, restoration projects may not take into account the value of nontimber forest products to local community health, welfare, and economy. The reasons stem from daunting challenges: dominating scientific methods, lack of formal TEK curriculum or applied experience, and TEK's oral tradition format. Many natural resource staff people working in tribal governments are trained and educated in Western universities. Unless they are tribal members actively immersed in the group's cultural traditions and practices, it cannot be assumed that they will adopt a TEK perspective in restoration projects.

This disconnect is further compounded when cultural data and program objectives are managed by one tribal agency or group, such as the tribal preservation office or elder cultural committees, while natural resource data and program objectives are managed by a different resource group, such as the tribal land program or natural resource department. At best, there is a good flow of information and compatibility of program objectives; at worst, the programs may work at cross purposes to one another.

Another observed reason for the lack of incorporation of TEK into modern restoration efforts is its specific, place-based nature in comparison to more widely applied methods of restoration ecology. TEK is derived from careful observations and adaptive interactions with a particular place over a very long period of time (Berkes 1999, Berkes et al. 2000). It does not lend itself to simplified one-size-fits-all or short-term restoration strategies. Most Western science restoration planning approaches draw on coarser scales of data with a few specific points and lack localized understandings of potential social or ecological barriers to implementation that reflect local conditions.

Many restoration projects are funded and implemented by government agencies or firms that typically work in the absence of local community or tribal involvement. Agencies or consultants may be unaware of TEK's usefulness and application and therefore overlook it at every phase, from planning to implementation. When these groups are required to consult with First Nations or American Indian tribes, this activity often is limited to a presentation of proposed actions in which little adjustment can be made to incorporate TEK after initial planning. In comparison, even when tribes have the opportunity to include TEK in the planning process, it may not find its way into restoration proposals because of prevailing trends in restoration prioritization and funding. In some cases First Nations and American Indian Tribes themselves are reluctant to incorporate TEK into their restoration proposals, fearing that the proposal, in comparison to "strict scientific" proposals, will be rejected by funding sources in a competitive funding allocation arena.

Planning processes for restoration treatments often are short, focusing on a perceived urgent call to action to save a species or a particular degraded habitat. It is rare that sufficient funding and qualified staff are allocated to restoration planning. The collection, analysis, and critique of appropriate application of TEK must take place in the planning phase of a restoration program or project. Yet many restoration ecologists and community-based practitioners who are guided by the Western science paradigm often move forward without consulting local indigenous people for their perspectives and data, even when these would be readily shared.

At the planning stage, TEK can identify issues such as particular species–habitat relationships, human–nature disturbance regimes, or long-term ecological cycles that may have been missed by purely Western science planning approaches (Berkes et al. 2000). Unfortunately, many agencies and organizations may not see the utility of adequately funding oral history interviews and archival research to collect and organize traditional knowledge related to a given restoration topic. Even agencies that do support documentation of ecocultural knowledge may miss the linkages or key points stressed by indigenous elders or subsistence practitioners because their methods are too compressed or because they do not understand what is being conveyed.

For example, a restoration focus placed strictly on a fish species and on instream flows may discard the points mentioned by tribal elders about historical practices that included burning vegetation in part to increase water yield. Yet the elders recall the importance of linking historical season and conditions of burning to decreased evapotranspiration, which increases stream flow during critical migration phases of key fish species. This critical connection is further illustrated in the prescribed burning and fish monitoring case studies described later in this chapter.

Finally, not all restoration projects are necessarily well suited to incorporating TEK. Attempting to draw on traditional knowledge and practice may be less effective in highly engineered projects where restoration practitioners are exclusively technical specialists whose experience is limited to megascale schemes involving large machinery and few individuals, where the goal is to complete the project and go home. TEK is better suited to projects that draw together social and ecological communities through long-term adaptive management strategies, where a long-term intergenerational community approach is used restore a degraded ecosystem or habitat or conserve threatened or endangered organisms (Berkes et al. 2000, Turner 2005).

Key Concepts in TEK Restoration

Kincentricity

A key concept in the indigenous world view is *kincentricity*, or a view of humans and nature as part of an extended ecological family that shares ancestry and origins (Salmón 2000, Martinez 1995). The kin or relatives include all the natural elements of an ecosystem; indigenous people sometimes refer to this interconnectedness as "all my relations." Kincentricity acknowledges that a healthy environment is achievable only when humans regard life around them as kin.

Kinship with plants and animals entails familial responsibilities; it tells us why we are on this earth and what our ecological role or niche is vis-à-vis our relatives in the natural world. To put it another way, it tells us that we are a legitimate part of nature, that we have responsibilities within nature, and that in exercising those responsibilities we are as "ecological" or "natural" as any other species. The indigenous land ethic holds that we can have a positive restoration effect in the very act of using natural resources. Kincentric ecology entails direct interaction with nature to promote enhanced ecosystem and cultural functioning. This is what sustainable practices are all about.

Our responsibility for participating in the "re-creation of the world" (as tribes on Klamath River in northwestern California call it) is never finished. Periodic intervention by humans in nature has long been part of ecosystem dynamics in the Pacific Northwest. Given the myriad ecological catastrophes we now face, the need for active restoration will only increase. There are no finished restoration projects. Nature has self-healing powers, but these engage only after a specific harmful disturbance (e.g., a dam or invasive species) is removed or modified. Humans also can lend a hand by restoring missing species, modifying structure, and so forth. As the SERI *Primer* notes, continuing management is necessary to "guarantee the continued well-being of the restored ecosystem thereafter" (Society for Ecological Restoration 2004:6).

Pioneering Western ecologists and restorationists have come to similar conclusions with respect to kincentricity, starting with Aldo Leopold (1949) and his land community ethic, which posited that an individual is a member of a community of interdependent parts and that each citizen is ethically bound to maintain cooperative relations with the biotic community. Restorationist Stephen Packard discovered that recovering degraded Midwest landscapes required corps of dedicated volunteers to restore several thousand acres of tallgrass prairie and oak savanna (Stevens 1995). This ambitious endeavor was not possible until people invested in their home places. Environmental philosopher Andrew Light (2005) has explored the personal, moral, and environmental dimensions of making amends to our kin through the act of restoration and considers how restoration provides a venue for ecological citizenship. Kincentricity provides a basis for considering restoration as a process of engagement with nature, a way to sustain or repair relations with the living world. In doing so we develop viable cultural, economic, and ecological practices that support and nurture our shared environment.

Reference Systems

When addressing the needs of restoration today, whether at the landscape, habitat, or species level, it is important to recognize indigenous peoples as an influence in shaping and maintaining the historical condition of many different ecosystems. In this sense, the effect of past indigenous management practices should be considered part of the reference ecosystem, or more generally as providing a set of reference processes to guide a restoration effort. In the Northwest, reference conditions influenced by indigenous land use practices of a pre-European era are the benchmark.

Any reference condition or design of future desired conditions must account for humans' use and management of the environment. The scale and intensity of human use and management of the environment are important to successful ecocultural restoration and to establishing a sustainable relationship to place. Our reference window is at least as large as 10,000 years, or the postglacial Holocene period, with particular attention paid to the last 4,000 years, during which a gradual cooling trend occurred that most resembles present climatic and ecological conditions (Figure 17.2). The last 10,000 years also is the period during which humans have exerted the most influence on North American ecosystems (Egan and Howell 2005).

The use of reference ecosystems in restoration is not without controversy. It can be expensive and time-consuming to use multiple disciplines and indirect proxy lines of ethnographic and scientific evidence to establish a reasonable probability of accuracy in identifying a site at a specific point in history (see Egan and Howell 2005 for technical information regarding reconstructing historical ecosystems).

Some restoration scientists and practitioners question the value of using historical baselines to

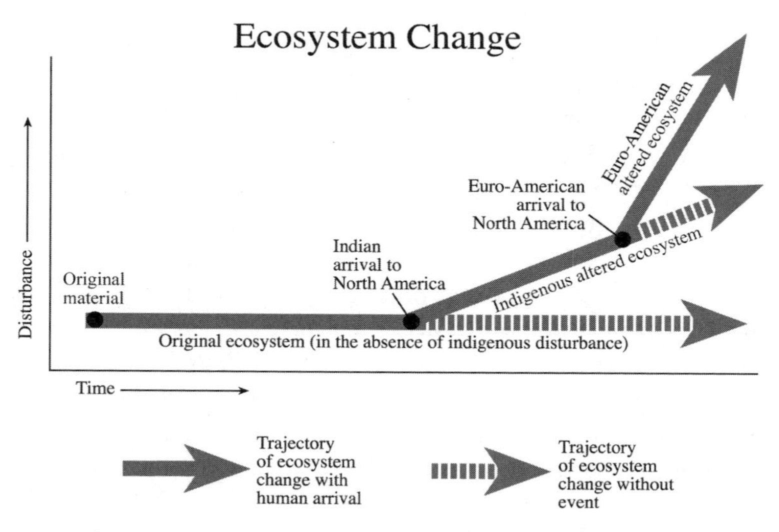

FIGURE 17.2. Ecosystem trajectory showing that humans represent an ecological force on the landscape. (*From Lewis and Anderson 2002. Copyright © 2002 by the University of Oklahoma Press, Norman. Reprinted with permission. All rights reserved.*)

guide restoration. Not only can it be difficult to reconstruct historical ecosystems because of severe changes in some environments, but why go back to an arbitrary point in history? Why choose, for example, a time before European contact as the reference? Why not just try to improve the function of degraded ecosystems? Isn't nature constantly changing?

However, TEK does not advocate that we stop change and freeze ecosystems in a particular timeframe or that we recreate a snapshot in time. After all, present conditions are also a snapshot in time. What we really need to do is connect the past with the present in order to reveal what kind of trajectory an ecosystem may be on and then nudge that trajectory just enough to restore key functions. History and function, then, are inseparable. Both Higgs and Martinez have written and spoken about balancing historical fidelity or authenticity with ecological functionality. Instead of fixing a snapshot in time, we need to rerun a moving picture, played out within boundaries determined by historical trends in disturbance regimes, including the kinds, intensities, and frequencies of disturbance with which an ecosystem has evolved.

For example, forest stand-level restoration projects are subject to constraints imposed by the greater landscape scale. Although a reference ecosystem can guide initial restoration efforts, these will be modified by larger landscape considerations (e.g., fragmentation, fire hazard, exotic species invasion, species losses), or even larger phenomena such as climate change. Anchoring the reference model in real historical time will help us to recover key features of ecosystem structure, composition, and processes within natural variability, with a look to designing the future desired condition.

Building sound reference models for ecocultural restoration requires the best Western science and the best of TEK, not one or the other. Each has the potential to reinforce the other and to compensate for inherent methodological limitations by considering history and function, quality and quantity, long term and short term, culture and ecology, economy and environment.

Successional Theory and Disturbance

Traditional ecological knowledge complements contemporary knowledge of fire ecology by providing information about historical and contemporary applications of fire by indigenous people, including fire effects on wildlife and vegetation in different environments. Indigenous knowledge of fire ecology includes but is not limited to variations in fire frequency, intensity, severity, and specificity of areas burned in different ecosystems or plant communities by indigenous people or by lightning ignitions. TEK provides knowledge about fire effects and ecosystem responses and about how physical and biological processes such as hydrology and forest succession respond to fire over time (Lewis and Anderson 2002).

Integrating multiple knowledge systems to understand the effects of fire on the remaining post-treatment vegetation or soils can lead to greater accomplishment of objectives. Thinning and spring season pile burning may be intermediate steps that help prepare the site for the reintroduction of fall season low-intensity burns that emulate Indian fire (Williams 2000). Ethnographic information and TEK may be instrumental at each treatment step, especially when one is considering restoration effects on wildlife, food plants, or non-timber forest products, resources that hold high social and ecological value to local communities (Anderson 2001).

Restoring and maintaining biocultural diversity of the landscape through integrated restoration planning involves an interdisciplinary as well as a multicultural approach. Fuel reduction projects that incorporate Indian fire will have higher levels of success in restoring and maintaining biodiversity, which in turn will support cultural diversity and local communities. This premise may hold true especially with native cultures that historically and currently rely on fire and fire-dependent landscapes for their sustenance and cultural survival (Boyd 1999). A community forestry approach can help local communities cope with likely future changes caused by climate change and intensified demands of natural resources.

Defining Scale

Issues of ecological and social scale are important considerations in restoration work. Any restoration program directed toward a given geographic area must carefully define the scale at which it will operate. For example, will projects focus on a single species or population, a particular habitat, or an entire watershed? Will restoration engage the collaboration of an individual, a community, or a national organization or institution?

Indigenous ways of understanding and relating to the environment provide useful models for framing restoration efforts at the appropriate scale. In coastal Pacific Northwest environs, individual families traditionally were responsible for a particular resource base at a specific location (e.g., shellfish beds). Villages were organized around specific places along stream reaches, and an affiliated tribal group (distinguished by common linguistics) managed a given bioregion (J. James, personal communication, 2001). Appropriate technologies and resource management practices were ritualized to maintain healthy functioning of the system at all social and ecological scales.

Defining the scale of operation provides context for our individual actions linking with others' actions across or up in scale. TEK provides an operational framework that addresses the integration of the various ecological and social scales, situated within temporal scales (Figure 17.3; Berkes et al. 2000). The perspective of scale can also be reflective in that the strengths and weaknesses of TEK and Western science are evaluated in the context of the restoration program or projects being planned, implemented, or monitored.

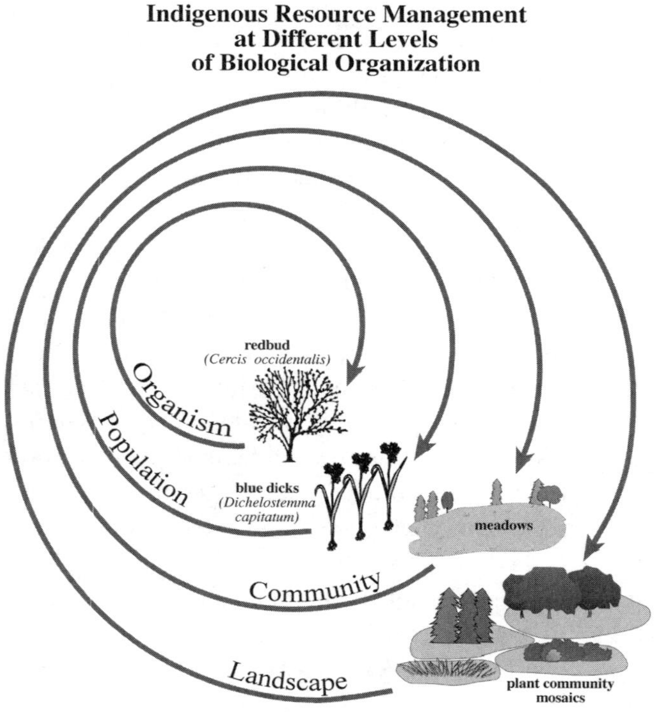

FIGURE 17.3. Scale of potential management effects. (*From Lewis and Anderson 2002. Copyright © 2002 by the University of Oklahoma Press, Norman. Reprinted with permission. All rights reserved.*)

Pacific Northwest TEK Restoration Projects

Cultural Fire Regimes and Prescribed Fire

Every ecosystem in the Pacific Northwest and throughout North America has been modified in some way by a fire regime implemented by indigenous people (Boyd 1999). Forest science, including vegetation classification, evolved from Gifford Pinchot–era observations of forests transitioning in condition from indigenous fire management to postsettlement fire suppression. Scientific understanding of forest processes therefore may be based largely on an atypical, transitional landscape (Kimmerer and Lake 2001).

Cultural fire regimes specific to certain ecosystems and plant communities (e.g., oak savanna or subalpine parkland) were intentionally created and maintained primarily by indigenous people. These indigenous fire regimes may or may not have occurred in conjunction with natural wildland fires ignited by lightning, volcanic eruptions, or other nonhuman causes (Lewis and Anderson 2002). Cultural fire regimes historically affected the composition, structure, function, and productivity of particular habitats, especially the culturally defined resources therein. The distinguishing features of cultural fire regimes include alternate seasons for burning under different kinds of settings; frequencies with which fires are set and reset over varying periods of time; corresponding intensities with which fuels can be burned; specific selection of sites fired and, alternatively, those that are not; and a range of natural and artificial controls that humans use in limiting the spread of human-set fires, such as time of day, winds, fuels, slope, relative humidity, location of streams and snowbanks, and natural and human-created firebreaks (Lewis 1982, in Bonnicksen et al. 1999).

FIGURE 17.4. Forest types created by different forest regimes: *(top)* a lightning fire regime; *(middle)* a cultural fire regime, or prescribed burning; *(bottom)* a fire suppression regime. Note that a fire regime that incorporates both cultural and natural fires results in increased diversity of understory species, decreased litter, and less dense tree canopy. (*Concept diagram by M. Kat Anderson and Michael Barbour. Courtesy of the University of Wisconsin Press*)

How have habitats been altered as a consequence of fire suppression and the cessation of indigenous land use practices? Consider the effects of fire suppression and the resulting changes in the composition, structure, function, and productivity of habitats that many indigenous people rely on for food, medicine, materials, and spiritual and cultural survival. The results of fire suppression and other contemporary forest management practices are at the heart of our national "healthy forests" debate today (Figure 17.4). In many areas of Pacific Northwest, First Nations and Native Americans historically and currently rely on fire-induced conditions of the environment through subsistence activities to support their cultures and livelihoods. Conversely, Western forests are deprived of the sustainable human interventions that promoted their former health and abundance and the expressions found in systems ranging from old-growth Douglas fir to oak savanna.

Do restoration prescriptions address the broader ecological role of plants for social, cultural, or spiritual uses? Land managers and the public should not view vegetation as merely fuel. The ecological and cultural significance of a plant varies but generally should be considered in light of its role as food, medicine, material, or habitat in relationship to the danger it may pose as fuel in its present setting. Many fuel reduction prescriptions do not account for the differences in the ecological or cultural function of specific plants. For example, many understory plant species may be considered dangerous ladder fuels, yet their complete removal may be ecologically or culturally undesirable. How a plant, shrub, or tree responds to fire, and the ecological and cultural services it provides, should be considered at a broad level with project prescriptions and at a specific level with the project implementation at a given site.

In general, fall burning was more widely practiced by indigenous people than spring burning. Fall burning was conducted because the response of culturally important plants and wildlife habitat vegetation was more in line with cultural needs. Fall burning favors nonsprouting brush species that are preferred by wildlife, and seeds of these species often need fire for germination. Fall fire also induces new growth on both sprouters and nonsprouters in early spring, a time when stored food supplies for people typically were running low (Lewis 1993, Boyd 1999). Indigenous fire regimes and selective harvesting probably modified the genetic structure and species dispersal of both cultural and noncultural plants and animals (Bonnicksen et al. 1999).

In many fuel reduction projects, intermediate treatment steps are taken to reduce the risk, intensity, and severity of fire on physical, biological, and social processes or conditions. For example, as a first treatment step mechanical thinning may reduce the connectivity and bulk density of fuels in areas near human settlements (USDA Forest Service 2001). Thinning may be followed by pile or broadcast burning in the winter or spring. Even if the historical seasonality of Indian burning was in the fall, the intensity of a fall burn in current forest conditions can be greater and more damaging to vegetation and soils. Thus a spring burn, after initial thinning, might be prescribed to reduce these risks. Later, when conditions are less conducive to a high-severity fire, the original pattern of fall burning may be resumed.

Case Study: The Karuk Tribe of Northern California and Local Fire Safe Councils—Fuel Reduction Projects, Wildland Urban Interface, and Fuel Breaks

> Lack of knowledgeable traditional stewardship has created landscape conditions that have ignored and devastated traditional resources and now threatens the well-being of both the forests and the forest based communities. The management of traditional resources through implementation of specific forest management practices requires addressing an intricate complex of political, cultural and technical issues. Building the Tribe's capacity to play an appropriate role in ecosystem management is the only means by which ecosystem restoration, cultural survival and community prosperity will be achieved. (Karuk Tribe, Department of Natural Resources Web page: karuk.us/)

In northwestern California, the Happy Camp and Orleans/Somes Bar Fire Safe Councils are working with the Karuk Tribe, the U.S. Forest Service, local native basket weavers, and private landowners to implement fuel reduction projects that achieve the objectives of reducing fuel loads while considering the importance of culturally significant native plants, wildlife habitat, and nontimber forest products. Vegetation is not viewed simply as fuel. Rather, during the planning and treatment phases, vegetation is judged from the perspective of fire danger as well as ecological and cultural importance. The key to incorporating TEK into the fuel reduction projects is to work closely with local communities. Collaboration between tribal elders, practitioners, and Western scientists is essential to restoring fire-adapted ecosystems from which many culturally valuable resources are used (USDA Forest Service 2001).

Information is collected on multiple perspectives held by local tribal members, landowners, and scientists on the functional role of trees, shrubs, forbs, and grasses. The various perspectives are incorporated in the prescription treatment. Local native basket weavers and tribal members may prefer that a combination of plant species be retained or removed and may favor spring and fall burning. Each project treatment is evaluated separately, given the social and ecological issues present at that site.

Acknowledging local tribal knowledge with respect to the importance of plants for wildlife habitat, cultural uses, and anticipated fire response is vital at all phases of the project. Because native cultural burning was a historically important agent in influencing the composition, structure, function, and productivity of low- to mid-elevation areas along the mid–lower Klamath River corridor (Pullen 1996, Lewis 1993), many members of the local community, both Indian and non-Indian, think it is appropriate to reinstate similar cultural burning practices as part of fuel reduction treatments (Orleans/Somes Bar Fire Safe Council 2005).

The most common integrated project elements include reducing hazardous fuels, restoring California black oak (*Quercus kelloggii*) and California hazel (*Corylus cornuta*) habitats, supporting basketry management practices, and improving wildlife habitat. Black oak- and hazel-dominated habitats were historically more prevalent before government fire suppression policies. Black oak–hazel habitats are ecologically and culturally significant in low- to mid-elevation communities (see Chapter 4). Yet shrub and conifer encroachment has threatened the black oak–hazel habitat quality and productivity. Current fuel reduction projects target removal of 2- to 50-year-old Douglas fir (*Pseudotsuga menziesii*), reduction of understory shrub density and crown area, and opening of forest floor.

After thinning operations and spring pile and broadcast burning, selected hazel clumps are spot or patched burned to promote shoot growth for basket materials (Anderson 1999). Native shrubs, forbs, and grasses that are important for wildlife habitat or forage and culturally useful for materials, foods, or medicines are rejuvenated after thinning and burning activities. Other habitats where TEK is integrated with fuel reduction and prescribed burning include sandbar willow (*Salix exigua*) riparian patches, pine (*Pinus ponderosa* and *P. lambertiana*) habitats, Oregon white oak (*Quercus garryana*) habitats, and higher-elevation meadows. Similar restoration projects are taking place on the Hoopa Valley Indian and Yurok Reservations in northwestern California.

In 1995 the Karuk Tribe presented the "Karuk Tribal Module for the Main Stem Salmon River Watershed Analysis" to the Klamath National Forest for inclusion in the agency's Final Forest Plan, as required by the Northwest Forest Plan. Ecosystem management was to be the guiding force behind forest management. Nearly all of the tribe's ceded ancestral lands were controlled by the Klamath, Six Rivers, and Siskiyou national forests. Consequently, the Karuk took the position that comanagement between the Klamath Forest and the tribe was imperative for two reasons: ecosystem management could not be accomplished without input from tribal members who used forest resources and therefore knew their turf better than the Forest Service, and the tribe could not stake its cultural and economic future on designated cultural resource areas that were too small to remain viable given continuing drought-driven catastrophic wildfires and other ecological degradation.

Although legally recognized comanagement has not yet been implemented, the Karuk work closely with the Forest Service on restoration projects. The key point is that for tribes to realize true sovereignty, they must have access to cultural resources on ceded ancestral lands, equity in treatment by the Forest Service (defined as comanagement), and the capacity to use traditional forest management practices such as prescribed fire. The Karuk are continuing their efforts to restore the fisheries and watersheds of the middle and lower Klamath River drainage in collaboration with the Klamath Restoration Council (formed in 2004), with participation from the Yurok and Hoopa tribes, environmental organizations, scientific and cultural consultants, the Indigenous Peoples Restoration Network, and local community members.

Case Study: Lomakatsi Restoration Project, Southern Oregon

The nonprofit Lomakatsi Restoration Project in southern Oregon is actively restoring sugar and ponderosa pine, black oak, and Oregon white oak habitats on public and private lands. Harvesting of mature conifers and fire suppression activities have allowed the expansion and encroachment of manzanita (*Arctostaphylos* spp.), ceanothus (*Ceanothus* spp.), and Douglas fir into former pine- and oak-dominated habitats. Manzanita, ceanothus, and Douglas fir encroachment has increased fuel loads and potential fire hazard. Former forb and grass openings associated with pine–oak habitats have been lost, endangering many native plants and degrading wildlife habitat quality.

The Lomakatsi Project has incorporated TEK into restoration prescriptions of pine–oak habitats. Restoration crews have undergone ecocultural workforce training, and native restoration practitioners have been hired to train crews. Fuel reduction projects involve hand equipment or chippers along access roads to remove small-diameter trees and shrubs before selective single or small-group tree harvesting.

Mature dense thickets of manzanita and ceanothus are thinned and pile burned in an effort to release pine and oaks and to restore openings (Figure 17.5). Retention patches of mature manzanita and ceanothus are left in clumps distributed across the project site in order to maintain overall diversity of species and habitats. Douglas firs, commonly 2–50 years old, are selectively thinned to favor sugar and ponderosa pines and mature and seedling black and white oaks across project sites. Native grasses and forbs locally harvested are also dispersed in burn piles and across the site. Native trees and shrubs reared at local nurseries are planted in project sites to restore diversity, and large conifer snags and downed wood are retained for wildlife habitat and soil formation.

The overall approach to restoration forestry thinning prescriptions by the Lomakatsi Project can be described as variable density or retention management (see Chapter 5). The intent is to restore enough variability in stand structure and composition to provide sufficient redundancy or repetition for suitable wildlife habitat where we lack knowledge of the lifeways of many species—a kind of risk spreading (Lindenmayer and Franklin 2002). Even if pines or oaks are favored over Douglas fir, some Douglas fir are preserved, and all potential native species and age classes are represented in the finished project. Another important feature of the Lomakatsi Project is the restoration of the herbaceous forest understory, because most cultural plants and wildlife habitat occur at the grass, forb, and shrub level, and many of these species are being lost to shading out by invasive conifers. (For more details see www.lomakatsi.org).

FIGURE 17.5. Lomakatski restoration project: thinning and burning manzanita and ceanothus to restore oaks and pines. (*Photo by Frank Lake*)

CASE STUDY: HUCKLEBERRY CROP MANAGEMENT

For First Nations and Native American people of the Pacific Northwest, several species of huckleberry (*Vaccinium* spp., especially *V. membranceum*, black mountain huckleberry) have always been important cultural foods (Mack and McClure 2002, Deur 2002). Historically, huckleberry patches were managed with fire and pruning to maintain and increase berry production. Cessation of Indian burning practices combined with suppression of natural ignitions has caused encroachment in many traditional huckleberry-gathering areas by conifers and other shrub species (Figure 17.6). Competition for light, nutrients, and water has reduced the abundance of berries and the former extent of berry patches (Minore 1972).

In a cooperative project between the U.S. Forest Service, the Mt. Hood National Forest, Oregon State University, and the Confederated Tribes of the Warm Springs, traditional ecological knowledge combined with Western forestry practices were used to examine the effects of different silvicultural treatments on the productivity of huckleberries (Communities Committee of the Seventh American Forest Congress 2000, Anzinger 2003). Silviculturists working with tribal members and elders are learning that huckleberry patches cannot be managed in isolation but are part of a complex, dynamic landscape mosaic of diverse and shifting successional patches. Elsewhere across the Pacific Northwest, other tribal and First Nations groups are incorporating traditional huckleberry management practices with forest restoration to increase the extent and production of huckleberries, a food source important for wildlife and human consumption.

FIGURE 17.6. Southern Washington Cascades huckleberry meadow historically maintained by indigenous burning is now encroached on by subalpine forest. (*Photo by René Senos*)

Traditional Ecological Knowledge and Wildlife Restoration

In both historic and contemporary cultures, wildlife species have occupied a major role in indigenous diet, clothing, tools, arts, stories, ecology, and cultural and spiritual life. As one Salish elder on the Flathead Indian Reservation in western Montana observed, "Wildlife aren't just animals. They are our medicine, our clothes, and our relatives. Wildlife are sacred" (P. Pierre, personal communication, 2003).

Traditional ecological knowledge offers several key practices and strategies that may be combined with scientific methods to enhance wildlife habitat and populations. For example, species monitoring is a practice shared by both disciplines. Traditional resource users regularly observe and communicate the status of a wildlife species and adjust habitat management and harvesting practices accordingly. Several indigenous practices entail the protection of vulnerable life stages of species or the preservation of key habitats. Temporal or spatial restriction of harvest is another common traditional practice, whereby hunting, fishing, and trapping areas are periodically rested to ensure healthy populations (Berkes 1999, Berkes et al. 2000). These seasonal or periodic restrictions on wildlife harvest typically are codified and ritualized into cultural and spiritual practices.

Resource rotation, or the activity of rotating harvest areas to allow wildlife species time to recover, is widely practiced. Unfortunately, the institution of a private property land system, prohibited access to usual and accustomed traditional resource areas, and habitat loss accelerated by human development all impose serious obstacles to resource managers' ability to rotate harvest areas. Greater pressure is thus placed on limited wildlife pools restricted to fragments of their former range. Thus restoration practice needs to include creative conservation strategies, such as enhancing wildlife corridors across multiple-owner boundaries.

Integrated system management is an important strategy that TEK offers to restoration practice. Rather than targeting a single species in the course of resource management, traditional practitioners use multiple-species management across multiple habitat types. This holistic world view of resource management often is expressed in phenological calendars maintained by many indigenous groups (Figure 17.7). Phenological cues are seasonal markers, such as the flowering of a particular plant species, that tell the observer when to engage in a resource activity, such as harvesting a specific ani-

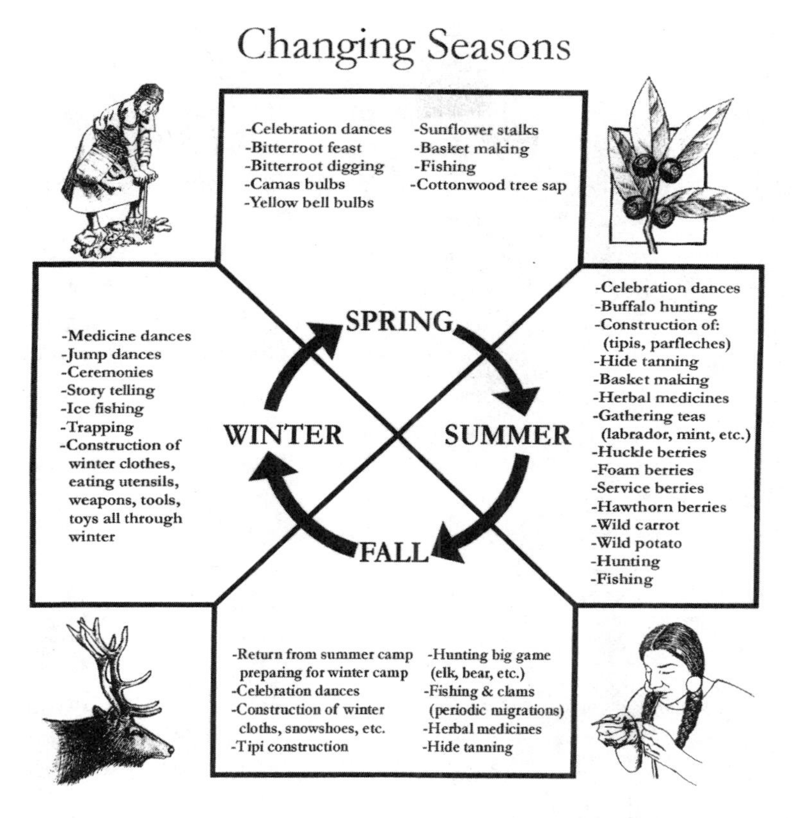

FIGURE 17.7. This Salish calendar illustrates the connection between seasonal markers, environmental practices, and cultural traditions. (*Courtesy of the Salish Cultural Committee*)

mal species. Connections are drawn between seemingly disparate natural events, management practices, and cultural traditions. Furthermore, phenological cues coupled with traditional rituals or ceremonies strengthen cultural cohesiveness and identity (Garibaldi and Turner 2004, Lantz and Turner 2003).

Northwest tribes are engaging in the restoration of many key wildlife species that are essential to their peoples' ecology and culture. For example, First Nations peoples in interior British Columbia are conducting oral histories to determine historical riparian conditions. These oral histories will help them incorporate TEK into cottonwood (*Populus balsamifera* spp. *trichocarpa*) and willow (*Salix* spp.) restoration planning to enhance riparian habitat for a key migratory bird, the yellow chat. The Nez Perce are leading wolf recovery efforts in Idaho and are also working to restore wildlife habitat in their traditional homeland, the Wallowa Valley in northeastern Oregon. An instrumental partnership with the Trust for Public Land enabled the Nez Perce Tribe to acquire a 10,000-acre cattle ranch to manage as a wildlife refuge in order to replace habitat lost with the construction of Columbia River dams (Mahler 2004). The Elakha Alliance, a partnership that includes the Confederated Indian Tribes of Siletz, is working to restore sea otters on the Pacific coast. Traditional ecological knowledge coupled with scientific testing was instrumental in distinguishing the Pacific coast otter as a unique and separate species and providing crucial information for their future recovery (see www.ecotrust.org/nativeprograms/elakha.html).

Case Study: Wildlife Crossing Design on the Flathead Indian Reservation

The Salish, Kootenai, and Pend d'Oreille peoples formerly occupied a vast territory that included northwest Montana and parts of Idaho, Wyoming, and southern British Columbia. They currently reside on a portion of that territory known as the Flathead Indian Reservation, a diverse landscape that contains part of the Continental Divide, the Bitterroot Range, and an extensive system of valleys, wetlands, and rivers. The land is also home to myriad bird, fish, amphibian, and wildlife species that range from painted turtle to grizzly bear. These species and their habitats are carefully managed by various agencies within the Confederated Tribes of the Salish, Kootenai, and Pend d'Oreille government and by cultural groups such as the Salish Cultural Committee and the Kootenai Cultural Committee. The tribes have engaged in restoration and management of several wildlife species and critical ecosystems, including the Jocko River.

U.S. Highway 93 is a north–south corridor that bisects the reservation. When the Montana Department of Transportation proposed widening the predominantly two-lane road to a four-lane highway, the tribes opposed the plan because of the project's potential impact to cultural and ecological values. The existing highway already impeded wildlife migration across the valley floor, disrupted hydrologic flows of dozens of streams and rivers, cut a swath through a half dozen rural communities, and affected several geological and ecological features directly tied to the peoples' origin stories.

Nonetheless, the highway was one of the most dangerous roads in the country, resulting in many fatalities. After a several-year impasse, the tribes, in collaboration with the state and federal highway departments, hired a design team to develop road alignment and design concepts and eventually to design a final project. The design team consisted of landscape architects Jones & Jones, engineers Skillings-Connolly, Inc., tribal resource agencies (both natural resource and tribal preservation agencies), and the Salish and Kootenai cultural committees. Additional assistance was provided by other restoration specialists, including the Western Transportation Institute, wildlife biologist Tony Clevenger (Banff National Park), Bitterroot Restoration, Inc., and other subconsulting landscape architecture and engineering firms.

Several important design concepts evolved out of this collaboration, which was driven by a context-sensitive approach to place and culture. First and foremost was the concept of respecting Spirit of Place and treating the road as a visitor. Out of this simple but powerful idea grew a project that better supported the ecological and social values inherent in the Flathead Indian Reservation. The process of rebuilding the road became one of restoring the lost or impaired functions that the original highway had compromised.

Instead of a four- or five-lane, straight highway in the present alignment, a variable-lane road was laid out in a curvilinear alignment that avoided sensitive resources. Tribal representatives, including both elders and resource managers, indicated the areas to avoid. A comprehensive restoration plan was developed in coordination with the elders, tribal resource managers, landscape architects, and biologists. The plan included restoration of sixteen different native plant communities, several wetland and riparian areas, and, most critically, the construction of nearly fifty wildlife crossings throughout the corridor length (Figure 17.8).

The location and individual design of each wildlife crossing were predicated on tribal wildlife data, traditional ecological knowledge, and independent scientific and design input. Tribal elders

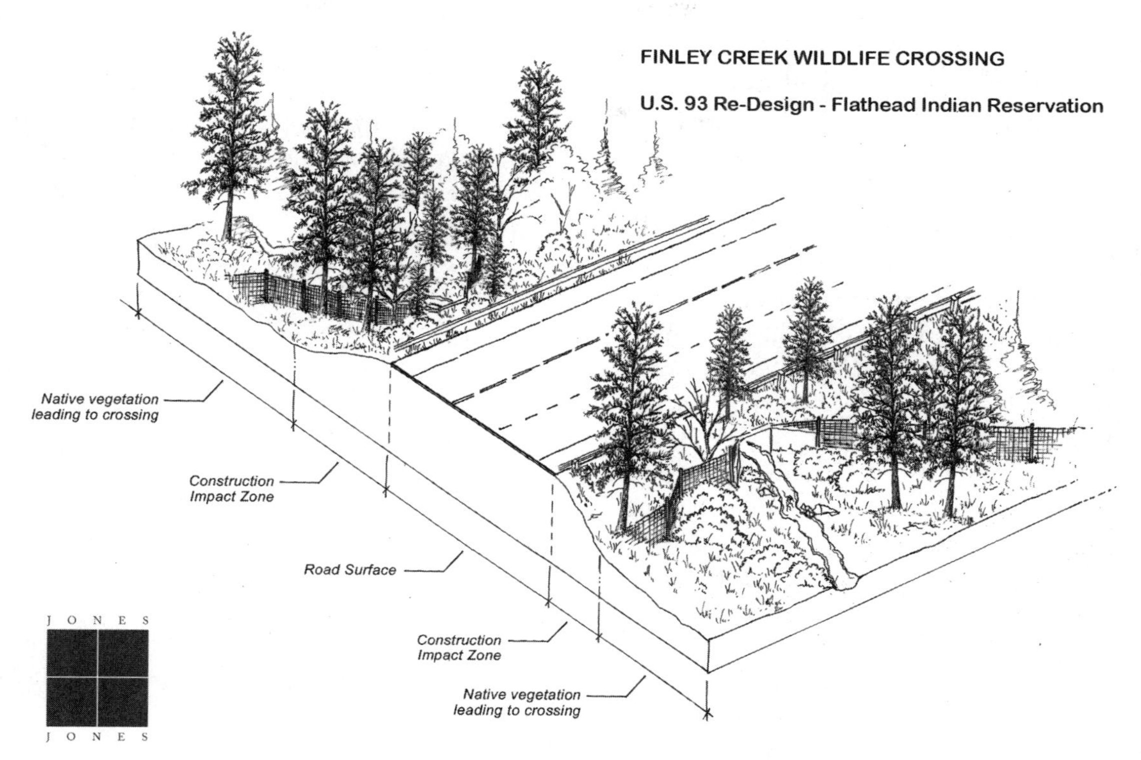

FIGURE 17.8. Conceptual drawing of one of nearly fifty wildlife crossings to be constructed as part of U.S. 93 redesign on the Flathead Indian Reservation. (*Drawing by Jones & Jones Architects and Landscape Architects*)

and the respective cultural committees have actively held and conveyed a very long record of cultural traditions, lifeways, and practices. This cumulative record of a people living in one place for several millennia is transmitted from one generation to the next through oral tradition, stories, language, arts, and specific cultural practices. On the scientific side, tribal wildlife managers have monitored populations and habitat conditions for several years and also maintained wildlife mortality records along U.S. 93. The scientific data of tribal wildlife managers, coupled with the traditional knowledge of the elders and guidance from the Tribal Preservation Office, identified key animal corridors and habitats.

The wildlife crossing structures range from 4 by 6 feet (1.2 by 1.8 meters) and 12 by 22 feet (3.7 by 6.7 meters) box culverts to bridges that span full floodplains and include one wildlife overpass. The structures accommodate a wide range of species, including painted turtles, fish, amphibians, numerous small mammals, deer, elk, coyote, wolf, moose, grizzly, and black bear. Critically, the crossing structures tie into restored plant habitats and key ecosystems. Construction began in the fall of 2004, and when completed the highway will contain more wildlife crossings than any other road in the United States. More importantly, the project design promotes the cultural and ecological values actively practiced by the Salish, Kootenai, and Pend d'Oreille people.

Ecocultural Restoration of Pacific Northwest Fishery Habitat and Populations

Fisheries are a mainstay of Northwest ecology and culture, and tribes throughout the region are fervently engaged in restoring fishery populations and habitats and sustaining the traditional cultures that coevolved with species such as the salmon. Like Pacific forests, fish have been thrown into the hotbed of politics and science. Hatcheries, dams, fish ladders, barges, and other artificial means of support are substituted for healthy, functioning rivers and streams. Still, many salmonid populations are declining to the brink of extinction.

In the midst of this crisis, tribes have organized individually and collectively to restore watersheds, estuaries, marine habitats, and fish and marine animal populations. Major coalitions provide a support network for tribal fishery restoration in the Northwest region: the Klamath River Intertribal Fish and Water Commission in the Klamath Basin; the Columbia River Inter-Tribal Fish Commission, based in the Columbia Basin; the Northwest Indian Fisheries Commission, based in Olympia, Washington; and the Fisheries Centre at the University of British Columbia. These organizations provide a visible forum for promoting indigenous fisheries and exchanging knowledge—both scientific and TEK-based—in restoration and conservation strategies.

The University of British Columbia Fisheries Centre strongly advocates the integration of TEK and science methods: "The Fisheries Centre at the University of British Columbia recognizes the tremendous value and importance of the Aboriginal perspective in fisheries conservation and management. . . . The partners are also working together to develop a framework for the inclusion of traditional ecological knowledge with quantitative fisheries management. This holistic approach will contribute to ecosystem restoration, habitat conservation of precious fisheries resources and pragmatic and feasible plans for sustainable fisheries use" (Endowed Chair 2002:2–3).

There are numerous case studies of Northwest fishery restoration; the ones that follow exemplify a restoration approach that creatively uses both traditional ecological knowledge and Western science, with promising results.

Case Study: Pacific Lamprey Research and Restoration

An oral history interview project conducted by the Confederated Tribes of the Umatilla Indian Reservation's Department of Natural Resources, in collaboration with the Columbia River Inter-Tribal Fish Commission, the Oregon Cooperative Fishery Research Unit, Oregon State University's Department of Fisheries and Wildlife for the U.S. Department of Energy, and Bonneville Power Administration's Division of Fish and Wildlife, has incorporated traditional ecological knowledge in a project to restore Pacific lampreys (eels; Figure 17.9). The restoration approach included a river subbasin method of collecting data on historical and present Pacific lamprey distribution and abundance. Collected indigenous knowledge will be used in an adaptive management process to make changes to factors inhibiting the passage and propagation of lampreys (Close et al. 2004).

In the central Coast Range of Oregon on the Siletz River, oral histories were conducted of local tribal members about factors leading to the decline of Pacific lampreys and about the extent and quality of spawning and rearing habitat. The Oregon Department of Fish and Wildlife is using this information in the management and restoration of habitat and populations of Pacific lamprey. Traditional knowledge was incorporated "by using the database queries to understand relationships between land

FIGURE 17.9. Elders and eels. Karuk elders Alme Allen and Eugene Coleman with lampreys harvested from a basket trap. (*Photo by Frank Lake*)

use and habitat or ecological parameters" in which "resource managers can focus their restoration techniques in the most cost-effective manner" (Chapin et al. 1998).

In northwestern California the Karuk and Yurok tribes have conducted oral history interviews with tribal elders and community members on the past and present distribution and abundance of Pacific lamprey and perspectives on the factors contributing to Pacific lamprey decline. The Yurok tribal fishery program completed a preliminary report on Pacific lamprey (Larson and Belchick 1998). Pacific lamprey monitoring and inventories by the Karuk and Yurok tribal fishery programs involve river habitat substrate sampling, outmigrant traps, and spawning surveys. Currently, traditional knowledge of tribal Pacific lamprey harvesters in the Klamath River and its tributaries about the historical abundance, food processing techniques, management practices, and habitat needs of Pacific lampreys is being gathered from local tribes. These data will be used to formulate restoration and conservation strategies (R. Peterson, personal communication, 2005).

Case Study: Salmon Restoration in the Pacific Northwest

Numerous federal and crown, state and provincial, American Indian and First Nations, and local community groups are actively working on salmon restoration in watersheds throughout the Northwest. Most approaches to salmon restoration are guided at two levels: Western scientific disciplines focusing at larger scales, such as watersheds, hydrologic units, and evolutionary significant units (ESUs), and local community-based practitioners at the river and stream system or reach level. Rarely, even with First Nations and American Indian fishery programs, has TEK been actively incorporated in restoration policies, planning, and project implementation.

Some of the best examples of salmon restoration not only integrate TEK and Western science methods but also work collaboratively across organizations. For example, the U.S. Fish and Wildlife Service and the Confederated Tribes of Warm Springs are working together for salmon recovery. Scott Aikin, a Native American liaison of the U.S. Fish and Wildlife Service, commented, "The tribe's traditional knowledge of the ecosystem and the salmon's role in it is vitally important to successful fish conservation in the Deschutes Basin, a tributary to the Columbia River. The service and tribe continually work together, melding their areas of knowledge into an effective process to protect this remaining wild stock of salmon" (Ikenson 2003). A similar approach is used by members of the Columbia River Inter-Tribal Fish Commission for the Columbia River basin and the Klamath Intertribal Fish and Water Commission in the Klamath River basin, working with other agencies and community groups to recover salmon species and habitat.

Case Study: Paleoecology and Salish Sea Restoration

The Center for the Study of Coast Salish Environments, based in Anacortes, Washington and encompassing the northern Puget Sound and northern Georgia Straits, has three main goals: to provide scientific support for sound stewardship of the Samish Historic Territory, including treaty rights; to prepare Samish tribal members for careers in the sciences and engineering; and to generate Samish tribal employment and income from research and conservation projects. Funding is provided by competitive sources, such as the National Science Foundation, and research contracts with state and federal agencies that share management responsibilities for the "Salish Sea."

The center actively recruits teams of Samish undergraduate science students to work year-round on its research projects. The program also attracts doctoral and postdoctoral science students as researchers and mentors for Samish scientists and creates partnerships and research-sharing arrangements with major Northwest universities and government research laboratories. Ongoing projects include archeological work on Fidalgo, Orcas, and Lopez islands; studies of native oysters in Fidalgo Bay and salmon at reef net sites on Orcas and Shaw islands; and estuary restoration activities (including genetics, geology, and botany) on Orcas and Lopez islands.

A major focus of the center is paleoecology, or the history of environmental change, including the ways in which Samish people shaped and were shaped by the ecosystems in which they lived, fished, and hunted. Paleoecology demonstrates that Samish ancestors successfully managed a complex and ever-changing marine environment for thousands of years. The center asserts, "Precise knowledge of how our ancestors took care of particular islands, bays, beaches, fish, and shellfish will give us guidance for the future protection and enjoyment of the resources that form our biological legacy" (Samish Research Center Web site: www.samishtribe.nsn.us/dnr/dnr_1.html).

Paleoecology entails the study of natural sediment deposits in estuaries and bays as well as cultural deposits, or vast shell middens that persist throughout the Northern Strait islands, which provide a record of what the Samish ancestors harvested, ate, manufactured, and traded. The center's mission is not only to survey and study the Samish archeological and cultural sites but also to help protect them. The center's SamArc database, created by Russell Barsh and Megan Jones, systematizes information on more than 500 historical and archeological sites in the Samish Historic Territory in Excel and geographic information system (GIS) formats.

The SamArc GIS database links archeology, ethnography, historical information, and fishery data in a comprehensive assessment of indigenous peoples' interaction with the marine environment across 300 islands and thousands of years. The center connects TEK with scientific research to study genetic evolution and coadaptation between humans and fisheries. The center has conducted a compelling study showing how human behavior has shaped salmon biology and demonstrating the intricate relationship between historic Samish reef net sites, Northern Straits kelp bed ecology, and sockeye salmon migratory patterns.

Northern Straits reef net fishing was a technology specialized to the Samish and other Straits Salish people (Stewart 1977, Claxton and Elliott 1994). Over millennia, Samish and Saanich reef net stewards cut paths in kelp beds for canoe access, a practice that influenced kelp bed growth, crustacean populations, and subsequent salmon migratory patterns. These reef net sites located offshore of Northern Strait islands were very important places connected to the field ecology of salmon, probably serving as way stations for genetically distinct salmon runs (Barsh and Hansen 2003, Turner and Berkes 2005).

To date there has been little scientific understanding of sockeye migration patterns and the importance and function of island systems. The Northern Straits region is rapidly developing, yet less than 5% is under protection. The center's research articulates the connection between traditional practices and fishery health and sheds light on the key elements that need to be in place for sockeye salmon restoration.

Case Study: Back to the Future

In the Strait of Georgia, British Columbia, the concept of "Back to the Future" encourages knowledge sharing between scientists, First Nations, fishers, historians, archeologists, and other interested parties to construct computer models of historical and contemporary ecosystem conditions. Past abundance and diversity form the basis for determining restoration goals that reflect productive potential rather than present scarcity. Research has included ecological studies and analyses of different fish species, marine mammals, birds, mollusks, and plankton from various parts of the ecosystem. Crucial information on the location, presence, and abundance of living organisms was obtained from historical records and documents, ethnographies, linguistic studies, archeological data, oral histories, and traditional ecological knowledge of the First Nations groups living around the Strait of Georgia.

Qualitative and quantitative data were used to construct mass-balanced models of the Strait of Georgia. The model was created with Ecopath, an ecosystem modeling software suite, and consists of two dozen or more functional groups. The epistemological, conceptual, and methodological issues brought forward by this interdisciplinary process served as a framework for ecosystem reconstructions and as a guide for future restoration. The "Back to the Future" model includes reconstruction of past and present ecosystems as a way to inform policy choices for fisheries. The assessment of alternative ecosystem management strategies, the design of restoration techniques, and the monitoring of ecosystem recovery are all factors that may generate powerful support among diverse stakeholders (Preikshot and Hearne 1998).

Invasive and Exotic Species Management

Because many First Nations and American Indian groups rely on multiple habitats or ecosystems in their aboriginal territory for ceremonial, spiritual, subsistence, and commercial purposes, they are often the first to recognize the presence, rates of spread, and reaction to disturbance by invasive and exotic species. Just as indigenous people observed and evaluated each native species for its potential habitat contribution as a food, medicine, or material or as wildlife habitat, the same criteria historically and currently are applied to exotic species. First Nations and American Indian groups adopted some exotic plants for use as medicines, foods, and materials. For example, in British Columbia, a number of elders recall eating watercress (*Rorippa nasturtium-aquaticum*) and dandelion greens (*Taraxacum officinale*), and broad-leaved plantain (*Plantago major*) is widely used as a medicinal poultice for burns, cuts, and bee stings. Exotic species that were observed exhibiting invasive traits were evaluated as to how they affected natural communities and cultural uses.

In northwestern California, tribes and tribal organizations working with fuel reduction, prescribed fire, and forest restoration projects prefer nonherbicide methods of invasive species control. Traditional knowledge of how different invasive native and exotic species respond to disturbance, the rates and methods of spread, and nonherbicide methods of control have been used to control invasive species (Salmon River Restoration Council 2005 and Mid-Klamath Watershed Council 2005). For example, the Karuk Indigenous Basketweavers organize with local fire safe and watershed councils, the Karuk tribe, and the U.S. Forest Service in mechanical

FIGURE 17.10. Karuk basket weavers and community members pulling Scotch broom at ceremonial sites to promote native plant production in the Klamath Basin. (*Photo by Luna Latimer*)

and hand pulling of Scotch broom (*Cytisus scoparius*) around tribal ceremonial grounds and highly infested areas (Figure 17.10).

Native Geophyte or Plant Bulb Restoration

Most terrestrial plant community restoration efforts in the Pacific Northwest focus on native tree, shrub, or grass species. Fewer projects focus on forbs, particularly native geophytes, plants with edible underground parts such as bulbs or tuberous roots. Many of the native geophytes that were culturally significant to Native Americans and First Nations present as understory forbs associated with threatened forest, prairie, and wetland habitats. Many native bulb plants belonging to the family Liliaceae (e.g., camas [*Camassia* spp.], onions [*Allium* spp.], and blue dicks [*Dichelostemma* spp.]) were managed and harvested by aboriginal peoples for food, medicine, and materials (Figure 17.11). At the foundation of this selective plant use was a well-developed

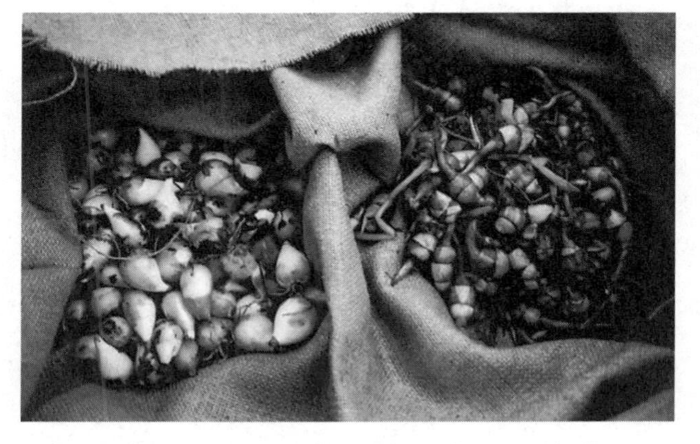

FIGURE 17.11. Camas bulbs *(left)* and wapato bulbs *(right)* were major staple foods to Western indigenous peoples. (*Photo by Nancy Turner*)

form of traditional knowledge particular to each species or assemblage of species occurring in similar habitats (Anderson and Rowney 1998). Traditional ecological knowledge held by aboriginal peoples or located in ethnographic literature provides substantial insights in support of restoration and establishment of native geophytes.

Camas was widely distributed throughout the Pacific Northwest, from coastal headlands to inland montane meadows, yet the ecology of camas production is a sensitive balance of moisture, soils, nutrients, and specific management practices such as burning, cultivation, and bulb division. Several projects in the interior Pacific Northwest focus on camas restoration. For example, the Kalispel and Nez Perce tribes are restoring camas to prairie lands or other areas after catastrophic wildfires (Marshall, in Goble and Hirt 1999). Other groups are working in the coastal regions of the Pacific Northwest to restore camas. Traditional ecological knowledge of Native American root harvesters is used to guide geophyte plantings in microsites that will facilitate improved establishment, as evidenced in the following case studies.

Case Study: Rekindling the Fire of Camas Production

Like with many contemporary First Nations, the Songhees Nation, of southern Vancouver Island (a Straits Salish community) developed a keen interest in their ecological heritage, not just for its historical value but also for cultural and practical reasons. The Songhees, like indigenous peoples all around the world, are negatively affected by the global "nutrition transition," the widespread conversion from a diverse, generally healthful diet of locally produced food to a diet of refined, marketed, mass-produced food, high in fat and simple carbohydrates. Although just a few generations ago diabetes was extremely rare in the Songhees Nation, today many people have developed or died from this disease (Songhees Nation 2004). Reintroducing some of their original food, the Songhees thought, would improve their health and strengthen their cultural identity.

Because of acculturation, Songhees people had largely forgotten most of their traditional plant foods; no one had eaten camas bulbs in any quantity since the eldest members of the community were children. Yet camas bulbs were once a staple—a cultural keystone species—central to people's livelihoods and cultural identity (Garibaldi and Turner 2004). Camas (*Camassia quamash* and *C. leichtlinii*) was featured in families' seasonal harvesting rounds, trade and reciprocity relations between communities and families, feasting, language, and stories. There was much traditional knowledge about its ecology and management. Formerly, camas patches were cleared and tended (Babcock 1967–1969), and people harvested the bulbs tremendous quantities: hundreds of thousands of bulbs annually from southern Vancouver Island (Deur and Turner 2005).

Camas, once a dominant element of the vegetation over most of the Greater Victoria landscape, has significantly diminished in both range and density and probably in productivity as well (Turner 1995, Beckwith 2004). Most of the remaining extensive patches of this liliaceous plant exist in local parks and protected areas, although some residual patches still remain on offshore islands, such as Discovery and Chatham islands, which include Songhees reserve lands. Decades of sheep grazing and invasive species introductions, coupled with fire suppression, have allowed the encroachment of coniferous forest where savanna prairie once existed (Moravets 1932, Norton 1979, Boyd 1999). Therefore, these patches were severely degraded. Nevertheless, Songhees land manager and historian Cheryl Bryce was determined to begin a process of reclaiming her heritage of camas tending and harvesting (Figure 17.12). She invited former residents of the islands, including her colleague

Joan Morris, to take part in the harvest on Discovery Island. Together with other band members and youth and accompanied by then–doctoral student Brenda Beckwith, Eric Higgs, Nancy Turner, and other nonindigenous participants, they first dug camas bulbs, pit-cooked them according to traditional methods, and ate them along with other root vegetables at a festive and memorable event.

The next step was to burn over areas of former camas meadow on the islands and ceremonially scatter camas seed over the charred ground; the Songhees accomplished this activity with the assistance of Brenda Beckwith and a group of volunteers (Beckwith 2004). Subsequently, there have been a number of pit-cooking events on Songhees lands and, as described in *Nature by Design*, Higgs was profoundly affected by the powerful message conveyed by the Songhees initiatives in ecocultural restoration (Higgs 2003).

These events are just the beginning of a series of focal restoration projects grounded in Coast Salish knowledge, practice, and beliefs that will include reintroducing regular burning regimes, promoting community participation and active experiential learning, reintegrating humans with the landscape, and once again tending and using the restored places and their resources in ways that sustain both people and their habitats. The students and faculty of the Restoration of Natural Systems Program and School of Environmental Studies at the University of Victoria have taken an active and ongoing interest in the Songhees community's reconnections to their traditional lands and resources. Through courses and individual student projects, the educational and participatory components of this ecocultural restoration endeavor are continually expanding.

In June 2005, a camas festival at the University of Victoria, organized by Cheryl Bryce and honors environmental studies student Pamela Tudge, continued the work of linking the Songhees with their traditional territory, this time on the university campus, with traditional food. In the presence of many youth and elders of neighboring First Nations, the participants ceremonially harvested camas bulbs using traditional methods and prepared the bulbs in an earth oven for tasting. Such activities are intended to reconnect and revive these peoples' relationship with their home place.

FIGURE 17.12. Songhees historian Cheryl Brice is determined to reclaim her people's tradition of tending camas. (*Photo by Nancy Turner*)

Case Study: Restoring Wapato on Shuswap Lake

Another example of ecocultural restoration of a former cultural keystone species is Ann Garibaldi's work with wapato (*Sagittaria latifolia* Willd.). Dr. Mary Thomas, a Secwepemc elder from Salmon Arm on Shuswap Lake, guided and helped inform the research. Born in 1917, Mary Thomas was raised in the vicinity of Salmon Arm, where she and her brother and sister spent much of their time learning from their grandmothers. She has witnessed tremendous environmental change over her lifetime. One of the most alarming transformations has been the deterioration of the wetland ecosystem at the delta where the Salmon River flows into Shuswap Lake, after its passage through the Neskonlith Reserve lands where Mary grew up.

Mary Thomas recalled going down to the mouth of the river with her grandmother "Macreet" (Marguerite, Mrs. Dick Andrew) as a child to dig the small egg-shaped tubers of *ckwalkwalul's* (wapato) in the spring, along with the finger-like roots of water parsnip (*Sium suave*), called *etsmáts* in the Shuswap language (cf. Turner 1997). Children gathered up the roots that the adults, wading out in the water, threw onto the bank. They rinsed and ate the water parsnip roots raw, but the wapato was placed into large baskets to be carried up and cooked for a much-enjoyed family meal (M. Thomas, personal communication, 1995). Extensive patches of these plants extended over wide areas together with many other culturally valued wetland species, such as highbush cranberry (*Viburnum opulus*), cattail (*Typha latifolia*), and, on higher ground, Indian hemp (*Apocynum cannabinum*). The songs of red-winged blackbirds (*Agelaius phoeniceus*) and western bluebirds (*Sialia mexicana*) filled the air, and many types of waterfowl and other wildlife were abundant (M. Thomas, personal communication, 2004).

Then, as Mary described to Ann Garibaldi, a whole series of changes took place: the CP Railway was rerouted and the trans-Canada highway built above the wetlands. To accommodate this development, some wetland areas were filled in, and the hydrologic regime was subsequently altered. Townspeople built houses along the shoreline of Shuswap Lake, and the number of boats and houseboats using the lake increased exponentially. The river became silted, in part because cattle grazing and farming along the riparian zone reduced vegetation and increased riverbank erosion. Logging on the upper slopes of the watershed also affected the ecosystem. The river was rechanneled through the delta for flood control and to allow construction of a wharf, and the open water was covered with oil every summer to reduce mosquitoes. Large numbers of cattle were allowed to graze over the area, and hayfields were planted to provide them with feed.

Today, much of the delta below Mary's house has dried out and become covered with invasive reed canarygrass (*Phalaris arundinacea*) and other weeds. The land is leased to a cattle owner who grazes his herd in the delta. Mary Thomas mourned, "There's not one plant left down there. Let alone a cattail where the birds used to sing beautiful music. You don't hear that anymore" (Thomas et al. unpublished manuscript).

Learning about the loss and apparent extirpation of these important species, particularly wapato, Ann Garibaldi designed a research experiment to assess the reintroduction of wapato in one of the remaining pond areas in the delta. She obtained tubers from a native plant nursery, grew them for one season, and then planted them in controlled, monitored plots across a gradient of water depth. Suspecting that the original wapato may have been destroyed in part by grazing or uprooting caused by ducks or introduced carp, which are known to have drastically reduced wapato populations on the Columbia River (Darby 1996), she established wire exclosures around a portion of the plantings, designating others as a control. Her results are reported in her thesis (Garibaldi 2004), and as of the summer of 2003, the wapato in the exclosures was flourishing.

Mary was delighted, and she used the plantings as a mechanism to introduce her own family and other members of the Secwepemc community to this forgotten root vegetable. People were able to taste wapato, and some have replanted it in other locations in Secwepemc territory. Again, as with the Songhees' efforts with camas, this example is just the beginning of a trend in ecocultural restoration in Mary's community. Plans are under way to build a Secwepemc center for ecocultural education in Salmon Arm, including a trail to the wapato experimental pond and an ethnobotanical garden to be planted around the main building. The garden is intended both for education and for production of plant materials, such as cattail, for use by schoolchildren, tourists, and any others wanting to learn about and experience the cultural and environmental importance of plants.

Mary Thomas is one of the last elders in her community to recall that wapato existed at this traditional gathering site and to remember its Secwepemc name, *ckwalkwalul's*, and associated harvesting and cooking methods. She is what Ann Garibaldi and Brenda Beckwith call a cultural refugium: a person who takes on a role metaphorically parallel to that of an ecological refugium, bringing cultural and ecological knowledge forward through times of great change and stress and then teaching that knowledge to others, fostering its renewed dissemination and the increased awareness that it brings. People such as Mary Thomas, Cheryl Bryce, Joan Morris, Brenda Beckwith, and Ann Garibaldi are all key players in ecocultural restoration. It is through the collective efforts of such people that the knowledge, practices, and approaches of indigenous peoples become the key ingredients in focal restoration.

Future Directions and Opportunities

Ecocultural restoration offers a strong model for restoring degraded environments to healthy, functioning systems that support social and ecological vitality. TEK is not folklore or a romantic way of viewing nature. Rather, this longstanding practice of cultures sustainably interacting with their bioregion over time is especially relevant in our rapidly changing world. Clearly science alone does not have all the answers to the alarming list of environmental and social problems. Traditional ecological knowledge not only offers creative, new solutions to the field of restoration but also asks different questions than would otherwise be posed.

Of course, restoration practitioners need to evaluate the appropriateness of any method to a specific project, and TEK is no exception. More often than not, however, restorationists may overlook the opportunity to include indigenous knowledge in a restoration project. We advocate a restoration approach that seeks out and engages the community of people who are invested in the place. If the community includes an indigenous group, it is the responsibility of the management agencies and restoration practitioners to solicit the input of that group through a consultation process. Community knowledge in general, and TEK specifically, will greatly enhance practitioners' understanding of a particular ecosystem and the subtle interactions that sustained it over multiple generations. Additionally, community investment is needed to care for a place over time, after the resource managers and scientists walk away.

Looking forward, there are numerous potential applications for TEK in the sphere of Pacific Northwest restoration. In addition to the possible avenues discussed in earlier sections, other opportunities for TEK-based restoration integrated with Western science include the following.

Climate Change

Given recent scientific understandings of potential global climate change and subsequent environmental changes in the Pacific Northwest (Rapp 2004), TEK probably will be more effective than strict science-based approaches to understanding local or finer-scale ecological changes. As vegetation communities shift in response to climate changes (e.g., trees and shrub migrate north) and corresponding management regimes such as prescribed fire are modified, Southern American Indian groups probably will exchange TEK with Northern groups.

Global warming is increasingly recognized as imposing some constraints on historical reference models. However, the restoration of as much biodiversity and redundancy as possible may contribute to sufficient genetic diversity for thermal adaptation by many species. Even so, a major problem will remain for high-elevation plant species, which cannot migrate much higher, and perhaps for many aquatic species as well.

Restoring old-growth trees not only will help sequester carbon but will cool temperatures in the understory enough to mitigate higher wind speeds and drying of live and dead fuels in forest openings. Moreover, long-established herbaceous species in these openings will contribute carbon to the soil through continual root decomposition. Thinning in some forest types to reduce crown-to-crown and ground-to-crown contact will lower fire risk and therefore prevent carbon from being lost to stand-replacing wildfires.

TEK regarding environmental change, adaptive management, and cultural practices that increase biodiversity can greatly enhance our ability to respond to climate change and allow us to actively tailor our restoration efforts accordingly.

Ecotones and Cultural Transitional Areas

Research of First Nations and American Indian tribes whose aboriginal territories encompass the boundaries of several ecosystems or physiographic provinces, described as ecotones or cultural transitional areas, have a greater breath of knowledge and inherent social-ecological resilience regarding species whose ranges converge (Turner et al. 2003). An ecotone is a transitional zone rich in species diversity and productivity that links major ecosystems or physiographic provinces. Indigenous peoples' territories encompassed such transitional zones, and through their cultural activities and practices they created or expanded these areas to enrich and diversify resources (Turner et al. 2003).

A major type of ecotone in several regions of the Pacific Northwest is the transition zone between forest and prairie. Native peoples purposefully set fires to enhance numerous patches and meadows of varying sizes and shapes in forest ecosystems (Boyd 1999). Prescribed burning and other management activities expanded the range of prairies or open meadows that facilitated high production of both plant resources (e.g., huckleberries and geophytes) and wildlife (e.g., deer and elk). These cultural ecotones were part of the overall mosaic of forest structure and composition and undoubtedly contributed to overall ecosystem diversity (Lewis 1993). Ecotone prairies enhanced opportunities for people and for other species and thereby increased their resilience.

The role of deliberate creation of ecological edges, an important aspect of indigenous knowledge and practice, often is disregarded today because of peoples' loss of access to their traditional territories, fire suppression, and other management practices, and it should be reinstated wherever appropriate. Landscape patches, corridors, and mosaics have constantly shifted in size, shape, and location while changing under the influence of factors such as weather patterns, moisture regimes, and successional cycles. Restoration cannot be a static, predetermined activity; instead it must flexibly incorporate transition zones if it is to reflect historic anthropogenic activities and cultural landscapes. Furthermore, restoration must enhance ecotones and ecocultural complexity at a range of temporal and spatial scales. TEK integrated with Western scientific modeling can provide a scaled, local to regional or global approach to defining the future desired conditions and the design of restoration treatments (Turner et al. 2003).

Local Knowledge and Community Participation

There are ample opportunities for applying TEK in contemporary restoration projects, both where science practitioners lack knowledge of ecosystem dynamics and species linkages and where fostering social approval and participation is important (Anderson and Barbour 2003, Ruppert 2003). Restoration researchers and practitioners are learning what communities have discovered on their own: Restoration is a vehicle for reconnecting with place. Scientists and practitioners now realize that community-based restoration may be a mutually beneficial activity whereby people develop stronger relations with each other and their local environment in the process of restoring a place. In many regions, restoration is an important process of community building. In the Pacific Northwest, for example, citizens' understanding of bioregional community health developed when they expanded their focus beyond saving single species to community-based restoration (Senos 2006, House 1999). As increasing numbers of degraded systems need our attention and intervention, ecological citizens are urgently needed to restore their respective home places.

TEK is not simply a repository of environmental information. It represents a diverse array of systems that integrate cultural and environmental knowledge, practice, and belief over various scales of time and space. Restoration practitioners and local communities can also draw on the social, cultural, and economic lessons implicit in TEK systems. Restoration specialists coming into a community need to foster social relationships and be as inclusive as possible while working with the community members who hold the resident knowledge. Exercising diplomacy while grasping the local political and cultural environments is key to successful restoration. TEK also shows us how we might reconceive restoration across nested social scales, from individual to family to affiliated group to the larger community. TEK challenges us to think beyond our own life spans and consider the implications of our actions in terms of at least seven generations.

TEK provides a model for how community members may interact with their local environments in ways that support or enhance social and ecological functioning. Stewards of TEK operate from an ethical point of view; similarly, accountability is required in our relations with each other, our bioregion, and our restoration work. We are called on to be attentive observers of our home places and to understand the intricate connections between individual actions and entire systems. To do this we must stand in place over time. Our role is not a passive one; instead we must actively intervene in nature to nurture its unique biotic relationships and landscapes. Our task as restorationists is not to simply put things back together the way they were in a technical sense but to apply our knowledge and awareness to creating dynamic, reciprocal relations with nature. As we re-story the landscape, we will perhaps develop new social codes, rituals, and language to describe and reinforce our cultural identity (Senos and Lake 2004).

Cultural Continuity

Restoration is an essential tool that can revitalize cultural traditions and practices in the course of repairing fragmented landscapes. As indigenous people reclaim their sovereignty, their land base, and their culture, restoration may assist the recovery of interrelated ecological, social, and cultural systems. Restoration serves a crucial role in activating traditional knowledge and practices that are at risk of disappearing. The process of restoring a lost prairie or a struggling species may become one of recovering knowledge, language, and environmental management techniques. As demonstrated in the geophyte case studies and numerous projects across the Northwest, as restoration practitioners and tribal resource managers seek out elders and community members for answers to environmental questions, the restoration project becomes a watershed event for the community.

Successfully restoring culturally significant resources and landscapes depends on how the goals and objectives of restoration are defined. Best available science must be relevant to local communities. Restoration programs should apply an intergenerational, cross-gender, and multicultural approach to planning, implementing, and evaluating projects. Youth ought to be involved with elders, men and women working to support each other, and various cultural values represented in each restoration project. Restoration practitioners should also consider how language, whether it is scientific jargon or local terms, Spanish, French, English, or indigenous, influences our understanding and ability to express a world view and relationship to place.

Language greatly influences the formation of one's world view and relationship to place. TEK is deeply entrenched in indigenous languages, and much of what is encoded in Native linguistics relates to environment and management practices (Salmón 1996). Ecological restoration can be strongly linked to indigenous language restoration. Interestingly, indigenous peoples in the Pacific Northwest who have remained in their ancestral territories have the highest indigenous language retention and use, continuity of cultural practices, and least degraded habitats and ecosystems (Schoonmaker et al. 1997). Indigenous languages describe ecological conditions and processes, cultural–ecological relationships, and a way of understanding the environment that can complement Western science.

Many First Nations and American Indian tribes believe that restoration projects should be inclusive of youth, elders, tribal practitioners, scientists, and non-Native community members. The Karuk involve local youth in their fish monitoring and restoration activities. Elder traditional fishers teach youth how to harvest fish, identify fish species, read local river hydrologic conditions, and understand how the river's environmental condition compares with past and likely future conditions. They also teach the indigenous language and terms used to describe these things. Karuk youth learn important scientific principles and techniques from the Western fishery biologists, including scientific vocabulary, proper measuring and aging techniques, and a broader context of how their monitoring studies are important in assessing fish health and abundance (Figure 17.13).

Partnerships and Collaboration

Collaboratively planned, designed, and implemented restoration projects between indigenous people and restoration practitioners that allow the reintroduction of traditional land management practices such as burning and horticultural techniques probably will show longer-term success both socioculturally and ecologically. For many indigenous groups, being recognized as contributors of the reference condition environment provides a critical opportunity for them to shape the future desired condition by participating in restoration activities today. Collaborative restoration of habitats historically significant to indigenous people facilitates the integration of knowledge systems, strengthens the human–nature relationship, and fosters sustainable use of the environment.

Formal agreements between American Indian and First Nations groups and government agencies that clearly define the roles and responsibilities of involved parties can greatly facilitate the integra-

FIGURE 17.13. Karuk youth learn both traditional dip-net harvesting and scientific monitoring techniques. (*Photo by Frank Lake*)

tion of cultural land use practices with restoration treatments. Working together, stakeholders can evaluate each style and technique of differing restoration practices and learn effective restoration strategies and practices (Ruppert 2003, Berkes et al. 2000). Joint endeavors acknowledge that social capital restores natural capital. The success of many restoration projects hinges on the ability of the parties involved to cooperatively build trust, prioritize recovery goals, and develop and assess restoration treatments.

Restoration projects are likely to be more successful when the restoration priorities, goals, objectives, and criteria used to measure project success are derived from the local community. Is the restoration ecologist's criterion the same as that of the local community, which depends on the successful outcome of restoration projects? Often the measure of success used to evaluate restoration projects differs between the managing agency and the community that relies on the cultural benefits and ecological services of restoration. Land managers and restoration practitioners need to align their restoration goals and objectives with the goals and objectives of local or tribal communities.

Each section of stream or river restored equates to higher fish numbers, which bolsters the opportunity for fishers to put local food on the table and has direct implications for indigenous peoples' diet and cultural sustenance (Harden 2005). Similarly, each unit area of forest or prairie successfully restored by fuel reduction and prescribed fire strategies may be measured by a biological index of vegetative species richness conducted via intensive survey methods. Yet local communities using these same sites may measure success in terms of the quantities of high-quality foods and medicinal plants harvested in the years after the restoration treatments. Project leaders come and go, but in the end our grandchildren will be our effectiveness monitors, and the landscape and waterways will be testimony to our collaborative efforts.

See Plate 12 in the color insert for a wildlife connectivity map.

REFERENCES

Anderson, M. K. 1999. The fire, pruning, and coppice management of temperate ecosystems for basketry material by California Indian tribes. *Human Ecology* 27(1): 79–113.

Anderson, M. K. 2001. The contribution of ethnobiology to the reconstruction and restoration of historic ecosystems. Pages 55–72 in D. Egan and E. A. Howell (eds.), *The Historical Ecology Handbook: A Restorationist's Guide to Reference Ecosystems*. The Society for Ecological Restoration and Island Press, Washington, DC.

Anderson, M. K. and M. G. Barbour. 2003. Simulated indigenous management: a new model for ecological restoration in national parks. *Ecological Restoration* 21(4): 269–277.

Anderson, M. K. and D. L. Rowney. 1998. California geophytes: ecology, ethnobotany, and conservation. *Fremontia* 26(1): 12–18.

Anonymous. 1992. What is TEK? *TEK TALK, A Newsletter on Traditional Ecological Knowledge* 1(1): 1.

Anzinger, D. L. 2003. *Big Huckleberry (*Vaccinium membranaceum*) Ecology and Forest Succession, Mt. Hood National Forest and Warm Springs Indian Reservation, Oregon*. Master's thesis, Oregon State University, Corvallis.

Babcock, M. 1967–1969. *Camas—Descriptions of Getting and Preparing—from Informants of Tsartlip Reserve (W. Saanich), Vancouver Island*. Unpublished manuscript, cited with permission of author; copy in possession of N. Turner.

Barsh, R. and K. C. Hansen. 2003. *A TEK-Based Marine Habitat Restoration Program: The Samish Indian Tribe of Puget Sound*. Paper presented at SER Symposium, Joint Conference of Society for Ecological Restoration and Ecological Society of America, Tucson, AZ.

Beckwith, B. R. 2004. *The Queen Root of This Clime: Ethnoecological Investigations of Blue Camas (*Camassia quamash, C. leichtlinii; *Liliaceae) Landscapes on Southern Vancouver Island, British Columbia*. Ph.D. dissertation, University of Victoria, Victoria, BC.

Berkes, F. 1993. Traditional ecological knowledge in perspective. In J. T. Inglis (ed.), *Traditional Ecological Knowledge: Concepts and Cases*. International Program on Traditional Ecological Knowledge and International Development Research Centre, Ottawa.

Berkes, F. 1999. *Sacred Ecology. Traditional Ecological Knowledge and Resource Management*. Taylor & Francis, Philadelphia, PA.

Berkes, F., J. Colding, and C. Folke. 2000. Rediscovery of traditional ecological knowledge as adaptive management. *Ecological Applications* 10(5): 1251–1262.

Blackburn, T. C. and M. K. Anderson. 1993. *Before the Wilderness: Native Californians as Environmental Managers*. Ballena Press, Banning, CA.

Bonnicksen, T. M., M. K. Anderson, H. T. Lewis, C. E. Kay, and R. Knudson. 1999. Native American influences on the development of forest ecosystems. Pages 439–470 in N. C. Johnson, A. J. Malk, W. T. Sexton, and R. Szaro (eds.), *Ecological Stewardship: A Common Reference for Ecosystem Management*. Elsevier Science Ltd., Oxford, UK.

Boyd, R., ed. 1999. *Indians, Fire and the Land in the Pacific Northwest*. Oregon State University Press, Corvallis.

Chapin, K., J. Vergun, and D. Wright. 1998. Development of an aquatic habitat geographical information system for pacific lamprey habitat in Rock Creek (Siletz), Oregon. In *Proceedings from Bridging Traditional Ecological Knowledge and Ecosystem Science Conference*, August 13–15, Northern Arizona University, Flagstaff.

Claxton, E. Sr. and J. Elliott Sr. 1994. *Reef Net Technology of the Saltwater People*. Saanich Indian School Board, Brentwood Bay, BC.

Close, D. A., A. D. Jackson, B. P. Conner, and H. W. Li. 2004. Traditional ecological knowledge of Pacific lamprey (*Entosphenus tridentatus*) in northeastern Oregon and southeastern Washington from indigenous peoples of the Confederated Tribes of the Umatilla Indian Reservation. *Journal of Northwest Anthropology* 38(2): 141–162.

Communities Committee of the Seventh American Forest Congress. 2000. Bringing back the huckleberries. *Communities and Forests* 4(1): 6.

Darby, M. C. 1996. *Wapato for the People: An Ecological Approach to Understanding the Native American Use of* Sagittaria latifolia *on the Lower Columbia River*. Master's thesis, Portland State University, Portland, OR.

Deur, D. August 2002. *Huckleberry Mountain Traditional-Use Study*. Final Report. National Park Service and USDA Forest Service (Rogue River National Forest), Seattle.

Deur, D. and N. J. Turner, eds. 2005. *"Keeping It Living": Indigenous Plant Management on the Northwest Coast*. University of Washington Press, Seattle.

Egan, D. and E. A. Howell, eds. 2005. *The Historical Ecology Handbook: A Restorationist's Guide to Reference Ecosystems*. 2nd ed. Island Press, Washington, DC.

Endowed Chair in Aboriginal Fisheries. February 2002. Draft discussion paper, pp. 2–3. Fisheries Centre at University of British Columbia, Vancouver.

Freeman, M. M. R. 1979. Traditional land users as a legitimate source of environmental experience. In J. G. Nelson, R. D. Needham, S. H. Nelson, and R. C. Scace (eds.), *The Canadian National Parks: Today and Tomorrow Conference II. Ten Years Later*. Vol. 1. Studies in Landscape History and Landscape Change, No. 7, University of Waterloo, Waterloo, Ontario.

Freeman, M. R. and L. N. Carbyn. 1988. *Traditional Knowledge and Renewable Resource Management in Northern Regions*. International Union for the Conservation of Nature, Commission on Ecology and Boreal Institute for Northern Studies, Edmonton, AB.

Garibaldi, A. and N. Turner. 2004. Cultural keystone species: implications for ecological conservation and restoration. *Ecology and Society* 9(3): 1. Available at www.ecologyandsociety.org/vol9/iss3/art1.

Goble, D. D. and P. W. Hirt, eds. 1999. *Northwest Lands, Northwest Peoples: Readings in Environmental History*. University of Washington Press, Seattle.

Harden, B. 2005. Tribe fights dams to get diet back: Karuks trying to regain salmon fisheries and their health. *Washington Post*, January 30, p. A03.

Higgs, E. 1997. What is good ecological restoration? *Conservation Biology* 11(2): 338–348.

Higgs, E. 2003. *Nature by Design: People, Natural Process and Ecological Restoration*. MIT Press, Cambridge, MA.

House, F. 1999. *Totem Salmon: Life Lessons from Another Species*. Beacon Press, Boston, MA.

Ikenson, B. 2003. Tribes work to restore traditional fisheries. In *Environmental News Network*. Interview with Scott Aikin, April 4. Available at www.enn.com/index.html.

Inglis, J. T., ed. 1993. *Traditional Ecological Knowledge. Concepts and Cases*. International Program on Traditional Ecological Knowledge, IDRC, Ottawa, ON.

Johnson, M., ed. 1992. *Lore: Capturing Traditional Environmental Knowledge*. Dene Cultural Institute and the International Development Research Center, Hay River, NWT.

Kimmerer, R. W. 2002. Weaving traditional ecological knowledge into biological education: a call to action. *BioScience* 52: 432–438.

Kimmerer, R. W. and F. K. Lake. 2001. Maintaining the mosaic: the role of indigenous burning in land management. *Journal of Forestry* 99: 36–41.

Lantz, T. and N. J. Turner. 2003. Traditional phenological knowledge of aboriginal peoples in British Columbia. *Journal of Ethnobiology* 23(2): 263–286.

Larson, Z. and M. Belchick. 1998. *A Preliminary Status Review of the Eulachon and Pacific Lamprey in the Klamath River Basin*. Report prepared for Yurok Tribe Fisheries Program, Klamath, CA.

Leopold, A. 1949. *A Sand County Almanac and Sketches Here and There*. Oxford University Press, New York.

Lewis, H. 1993. Patterns of Indian burning in California: ecology and ethnohistory. Pages 55–116 in T. C. Blackburn and M. K. Anderson (eds.), *Before the Wilderness: Environmental Management by Native Californians*. Ballena Press, Menlo Park, CA.

Lewis, H. T. and K. M. Anderson, eds. 2002. *Forgotten Fires: Native Americans and the Transient Wilderness*. University of Oklahoma Press, Norman.

Light, A. 2005. Ecological citizenship: the democratic promise of restoration. In R. Platt (ed.), *The Humane Metropolis: People and Nature in the 21st Century City*. University of Massachusetts Press, Amherst.

Lindenmayer, D. B. and J. F. Franklin. 2002. *Conserving Forest Biodiversity: A Comprehensive Multiscaled Approach*. Island Press, Washington, DC.

Mack, C. A. and R. H. McClure. 2002. Vaccinium processing in the Washington Cascades. *Journal of Ethnobiology* 22(1): 35–60.

Mahler, R. 2004. Restoring homelands: an innovative TPL program helps reverse centuries of Native American land loss. *Lands & People* 16(1): 44–49.

Martinez, D. 1995. *Karuk Tribal Module of Mainstem Salmon Watershed Analysis: Karuk Ancestral Lands and People as Reference Ecosystem for Eco-Cultural Restoration in Collaborative Ecosystem Management*. Unpublished report. On file with the U.S. Department of Agriculture, Forest Service, Klamath National Forest, CA.

McDonald, T., ed. 2003. Restoration practice: art, toil and focal practice. *Ecological Management and Restoration* 4(3): 160.

Mid-Klamath Watershed Council. 2005. *Restoration Applications for Native Plants in the Mid-Klamath Sub-Basin*. Unpublished document, Somes Bar, CA.

Minore, D. 1972. *The Wild Huckleberries of Oregon and Washington: A Dwindling Resource*. USDA Research Paper 143. Pacific Northwest Forest and Range Experiment Station, U.S. Department of Agriculture, Portland, OR.

Moravets, F. L. 1932. Second growth Douglas fir follows cessation of Indian fires. *Service Bulletin, USDA Forest Service* 16(20): 3.

Norton, H. H. 1979. The association between anthropogenic prairies and important food plants in western Washington. *Northwest Anthropological Research Notes* 13(2): 175–200.

Orleans/Somes Bar Fire Safe Council. *Orleans/Somes Bar Community Wildfire Protection Plan. 2005 Draft Plan*. Unpublished document, Somes Bar, CA.

Preikshot, D. and J. Hearne, eds. 1998. Back to the future: reconstructing the Strait of Georgia ecosystem. *Fisheries Centre Research Reports* 6(5).

Pullen, R. 1996. *Overview of the Environment Native Inhabitants of Southwestern Oregon, Late Prehistoric Era*. Contracted document compiled for USFS, Siskiyou and Rogue River National Forest and BLM, Medford District.

Rapp, V., ed. January 2004. *Western Forest, Fire Risk, and Climate Change*, Issue 6. Pacific Northwest Research Station, USDA Forest Service, Portland, OR.

Ruppert, D. 2003. Building partnerships between American Indian tribes and the National Park Service. *Ecological Restoration* 21(4): 261–268.

Salmón, E. 1996. Language affects knowledge. *Winds of Change* Summer 1996: 70–72.

Salmón, E. 2000. Kincentric ecology: indigenous perceptions of the human–nature relationship. *Ecological Applications* 10(5): 1327–332.

Salmon River Restoration Council. 2005. *Salmon River Cooperative Noxious Weed Management Plan*. Salmon River Restoration Council, Sawyers Bar, CA. Available at www.srrc.org.

Senos, R. 2006. Rebuilding salmon relations: participatory ecological restoration as community healing. In R. France (ed.), *Handbook of Restoration Design*. Taylor & Francis CRC Press, Boca Raton, FL.

Senos, R. and F. Lake. 2004. Traditional ecological knowledge and restoration practice in the Northwest. *Proceedings from Society for Ecological Restoration Conference*, August 24–26, Victoria, BC.

Schoonmaker, P. K., B. V. Hagen, and E. C. Wolf. 1997. *The Rainforests of Home: Profile of a North American Bioregion*. Ecotrust publication. Island Press, Washington, DC.

Society for Ecological Restoration International Science & Policy Working Group. October 2004. *The SER International Primer on Ecological Restoration*. Society for Ecological Restoration International, Tucson, AZ. Available at www.ser.org.

Songhees Nation. 2004. *Diabetes*. An educational video produced by Cheryl Bryce and other Songhees participants. Vancouver Island, BC.

Stevens, W. K. 1995. *Miracle under the Oaks: the Revival of Nature in America*. Pocket Books, New York.

Stewart, H. 1977. *Indian Fishing: Early Methods of the Northwest Coast*. Douglas and McIntyre, Ltd., Vancouver, BC.

Thomas, M., N. J. Turner, and A. Garibaldi. Unpublished ms. "Everything Is Deteriorating": Environmental and Cultural Loss in Secwepemc Territory. In K. P. Bannister, N. Turner, and M. B. Ignace (eds.), *Secwepemc People and Plants: Research Papers in Shuswap Ethnobotany*. Secwepemc Cultural Education Society, Kamloops. Western Geographic Series, Department of Geography, University of Victoria.

Turner, N. J. 1995. *Food Plants of Coastal First Peoples*. Royal British Columbia Museum Handbook. University of British Columbia Press, Victoria.

Turner, N. J. 1997. *Food Plants of Interior First Peoples*. University of British Columbia Press, Vancouver and Royal British Columbia Museum, Victoria.

Turner, N. J. 2005. *The Earth's Blanket. Traditional Teachings for Sustainable Living*. Douglas and McIntyre Press, Vancouver and University of Washington Press, Seattle.

Turner, N. J. and F. Berkes. 2005. Coming to understanding: developing conservation through incremental learning. *Human Ecology* 31(3): 439–463.

Turner, N. J., I. J. Davidson-Hunt, and M. O'Flaherty. 2003. Living on the edge: ecological and cultural edges as sources of diversity for social-ecological resilience. *Human Ecology* 31(3): 439–461.

USDA Forest Service. August 2001. *National Fire Plan, A Collaborative Approach for Reducing Wildland Fire Risks to Communities and the Environment: 10 Year Comprehensive Strategy*. "Goal 3, Restore Fire Adapted Ecosystems," p. 10. Available at fs.fed.us/nfp.

Williams, G. W. 2000. Introduction to aboriginal fire use in North America. *Fire Management Today* 60: 8–11.

Williams, N. M. and G. Baines. 1993. *Traditional Ecological Knowledge. Wisdom for Sustainable Development*. Based on Traditional Ecological Knowledge Workshop, Centre for Resource and Environmental Studies, Australian National University, Canberra, April 1988.

Conclusion: The Status and Future of Restoration in the Pacific Northwest

Dean Apostol and Marcia Sinclair

Ecological restoration in the Pacific Northwest has come a long way over the past 10–15 years. The preceding chapters make it evident that practitioners are working on all major ecosystem types, from city centers to alpine wilderness, and are applying restoration at scales that range from schoolyards to river basins. The growth in restoration practice has run ahead of policy development and educational offerings, and the literature on restoration theory is just beginning to be published (see, for example, other volumes in this SERI-sponsored book series). This book represents an attempt to catch a collective regional breath, take stock of achievements, and consider future challenges and opportunities.

Many influences sparked the initial interest in restoration in the Northwest and probably will continue to drive it forward. Chief among these is the widening recognition of the extent of degradation and loss of ecosystems that were once taken for granted. The Pacific Northwest still has an impressive portfolio of natural ecosystems compared with most regions of the world, but as this book has made clear, restoration is needed from sea level (or even below) to the highest mountains. Government entities at nearly every level that are responsible for land management or regulation now include restoration in their work portfolios. Local elementary schools adopt and restore wetlands and streams, as do state and county highway departments. Federal agencies have radically shifted the practice of forestry, remove more roads than they build, replant trampled alpine meadows, and increasingly harness fire in the cause of restoration. Nongovernment organizations, particularly The Nature Conservancy, have taken lead roles in eradicating invasive weeds, replanting prairies, training managers in fire ecology, and building landscape-level strategies. Citizen-based watershed councils have assumed coordinating roles in restoring aquatic ecosystems, particularly on private lands.

Three interrelated issues have shaped the practice of restoration and will influence its continued development: public policy, technical education and knowledge, and culture.

Public Policies and Programs

In general, public policies tend to be more crisis driven than proactive. Typically, a problem builds until it can no longer be ignored, a constituency organizes and gains strength, and policymakers eventually take action, ideally before the crisis is beyond the point of solution. In the case of ecological restoration, the growing problem is the degradation of natural ecosystems to the point at which biological diversity is threatened. The constituency for action has been an environmental movement allied with scientists who have studied, uncovered, and become engaged by the problem. In the Pacific Northwest a strong public constituency has long supported conservation. For example, Oregon created the nation's first statewide land use planning system, developed the first state regulation of forest practices, and designated its entire Pacific shoreline as public property. Washington was the first state to create a Department of Ecology and the first to adopt a coastal management program. British Columbia has developed one of the best ecological classification systems in North America. The region's rugged natural beauty has attracted residents who love it. Yet citizens who want to restore the land find themselves competing with those who want to convert its natural capital to financial capital. Additionally, many residents who are sympathetic to the need for conservation mistrust government-led initiatives.

Because so much of the land in the region is publicly owned and managed by government entities (more than 50% in most states and more than 90% in British Columbia), national politics and policies strongly influence conservation (Fairbank, Maslin, Maullin & Associates 2004). The two key federal policies with the greatest influence over the practice of restoration in the United States are the Endangered Species Act and the Clean Water Act. Listing a species under the federal Endangered Species Act triggers a series of actions that cross ownerships, from federal to state, local, and private lands. Although these actions do not compel restoration directly, the Endangered Species Act requires identification of critical habitat for each listed species, setting the stage for habitat conservation and restoration. The Clean Water Act sets requirements for surface water quality. Initially focused on reducing point source chemical and sewage pollution, the act has evolved over the past decade, placing greater focus on the physical and biological integrity of watersheds and thus greatly influencing wetland, stream, watershed, and riparian management, conservation, and restoration.

The listing of the northern spotted owl and numerous stocks of salmonids created momentum for restoration that is reshaping much of the regional landscape. Most endangered species listings have limited geographic scope, even when species are wide ranging. For example, one of the first species to be listed as endangered, the bald eagle, inhabits a much wider area than the northern spotted owl but is less dependent on a specific landscape condition for its survival. It needs a few trees large enough to build nests in areas that have abundant food (fish and waterfowl). It needs food that is not laced with chemicals that weaken its eggshells (the lesson learned from decades of use of DDT insecticide that led to establishment of the Endangered Species Act). But it can and does take up residence in highly altered ecosystems, including the middle of large cities. The spotted owl and salmon are different. The owl depends on old, complex conifer forests in a territory that encompasses more than 57 million acres (FEMAT 1993). It does not adapt to intensively managed forests, so its protection entails significant change and economic sacrifice across a vast area.

Until the spotted owl was listed, complex old forests were being rapidly converted to simplified young ones. Wilderness areas were concentrated in the high mountains on lands with little timber value. But the spotted owl needs lower-elevation forests, and its listing has forced many foresters to rethink management strategies. In addition, the owl was the proverbial canary in the coal mine. Researchers have now identified nearly 1,100 terrestrial wildlife species as closely associated with Northwest old-growth conifer forests (FEMAT 1993). This discovery caused a shift in focus from a single species to an entire ecosystem.

It also shifted forest practice in British Columbia, even though Canada lacks an endangered species act and the owl barely ranges north of Washington. An increasing cultural and political awareness of the uniqueness and ecological importance of Northwest temperate old-growth forests put British Columbia under market pressure as Europeans began to shun their forest products. As a consequence, the province has increased old-growth conservation and shifted harvest practices to natural forest structure (see Chapter 5). Although British Columbia still probably harvests old growth faster than it grows back, they have moved away from wholesale conversion to plantations.

The spotted owl listing also altered forest management on state and private lands and has even begun to shift forestry internationally as lessons learned here are applied elsewhere. Nest sites are protected at least for a time, and more old trees and snags are being left behind after clearcut harvests. A key point is that the psychology of Pacific Northwest forestry shifted from abundance to scarcity. Old-growth conifer forests, which had been viewed as unproductive and inferior to young plantations, are now widely viewed as valuable resources with unique attributes.

In similar fashion, the listing of multiple stocks of salmon forced land managers, policymakers, and the public to begin viewing streams and rivers as dynamic and complex ecosystems rather than as simple channels that carry water away or as sources for electricity. Whereas the owl and its old-growth cohorts occupied only the western part of the Northwest, the salmon had a much bigger range, reaching well into Idaho, 900 miles upriver from the Pacific Ocean. People eat fish, and salmon has held a special place in the diet and culture of Northwest people for thousands of years. Although decline of native salmon has been noted since at least the late nineteenth century, until the Endangered Species Act listings many still believed that the region had largely made up for this loss through the rearing of hatchery fish. When it turned out that hatchery fish probably were just one more contributor to the decline of native fish, it was a shock to many (Lichatowich 1999).

Additionally, restoring native salmon cannot be isolated to a specific location or type of habitat because they range nearly everywhere in the region and occur in watersheds from sagebrush steppe to coastal rainforests. It was the listing of the salmon, not the owl, that spurred local communities to form watershed councils and initiated the larger culture of restoration that is now taking root (see Chapters 1, 13, and 14).

Along with the Endangered Species Act, the U.S. Federal Clean Water Act of 1972 also has had an increasing influence on ecological restoration. The best-known aspect of this act is wetland protection, which has caused the creation of a wetland mitigation industry that overlaps with restoration now and then. A second aspect has been the effect on streams. As streams become listed as "water quality limited," meaning that they have become polluted with sediment or waste, or their temperatures are above a key threshold, riparian restoration efforts often ensue. Much of the riparian restoration on streams that flow through agricultural land is in response to the Clean Water Act rather than the Endangered Species Act.

Restoration on U.S. Federal Lands

U.S. federal lands, including national forests, national parks, and lands managed by the Bureau of Land Management and the Fish and Wildlife Service, make up most of the land surface of the Northwest, excluding British Columbia. One consequence of the endangered species listings has been that all four federal land management agencies have embraced ecological restoration, although at varying levels of commitment and effectiveness. The Forest Service has focused primarily on aquatic ecosystems including streams and watersheds. The Northwest Forest Plan pioneered the idea of watershed analysis, which has been adopted and modified by many agencies and groups to assess causes of ecosystem decline and prioritize conservation and restoration actions. Despite continuing budget cuts resulting from the decline of timber harvest, the Forest Service has been able to reprogram appropriated funds and build partnerships with other state and federal agencies to remove hundreds of miles of roads (Figure 18.1), open new habitat by replacing dysfunctional culverts, and reforest riparian areas. The Forest Service has prioritized watersheds for restoration in order to make better use of scarce resources, with a focus on the best-quality habitats first (J. Uebels, personal communication, 2005). The Siuslaw National Forest along the Oregon Coast, once among the leading

Figure 18.1. A young citizen steward monitoring aquatic habitat in the Siuslaw Watershed. *(Photo courtesy of the Karnowsky Creek Restoration team and Siuslaw Watershed Council)*

timber-producing forests in the nation, has become a leader in restoration. They have partnered with many other agencies, foundations, and citizen groups to remove old roads, replace dysfunctional culverts, thin overcrowded young forests, repair estuaries, and remove invasive species from dunes and prairies. They won the prestigious International Thiess Riverprize in 2004, in recognition of their restoration efforts.

The Bureau of Land Management has also begun restoring aquatic and old-growth ecosystems, often in partnership with the Forest Service. The Bureau of Land Management also restores oak woodlands in concert with fire hazard reduction projects, particularly in southern Oregon, and has taken a leading role in sagebrush steppe restoration east of the Cascade Mountains (Chapter 10). One impediment to restoring fire-dependent forests is the history of fire suppression and the many and overlapping environmental regulations that make it difficult for land managers to use prescribed fire more frequently and effectively (S. Arno, personal communication, 2004).

The historical mission of the U.S. Fish and Wildlife Service has been strongly tied to conserving game populations through habitat manipulation. Typical Northwest refuges are far from natural habitats and often include highly engineered water control systems and grain fields. But over the past few years the agency has turned increasing attention to restoring bunchgrass prairies, oak woodlands, floodplain forests, wetlands, and sagebrush steppe. The new Tualatin River National Wildlife Refuge, largely carved from an agricultural floodplain, has become a showcase of restoration in the Portland, Oregon metropolitan area.

The National Park Service has long been the regional leader in restoring high mountain ecosystems (Chapter 11) but also restores oak and prairie ecosystems in California at Redwood National Park and in Washington State at Ebey's Landing National Historic Reserve. Yellowstone National Park has become the nation's and perhaps the world's most intriguing experiment in restoration of complex ecosystems through reintroduction of key wildlife species (Chapter 15). This is the reverse of the way most restoration takes place.

Another key aspect of restoration on federal lands is the return of fire as a restoration agent. Some Forest Service, Bureau of Land Management, and Park Service land holdings are large enough and remote enough to allow land managers to experiment with prescribed natural fire. Land managers have seen important restorative results in areas such as mountain wilderness lands of Idaho, where they have allowed some fires ignited by lighting to burn (Chapter 9).

Despite continuing budget cuts and limitations in adding new staff with restoration training and skills, all four federal land management agencies probably will continue to increase their focus on ecological restoration. This work will reshape tens of thousands, possibly millions of acres of regional real estate.

A key unknown is the future of the Endangered Species and Clean Water acts. There is a lot of pressure in Congress and the Bush Administration to weaken or ignore these laws in order to unleash more logging, grazing, drilling, and other development on federal and private lands. If they are successful, this could have a profound impact on the support for and pace of restoration in the Pacific Northwest states.

Additional Programs

In addition to the work of federal land management agencies, the Northwest Power and Conservation Council has a profound influence on regional restoration. The council was established by an act of Congress in 1980 to help mitigate the effects of hydroelectric dams on fish and wildlife. In recent years it has played a key role in allocating resources for Columbia River Basin fish and wildlife conservation and restoration. This has become the largest regional fish and wildlife conservation effort in the United States, covering the entire area that drains to the Columbia River, a territory the size of France. Beginning in 2000 the council began developing fifty-nine Columbia River tributary plans (called subbasin plans) to better prioritize and guide investments and tie them to expected outcomes. Habitat restoration is only one component, along with managing dams (e.g., adjusting flows) and fish hatcheries and regulating fishing levels. The overall conservation program provides a vision, a description of ecological conditions needed to realize the vision, and a set of implementation strategies intended to help create the ecological conditions. An independent scientific review panel provides technical advice and has established a conceptual foundation based on natural ecological functions and processes. Members are nominated by the National Academy of Sciences and appointed by the council (Northwest Power and Conservation Council 2005).

Restoration funding is provided through electricity revenues of the Bonneville Power Administration and averages $139 million per year. This amount funds 200 ongoing projects and a number of new projects selected from proposals. According to the council, funding is expected to increase in the years ahead.

At the state level, the Oregon Watershed Enhancement Board (OWEB) has become an important fund provider for watershed, stream, and riparian restoration. In the late 1990s Oregon voters dedicated a portion of state lottery funds to park development and salmon restoration, much of which has been used to fund watershed councils and their activities. In just one recent 4-month grant cycle OWEB provided more than $6.5 million for restoration projects, including riparian planting, instream enhancement, culvert replacements, and dam removal (OWEB 2005). In 1999 Washington created the Salmon Recovery Funding Board to route state and federal grant funds to the most promising conservation and restoration projects (Chapter 14). From 2000 to 2004 more than $195 million was distributed for projects including conservation easements, culvert replacement, riparian planting, and dike removal. Funding sources include state general obligation bonds and the Federal Pacific Coast Salmon Recovery Fund, to which Congress provided more than $375 million in the most recent 4-year period (2005 *Pacific Coast Salmon Recovery Fund Report to Congress*).

Proactive planning is now seen as essential to moving up the line for grant funding in Washington, so new salmon recovery plans are emerging continuously. One of the more recent (July 2005) is the Shared Strategy for Puget Sound, a coalition of local watershed organizations, tribes, and state and federal agencies.

Local governments also are increasingly promoting restoration. The City of Eugene, Oregon and Lane County initiated the West Eugene Wetlands Plan (Chapter 3) to manage development of a proposed industrial area, leveraging multiple resources to fund restoration of prairie, wetland, and riparian ecosystems as an interconnected habitat network. Portland, Seattle, and other towns and cities throughout the Northwest have ecologists on staff and regularly engage in restoration in parks and along streams and rivers. Portland's Bureau of Environmental Services runs a watershed stewardship grant program that disbursed $362,000 to ninety-two projects over a 10-year period (see the Portland Bureau of Environmental Services Web site at www.portlandonline.com/bes). Portland Metro, the only elected regional government in the United States, has used public bond measures to purchase more than 8,000 acres of natural areas and has hired a technical staff of ecologists who plan and carry out restoration projects in upland and riparian forests, oak woodlands, wetlands, and floodplains. An array of financial resources are used by local governments for restoration, including allocated park management funds, garbage tipping fees, system development charges (fees paid by developers), and partnerships with organizations including Ducks Unlimited and the Oregon Watershed Enhancement Board.

In British Columbia restoration funding is provided through the Forest Renewal BC program. Once well funded, this program has been cut back drastically over the past few years because of a decline in timber harvest receipts. A number of small grant programs are available through the Vancouver Foundation, Habitat Conservation Trust Fund, Pacific Salmon Foundation, and other sources. Regional districts (similar to counties in the United States) and the federal government also provide technical assistance or small grants for restoration (W. Warttig, personal communication, 2005). The Pacific Salmon Commission and a federal program called Eco-Action also provide restoration funding, but available resources are far below the need and are stagnant or declining.

The Canadian Species at Risk program, similar to the U.S. Endangered Species Act, includes a Federal Habitat Stewardship Program established to support projects aimed at assisting restoration. This program recently provided $260,000 (Canadian) for inventory, mapping, assessment, and restoration of Garry oak ecosystems.

Private Land Restoration

Nearly half of the land in Oregon and Washington is in private hands. These lands are low elevation and historically tended to be the most biologically productive in the region. Many wildlife species use both public and private lands, and some habitats such as lowland riparian and Garry oak woodlands are almost exclusively in private hands. An unknown amount of privately owned land is being restored in the Northwest. Some landowners initiate restoration at their own expense and on their own time without seeking support or advice. Many take advantage of tax incentives, cost share grants, and technical advice from university extension or soil and water conservation districts. Others use farm conservation set-aside programs. Some certification and marketing programs such as Oregon Tilth, Salmon Safe, and Oregon Country Beef for agriculture and the Forest Stewardship Council for wood products encourage restorative practices (Oregon Department of Fish and Wildlife 2005).

Portland General Electric (PGE) has agreed to remove two privately owned hydroelectric dams in the Sandy River Basin, near Portland, Oregon. The federal license to operate the dams expired in 2004, but PGE and other utilities normally fight hard to renew their licenses. In this case, the dams do not produce electricity efficiently, and it would have been expensive to upgrade fish passage facilities. Interestingly, plans for their removal call for building temporary dams using river rocks and boulders, then removing the existing dams and letting naturally high spring flows flush out accumulated sediments and redeposit them downstream. Fifteen hundred acres of PGE land in the area will be transferred to a nonprofit organization as part of a nature reserve (Brinkman 2002).

Wetland mitigation banking sets aside some land for conservation or restoration as compensation for wetlands destroyed elsewhere. It is used to stitch together larger, more functional wetland ecosystems instead of creating scattered small wetlands. A multiagency review team evaluates proposals by landowners to establish mitigation banks. Credits are allocated based on various wetland functions and total area (P. Agrimis, personal communication, 2005). At this writing, Oregon has sixteen mitigation banks either established or in process and is actively seeking to create more. Acreage ranges from thirteen to several hundred at the West Eugene area (see the Oregon Division of State Lands Web site at egov.oregon.gov/DSL/WETLAND/). Mitigation banks often are controversial because they allow existing wetlands to be destroyed without adding more wetland acreage. Restoration locations often are some distance from the impact area, although they are usually in the same county. The advantage of mitigation banks lies in the opportunity to gain ecological function through larger scale and to provide more restoration expertise and monitoring than is often available on smaller projects. In the Willamette Valley of western Oregon, efforts are under way to create conservation banks that would broaden the scope of banking beyond wetlands to additional habitats, develop an active credit trading system to secure funding for larger-scale restoration, and entice more private investment in conservation.

Are existing policies and programs that guide and fund ecological restoration sufficient to meet the many emerging needs? Generally the answer is "no," although program effectiveness appears to be increasing as experience and knowledge increase. The total amount of resources going to restoration appears to have topped out in the past few years and may even be declining as budgets are tightened to cope with war, hurricanes, and tax cuts for the wealthy in the United States. Practitioners remain hopeful (but not optimistic) that resources will be increased over time, particularly as the effectiveness and accountability of restoration are further demonstrated.

Technical Education and Knowledge

A second key influence on the development of ecological restoration is technical education, training, and knowledge. Still a new field, ecological restoration is just beginning to establish an academic base. The primary means of sharing information and learning about restoration in the region is still through conferences sponsored by various organizations, including the Northwest and British Columbia chapters of the Society for Ecological Restoration and the Society for Wetland Scientists. Professional associations, including those affiliated with landscape architects, wildlife biologists, foresters, and especially aquatic ecologists, have also included restoration presentations and seminars at their conferences and gatherings. Interest in restoration is strong, and regional conferences have drawn as many as 1,000 participants.

The flagship academic offering in the region is the Restoration of Natural Systems program at the University of Victoria in British Columbia. First initiated in 1995, it is an interdisciplinary program under the auspices of the School of Environmental Studies and the Division of Continuing Studies. It is designed to appeal to working professionals or those with full-time jobs seeking a career change. A diploma option requires twelve courses and full admission to the university. The certificate program is noncredit, covers eight courses, and does not require full admission. Students can also take individual courses if space is available without being enrolled in the program. Courses include "Principles and Concepts of Ecological Restoration Field Study and Practicum in Environmental Restoration," "Field Study and Practicum in Environmental Restoration, Biodiversity and Conservation Biology," "Ethical, Legal, and Policy Aspects of Environmental Restoration," and a wide variety of electives. For example, courses are offered in forest restoration, aquatic restoration, traditional systems of land management, and ecosystems of British Columbia, among many others. Plans are under way to initiate a graduate program as well, and further expansion of course offerings is anticipated.

The University of Washington Restoration Ecology Network offers a certificate program in ecological restoration at its Bothell, Seattle, and Tacoma campuses. The requirement is twenty-five total credits, of which ten are earned in a real-world capstone project in which students work through all phases of restoration in a multidisciplinary framework. Additional courses include "Introduction to Restoration Ecology," "River Restoration,"

"Native Plant Production," "Invasive Species," "Wetland Ecology," and a number of other electives offered through several cooperating departments. Student demand is running far ahead of available space in the graduate program at University of Washington. Budget limitations and academic inertia limit the rate of growth (K. Ewing, personal communication, 2005).

Washington State University in Pullman has a new native plant and landscape restoration nursery and sponsors courses and research on the Palouse Prairie ecosystem. The Department of Natural Resource Sciences offers an emphasis in landscape and restoration ecology under its natural resource bachelor of science degree program.

Portland State University offers a watershed management professional program under the auspices of the Hatfield School of Government that includes four core and three elective courses that focus on law, policy, and science of watersheds. Students can earn the certificate by taking at least five courses, either for credit or noncredit. The program is designed primarily for professionals who want to broaden their knowledge.

Other regional colleges and universities offer the occasional course in ecological restoration but no full programs. There is a clear trend toward increasing education opportunities, and we should expect that programs in restoration will continue to grow and expand, albeit in fits and starts. The tendency is to treat restoration as an interdisciplinary subject that is shared by several departments, and ideally that will remain the case. It will be challenging to integrate busy practitioners into the classroom as teachers and lecturers because academic departments are infamous for defending their turf.

The University of California at Davis may be creating the future model for academic programs in restoration. They offer both master's and doctoral programs in restoration ecology as an area of emphasis in the Graduate Group in Ecology. This is a coalition of twenty-four departments with 125 faculty members participating (K. Ewing, personal communication, 2005, University of California 2005).

A key question in restoration education is the extent to which practitioners learn to improve their methods and efficiency, which along with increased funding is the key to gaining net ecosystem acres. Ed Alverson (personal communication, 2005) says that although knowledge is improving, the gap between what we know and what we would like to know may be getting wider because the engagement of restoration often generates more questions than it answers.

A Culture of Restoration

The third and probably the most important influence on how ecological restoration develops into the future is the culture of restoration. The word *culture* means both "to grow or cultivate" and "the sum of human achievements that can be passed on to future generations." Culture reflects what is of value to a society. To the extent that people of the Northwest continue to consider ecological restoration an honored and valued activity, they will support it, participate in it, and even expect it to be done. Clearly people derive great satisfaction from collective accomplishments, and restoration is nothing if not an accomplishment by many hands working together.

As already demonstrated, there has been a shift in the culture of land management agencies toward embracing restoration as a key conservation tool. Popular media increasingly include articles and stories about restoration. For example, the *Oregonian* newspaper ran an article about a couple who own 1,300 acres near Ashland, Oregon. They are restoring logged-over forest at the junction of the Siskiyou and Cascade Mountain ranges, a key biological link for many species (Larabee 2005). Words that would have been foreign in the 1980s—*restoration, riparian zones, watersheds,* and even *biodiversity*—are becoming part of the common language of average citizens across the region.

Much of the cultural development of restoration is occurring in elementary and high schools. Many schools have adopted wetlands or streams, disconnected downspouts, and constructed native plant gardens. Schoolchildren not only learn about restoration but also teach their parents about it.

Catlin Gable, a private K–12 school in Portland, has one of the longest-running school restoration programs. Beginning in 1978, students began restoring degraded riparian areas in eastern Oregon rangelands in cooperation with the U.S. Forest Service, under the leadership of science teacher Dave Corkran. Each year the school sends three groups to participate in restoration projects. Freshmen do 1 day of work as part of their orientation trip. Seniors work 3 days on their senior trip, which is a final community service activity before graduation. In addition, about twenty-five students and alumni volunteer to spend 4 days on the Elana Gold Environmental Restoration Project, a living restoration memorial to a deceased student (D. Corkran, personal communication, 2005). The school's restoration projects have included fencing cows and off-road vehicles out of riparian areas and thinning dense pine stands to restore their resistance to fire. Some of the thinned material is used for fencing, and some is placed in streams to restore woody debris. Other projects include planting riparian vegetation, stabilizing banks, building fire lines for controlled burns, installing bird boxes, and monitoring ecological conditions.

These projects have had significant effects, By 2001, a degraded stream bed restored by the 1982 seniors had risen 18 inches, its formerly sheer banks replaced by gentle vegetated slopes. A foot of cool water was present where there had been only dry stream bed. Ten years after restoration work began on Rock Creek, riparian vegetation had grown 25 feet, summer water temperatures were reduced, and fish populations had risen measurably. The class of 1997 built fencing to keep cows, campers, and off-road vehicles out of a severely degraded floodplain on Gate Creek. Just 6 years later beavers used new vegetation there to build three dams, creating new habitat for wetland vegetation, wood ducks, and breeding chorus frogs.

The projects have also had significant effects on the students. Catlin Gable students won acceptance for restoration from skeptical rancher neighbors through their hard work and idealism. A member of the class of 1978 used his experience planting willows for stream restoration to restore habitat on his own property. A 1992 graduate persuaded his college fraternity brothers to volunteer to help restore habitat of the red-cockaded woodpecker in a Texas national forest. The Catlin Gable experience supports William Jordan's theory that practicing ecological restoration is an engagement with nature that inspires hope for the future and prompts greater concern for nature (Jordan 2003). The future impact of thousands of Northwest students coming into contact with restoration on and off their campuses is yet to be known but has to be positive.

Another key has been the establishment of citizen-based watershed councils. Salmon, as Freeman House describes them, are a totem species that live in the rivers that course through the veins of the entire Northwest ecosystem (House 1999). The goal to bring back wild salmon cuts across political divisions and the urban–rural divide. Watershed councils include environmental advocates, timber and farm landowners, and agency representatives. At first many councils were dysfunctional, Monty Pythonesque argument clinics with no funding and ridiculously impossible missions. Factions wary of each other's motives and world views disagreed over fundamental causes of salmon decline and possible solutions ("It's the dams." "No, it's the ocean conditions." "It is the return of the seals." "It is overfishing." "It is logging." "No, it is urban sprawl."). But gradually councils have begun to make real progress. Council staff have increasingly become community organizers and institution builders rather than technical experts. Although watershed councils have no regulatory authority and little base funding and rely on voluntary cooperation, many landowners need only technical help and small amounts of support. All across the region, councils are building fences, planting trees, sowing native grasses, filling unused ditches, and, most importantly, spreading acceptance of restoration in local communities.

It is hard to tell how wide or deep restoration will spread into the culture of the Northwest. It may be that only nature nerds and their immediate friends and families will learn, pitch in, and provide support, and the effort will be limited to a small subculture of a consumer-driven society (E. Alverson, personal communication, 2005). To be sure, there are still plenty of rural Northwesterners who will never be persuaded to fence cattle out of a stream, and shopping malls will continue to attract more people on weekends than ivy-pulling contests. But as more and more of their neighbors invest in better stewardship, peer pressure builds. As landscape conditions change and knowledge builds, there is a trend toward better land responsibility and stewardship (Figure 18.2), but it is countered by the drive to use it all up or cash it in now.

Assisted Renewal and the Restoration of Hope

A few weeks before the final manuscript due date, we were on a short hike along Santiam Pass in the Mt. Jefferson Wilderness of Oregon's Cascade Mountains. A huge stand-replacing fire had swept through the area 2 years earlier, and nearly all the overstory trees were dead. A few clumps of surviv-

FIGURE 18.2. A former road in Mt. Hood National Forest, ripped and replanted. The Forest Service and Bureau of Land Management have closed many more miles of road than they have built in the past decade. (*Photo by Dean Apostol*)

ing trees lined the road that led to the trailhead, and a profusion of wildflowers carpeted the former forest floor. A view from a rocky knoll showed a burned forest as far as the eye could see. We talked about how, not so long ago, nearly all Northwest land managers and residents would have viewed this scene as devastation, and the media would describe the forest as having been "destroyed." But now, many see renewal.

Forests of the high Cascades are among the least altered ecosystems in North America, and as Aldo Leopold (1949) would say, they have all their parts except for wolves and grizzlies. Because these forests have an infrequent, high-intensity stand replacement fire regime, they have not been as affected by well-meaning but short-sighted fire suppression policies as have lower-elevation, drier forests. Ecologists now have confidence that once burned, these high montane forests will renew themselves. Some introduced invasive species may join early-successional communities, but this should not be a long-lasting effect as competition increases. So if nature can renew itself, why do we need ecological restoration?

An interesting debate in restoration circles over the past 15 years has been the issue of natural- versus human-influenced landscapes. Eric Katz (1992) and other environmental philosophers have challenged restorationists on ethical grounds. They say landscapes cannot be natural if humans are tending them. Restorationists argue back that humans are part of nature and that indigenous people have helped shape most landscapes on Earth, certainly in temperate zones. The counter is that we have to distinguish culture from nature, or anything goes. Replacing a wilderness with a strip mine would be easy to accept if we define all human works as natural because humans are natural.

But the new working definition—"ecological restoration is the assisted natural recovery of damaged ecosystems"—means that anything does *not* go. As William Jordan (2003) has stated, restorationists have business with ecosystems. This business is the work of helping ecosystems heal themselves. It is both a natural and a cultural activity. Restored landscapes are not diminished simply because humans had a hand in shaping them.

At Santiam Pass, the human community could take responsibility for monitoring the forest's recovery and yank out invasive weeds for a few years, thus giving natives the edge in the successional race. They could make sure erosion is controlled along trails and campsites, our means of access to wild areas. They could even reintroduce wolves and grizzly bears, which are missing from the Cascades only because of previous human interventions. These are all restorative actions that may assist natural recovery. None of them would diminish the naturalness or value of the forest as it recovers.

There are numerous examples of Northwest ecosystems beginning to renew themselves, sometimes with only a gentle nudge from humans, at other times with a much firmer hand. Large dams across the Northwest are scheduled for removal, with almost no new ones being built. Tidal wetlands are recovering once dikes are breached. All-out assaults on invasive species are helping to stabilize and restore degraded prairies, oak woodlands, and sagebrush steppe. Restorationists are learning how to work in tandem with nature's self-healing power. A recent report by the Pacific Northwest Research Station notes that the acreage of late-successional and old-growth forests has increased by about 600,000 acres over the past 10 years, despite several large fires and continued timber harvest (Haynes et al. in press). This increase can be attributed mostly to protection rather than restoration, yet road removal and careful thinning will strengthen and accelerate this renewal if land managers are careful and continue to adapt their methods and approaches. Yet even with this increase in habitat, spotted owl populations continue to decline (Figure 18.3). This may be a temporary trend in response to habitat depletion, or it may be that habitat alterations have provided an advantage for competing nonnative species such as the barred owl. For all we know, this may be the case for other regional ecosystems and species; thus, even as habi-

FIGURE 18.3. The northern spotted owl (*Strix occidentalis*). Conservation of this bird has initiated widespread conservation and restoration of Pacific Northwest forests. (*Photo courtesy of the U.S. Forest Service*)

tat is recovered, populations may continue to diminish for a time (E. Alverson, personal communication, 2005).

Restoration practitioners have chosen a demanding, generally low-paying, technically challenging field of work. They are a small but growing part of the worldwide team intent on rescuing the planet's biodiversity. But in addition to restoring ecosystems, restoration must restore hope. The environmental studies section at Powell's Books in Portland (one of the world's greatest bookstores) has stacks of titles on industrial agriculture, global warming, loss of rainforests and biodiversity, ocean depletion, and much more—the gloom and doom of environmental degradation. This literary genre started with *Silent Spring* (Carson 1962) and moved to *Limits to Growth* (Meadows et al. 1972), *The End of Nature* (McKibben 1990), and many more recent titles. These are important books that have roused environmentalists and many restorationists into taking action. But they can also drain hope because in many cases they focus more on problems than on solutions.

Ecological restoration is part of a wider political, cultural, and technical effort to turn the wheel of the great ship of human progress before it runs smack into rocks that lie at shallow depth, possibly just ahead. Environmental authors have returned from scouting missions. They have seen the rocks ahead and have reported back to the ship. The chart on the bridge says there are no rocks, or if there are they are at safe depths. The captain says changing course is very expensive and could lead us to more dangerous waters. Better to continue full steam ahead. Meanwhile, restorationists do their work mostly belowdecks, repairing damage already inflicted when the ship hit previously unseen or ignored rocks (DDT, draining the Everglades, old-growth logging). They are learning critical skills that will be needed far into the future. Most are so busy repairing damage, they have little time to lobby the bridge to change course. Deep down, most know that the few repairs made will not be sufficient if the ship hits something really big, such as rapid and extreme global warming.

Nevertheless, the products of restoration cocreated with nature have real value to our fellow species as well as to humans. Perhaps more importantly, this work shows others that humans can coexist with nature and even be beneficial, not inherently harmful. In the Pacific Northwest, a lot has been accomplished in the past few decades. The future is not at all clear, but if hope can be restored, then the repair of ecosystems will not be far behind.

REFERENCES

Brinkman, J. 2002. PGE will remove 2 dams in basin of Sandy River. *The Oregonian*, October 12.

Carson, R. 1962. *Silent Spring*. Houghton Mifflin, Boston.

Fairbank, Maslin, Maullin & Associates. 2004. *Washington Public Lands Survey, 2004*. Unpublished findings.

FEMAT. 1993. *Forest Ecosystem Management: An Ecological, Economic, and Social Assessment*. U.S. GPO 1993-793-071. U.S. Government Printing Office, Washington, DC.

Haynes, R. W., B. T. Bormann, D. C. Lee, and J. R. Martin. In press. *Northwest Forest Plan: The First 10 Years (1994–2003): Synthesis of Monitoring and Research Results*. Gen. Tech. Rep. PNW-GTR-651. U.S. Department of Agriculture, Forest Service, Pacific Northwest Research Station, Portland, OR.

House, F. 1999. *Totem Salmon: Life Lessons from Another Species*. Beacon Press, Boston, MA.

Jordan, W. R. III. 2003. *The Sunflower Forest: Ecological Restoration and the New Communion with Nature*. University of California Press, Berkeley.

Katz, E. 1992. The big lie: human restoration of nature. *Research in Philosophy and Technology* 12: 231–242.

Larabee, M. 2005. Trail winds through a rare wildlife link on private land. *The Oregonian*, July 28.

Leopold, A. 1949. *A Sand County Almanac, and Sketches Here and There*. Oxford University Press, New York.

Lichatowich, J. 1999. *Salmon without Rivers: A History of the Pacific Salmon Crisis*. Island Press, Washington, DC.

McKibben, B. 1990. *The End of Nature*. Anchor Books/Doubleday, New York.

Meadows, D. H., D. I. Meadows, J. Randers, and W. W. Behrens III. 1972. *The Limits to Growth. A Report to the Club of Rome*. Potomac Associates, New York.

Northwest Power and Conservation Council. 2005. *Briefing Book*. Available online at www.nwcouncil.org/library/2005/2005-1.htm.

Oregon Department of Fish and Wildlife. 2005. *Draft State Wildlife Habitat Conservation Strategy*. Unpublished manuscript.

Oregon Watershed Enhancement Board Web site. 2005. www.oweb.state.or.us/OWEB/GRANTS.

Pacific Coast Salmon Recovery Fund Report to Congress. 2005. Available online at www.nwr.noaa.gov/pcsrf/.

University of California at Davis. 2005. Web site: ecology.ucdavis.edu/AOE/Restoration/rest_home.htm.

About the Contributors

Bob Altman works for the American Bird Conservancy as the Northern Pacific Rainforest Bird Conservation Region Coordinator. He also is a science coordinator for the Pacific Coast Joint Venture. In these roles he works with all the bird initiatives and numerous other partners from northern California to western Alaska to advance bird and habitat conservation through a variety of activities including research and monitoring, planning, management, and outreach. Bob has been active in the Partners in Flight Initiative for landbird conservation at regional, national, and international levels, including authoring five Oregon–Washington Bird Conservation Plans, and is a current member of the National Science Team.

Ed Alverson has more than 25 years experience as a field botanist in the Pacific Northwest and has been The Nature Conservancy's Willamette Valley Stewardship Ecologist since 1991. He has bachelor's degrees from Evergreen State College and a master's degree in botany from Oregon State University. His interests include ecological restoration and management, field inventory and surveys, plant systematics and floristics, and ecological history.

Dean Apostol is a landscape architect, natural resource planner, writer, and teacher. He graduated from Iowa State University with a B.S. in landscape architecture and did graduate work in geography at Portland State University. Dean coauthored with Nancy Diaz the book *Forest Landscape Analysis and Design*, published by the Pacific Northwest Research Station in 1992. He has nearly completed (with Simon Bell) an updated volume titled *Designing Sustainable Forest Landscapes* (E&F Spon Press). His professional practice includes watershed assessment, open space master planning, urban design, forest ecology and certification, landscape aesthetics, trail design, and ecological restoration.

Stephen F. Arno is a forest ecologist who retired from the USDA Forest Service's Rocky Mountain Research Station after 31 years. He has studied the effects of fire and the use of prescribed fire and fuel reduction treatments and has authored more than 100 scientific publications. He holds a Ph.D. in forestry and plant science from the University of Montana and has practiced restoration forestry on his family's ponderosa pine forest for 30 years. In 2002, his book *Flames in Our Forest: Disaster or Renewal?* coauthored by science writer Steve Allison-Bunnell, was published by Island Press. He has authored numerous other papers and books.

Edward E. Bangs has been the U.S. Department of Interior Fish and Wildlife Service wolf recovery coordinator for the Northwest since 1988. He led efforts to reintroduce wolves to central Idaho and Yellowstone National Park in 1995. Before moving to Helena, Montana, Ed was a wildlife biologist on the Kenai National Wildlife Refuge in Alaska for 13 years, where he worked on a large variety of species including wolves, brown bear, lynx, moose, caribou, bald eagles, and red-backed voles. His professional interests include restoration of ecological processes and human values toward wildlife management.

Dean Rae Berg is a consulting silviculturist and forest engineer based near Everett, Washington. His graduate studies and experience are in forestry, forest engineering, ecology, riparian management, and fisheries. Dean has 26 years of experience in forest management, most recently in variable retention silviculture and ecological engineering. He has worked as a forest consultant and currently maintains a private practice in forest engineering, with a focus on ecologically based harvest practices. Dean has been widely published in forestry journals and books.

Robert W. Butler is a research scientist with the Canadian wildlife service. He is one of the leading authorities on great blue herons in western Canada. He has published a number of books on wildlife conservation.

Bruce H. Campbell is a biologist with the Oregon Department of Fish and Wildlife. He grew to love the outdoors and wildlife while growing up on farms in northeastern Oregon, the Willamette Valley, Idaho, and the Oregon coast. Bruce earned his bachelor's degree in biology from Central Oklahoma State University. He attended graduate school at Northern Arizona University, where he concentrated on botany and bird and mammal ecology. Bruce's career has taken him from the Sonora Desert to the Arctic. Bruce has authored more than fifteen journal articles and major symposium papers. His current professional interests include working with private landowners to restore wildlife habitats and human dimensions in wildlife management. He recently authored *Restoring Rare Habitats in the Willamette Valley*, published by Defenders of Wildlife.

Andrew B. Carey is chief research biologist and leader of the Ecological Foundations of Biodiversity Team of the Pacific Northwest Research Station. He has a B.S. in forestry and wildlife, an M.S. in wildlife management, an M.S. in organization development, and a Ph.D. in zoology and entomology. He has been leading the Forest Ecosystem Study, which emphasizes the effects of management history, biological legacies, and induced canopy heterogeneity on vascular plants, fungi, soil food webs, litter invertebrate communities, small mammal communities, arboreal rodent communities, winter bird communities, and spring bird communities in second-growth forests in the Puget Trough of Washington. He has published his research widely in major forest, wildlife, and ecology journals.

Bob Carey has been the Skagit River program director for The Nature Conservancy since 1998. He has an M.A. in political ecology from Western Washington University and a B.A. in economics from Rutgers University. Before his current position he spent 3 years as a community conservation specialist for The Nature Conservancy in Nicaragua and 5 years as a wilderness ranger for the National Park Service and U.S. Forest Service.

Tara W. Carolin is an ecologist and research coordinator at Glacier National Park. She earned a B.S. and M.S. in wildlife and rangeland resources at Brigham Young University. She completed additional graduate work at Texas A&M University. She began work at Glacier National Park in 1996 and before that worked as a biologist for the National Park Service Southeast Utah Group, including Arches National Park, Canyonlands National Park, and Natural Bridges National Monument. Her work has focused on restoration and long-term monitoring, inventory and monitoring of sensitive resources, whitebark and limber pine restoration, vegetation mapping, research coordination, and environmental analysis and compliance.

Blair Csuti is the conservation program coordinator for the Oregon Zoo.

Patrick Dunn has been restoring prairies with a range of partners from the Department of Defense to individual landowners in the South Puget Sound region for more than a decade. He is a program manager for The Nature Conservancy, based in Olympia, Washington. He has an M.S. from California State University, Los Angeles. In his 25 years of conservation work he has also led a private company and worked for public agencies. His restoration projects have occurred in habitats ranging from coastal salt marshes to alpine meadows, tropical rain and dry forests, and prairies and oak woodlands.

Peter Dunwiddie is the director of research programs in the Washington Field Office of The Nature Conservancy. He earned his doctorate in plant ecology at the University of Washington and has spent much of his career as an ecologist with conservation organizations. He has published extensively in a variety of fields, including fire ecology, dendrochronology, conservation biology, paleoecology, and land use history.

Carl E. Fiedler earned a Ph.D. in silviculture and forest ecology from the University of Minnesota and is currently a research professor at the University of Montana. His research focuses on uneven-aged silvicultural applications, forest restoration, and old-growth stand dynamics. He has testified to congressional subcommittees on forest health and wildfire hazard issues and was lead scientist in the recently completed statewide analyses of fire hazard in Montana and New Mexico. He also collaborates with the Bureau of Indian Affairs in developing and presenting silviculture and fire training courses on Indian reservations throughout the United States. He coauthored *Mimicking Nature's Fire: Restoring Fire-Prone Forests in the West*, published by Island Press in the spring of 2005.

Jerry F. Franklin is a professor of ecosystem analysis at the University of Washington. He is among the nation's foremost experts on the ecology of old-growth conifer forests of the Pacific Northwest. A native of the Northwest, he began his career as a research forester with the Pacific Northwest Research Station, where he eventually became chief plant ecologist. His publications include the classic *Natural Vegetation of Oregon and Washington*, first published in 1969, and *Creating a New Forestry for the 21st Century*, a book that has profoundly influenced the way we think about and manage forests across the world.

Kimberly Frappier is a botanist and restoration ecologist from Seattle, Washington. She earned an M.S. in environmental horticulture from the University of Washington College of Forest Resources. She currently works for the University of Washington Botanic Gardens' rare plant recovery and education programs. She has worked as a botanist for Mt. Rainier National Park and the Washington Department of Natural Resources.

Ralph J. Garono is president of Earth Design Consultants, Inc. in Corvallis, Oregon, where he heads the Wetland & Watershed Assessment Group. He holds a faculty (courtesy) appointment in the College of Oceanic and Atmospheric Sciences at Oregon State University and advises graduate students in the Marine Resource Management Program. He is trained in biochemical limnology and aquatic entomology. His current research focuses on assessments of aquatic and marine ecosystems. His recent research projects include developing restoration scenarios for the restoration of Deschutes River Estuary and Capitol Lake (Olympia, Washington), the use of aquatic insects as a wetland assessment tool, GIS-based watershed assessments, the use of GIS-based models to evaluate alternative land use scenarios, and the use of hyperspectral imagery to map landscape patterns in estuarine vegetation.

Jennifer Goodridge is a wetland specialist at Pacific Habitat Services Inc. in Wilsonville, Oregon. She has an M.S. in plant ecology from Oregon State University and teaches a native wetland plant course at Portland State University.

Elizabeth Gray is the director of conservation science for The Nature Conservancy of Washington. In this role, Elizabeth oversees ecoregional and site planning, on-the-ground stewardship, restoration, and research activities, including a large-scale experiment on prairie restoration in the Pacific Northwest. Elizabeth earned her Ph.D. from the University of Washington and her A.B. from Harvard. Her research specialties include conservation biology, behavioral ecology, and conservation genetics, with a focus on how birds can be used as indicators of successful habitat restoration.

Fritzi Grevstad is the biocontrol specialist for the University of Washington's Olympic Natural Resources Center. She has a B.S. in zoology from the University of Washington and a Ph.D. in ecology, with emphases in insect ecology and biological control, from Cornell University. Since 2000 she has led a biocontrol program against *Spartina* in Willapa Bay and Puget Sound, Washington. She has also worked with the purple loosestrife and Japanese knotweed biocontrol systems. Her research interests include optimal release strategies for establishing agent populations, interactions between agent species, and the role of the agent geographic source on performance in the introduced environment.

Dale Groff is a horticulturist and geologist for Pacific Habitat Services, Inc. in Wilsonville, Oregon. He is an expert in soil and water chemistry, plant physiology, and native plant growth needs. Dale is a restoration specialist who provides technical support on wetland mitigation, constructed wetlands for wastewater treatment, erosion control, and stream restoration projects.

Richard A. Hardt is a forest ecologist with the Eugene District Office of the Bureau of Land Management. He has worked for the Bureau of Land Management for 11 years, specializing in forest ecology, planning, and environmental analysis. He has a B.A. in natural sciences from the Johns Hopkins University, an M.L.A. from Harvard University, and a Ph.D. in forest resources from the University of Georgia.

David W. Hays is with the Washington Department of Fish and Wildlife in Olympia.

O. Eugene (Gene) Hickman is a consulting plant ecologist and range management specialist, previously retired from a career with the U.S. Department of Agriculture Natural Resources Conservation Service. His experience includes extensive vegetation inventory and mapping for range management plans, watershed planning, and soil survey publications. Gene has written papers on rangeland, historic vegetation, and the ecology of western and central Oregon.

Eric Higgs is a professor and director of the School of Environmental Studies at the University of Victoria. He served as chair of the Board of the Directors for the Society for Ecological Restoration International from 2001 to 2003 and is the author of *Nature by Design: People, Process and Ecological Restoration.*

Randal W. Hill is a wildlife biologist with the U.S. Department of the Interior Fish and Wildlife Service in Othello, Washington. As the biologist for Columbia National Wildlife Refuge for 15 years he is responsible for management and restoration of shrub steppe, wetland, and riparian habitats. He has a special interest in controlling cheatgrass, Russian-olive, and other invasive plants to benefit native plant communities in the Columbia Basin.

Paul E. Hosten started his botanical training in utero as intern to his botanist mother. He completed his undergraduate training in botany at the University of Port Elizabeth and Rhodes University and an M.S. degree at the University of

Natal in South Africa. He worked at the University of Ben Gurion, examining the influence of Bedouin goat and sheep grazing on the sand field vegetation of the Negev desert in Israel and then moved to Utah to study the influence of livestock on soil microphytic crust. He completed his Ph.D. at Utah State University, examining the influence of fire and livestock grazing on long-term vegetation dynamics of sagebrush steppe vegetation. He has lived in southwest Oregon for the past 10 years, studying local grasslands, chaparral, and woodlands.

Laurie L. Kurth earned a B.S. in botany from Ohio University and an M.S. in biology from the University of Chicago. She started working in the Pacific Northwest as a backcountry ranger in Olympic National Park in 1984 and 1985. After working as a biologist and plant ecologist at Indiana Dunes National Lakeshore, the U.S. Army Corps of Engineers Chicago District, and Glacier and Zion national parks, she returned to the Pacific Northwest in 1999 as a plant ecologist at Mt. Rainier National Park. At Mt. Rainier she manages the restoration, rare plant, human impacts, whitebark pine, and introduced species programs. Previously she established restoration and fire monitoring programs at Glacier National Park and established the restoration and invasive species control programs at Zion National Park.

Frank K. Lake was raised in northwestern California, learning subsistence activities and traditional ecological knowledge from the local tribes and family. He graduated in 1995 with a B.S. from University of California, Davis. He has worked as a fishery habitat biologist, ethnoecologist, and restoration ecologist, incorporating traditional ecological knowledge with Western science to restore and conserve biodiversity in the Klamath–Siskiyou and Pacific Northwest bioregions. Currently, he is a Ph.D. candidate in the Environmental Sciences Program at Oregon State University and works for the U.S. Department of Agriculture Forest Service Pacific Southwest Research Station on tribal and community forestry issues, fuel reduction and prescribed burning, and fire effects on riparian areas.

Frank A. Lang is an emeritus professor of biology at Southern Oregon University. He holds a B.S. from Oregon State College (botany), an M.S. from the University of Washington, and a Ph.D. from the University of British Columbia. Retirement activities include a weekly radio spot on Jefferson Public Radio called *Nature Notes*, associate editor of *Kalmiopsis Journal* of the Native Plant Society of Oregon, a continuing interest in the vegetation and flora of the Klamath Ecoregion and the history of the botanical exploration of the Pacific Northwest, service on the Jackson County Natural Resources Advisory Council and the Medford District Bureau of Land Management Natural Resources Advisory Committee, and service on the Board of Directors for the Klamath Bird Observatory and the Crater Lake Natural History Association.

Steven O. Link earned a B.S. in biology and a B.M. in mathematics at the University of Minnesota in 1977 and a Ph.D. in botany at Arizona State University in 1983. He did postdoctoral work in the Range Science Department at Utah State University until 1985. Steve has been active in environmental associations, serving as president of the Benton County Noxious Weed Control Board, president of the Columbia Basin Chapter of the Washington Native Plant Society, and president of the Society for Ecological Restoration Northwest Chapter. He has authored more than 400 articles, book chapters, and proceedings. Steve joined the faculty of Washington State University in 1996 and currently does research on restoration ecology, fire risk, invasive plant species management, and plant–water relationships.

Dennis Martinez is of mixed O'odham, Chicano, and Anglo heritage and has 36 years of restoration experience. Over the past 20 years Dennis has been building bridges between Western science and traditional ecological knowledge by bringing indigenous scholars to scientific conferences. Dennis is founder and chair of the Indigenous Peoples' Restoration Network and has served on the Society for Ecological Restoration International Board of Directors. He has also served on the Traditional Knowledge Council of the American Indian Science and Engineering Society. Dennis coauthored the *Educators' Guide to American Indian Resource Management* and has delivered more than 150 invited papers to major scientific, environmental, and indigenous conferences in the United States and several foreign countries. He has worked as a restoration contractor and cofounded Design Associates Working with Nature, the first major restoration contracting, consulting, and native plant growing organization on the West Coast. His academic background is in the history and philosophy of science (University of California at Berkeley). Dennis received the John Reiger Service Award in ecological restoration, two fellowships from Bioneers Collective Heritage Foundation for environmental justice, and Ecotrust's Buffett Award for Indigenous Conservation Leadership in N.W. North America and is noted in *Biography of American Environmental Leaders*.

William H. Mast is a principal at Wildlands Inc. in Richland, Washington. He specializes in design and implementation of environmental restoration. He has managed and completed complex environmental restoration projects in Washington, Utah, Idaho, Montana, Colorado, and Oregon. William's work includes major projects on pipelines, linear projects, desert shrub steppe restoration, wetlands, landfills, highway rights of way, and dust control. He also has extensive experience with production agriculture.

Robert R. Mierendorf is the park archeologist at North Cascades National Park Service Complex. He holds degrees in sociology and anthropology from Iowa State and Washington State universities. For the last 21 years he has focused on the management and research of archeological sites representing the precontact history of mountain-oriented indigenous peoples. Robert established the archeological program at North Cascades National Park as the first park archeologist in the Pacific Northwest. He also serves as the park's liaison with indigenous communities and is active in public education as an instructor and board member at North Cascades Institute, a nonprofit environmental education organization based in the Pacific Northwest.

Judith Myers is with the University of British Columbia and is a leading expert on invasive plant management and biological control methods.

Ben Perkowski has been technical coordinator for the Skagit Watershed Council in Mt. Vernon, Washington since 1998. Before coming to the council, he was a restoration ecologist for an environmental consulting firm in Seattle for 5 years. He also has worked as a newspaper editor and reporter in the Carolinas. Ben earned a master's degree in environmental management from Duke University and a bachelor's degree from the University of North Carolina at Chapel Hill.

David F. Polster is a plant ecologist with more than 27 years of experience in vegetation studies, reclamation, and invasive species management. He graduated from the University of Victoria with an honors B.S. degree in 1975 and an M.S. degree in 1977. He has developed a wide variety of reclamation techniques for unstable slopes and techniques for reestablishing riparian and aquatic habitats. David is president of the Canadian Land Reclamation Association and treasurer for the British Columbia Chapter of the Society for Ecological Restoration, and he serves on the board of the Provincial Invasive Plant Council.

Gordon H. Reeves has worked as a research fish biologist with the Aquatic and Land Interaction Program of the U.S. Department of Agriculture Forest Service, Pacific Northwest Research Station in Corvallis, Oregon since 1984. He has an M.S. from Humboldt State University, Arcata, California (1978) and a Ph.D. from Oregon State University, Corvallis in 1984. His research focuses on the impact of land management practices on juvenile anadromous salmon and trout and their freshwater habitats, the dynamics of aquatic ecosystems, and the role of disturbances in creating and maintaining fish habitats in the Pacific Northwest and Alaska, and the development of monitoring plans. He has participated in several efforts that developed and evaluated options for managing federal lands in the Pacific Northwest and Alaska. He was the co-leader of the aquatic group in the Federal Ecosystem Management Assessment Team that produced the Northwest Forest Plan. He is a courtesy professor at both Oregon State University and Humboldt State University.

Jon L. Riedel is a geologist at North Cascades National Park Service Complex. Jon earned B.S. degrees in biology and geography from the University of Wisconsin at LaCrosse in 1982 and an M.S. in physical geography from the University of Wisconsin at Madison in 1987 and is completing a Ph.D. at Simon Fraser University. He has published papers on recent glacial fluctuations and landslides in the North Cascades and is studying the glacial history of the upper Skagit Valley. Jon is also leading an effort to monitor the mass balance of six glaciers in North Cascades and Mt. Rainier national parks and is engaged in mapping surficial geology and soils for all of the national parks in Washington.

Regina M. Rochefort is a plant ecologist and science advisor at North Cascades National Park Service Complex. She earned a B.S. in biology from Northeastern University, an M.F.S. from Yale University, and a Ph.D. from the University of Washington. Regina began working in the Pacific Northwest in 1984 as botanist for Mt. Rainier National Park. While at Mt. Rainier she established a parkwide subalpine restoration program and developed programs to monitor rare plants, whitebark pine, and human impacts and to survey and control introduced invasive plants. Before coming to the Pacific Northwest, she worked in Florida as a botanist at Everglades National Park and fire ecologist in Big Cypress National Preserve.

Emily Roth is an ecologist at Jones and Stokes in Portland, Oregon and has extensive expertise in wetlands, natural area management, and environmental analysis.

Cheryl Schultz is a conservation biologist at Washington State University in Vancouver. Her research focuses on ecology and conservation of at-risk butterflies in the Pacific Northwest and restoration of their prairie habitats.

René Senos, of Jones & Jones Architects and Landscape Architects in Seattle, earned M.L.A. and B.L.A. degrees in landscape architecture from the University of Oregon and a B.S. in sociology from the University of Delaware. Her work at Jones & Jones focuses on integrating cultural and ecological issues in a holistic approach to landscape planning and design. She has also worked with Native American tribes to identify and protect traditional resources, including collaboration with the Confederated Salish and Kootenai to identify opportunities for cultural landscape restoration along the U.S. 93 scenic corridor. Her ongoing research addresses ecological restoration as a model for healing both natural and social communities, including a comprehensive study on the role of citizen-based action related to restoring wild salmon.

Marcia Sinclair is a writer, editor, and public involvement consultant. She is the outreach specialist for the Willamette Partnership, a nonprofit organization building a conservation marketplace to increase the pace, scope, and effectiveness of conservation actions in the Willamette River Basin of western Oregon. She earned a B.A. in public relations from San Jose State University. Marcia is the author of *Hiking Mount Hood National Forest*, an interpretive natural and cultural history hiking guide. Recent publications include *Challenge of Change*, a summary of the key findings from the *Willamette River Basin Planning Atlas* (OSU Press). She is a self-taught naturalist who has translated reams of government natural resource documents and technical materials into clear English, given talks, and led tours on habitat restoration, and she writes a nature column for her local newspaper. She focuses on restoring people's sense of wonder in the natural world and identifying ways for human society to work in harmony with natural systems.

Frank Slavens is the former curator of reptiles for the Woodland Park Zoological Society and is currently a field biologist for the Western Pond Turtle Project at Woodland Park Zoological Gardens.

Kate Slavens is a field biologist with the Washington Department of Fish and Wildlife, specializing in recovery of the western pond turtle.

Doug W. Smith has conducted research on wolves in Yellowstone National Park since 1994 and is currently the park's project leader for wolf research. Doug previously worked on various wolf-related programs on Isle Royale National Park, where he also conducted work on beaver ecology for his Ph.D. Doug's professional interests include discovering the mechanisms involved in top-down trophic cascades and predator–prey theory.

Jonathan Soll is a Willamette Basin conservation director with The Nature Conservancy in Portland, Oregon. He has investigated and managed invasive species since 1989, with emphasis on urban natural areas since 1999.

David L. Steensen is a geologist with the National Park Service, Geologic Resources Division. He earned a B.S. from Western Washington University and an M.S. from California State University, Humboldt, both in geology. He has been working as an engineering geologist and geomorphologist for the past 20 years, concerned primarily with correcting the physical effects of land uses. He has worked for the National Park Service since 1986, where he began with the Watershed Restoration Program at Redwood National Park. He now works for the Geologic Resources Division, where he coordinates a program of restoring resources and processes disturbed by human activities.

Erin Thompson is an assistant research ecologist with Earth Design Consultants. She has a master's degree from the University of Michigan's School of Natural Resources and Environment in forest ecology and entomology. Her training includes forest insects and diseases, with a special interest in wood-boring beetles. Her experience with aquatic ecosystems comes from growing up very near the Great Lakes and her summer spent as stream survey technician for the Wind River Ranger District in Carson, Washington.

Nancy Turner is an ethnobotanist and professor in the School of Environmental Studies at the University of Victoria in British Columbia. While working on her thesis, she collaborated with Saanich First Nations elders to learn about the significance of plants to their culture. Her postgraduate work concentrated on plant classification systems among the Haida, Nuxalk (Bella Coola), and Stl'atl'imx (Lillooet) people. Her major research contributions have been in demonstrating the pivotal role of plant resources in past and contemporary aboriginal cultures and languages, as an integral component of traditional knowledge systems, and the ways in which traditional management of plant resources has shaped the landscapes and habitats of western Canada. Nancy has authored many papers and several books, including *Plant Technology of British Columbia First Peoples* (1998), a classic among Northwest restoration ecologists.

John van Staveren is president and senior scientist at Pacific Habitat Services Inc., located in Wilsonville, Oregon. John is one of the leading wetland specialists in the Northwest. As senior scientist, he directs the firm's environmental and regulatory compliance activities. As a professional wetland scientist, he has extensive experience in designing freshwater and estuarine mitigation projects, wetland and riparian assessments, environmental regulations, and public presentations.

David Vesely is a wildlife biologist and partner with Pacific Wildlife Research, an ecological consulting team based in Corvallis, Oregon.

Warren Warttig is a registered professional biologist for International Forest Products Ltd. at Campbell River, British Columbia. Warren has completed several ecosystem-based watershed restoration plans on the coast of British Columbia and Vancouver Island over the last 10 years, including the Kennedy Flats Restoration Plan in Clayoquot Sound. The Kennedy Flats restoration has been implemented since 1994, with dramatic results in improved habitat conditions and subsequent increases in salmon returns. Warren is also intimately involved in the forest resource planning in Clayoquot Sound, following the Clayoquot Sound Scientific Panel report's five recommendations. In Warren's words, "Restoration is pointless unless all development follows the same ecosystem-based planning principles that the restoration does."

Jack E. Williams is the executive director of Trout Unlimited. Jack served as the forest supervisor of the Rogue River and Siskiyou national forests in Oregon and the deputy forest supervisor of the Boise National Forest in Idaho. He also has worked for the Bureau of Land Management, including serving as the national fishery program manager and science advisor to the director in Washington, D.C. Jack has written more than 100 technical scientific articles, popular science articles, books, and book chapters on the subjects of conservation, watershed restoration, fisheries, ecosystem management, and endangered species. His education includes a B.S. in wildlife biology from Arizona State University, an M.S. in biology from the University of Nevada in Las Vegas, and a Ph.D. in fishery science from Oregon State University.

Mark Griswold Wilson is a restoration ecologist with the City Nature Program of Portland, Oregon, where he designs, finds funding for, and implements restoration and revegetation projects in urban natural areas. Previously Mark was an independent consulting ecologist and an environmental educator in high school and community college settings. He also served on the board of directors for the Society for Ecological Restoration Northwest Chapter.

Supporters and Partners

We gratefully acknowledge and thank the foundations, individuals, businesses, and agencies who contributed funds, staff, or other assistance critical to the making of this book.

Ann Lennartz has been a Seattle resident for more than 20 years. Her varied interests include ecology and ecological restoration, science, geography, music, and support for strong communities. She has served as a community volunteer in many of these areas of interest.

The Bullitt Foundation's mission is to protect, restore, and maintain the natural environment of the Pacific Northwest for present and future generations. The foundation supports nonprofit organizations that serve Washington, Oregon, Idaho, British Columbia, western Montana (including the Rocky Mountain Range), and coastal Alaska from the Cook Inlet to the Canadian border.

The Spirit Mountain Community Fund promotes Native American culture, history, and values through charitable support of education, arts and culture, historic preservation, and the environment. Benevolence and compassion are primary Native American values. Spirit Mountain promotes supporting aspirations of youth, respecting and honoring elders, and developing self-sufficiency while sustaining and preserving the air, water, and land and its inhabitants.

Portland State University's Mark O. Hatfield School of Government offers students the latest technological capabilities and exposure to a wide variety of courses in political science, public administration, and criminal justice. In addition, the Hatfield School draws significantly from its five institutes and one center, which conduct research and offer classes specializing in environmental studies, executive leadership, nonprofit management, criminal justice policy, tribal government, and U.S.–Turkish affairs.

Trout Unlimited's mission is to conserve, protect, and restore North America's trout and salmon fisheries and their watersheds. They accomplish this mission on local, state, and national levels with an extensive and dedicated volunteer network. Their national and regional offices employ professionals who testify before Congress, publish a quarterly magazine, intervene in federal legal proceedings, and work with 142,000 volunteers in 450 chapters nationwide to keep them active and involved in conservation issues.

The South Slough Estuarine Reserve is a 4,700-acre natural area on the Oregon coast dedicated to research, education, and stewardship. Their support for this book was in collaboration with the **Oregon Department of State Lands**. The mission of the Department of State Lands is to ensure a legacy for Oregonians and their public schools through sound stewardship of lands, wetlands, waterways, unclaimed property, estates, and the Common School Fund.

Walker Macy is an award-winning Portland landscape architecture, planning, and sustainable design firm that specializes in parks and open spaces, waterfronts, roof gardens, campuses, stormwater management, and similar projects.

Portland Metro is the directly elected regional government that serves more than 1.3 million residents in Clackamas, Multnomah, and Washington counties and the twenty-six cities in the Portland, Oregon metropolitan area. Their Parks, Trails, and Greenspaces program is a leader in restoration and management of urban natural areas, including wetlands, oak woodlands, and old-growth conifer forests.

Vigil–Agrimis is a professional service consulting firm specializing in water and natural resource planning, analysis, and design. Staff includes water resource engineers, landscape architects, and environmental scientists.

Greenworks, PC is an award-winning Portland landscape architecture firm with a mission to integrate people and nature through creative and sustainable design.

The Society for Ecological Restoration Northwest Chapter is dedicated to the art and science of restoration. Members actively protect and restore ecosystems throughout the Cascadia bioregion, which includes Washington, Oregon, Montana, Idaho, Alaska, and northern California.

The Rivers Foundation of the Americas (RFA) is a public operating foundation dedicated to promoting and funding the conservation, protection, and restoration of rivers and their watersheds throughout the Americas. RFA uses its expert financial and technical support to promote clean water, biodiversity, and human health through watershed protection and restoration. Contributors to RFA, matching grants for this project, include The Paulus Foundation, Pamela Hyde, Mary Mills Dunea, Peter Lavigne and Nancy Parent, Moshe Lenske, Dean Marriott, Jennifer P. Speers, Michael Biehler, Rebecca Miller, Mike Fremont, Mark Dubois, Jacqui Reisner Bostrom, Jeremy O'Leary, Bill Hutchison, Sylvia Schultz, Copper River Watershed Project, Riki Ott, Shannon Gardner, Ann Christensen, Coralie Brown, Pete Richardson, Ed Pembleton, Hanna Cortner,

Jay Austin, Rick Deats and Kirsten Day, Prof. Scott Burns, Prof. Jack Corbett, Prof. Marcus Ingle, Jane and David Schue, Hawthorne Auto Clinic, Watercycle LLC, Dennis Burkhart, Jackie Smith, Gil Kelley, Cheron Calder, Bill Lacey, David E. Cooke, Joanne and Ed Harris, Daniel Lichtenwald, Jean Mosely, Matt Leidecker, Diana Bartlett, Kathryn Cahoon, Marta Boyett, Jenny Holmes, Albert Kern, Paul Whitefield, Neal Schwierterma, David Koepping, Steve Blackmer, Lysa Leland, Ross and Sheila Lienhart, Langdon Marsh, Allison L. Handler, W. M. Morton, Chris Brown, Jerry Hess, Bill and Ann Beverly, and Liz and Chuck Cole.

The Society of Wetland Scientists Northwest Chapter promotes wetland science and the exchange of information for members in Washington, Oregon, and Idaho. It is a branch of the National Society of Wetland Scientists, which has more than 4,000 members from the United States, Canada, Mexico, and many other countries.

Defenders of Wildlife West Coast Office emphasizes alternative approaches to environmental decision making through partnerships that engage a broad spectrum of participants to find common ground and constructive solutions. Their programs focus on the conservation of biodiversity in the context of human activity on the landscape.

The Nature Conservancy of Oregon and Washington works to preserve the plants, animals, and natural communities that represent the diversity of life in the Pacific Northwest by protecting the lands and waters they need to survive.

The Society for Ecological Restoration British Columbia Chapter is the association of organizations and individuals working in the field of ecological restoration in British Columbia. Membership includes scientists, educators, community organizers, designers, land managers, and people from many other areas of restoration practice.

INDEX

Island Press Board of Directors